Electric Power Distribution Reliability

Second Edition

POWER ENGINEERING

Series Editor
H. Lee Willis
Quanta Technology
Raleigh, North Carolina

Advisory Editor
Muhammad H. Rashid
University of West Florida
Pensacola, Florida

1. Power Distribution Planning Reference Book, *H. Lee Willis*
2. Transmission Network Protection: Theory and Practice, *Y. G. Paithankar*
3. Electrical Insulation in Power Systems, *N. H. Malik, A. A. Al-Arainy, and M. I. Qureshi*
4. Electrical Power Equipment Maintenance and Testing, *Paul Gill*
5. Protective Relaying: Principles and Applications, Second Edition, *J. Lewis Blackburn*
6. Understanding Electric Utilities and De-Regulation, *Lorrin Philipson and H. Lee Willis*
7. Electrical Power Cable Engineering, *William A. Thue*
8. Electric Systems, Dynamics, and Stability with Artificial Intelligence Applications, *James A. Momoh and Mohamed E. El-Hawary*
9. Insulation Coordination for Power Systems, *Andrew R. Hileman*
10. Distributed Power Generation: Planning and Evaluation, *H. Lee Willis and Walter G. Scott*
11. Electric Power System Applications of Optimization, *James A. Momoh*
12. Aging Power Delivery Infrastructures, *H. Lee Willis, Gregory V. Welch, and Randall R. Schrieber*
13. Restructured Electrical Power Systems: Operation, Trading, and Volatility, *Mohammad Shahidehpour and Muwaffaq Alomoush*
14. Electric Power Distribution Reliability, *Richard E. Brown*
15. Computer-Aided Power System Analysis, *Ramasamy Natarajan*

16. Power System Analysis: Short-Circuit Load Flow and Harmonics, *J. C. Das*
17. Power Transformers: Principles and Applications, *John J. Winders, Jr.*
18. Spatial Electric Load Forecasting: Second Edition, Revised and Expanded, *H. Lee Willis*
19. Dielectrics in Electric Fields, *Gorur G. Raju*
20. Protection Devices and Systems for High-Voltage Applications, *Vladimir Gurevich*
21. Electrical Power Cable Engineering, Second Edition, *William Thue*
22. Vehicular Electric Power Systems: Land, Sea, Air, and Space Vehicles, *Ali Emadi, Mehrdad Ehsani, and John Miller*
23. Power Distribution Planning Reference Book, Second Edition, *H. Lee Willis*
24. Power System State Estimation: Theory and Implementation, *Ali Abur*
25. Transformer Engineering: Design and Practice, *S.V. Kulkarni and S. A. Khaparde*
26. Power System Capacitors, *Ramasamy Natarajan*
27. Understanding Electric Utilities and De-regulation: Second Edition, *Lorrin Philipson and H. Lee Willis*
28. Control and Automation of Electric Power Distribution Systems, *James Northcote-Green and Robert G. Wilson*
29. Protective Relaying for Power Generation Systems, *Donald Reimert*
30. Protective Relaying: Principles and Applications, Third Edition, *J. Lewis Blackburn and Thomas J. Domin*
31. Electric Power Distribution Reliability, Second Edition, *Richard E. Brown*

Electric Power Distribution Reliability

SECOND EDITION

Richard E. Brown

CRC Press
Taylor & Francis Group
Boca Raton London New York

CRC Press is an imprint of the
Taylor & Francis Group, an **informa** business

CRC Press
Taylor & Francis Group
6000 Broken Sound Parkway NW, Suite 300
Boca Raton, FL 33487-2742

Printed in the United States of America on acid-free paper
10 9 8 7 6 5 4 3 2 1

International Standard Book Number-13: 978-0-8493-7567-5 (Hardcover)

Visit the Taylor & Francis Web site at
http://www.taylorandfrancis.com

and the CRC Press Web site at
http://www.crcpress.com

Contents

Series Introduction

Power engineering is the oldest and most traditional of the various areas within electrical engineering, yet no other facet of modern technology is currently undergoing a more dramatic revolution in technology or business structure. Perhaps the most fundamental change taking place in the electric utility industry is the move toward a quantitative basis for the management of service reliability. Traditionally, electric utilities achieved satisfactory customer service quality through the use of more or less "one size fits all situations" standards and criteria that experience had shown would lead to no more than an acceptable level of trouble on their system. Tried and true, these methods succeeded in achieving acceptable service quality.

But evolving industry requirements changed the relevance of these methods in two ways. First, the *needs* of modern electric energy consumers changed. Even into the early 1980s, very short (less than 10 second) interruptions of power had minimal impact on most consumers. Then, utilities routinely performed field switching of feeders in the early morning hours, creating 10-second interruptions of power flow that most consumers would not even notice. But where the synchronous-motor alarm clocks of the 1960s and 1970s would just fall a few seconds behind during such interruptions, modern digital clocks, microelectronic equipment and computers cease working altogether. Homeowners of the 1970s woke up the next morning—not even knowing or caring—that their alarm clocks were a few seconds behind. Homeowners today wake up minutes or hours late, to blinking digital displays throughout their home. In this and in many other ways, the widespread use of digital equipment and automated processes has redefined the term "acceptable service quality" and has particularly increased the importance of interruption frequency as a measure of utility performance.

Second, while the traditional standards-driven paradigm did achieve satisfactory service quality in most cases, it did not do so at the lowest possible cost. In addition, it had no mechanism for achieving reliability targets in a demonstrated least-cost manner. As a result, in the late 20th century, electric utility management, public utility regulators, and energy consumers alike realized there had to be a more economically effective way to achieve satisfactory reliability levels of electric service. This was to *engineer* the system to provide the type of reliability needed at the lowest possible cost, creating a need for rigorous, quantitative reliability analysis and engineering methods—techniques capable of "engineering reliability into a system" in the same way that capacity or voltage regulation targets had traditionally been targeted and designed to.

Many people throughout the industry contributed to the development of what are today the accepted methods of reliability analysis and predictive design. But none contributed as much to either theory, or practice, as Richard Brown. His work is the foundation of modern power distribution reliability engineering. It is therefore with great pride that I welcome this book as the newest addition to the CRC Press series on Power Engineering. This is all the more rewarding to me because for the past decade Richard Brown has been one of my most trusted co-workers and research collaborators, and a good friend.

Dr. Brown's book lays out the rules and structure for modern power distribution reliability engineering in a rigorous yet accessible manner. While scrupulously correct in theory and mathematics, his book provides a wealth of practical experience and useful knowledge that can be applied by any electric power engineer to improve power distribution reliability performance. Thus, *Electric Power Distribution Reliability* fits particularly well into the theme of the Power Engineering series, which focuses on providing modern power technology in a context of proven, practical application—books useful as references as well as for self-study and classroom use. I have no doubt that this book will be *the* reference in power delivery reliability engineering for years to come.

Good work, Richard.

H. Lee Willis

Preface

Distribution reliability is one of the most important topics in the electric power industry due to its high impact on the cost of electricity and its high correlation with customer satisfaction. The breadth and depth of issues relating to this subject span nearly every distribution company department including procurement, operations, engineering, planning, rate making, customer relations, and regulatory. Due in large part to its all-encompassing nature, distribution reliability has been difficult for utilities to address in a holistic manner. Most departments, if they address reliability at all, do so in isolation without considering how their actions may relate to those in different parts of the company—an understandable situation since there has been no single reference that covers all related issues and explains their interrelationships. This book is an attempt to fill this void by serving as a comprehensive tutorial and reference book covering all major topics related to distribution reliability. Each subject has been extensively researched and referenced with the intent of presenting a balance of theory, practical knowledge, and practical applications. After reading this book, readers will have a basic understanding of distribution reliability issues and will know how these issues have affected typical utilities in the past. Further, readers will be knowledgeable about techniques capable of addressing reliability issues and will have a basic feel for the results that can be expected from their proper application.

Electric Power Distribution Reliability is intended for engineering professionals interested in the topic described by its title. Utility distribution planners and reliability engineers will find it of greatest use, but it also contains valuable information for design engineers, dispatchers, operations personnel, and maintenance personnel. Because of its breadth, this book may also find use with distribution company directors and executives, as well as with state regulatory authorities. It is intended to be a scholarly work and is suitable for use with senior or graduate level instruction as well as for self-instruction.

This book is divided into eight chapters. Although each is a self-contained topic, the book is written so that each chapter builds upon the knowledge of prior chapters. As such, this book should be read through sequentially upon first encounter. Terminology and context introduced in prior chapters are required knowledge to fully comprehend and assimilate subsequent topics. After an initial reading, this book will serve well as a refresher and reference volume and has a detailed index to facilitate the quick location of specific material.

The first chapter, "Distribution Systems," presents fundamental concepts, terminology, and symbology that serve as a foundation of knowledge for reliability-specific topics. It begins by describing the function of distribution systems in the overall electric power system. It continues by describing the component and system characteristics of substations, feeders, and secondary systems. The chapter concludes by discussing issues associated with load characteristics and distribution operations.

The second chapter, "Reliability Metrics and Indices," discusses the various aspects of distribution reliability and defines terms that are frequently used later in the book. It begins at a high level by discussing power quality and its relationship to reliability. Standard reliability indices are then presented along with benchmark data and a discussion of their benefits and drawbacks. The chapter continues by discussing reliability from the customer perspective including the customer cost of interrupted electrical service and the customer surveys used to obtain this information. The chapter ends with a discussion of reliability targets and the industry trend towards performance-based rates, reliability guarantees, and customer choice.

Remembering that reliability problems are caused by real events, Chapter 3 provides a comprehensive discussion of all major causes of customer interruptions. It begins by describing the most common types of equipment failures and their associated failure modes, incipient failure detection possibilities, and failure prevention strategies. It then discusses reliability issues associated with animals, presents animal data associated with reliability, and offers recommendations to mitigate and prevent animal problems. The chapter continues by discussing severe weather including wind, lightning, ice storms, heat storms, earthquakes, and fires. Human causes are the last interruption category addressed, including operating errors, vehicular accidents, dig-ins, and vandalism.

To place all of this information in perspective, the chapter concludes by discussing the most common interruption causes experienced by typical utilities.

The analytical section of this book begins in Chapter 4, "Component Modeling." The chapter starts by defining the component reliability parameters that form the basis of all reliability models. It then discusses basic modeling concepts such as hazard functions, probability distribution functions, and statistics. It ends by providing component reliability data for a wide variety of distribution equipment, which can be used both as a benchmark for custom data or as generic data in lieu of custom data.

The topic of component reliability modeling leads naturally into the next chapter, "System Modeling." This chapter begins with a tutorial on basic system analysis concepts such as states, Venn diagrams, network modeling, and Markov modeling. The bulk of the chapter focuses on analytical and Monte Carlo simulation methods, which are the recommended approaches for most distribution system reliability assessment needs. Algorithms are presented with detail sufficient for the reader to implement models in computer software, and reflect all of the major system issues associated with distribution reliability. For completeness, the chapter concludes by presenting reliability analysis techniques commonly used in other fields and discusses their applicability to distribution systems.

The sixth chapter, "System Analysis," focuses on how to use the modeling concepts developed in the previous two chapters to improve system reliability. It begins with the practical issues of actually creating a system model, populating it with default data and calibrating it to historical data. It then presents techniques to analyze the system model including visualization, risk analysis, sensitivity analyses, root-cause analysis, and loading analysis. One of the most important topics of the book comes next: strategies to improve reliability and how to quantify their impact by incorporating them into component and system models. This includes the nontraditional topics of underground conversion and storm hardening. The chapter then discusses how to view reliability improvement projects from a value perspective by presenting the basics of economic analysis and the prioritization method of marginal benefit-to-cost analysis. The chapter concludes with a comprehensive example that shows how system analysis techniques can be applied to improve the reliability of an actual distribution system.

Since most distribution companies would like to optimize the reliability of their distribution system, this book continues with a chapter on system optimization. It begins by discussing common misconceptions about optimization and continues by showing how to properly formulate an optimization problem. It then presents several optimization methods that are particularly suitable for distribution system reliability. Finally, the chapter presents several practical applications of reliability optimization and discusses potential barriers that might be encountered when attempting to implement a reliability optimization initiative that spans many distribution company departments and budgets.

This book concludes with a chapter on aging infrastructure and the impact of aging infrastructure on reliability. It begins by discussing equipment and population aging, and when age can be used as a reasonable proxy for equipment condition. The chapter continues by discussing how failure rates increase as a function of age. This includes techniques to develop age-versus-failure models using data available at most utilities. The book concludes by presenting the state of the industry in terms of equipment age; US distribution systems are surprisingly old, are getting older, and will become less reliability as a result. As such, the topics covered in this book will become increasingly important in the next decade. Utilities will have to spend increasingly more money just to keep reliability from getting worse. Using the techniques described in this book, utilities can ensure that this reliability spending is done so that the highest level of reliability can be attained for the lowest possible cost.

The second edition of *Electric Power Distribution Reliability* is the product of approximately fifteen years of effort in various aspects of electric power distribution reliability. I would like to thank the following people for teaching, collaborating, and supporting me during this time. In the academic world, I would like to thank Dr. Mani Venkata, Dr. Richard Christie, and Dr. Anil Pahwa for their insight, guidance and support. In industry, I would like to acknowledge the contributions and suggestions of my co-workers at with special thanks to Dr. Damir Novosel, Mr. Lee Willis, and Mr. Jim Burke, all IEEE Fellows. Last, I would like to offer special thanks to my wife Christelle and to our four children for providing the inspiration and support without which this book would not be possible.

Richard E. Brown

Author

Richard E. Brown is the Vice President of Operations and co-founder for Quanta Technology, a firm specializing in technical and management consulting for electric utilities. He has previously worked at Jacobs Engineering, ABB, and KEMA. During his career, Dr. Brown has developed several generations of reliability assessment software programs, has provided consulting services to most major utilities in the United States and many around the world, and has published more than 90 technical papers. In 2007, Dr. Brown was made an IEEE Fellow for "contributions to distribution system reliability and risk assessment." He earned his BSEE, MSEE, and PhD degrees from the University of Washington in Seattle, and his MBA from the University of North Carolina at Chapel Hill. He is a registered professional engineer.

Dr. Brown lives in Cary, North Carolina with his wife and four children.

1
Distribution Systems

Since distribution systems account for up to 90% of all customer reliability problems, improving distribution reliability is the key to improving customer reliability. To make effective improvements, a basic understanding of distribution system functions, subsystems, equipment, and operation is required. This chapter presents fundamental concepts, terminology, and symbology that serve as a foundation of knowledge for reliability-specific topics. Careful reading will magnify the clarity and utility of the rest of this book.

1.1 GENERATION, TRANSMISSION, AND DISTRIBUTION

Electricity, produced and delivered to customers through generation, transmission and distribution systems, constitutes one of the largest consumer markets in the world. Electric energy purchases are 3% of the US gross domestic product and are increasing faster than the US rate of economic growth (see Figure 1.1). Numbers vary for individual utilities, but the cost of electricity is approximately 50% fuel, 20% generation, 5% transmission, and 25% distribution.

Reliable electric power systems serve customer loads without interruptions in supply voltage. Generation facilities must produce enough power to meet customer demand. Transmission systems must transport bulk power over long distances without overheating or jeopardizing system stability. Distribution systems must deliver electricity to each customer's service entrance. In the context of reliability, generation, transmission, and distribution are referred to as functional zones.[1]

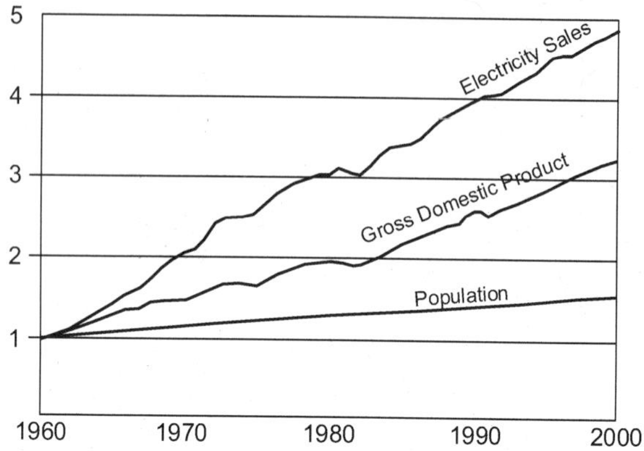

Figure 1.1. Growth of electricity sales in the US as compared to growth in gross domestic product and population (normalized to 1960 values). Electricity sales growth consistently outpaces population growth and GDP. Absolute energy usage is increasing as well as per-capita energy usage.

Each functional zone is made up of several subsystems. Generation consists of generation plants and generation substations. Transmission consists of transmission lines, transmission switching stations, transmission substations, and subtransmission systems. Distribution systems consist of distribution substations, primary distribution systems, distribution transformers, and secondary distribution systems. A simplified drawing of an overall power system and its subsystems is shown in Figure 1.2.

Generation Subsystems

Generation Plants produce electrical energy from another form of energy such as fossil fuels, nuclear fuels, or hydropower. Typically, a prime mover turns an alternator that generates voltage between 11 kV and 30 kV.

Generation Substations connect generation plants to transmission lines through a step-up transformer that increases voltage to transmission levels.

Transmission Subsystems

Transmission Systems transport electricity over long distances from generation substations to transmission or distribution substations. Typical US voltage levels include 69 kV, 115 kV, 138 kV, 161 kV, 230 kV, 345 kV, 500 kV, 765 kV, and 1100 kV.

Transmission Switching Stations serve as nodes in the transmission system that allow transmission line connections to be reconfigured.

Transmission Substations are transmission switching stations with transformers that step down voltage to subtransmission levels.

Subtransmission Systems transport electricity from transmission substations to distribution substations. Typical US voltage levels include 34.5 kV, 46 kV, 69 kV, 115 kV, 138 kV, 161 kV, and 230 kV.

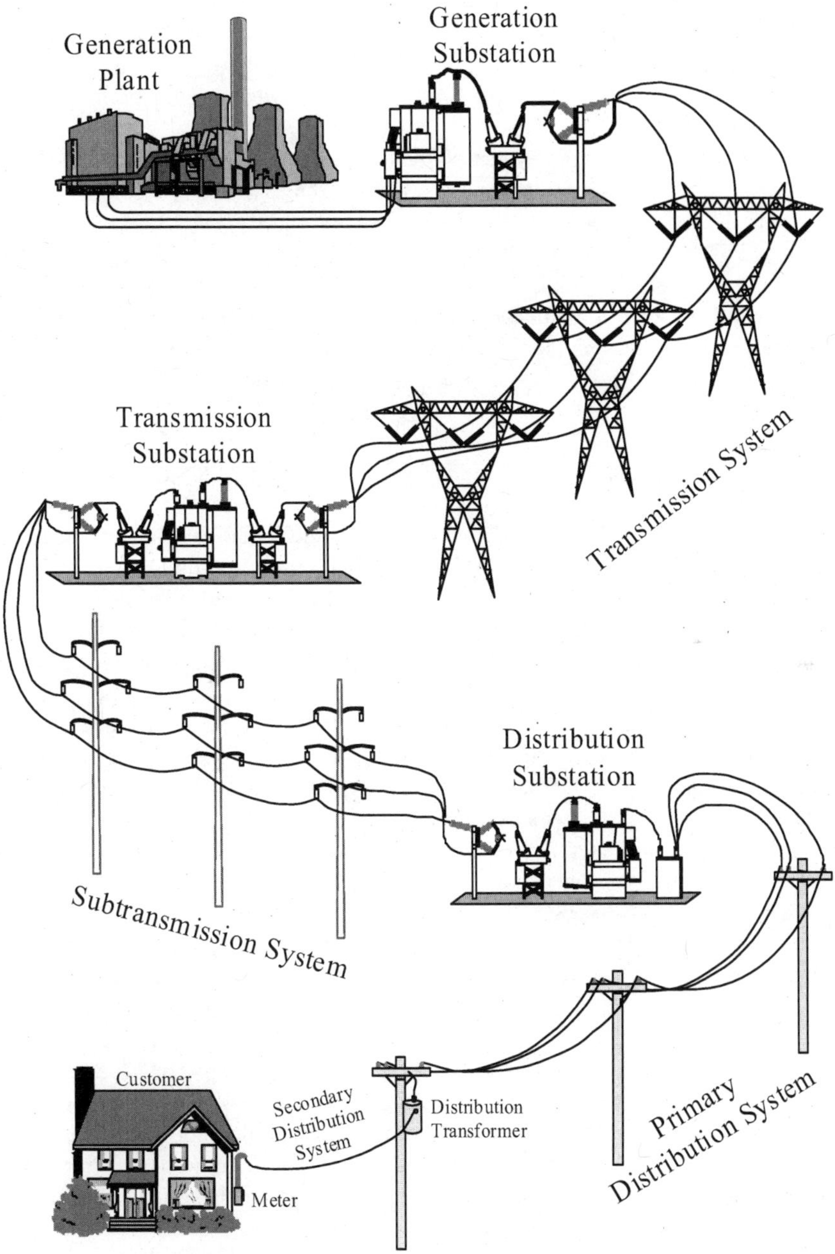

Figure 1.2. Electric power systems consist of many subsystems. Reliability depends upon generating enough electric power and delivering it to customers without any interruptions in supply voltage. A majority of interruptions in developed nations result from problems occurring between customer meters and distribution substations.

Distribution Subsystems

Distribution Substations are nodes for terminating and reconfiguring subtransmission lines plus transformers that step down voltage to primary distribution levels.

Primary Distribution Systems deliver electricity from distribution substations to distribution transformers. Voltages range from 4.16 kV to 34.5 kV with the most common being 15-kV class (e.g., 12.47 kV, 13.8 kV).

Distribution Transformers convert primary distribution voltages to utilization voltages. Typical sizes range from 5 kVA to 2500 kVA.

Secondary Distribution Systems deliver electricity from distribution transformers to customer service entrances. Voltages are typically 120/240V single phase, 120/208V three phase, or 277/480V three phase.

1.1.1 Generation

Generation plants consist of one or more generating units that convert mechanical energy into electricity by turning a prime mover coupled to an electric generator. Most prime movers are driven by steam produced in a boiler fired by coal, oil, natural gas, or nuclear fuel. Others may be driven by nonthermal sources such as hydroelectric dams and wind farms. Generators produce line-to-line voltages between 11 kV and 30 kV.[2]

The ability of generation plants to supply all of the power demanded by customers is referred to as *system adequacy*. Three conditions must be met to ensure system adequacy. First, available generation capacity must be greater than demanded load plus system losses. Second, the system must be able to transport demanded power to customers without overloading equipment. Third, customers must be served within an acceptable voltage range.

System adequacy assessment is probabilistic in nature.[3] Each generator has a probability of being available, a probability of being available with a reduced capacity, and a probability of being unavailable. This allows the probability of all generator state combinations to be computed. To perform an adequacy assessment, each generation state combination is compared to hourly system loads for an entire year. If available generation cannot supply demanded load or constraints are violated, the system is inadequate and load must be curtailed.

Generation adequacy assessments produce the following information for each load bus: (1) the combinations of generation and loading that require load curtailment, and (2) the probability of being in each of these inadequate state combinations. From this information, it is simple to compute the expected number of interruptions, interruption minutes, and unserved energy for each load bus. Load bus results can then be easily aggregated to produce the following system indices:

LOLE (Loss of Load Expectation) — The expected number of hours per year that a system must curtail load due to inadequate generation.

EENS (Expected Energy Not Served) — The expected number of megawatt hours per year that a system must curtail due to inadequate generation.

Most generation plants produce electricity at voltages less than 30 kV. Since this is not a sufficiently high voltage to transport electricity long distances, generation substations step up voltages to transmission levels (typically between 115 kV and 1100 kV). Current research utilizing high voltage cables in generators is able to produce electricity directly at transmission voltages and may eliminate the need for generation substations.

1.1.2 Transmission

Transmission systems transport electricity over long distances from bulk power generation facilities to substations that serve subtransmission or distribution systems. Most transmission lines are overhead but there is a growing trend towards the use of underground transmission cable (oil-filled, SF_6 filled, extruded dielectric, and possibly superconducting).

To increase flexibility and improve reliability, transmission lines are interconnected at transmission switching stations and transmission substations. This improves overall performance, but makes the system vulnerable to cascading failures. A classic example is the Northeastern Blackout of November 9th, 1965, which left an entire region without electricity for many hours.

The North American Electric Reliability Council (NERC) was formed in 1968 as a response to the 1965 blackout to provide planning recommendations and operating guidelines for electric utilities. In 2007, NERC become the federal National Reliability Organization (NRO) under the Federal Energy Regulatory Commission (FERC). It changed its name to the North American Electric Reliability Corporation, and now has the authority to create and enforce federal reliability standards (including fines for noncompliance). The territory covered by NERC is divided into eight regions, but there are only four major transmission grids in the United States and Canada. Each grid is highly interconnected within its boundaries, but only has weak connections to adjacent grids. NERC regions are shown in Figure 1.3. Abbreviations for NERC regions are shown in Table 1.1.

Each NERC region insures transmission system reliability by performing transmission security assessments. Transmission security assessment determines whether a power system is able to supply peak demand after one or more pieces of equipment (such as a line or a transformer) are disconnected. The system is tested by removing a piece (or multiple pieces) of equipment from the normal power flow model, re-running the power flow, and determining if all bus voltages are acceptable and all pieces of equipment are loaded below emergency ratings. If an unacceptable voltage or overload violation occurs, load must be shed and the system is *insecure*. If removing any single component will not result in the loss of load, the system is *N-1 Secure*. If removing any *X* arbitrary components will not result in the loss of load, the system is *N-X Secure*. *N* refers to the number of components on the system and *X* refers to the number of components that can be safely removed.

Table 1.1. North American Electric Reliability Corporation (NERC) regions.

Abbreviation	NERC Region
ERCOT	Electric Reliability Council of Texas, Inc. (ERCOT)
FRCC	Florida Reliability Coordinating Council (FRCC)
MRO	Midwest Reliability Organization (MRO)
NPCC	Northeast Power Coordinating Council (NPCC)
RFC	ReliabilityFirst Corporation (RFC)
SERC	SERC Reliability Corporation (SERC)
SPP	Southwest Power Pool, Inc. (SPP)
WECC	Western Electricity Coordinating Council (WECC)

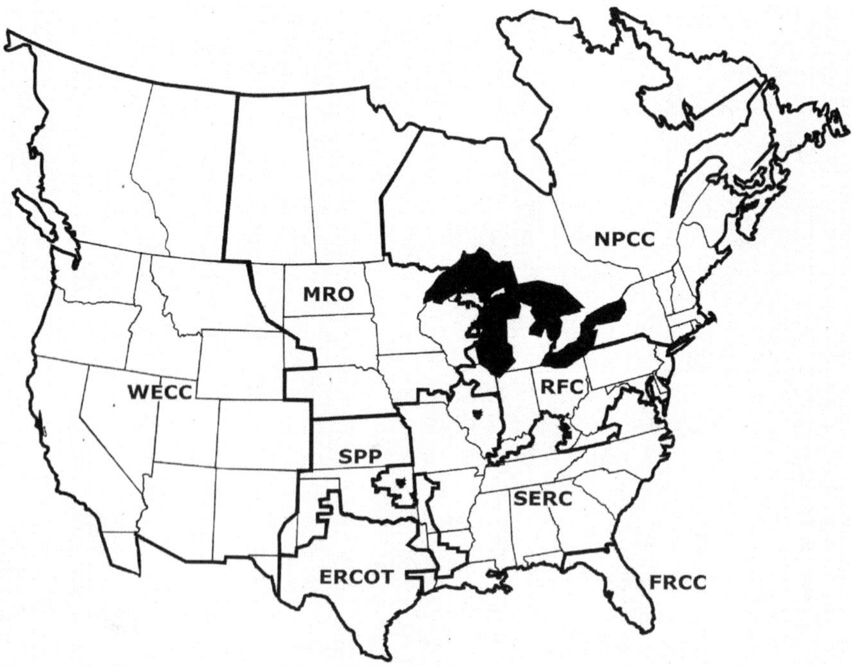

Figure 1.3. The territory covered by NERC is divided into eight operating regions. Despite this, there are only four strongly interconnected transmission networks: WECC, ERCOT, Quebec, and the remainder (typically called the Eastern Interconnection).

1.1.3 Distribution

Distribution systems deliver power from bulk power systems to retail customers. To do this, distribution substations receive power from subtransmission lines and step down voltages with power transformers. These transformers supply primary distribution systems made up of many distribution feeders. Feeders consist of a main 3φ trunk, 2φ and 1φ laterals, feeder interconnections, and distribution transformers. Distribution transformers step down voltages to utilization levels and supply secondary mains or service drops.

Distribution planning departments at electric utilities have historically concentrated on capacity issues, focusing on designs that supply all customers at peak demand within acceptable voltage tolerances without violating equipment ratings. Capacity planning is almost always performed with rigorous analytical tools such as power flow models. Reliability, although considered important, has been a secondary concern usually addressed by adding extra capacity and feeder ties so that certain loads can be restored after a fault occurs.

Although capacity planning is important, it is only half of the story. A distribution system designed purely for capacity (and minimum safety standards) costs between 40% and 50% of a typical US overhead design. This minimal system has no switching, no fuse cutouts, no tie switches, no extra capacity, and no lightning protection. Poles and hardware are as inexpensive as possible, and feeders protection is limited to fuses at substations. Any money spent beyond such a "minimal capacity design" is spent to improve reliability. Viewed from this perspective, about 50% of the cost of a distribution system is for reliability and 50% for capacity. To spend distribution reliability dollars as efficiently as capacity dollars, utilities must transition from capacity planning to integrated capacity and reliability planning.[4] Such a department will keep track of accurate historical reliability data, utilize predictive reliability models, engineer systems to specific reliability targets, and optimize spending based on cost per reliability benefit ratios.

The impact of distribution reliability on customers is even more profound than cost. For a typical residential customer with 90 minutes of interrupted power per year, between 70 and 80 minutes will be attributable to problems occurring on the distribution system.[5] This is largely due to the radial nature of most distribution systems, the large number of components involved, the sparsity of protection devices and sectionalizing switches, and the proximity of the distribution system to end-use customers. The remaining sections of this chapter address these distribution characteristics in more detail.

1.2 DISTRIBUTION SUBSTATIONS

Distribution systems begin at distribution substations. An elevation and corresponding one-line diagram of a simple distribution substation is shown in Figure 1.4. The substation's source of power is a single overhead subtransmission line that enters from the left and terminates on a take-off (dead-end) structure. The line is connected to a disconnect switch, mounted on this same structure, capable of visibly isolating the substation from the subtransmission line. Electricity is routed from the switch across a voltage transformer through a current transformer to a circuit breaker. This breaker protects a power transformer that steps voltage down to distribution levels. High voltage components are said to be located on the "high side" or "primary side" of the substation.

The medium voltage side of the transformer is connected to a secondary breaker. If a transformer fault occurs, both the primary and secondary breaker will open to isolate the transformer from the rest of the system. The secondary breaker is connected to a secondary bus that provides power to four feeder breakers. These breakers are connected to cables that exit the substation in an underground ductbank called a "feeder get-away." Medium voltage components are said to be located on the "low side" or "secondary side" of the substation. Confusingly, substation secondary components supply power to primary distribution systems.

The substation in Figure 1.4 may cause reliability concerns due to its simple configuration. If any major component fails or is taken out of service, there will be no electrical path from the subtransmission source to the secondary bus and all feeders will become de-energized. Consequently, many distribution substations are designed with redundancy allowing a portion of feeders to remain energized if any major component fails or is taken out of service for maintenance.

Figure 1.5 shows a common substation layout to the left and a much more complicated (and reliable) substation to the right ("n.o." refers to a normally open switch). The substation to the left is sometimes referred to as an "H-station" or a "transmission loop-through" design. It is able to supply both secondary buses after the loss of either transmission line or either transformer. Faults, however, will generally cause one of both secondary buses to be de-energized until switching can be performed. The substation to the right further increases reliability by having an additional transmission line, an energized spare power transformer, primary ring-bus protection, motor-operated switches, and a secondary transfer bus.

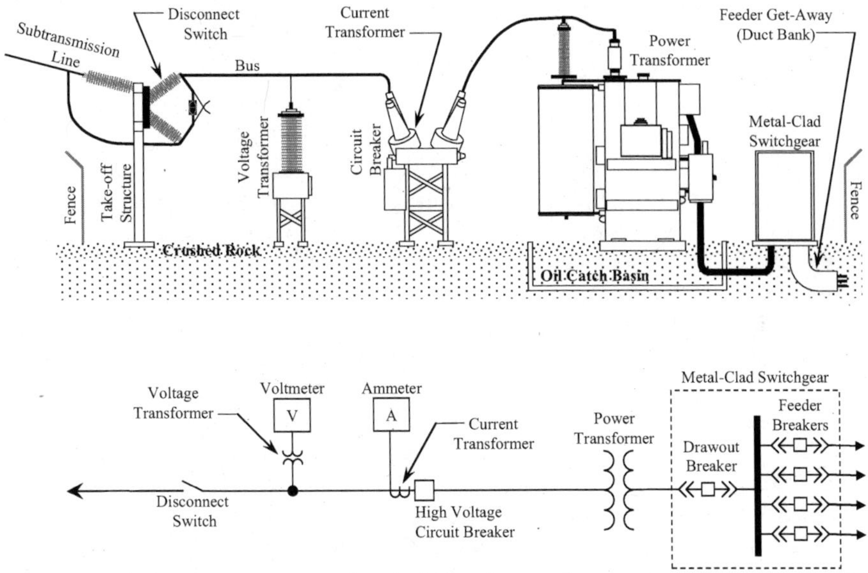

Figure 1.4. An elevation and corresponding single-line diagram illustrates the basic components of a distribution substation. This substation has a single source, a single transformer and four feeders.

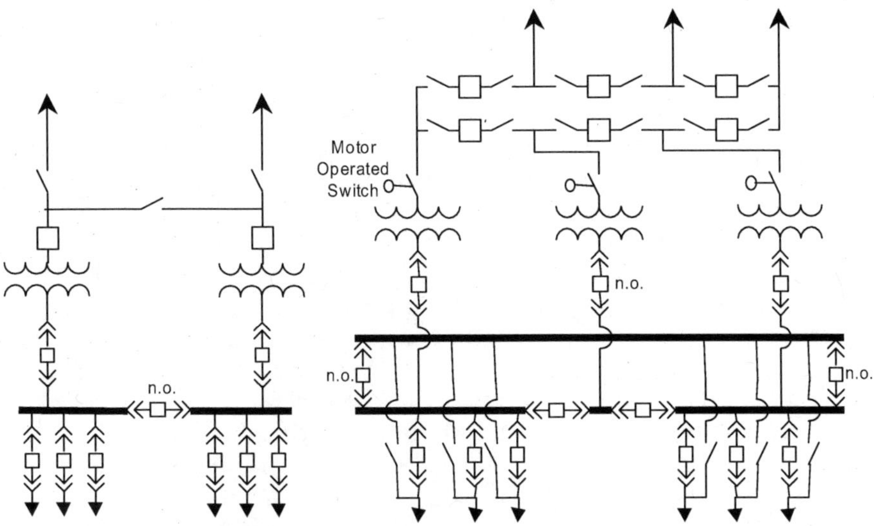

Figure 1.5. The left substation is a typical design with two subtransmission lines and two transformers. The right substation is a very reliable design with a primary ring bus, motor operated switches, an energized spare power transformer, and a secondary transfer bus.

1.2.1 Substation Components

Many different types of components must be interconnected to build a distribution substation. Understanding these components is the first step in understanding substation reliability. A brief description of major substation components is now provided (see Figure 1.6 for pictures of common equipment).

High Voltage Disconnect Switches — switches used to visibly isolate parts of a substation during maintenance or repair periods. They can also be used to reconfigure connections between subtransmission lines and/or power transformers. Disconnect switches are classified as either *load break* or *no-load break.* Load break switches can open and close with normal load current flowing through them. No-load break switches can only open and close if there is no current. Disconnect switches are not able to interrupt fault currents.

High Voltage Bus — rigid conductor used to interconnect primary equipment. It is made out of extruded metal (such as aluminum pipe) and is supported by insulator posts. A high voltage bus must be mechanically braced to withstand mechanical forces caused by high fault currents.

High Voltage Circuit Breakers — switches that can interrupt fault current. Classification is based on continuous rating, interruption rating, insulating medium, and tank potential. Continuous rating is the continuous current that can flow through the device without overheating (typically from 1200 A to 5000 A). Interruption rating is the largest amount of fault current that can be interrupted (e.g., 50 kA or 64 kA). The most common insulating mediums are oil, SF_6 (sulfur hexafluoride gas), and vacuum. In the US, most breakers have a grounded tank—referred to as a dead tank—enclosing the breaker contacts. In Europe, most breakers have the tank at line potential—referred to as a live tank.

Circuit Switchers — combination devices consisting of a visible disconnect switch and circuit breaker. They typically do not have a high short circuit interruption rating, but save space and cost less than purchasing a switch and breaker separately. Circuit switchers are typically used in rural areas or other parts of the system where available fault current is low.

Voltage and Current Transformers — these devices step down high voltages and currents to levels usable by meters and relays. Voltage transformers and current transformers are commonly referred to as VTs and CTs, respectively. Voltage transformers are sometimes referred to as potential transformers (PTs).

Power Transformers — devices that step down transmission and subtransmission voltages to distribution voltages. The ratio of primary windings to secondary windings determines the voltage reduction. This ratio can be adjusted up or down with tap changers located on primary and/or secondary windings. A no-load tap changer can only be adjusted when the transformer is de-energized while a load tap changer (LTC) can be adjusted under load current. Nearly all power transformers are liquid filled, but new applications using extruded cable technology have recently made dry-type power transformers available.

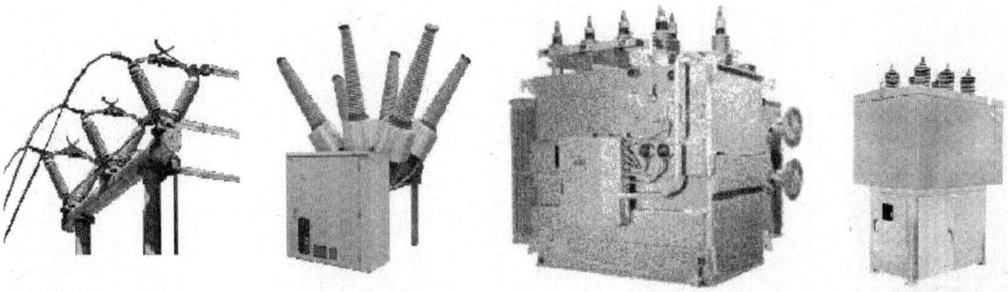

Figure 1.6. Typical components found in an air-insulated substation (AIS). From left to right: v-break sectionalizing switch, 115-kV SF_6 circuit breaker, power transformer, and 15-kV vacuum circuit breaker.

Power transformers are characterized by a base MVA rating based on a maximum hot-spot temperature at a constant load at a specific ambient temperature. Since this rating uses ambient air to cool oil, it is labeled OA. Ratings are increased by installing oil pumps and/or radiator fans. A radiator fan stage is labeled FA for forced air. A pumped oil stage is labeled FO for forced oil. Each stage will typically increase the transformer rating by 33% of its base rating. For example, a power transformer having a base rating of 15 MVA, a first stage of radiator fans and a second stage of pumps has a rating of 15/20/25 MVA, OA/FA/FOA (FOA refers to forced oil and air).

Power transformers are also characterized by an impedance expressed as a percentage of base ratings. The following equations show the relationships between base transformer ratings.

$$\text{MVA}_{\text{base}} \qquad 3\phi\ \text{MVA} \tag{1.1}$$

$$\text{kV}_{\text{base}} \qquad \text{line-to-line kV} \tag{1.2}$$

$$I_{\text{base}} = 1000 \cdot \frac{\text{MVA}_{\text{base}}}{\sqrt{3} \cdot \text{kV}_{\text{base}}} \qquad \text{amps} \tag{1.3}$$

$$Z_{\text{base}} = \frac{\text{kV}_{\text{base}}^2}{\text{MVA}_{\text{base}}} \qquad \text{ohms} \tag{1.4}$$

Impedance is an important tradeoff because high impedances limit fault current (less damage to the transformer) but cause a larger voltage drop. It can also be used to compute resistance and reactance if an X/R ratio is given. Typical power transformers can have impedances ranging from Z = 6% to Z = 16%.

Autotransformers — power transformers with electrical connections between primary and secondary windings. Autotransformers are characterized by low cost and low impedance. Due to low impedances, autotransformers are subject to higher fault currents and tend to fail more frequently than standard power transformers.

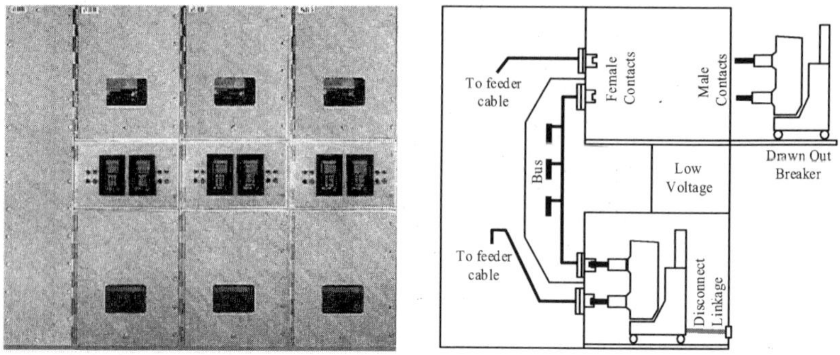

Figure 1.7. Metal-clad switchgear and cross section of a compartment fitted with two drawout feeder breakers. After a breaker is disconnected by a closed-door racking system, the door can be opened and the breaker can be rolled out of its compartment.

Medium Voltage Switchgear — refers to switches, breakers, and interconnecting buswork located downstream of power transformers. These devices can be free standing, but are often enclosed in a lineup of cabinets called *metal-clad switchgear*. Breakers in metal-clad switchgear are typically mounted on wheels and can be removed by rolling them out of their compartment (referred to as *drawout breakers*). A metal-clad switchgear lineup and corresponding cross section is shown in Figure 1.7.

Protective Relays — these devices receive information about the system and send signals for circuit breakers to open and close when appropriate. Older relays are based on electromechanical principles (e.g., the spinning of a disc or the movement of a plunger) but most modern relays are based on digital electronics.

Overcurrent relays send trip signals when high currents (typically caused by faults) are sensed. Instantaneous relays send this signal without any intentional delay. Time overcurrent relays send a signal that is delayed longer for lower currents, allowing breakers in a series to be coordinated. Delay versus current for each device is characterized by a time current curve (TCC).

Differential relays send a trip signal if the current flowing into a zone is not equal to current flowing out of a zone. Common applications include transformer differential protection and bus differential protection.

Reclosing relays tell circuit breakers to close after clearing a fault in hope that the fault has cleared. Reclosing will typically occur multiple times with increasing delays. If the fault fails to clear after the last reclose, the circuit breaker *locks out*.

Several common relays are shown in Figure 1.8. There are many other types and comprehensive treatment is beyond the scope of this book. For more detailed information, the reader is referred to References 6 and 7.

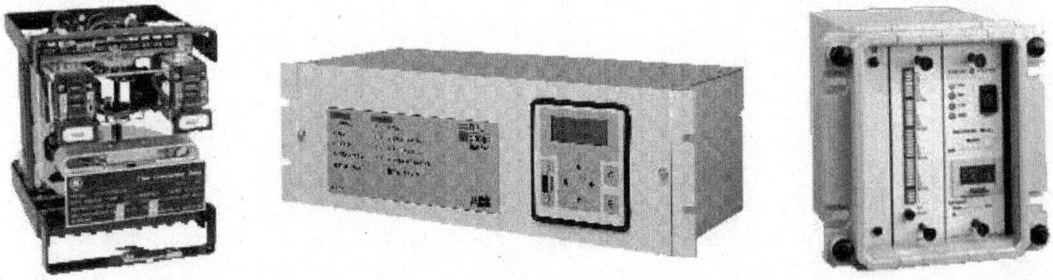

Figure 1.8. From left to right: an electromechanical time-overcurrent relay, a multifunction digital relay and a digital reclosing relay.

Substation Automation — this term refers to supervisory control and data acquisition (SCADA) equipment located in distribution substations. Typically, substation automation allows transformer and feeders to be monitored and circuit breakers to be remotely opened and closed. Substations can also serve as communication links to automated equipment located on feeders.

Gas Insulated Substations — substations that enclose high voltage bus, switches, and breakers in containers filled with SF_6 gas. GISs greatly reduce the substation footprint and protects equipment from many causes of equipment failures. GISs are sold in modular units, GIS bays, consisting of a transition component, a circuit breaker, and one or more buses. A GIS bay and a GIS lineup are shown in Figure 1.9.

Mobile Substations — substations that have a primary circuit breaker or fuse, a transformer and secondary switchgear mounted on a trailer. These are used to temporarily replace permanent substations following severe events such as the loss of multiple power transformers. A mobile substation can be used to support multiple substations but are typically limited to a maximum capacity of 25 MVA due to size and weight constraints.

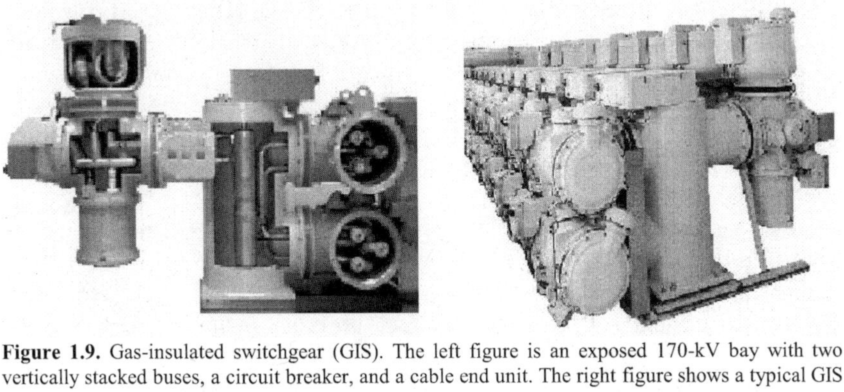

Figure 1.9. Gas-insulated switchgear (GIS). The left figure is an exposed 170-kV bay with two vertically stacked buses, a circuit breaker, and a cable end unit. The right figure shows a typical GIS lineup.

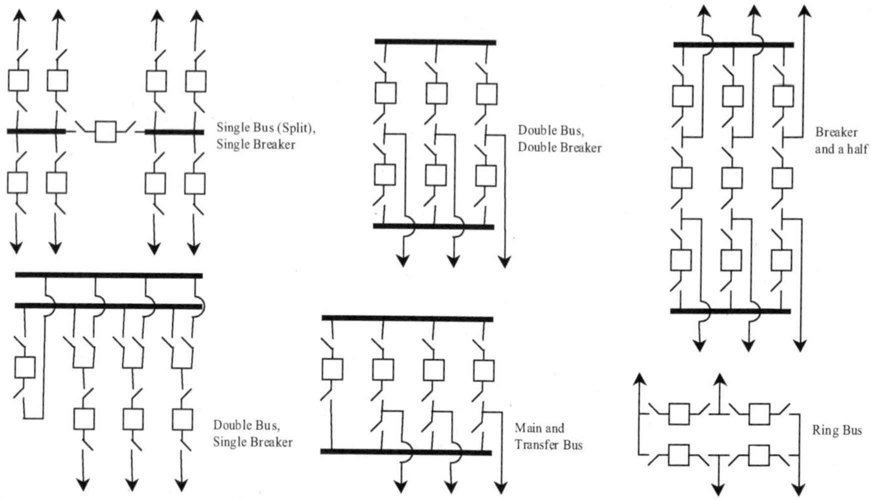

Figure 1.10. Typical substation bus configurations (see main text for more detailed descriptions). Arrows are connected to either subtransmission lines or power transformers.

1.2.2 Bus Configurations

The ability of subtransmission lines and power transformers to be electrically connected is determined by bus connections, disconnect switches, circuit breakers, circuit switchers, and fuses. Together, these components determine the bus configuration of distribution substations. Bus configurations are an important aspect of substation reliability, operational flexibility, and cost.

An infinite number of possible substation configurations exist. The six most commonly encountered are shown in Figure 1.10. The reliability of substation configurations will be addressed in detail by future chapters, but a brief description of each is now provided.

Single Bus, Single Breaker — all connections terminate on a common bus. They are low in cost, but must be completely de-energized for bus maintenance of bus faults. To improve reliability, the bus is often split and connected by a switch or breaker.

Main and Transfer Bus — a transfer bus is connected to a main bus through a tie breaker. Circuits are normally connected to the main bus, but can be switched to the transfer bus using sectionalizing switches. Since circuits on the transfer bus are not protected by circuit breakers, faults on one transferred circuit result in outages for all transferred circuits.

Double Bus, Single Breaker — utilizes a single breaker per circuit that can be connected to either bus. A tie breaker between buses allows circuits to be transferred without being de-energized. Since this configuration requires four switches per circuit, space, maintenance, and reliability are concerns for AIS applications. Double bus, single breaker configurations are well suited for GIS applications where space, maintenance, and reliability of switches are significantly less of a concern.

Double Bus, Double Breaker — each circuit is connected to two buses through dedicated circuit breakers. The use of two breakers per circuit makes this configuration reliable, flexible, and expensive.

Breaker and a Half — utilizes legs consisting of three series breakers connected between two buses. Since two circuits are connected on each leg, 1.5 breakers are required for every circuit. This configuration is more expensive than other options (except double bus, double breaker), but provides high reliability, maintainability, and flexibility. Protective relaying is more complex than for previously mentioned schemes.

Ring Bus — arranges breakers in a closed loop with circuits placed between breakers. Since one breaker per circuit is required, ring buses are economical while providing a high level of reliability. For AIS applications, ring buses are practical up to five circuits. It is common to initially build a substation as a ring bus and convert it to breaker and a half when it requires more than this amount.[8] Ring buses are a natural configuration for GIS applications with any number of circuits. Like the breaker and a half configuration, ring bus relaying is relatively complex.

1.3 PRIMARY DISTRIBUTION SYSTEMS

Primary distribution systems consist of feeders that deliver power from distribution substations to distribution transformers. A feeder begins with a feeder breaker at the distribution substation. Many will exit the substation in a concrete ductbank (feeder get-away) and be routed to a nearby pole. At this point, underground cable transitions to an overhead three-phase main trunk. The main trunk is routed around the feeder service territory and may be connected to other feeders through normally-open tie points. Underground main trunks are possible—even common in urban areas—but cost much more than overhead construction.

Lateral taps off of the main trunk are used to cover most of a feeder's service territory. These taps are typically 1ϕ, but may also be 2ϕ or 3ϕ. Laterals can be directly connected to main trunks, but are more commonly protected by fuses, reclosers, or automatic sectionalizers (see Section 1.3.1). Overhead laterals use pole-mounted distribution transformers to serve customers and underground laterals use padmount transformers. An illustrative feeder showing different types of laterals and devices is shown in Figure 1.11.

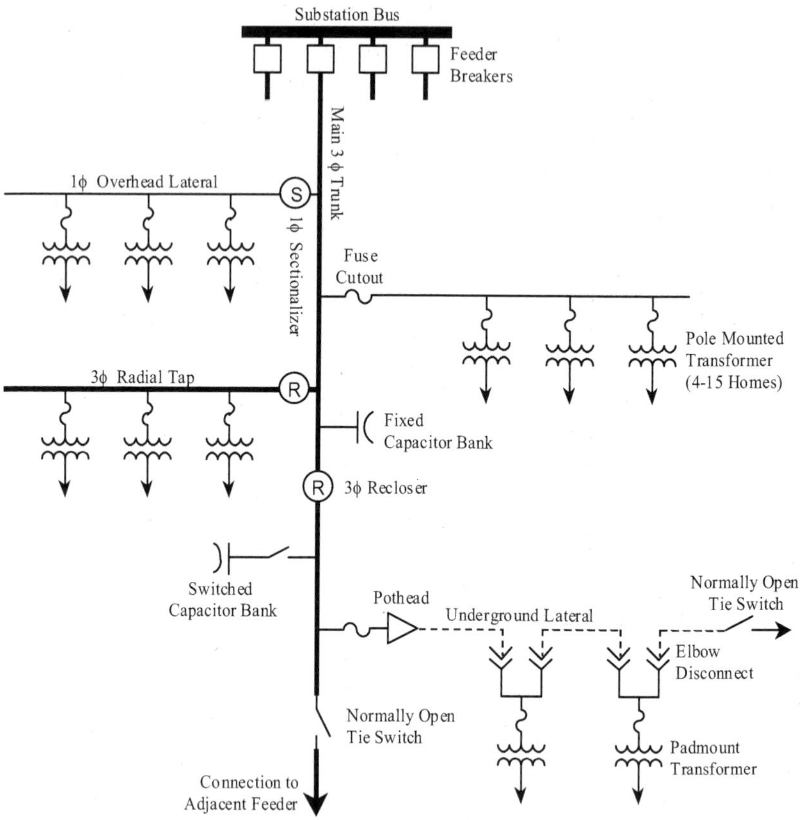

Figure 1.11 A primary distribution feeder showing major components and characteristics.

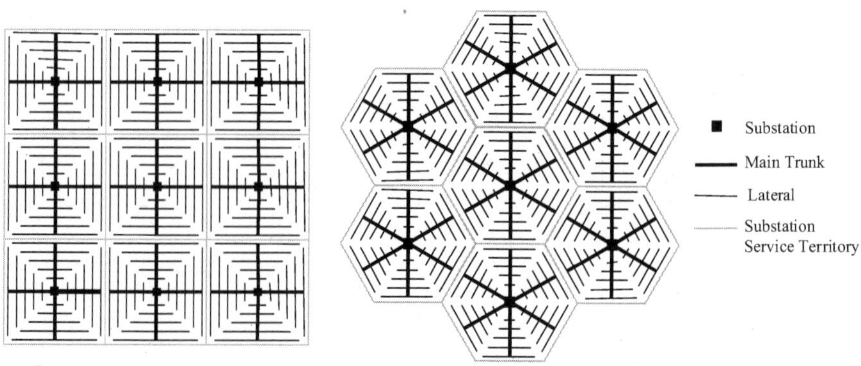

Figure 1.12. Substations supply a number of feeders to cover their service territories. The left is organized into square service territories with four feeders per substation. The right is organized into hexagonal service territories with six feeders per substation.

Feeder routes must pass near every customer. To accomplish this, each substation uses multiple feeders to cover an allocated service territory. Figure 1.12 illustrates this with (1) square service territories and four feeders per substation, and (2) hexagonal service territories and six feeders per substation.[9] In most cases, feeder routings and substation service territories evolve with time, overlap each other and are not easily categorized into simple geometric configurations.

1.3.1 Overhead Feeder Components

Feeders consist of many types of components—all playing an interconnected role in distribution reliability. This section provides a brief description of major components and construction practices. It is not intended to be exhaustive in breadth or depth, but to provide a basis for future discussions in this book. Readers wishing a more comprehensive treatment of distribution equipment are referred to Reference 10.

Feeder components can be broadly categorized into overhead and underground. Overhead equipment is less expensive to purchase, install and maintain, but is more exposed to weather and has poor aesthetic characteristics. Figure 1.15 shows pictures of some common overhead feeder components.

Poles — Poles support overhead distribution equipment and are an important part of all overhead systems. Most poles are treated wood, but concrete, steel, composite, and other materials are also used. Typical distribution pole constructions are shown in Figure 1.13 (these examples are by no means exhaustive).

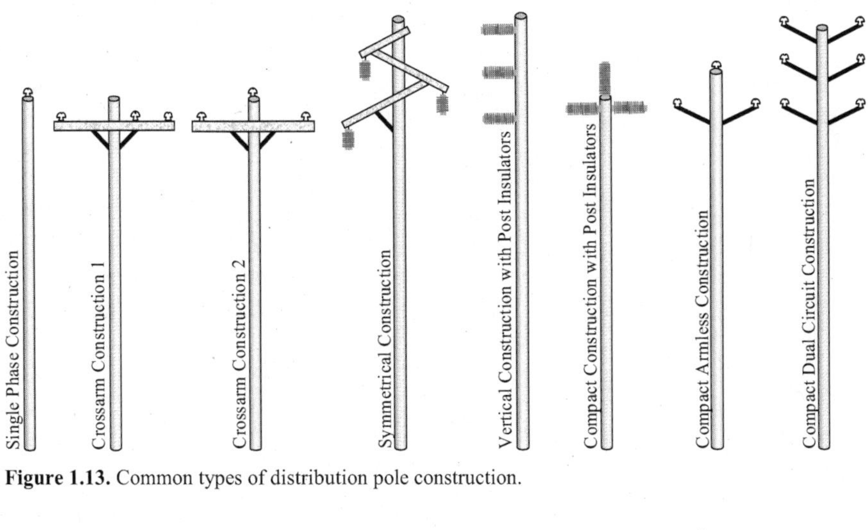

Figure 1.13. Common types of distribution pole construction.

Wood distribution poles must be specified based on height and strength. Generally, height is specified in feet, and strength is specified by an ANSI pole class. ANSI pole classes used for distribution are as follows (from weakest to strongest): 7, 6, 5, 4, 3, 2, 1, H1, H2, H3, and possibly higher H-class poles. For example, a 45' distribution pole of Class 3 would be referred to as a 45/3 pole. Common poles for distribution application are from 30' to 50' in height and from Class 5 to Class 2 in strength.

The governing safety standard for distribution pole strength in the US is the National Electrical Safety Code (NESC). A pole meeting the NESC requirements can be considered safe, but may or may not be desirable from an economic or reliability perspective.

The NESC defines three different grades of safety requirements depending upon the public safety issues related to a particular installation. These are termed Grade B, Grade C, and Grade N, with Grade B being the highest requirement. In general, the NESC requires distribution structures to meet Grade C construction except when crossing railroad tracks or limited-access highways (these require Grade B construction).

Overhead Lines — wires that carry load current in an overhead system. Major classifications are by insulation, size, stranding, material, impedance, and ampacity.

Lines without an insulated cover are called *bare conductors* and all other lines are referred to as *insulated conductors*. Insulated conductors are further classified into *covered conductor, tree wire, spacer cable,* and *aerial cable*. Covered conductor and tree wire have a thin covering of insulation that cannot withstand phase to ground voltages, but reduce the probability of a fault if vegetation bridges two conductors. Spacer cable has increased insulation that allows conductors to be arranged in a small triangular configuration. Aerial cable has fully rated insulation capable of withstanding phase to ground voltages.

Line sizes are measured by American Wire Gauge (AWG) and by circular mills (CM). AWG (also known as Brown & Sharpe gauge) numbers each wire size and bases successive diameters on a geometrical progression. AWG 14 is used for small house wire and AWG 6 is considered very small utility wire. The largest sizes are AWG 0, 00, 000 and 0000. AWG 0000 (also referred to as 4/0 or "four ought") is the largest and has a diameter of 0.46 inches.

A circular mil is the area of a circle with a one mil (one thousandth of an inch) diameter. Large utility wire is measured in thousands of circular mils, labeled in the metric-based *kcmil* or the Roman-based *MCM*. Sizes start at 250 kcmil (diameter ≈ 0.6 inches) and are usually limited to 1000 kcmil due to handling and termination considerations.

Most lines consist of steel strands surrounded by aluminum strands. Referred to as *Aluminum Conductor Steel Reinforced (ACSR),* this wire type uses steel to achieve high tensile strength and aluminum to achieve high conductivity at a low weight. There are many other wire types with a range of advantages and

Table 1.2. Common types of overhead lines (ranges reflect various stranding options). ACSR is the most popular choice due to its high strength-to-weight and conductivity-to-weight ratios.

Type	Description	Conductivity (%)	Strength (%)	Weight (%)
CU	Stranded Copper	100	100	100
AAC	All Aluminum Conductor	64	39	30
AAAC	All Aluminum Alloy Conductor	55	75	30
ACSR	Aluminum Conductor Steel Reinforced	64-65	55-111	35-51
ACAR	Aluminum Conductor Alloy Reinforced	71-73	55-63	36-37
ACSS	Aluminum Conductor Steel Supported	66-67	50-98	39-51

disadvantages (see Table 1.2). Full treatment is beyond the scope of this book and the curious reader is referred to Reference 11.

Two important line characteristics are impedance and ampacity. Impedances are series resistances and reactances that determine Ohmic losses and voltage drops. Resistance is primarily determined by line material, but is also influenced by conductor temperature and current waveform frequency. Reactance is primarily determined by construction geometry, with compact designs having a smaller reactance than designs with large phase conductor separation.

When temperature increases, lines expand and sag. The maximum current that a line can support without violating safe ground clearances is referred to as *ampacity*. The ampacity of a line is usually based on a specified ambient temperature and wind speed. Because of this, many utilities have separate summer and winter ratings. It is also common to have emergency ratings that allow lines to exceed their normal ratings for a limited amount of time. This is possible because thermal inertia results in a time lag between line overloading and line sag. Typical ampacities for ACSR lines are provided in Table 1.3.

Table 1.3. Typical conductor ratings based on 75°C conductor temperature, 1-ft spacing and 60 Hz.

Codeword	Size (AWG or kcmil)	Stranding (Al/St)	R (Ω/mi)	X (Ω/mi)	Ampacity (amps)
Turkey	6	6/1	4.257	0.760	105
Swan	4	6/1	2.719	0.723	140
Sparrow	2	6/1	1.753	0.674	184
Raven	1/0	6/1	1.145	0.614	242
Quail	2/0	6/1	0.932	0.599	276
Pigeon	3/0	6/1	0.762	0.578	315
Penguin	4/0	6/1	0.628	0.556	357
Partridge	266.8	26/7	0.411	0.465	457
Oriole	336.4	30/7	0.324	0.445	535
Lark	397.5	30/7	0.274	0.435	594
Hen	477	30/7	0.229	0.424	666
Eagle	556.5	30/7	0.196	0.415	734
Egret	636	30/19	0.172	0.407	798
Redwing	715	30/19	0.153	0.399	859
Mallard	795	30/19	0.138	0.393	917

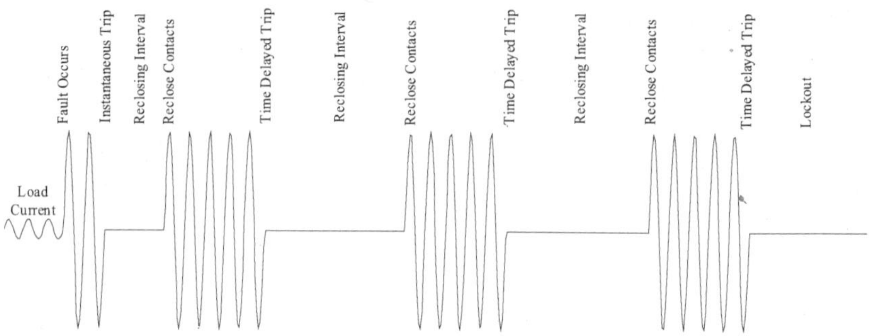

Figure 1.14. Typical reclosing sequence. After a fault occurs, an instantaneous relay clears the fault without any intentional time delay. After a short interval, the recloser closes its contacts to see if the fault has cleared. If not, a time delay relay trip will occur if no other device operates first. If the fault still persists after several longer reclosing intervals, the recloser locks out.

Sectionalizing Switches — devices that can be opened and closed to reconfigure a primary distribution system. Like substation disconnect switches, they are rated as either load break or no-load break.

Fuse Cutouts — Fuse cutouts are containers that hold expulsion fuse links. Since expulsion fuses are not able to interrupt high fault currents, fuse cutouts may be fitted with a backup current-limiting fuse. Since current-limiting fuses will clear faults quickly by forcing a current zero, they have the additional advantage of greatly reducing the energy of a fault.

Reclosers — self-contained protection devices with fault interrupting capability and reclosing relays. Interruption capability is lower than for a circuit breaker, and applications are typically away from substations where fault current is lower. Most reclosers utilize one or two instantaneous operations that clear a fault without any intentional time delay. This allows temporary faults on fused laterals to clear. If the fault persists, time-delayed operations allow time overcurrent devices to operate in a coordinated manner. Reclosers typically utilize a short reclosing interval followed by several longer intervals. A typical reclosing sequence is shown in Figure 1.14.

Historically, reclosers have been used in areas where fault current is too low for the feeder breaker to protect. This is certainly a valid application, but underutilizes the ability of reclosers to improve distribution system reliability.

Sectionalizers — protection devices used in conjunction with reclosers (or breakers with reclosing relays) to automatically isolate faulted sections of feeders. Sectionalizers do not have fault interrupting capability. Rather, they count the number of fault currents that pass and open after a specified count is reached.

Consider a sectionalizer directed to open after a count of 2. If a fault occurs downstream of the sectionalizer, it will increment its counter to "1" and an upstream reclosing device will open. The device will pause and then reclose. If the fault persists, the sectionalizer will increment its counter to "2" and the reclosing device will open a second time. At this point, the sectionalizer will automatically open. When the reclosing device closes, the fault will be isolated and all customers upstream of the sectionalizer will have power restored.

Capacitors — devices used to provide reactive current to counteract inductive loads such as motors. Properly applying capacitors will reduce losses, improve voltage regulation and allow a distribution system to deliver increased kilowatts. Typical distribution systems utilize *fixed capacitors* (unable to be switched on and off) during light loading conditions and turn on *switched capacitors* during heavy loading conditions. Switched capacitors are typically turned on and off automatically based on temperature, timers, current, voltage, reactive power, or power factor measurements.

Voltage Regulators — transformers with load tap changers used on feeders to support voltage. They are becoming less common as the use of higher voltages, large wire, and capacitors becomes more common.

Pole-Mounted Transformers — step down voltage to utilization levels and are characterized by voltage and kVA rating. Since typical distribution transformers only experience peak loading several hours per year, many utilities do not replace them until peak loading exceeds 200% of rated kVA. They can be single phase or three phase, and can be conventional or completely self-protected (CSP). Conventional transformers require separate overcurrent and lightning protection. CSP transformers provide overvoltage protection with a primary arrester and air gaps on secondary bushings, and provide overcurrent protection with a secondary circuit breaker. Standard kVA ratings range from 5 kVA to 500 kVA for 1ϕ units.[12]

Lightning Protection — refers to voltage transient protection accomplished with surge arresters and static ground wire. Surge arresters are nonlinear resistors that have high impedances at normal voltages and near-zero impedances at higher voltages. They protect equipment by clamping phase to ground voltages. Static ground wires are strung above phase wires to intercept lightning strokes. If ground impedance is too high, high current flowing into the ground can cause a large ground potential rise resulting in a backflash.

Figure 1.15. Typical overhead feeder components. From left to right: stranded ACSR wire, a pole-mounted distribution transformer, a vacuum recloser, a capacitor, and a fused cutout.

Feeder Automation — refers to SCADA and locally controlled devices on feeders. This includes faulted circuit indicators (FCIs), remote terminal units (RTUs), intelligent electronic devices (IEDs), automatic meter reading (AMR), capacitor control, automatic reconfiguration, and a host of other functions. In the context of distribution reliability, feeder automation usually refers to switches that can automatically open and close after a fault has occurred.

1.3.2 Underground Feeder Components

In the past, underground feeders have not been as common as overhead feeders due to high initial cost and maintenance difficulties. As the cost differential between overhead and underground continues to decline, underground systems are becoming increasingly popular. Public concern over aesthetics often drives the decision to build underground systems, and laws and statutes mandating underground construction are becoming common. A brief description of major underground feeder components is now provided. Figure 1.17 shows pictures of some common underground feeder components.

Riser Poles — poles that transition overhead wire to underground cable. Wire will typically terminate at a pothead, transition to cable and be routed down the riser pole in conduit.

Cable — wires covered with a dielectric insulation characterized by voltage rating, current rating, and material. Typical voltage ratings for distribution cable are 5 kV, 15 kV, 25 kV, and 35 kV. The most common insulating materials are paper-insulated lead-covered (PILC), ethylene propylene rubber (EPR), and cross-linked polyethylene (XLPE).

The simplest cable consists of one phase conductor. Adding a concentric neutral eliminates the need for a separate neutral cable (most new concentric neutral cable have an outer jacket to minimize neutral corrosion). More complicated cables contain three phase conductors and possibly a neutral wire. Some typical cable configurations are shown in Figure 1.16.

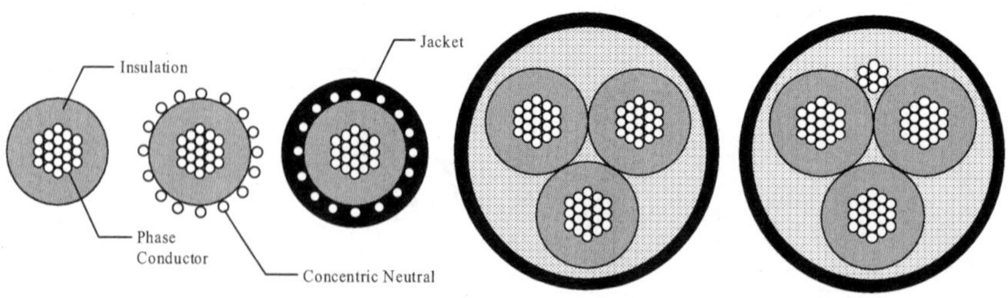

Figure 1.16. Typical cable cross sections. Simple cables consist of a single insulated conductor (far left). Adding a concentric conductor allows the cable to serve single-phase loads without requiring a separate neutral cable (second to left). Since bare concentric neutrals can be subject to undesirable corrosion, they are often covered by a waterproof jacket (middle). If three-phase service is required, three single-phase conductors can be used. Alternatively, a single cable with three fully-insulated phase conductors can be used. These multiphase cables are available both with and without a neutral conductor (second to right and right, respectively).

Table 1.4. Ampacities for three-conductor cable (with an overall covering) in underground electrical ducts. Ratings assume one cable per duct, ambient earth temperature of 20°C, 100% load factor, and a soil resistivity of 90°C–cm/watt. This table is valid for cables rated between 5001 kV and 35000 kV.

AWG	90°C Rated Insulation			105°C Rated Insulation		
kcmil	1 Circuit	3 Circuits	6 Circuits	1 Circuit	3 Circuits	6 Circuits
6	88	75	63	95	81	68
4	115	97	81	125	105	87
2	150	125	105	160	135	110
1	170	140	115	185	155	125
1/0	195	160	130	210	175	145
2/0	220	185	150	235	195	160
3/0	250	205	170	270	220	180
4/0	285	230	190	305	250	200
250	310	255	205	335	270	220
350	375	305	245	400	325	275
500	450	360	290	485	385	305
750	545	430	340	585	465	365
1000	615	485	380	660	515	405

Based on Table 310-79 of the National Electric Code.[14]

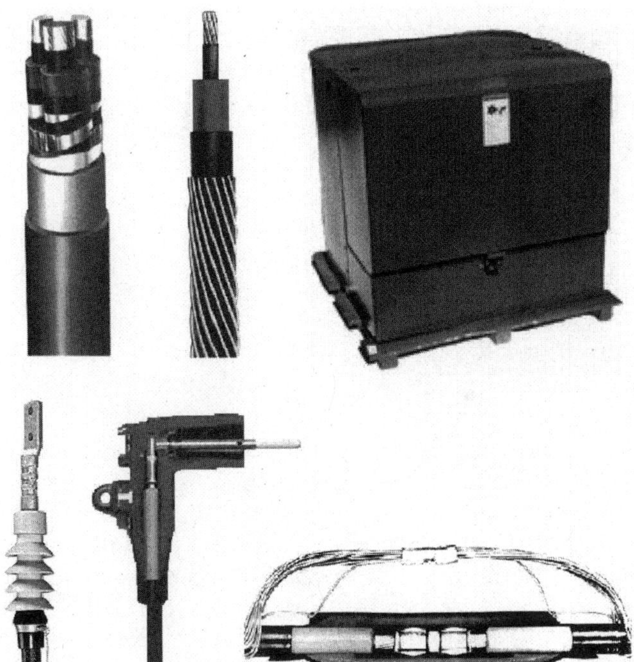

Figure 1.17. Common underground feeder equipment. Top left: three-conductor XLPE cable. Top center: single-conductor URD cable with concentric neutral. Top right: pad-mounted distribution transformer. Bottom left: cable termination. Bottom center: load break elbow. Bottom right: cable splice.

Like overhead wire, underground cables have normal and emergency ampacity ratings. These ratings are based on thermal degradation to insulation and depend on the geometry of adjacent cables, soil temperature and the ability of soil to dissipate heat (thermal resistivity). Ampacity calculations are beyond the scope of this book, but sample ratings are provided in Table 1.4. See Reference 13 for more detailed information on this and other aspects of cables.

URD Cable — abbreviation for underground residential distribution cable. Typically single phase XLPE or EPR cable with a concentric neutral that is direct buried to supply pad-mounted distribution transformers in residential neighborhoods.

Cable Terminations — devices placed on the end of cables so that they can be connected to other equipment. Cable terminations are connected with bolts and must be de-energized to connect and disconnect.

Cable Splices — devices used to connect two cables together and must be compatible with the cable insulation material. *Transition splices* are designed to connect cables with different types of insulation (e.g., PILC and XLPE).

Load Break Elbows — cable terminations that can be connected and disconnected while energized (typically up to 200 amps). These can be properly referred to as 1ϕ load break switches, but are almost always called elbows due to their "L" shape.

Pad-Mounted Distribution Transformers — metal containers containing distribution transformers, cable terminations, and switching equipment (non-metallic containers are also possible). Pad refers to the concrete slab that the devices are installed on. For public safety reasons, pad-mounted equipment must be tamper resistant and have no exposed energized parts. Typical sizes range from 10 kVA to 500 kVA for single-phase units and from 75 kVA to 5000 kVA for three-phase units.

1.3.3 Typical Configurations

Some typical primary distribution system configurations are shown in Figure 1.18. The simplest primary distribution system consists of independent feeders with each customer connected to a single feeder. Since there are no feeder interconnections, a fault will interrupt all downstream customers until it is repaired. This configuration is called a *radial system* and is common for low-density rural areas where more complex systems are cost prohibitive.

A slightly more common configuration connects two feeders together at their endpoints with a normally open tie switch. This *primary loop* increases reliability by allowing customers downstream of a fault to receive power by opening an upstream switch and closing the tie switch. The only customers that cannot be restored are those in switchable section where the fault occurred.

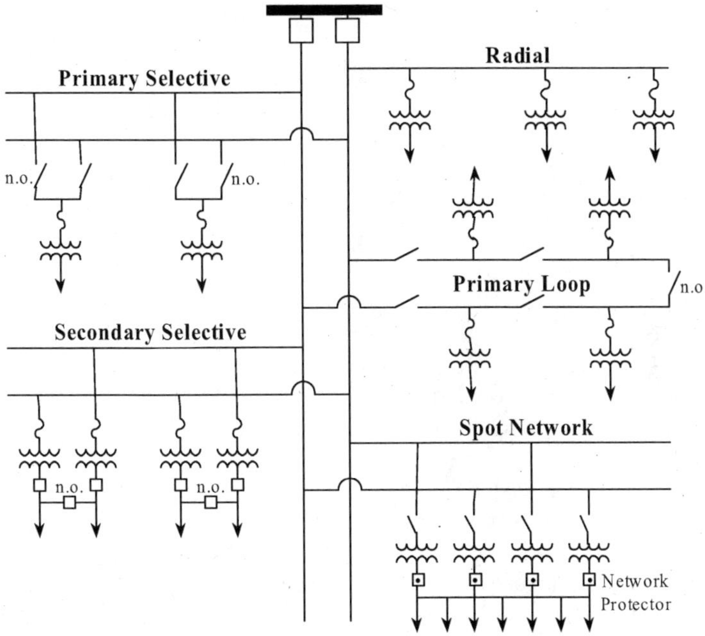

Figure 1.18. Typical primary distribution systems. Each offers different reliability characteristics with radial systems having the lowest inherent reliability and spot networks having the highest.

Many distribution systems have multiple tie switches between multiple feeders. Reliability benefits are similar to a primary loop with greater switching flexibility. These highly interconnected primary distribution systems are referred to as *radially operated networks*.

Certain classes of customers require higher reliability than a single feeder can provide. *Primary Selective* service connects each customer to a preferred feeder and an alternate feeder. If the preferred feeder becomes de-energized, a transfer switch disconnects the preferred feeder and connects the alternate feeder. *Secondary Selective* service achieves similar results by using switches on secondary voltages rather than primary voltages. With secondary selective service, each distribution transformer must be able to supply the entire load for maximum reliability benefits.

Spot Networks are used for customers with the highest reliability requirements. This configuration connects two or more transformers (fed from at least two feeders) in parallel to energize the secondary bus. To prevent reverse power flow through the transformers, special network protectors with sensitive reverse power relays are used. Spot networks allow multiple component failures to occur without any noticeable impact on customers. They are common in central business districts and high-density areas and are being applied frequently in outlying areas for large commercial services where multiple supply feeders can be made available.[15]

1.4 SECONDARY DISTRIBUTION SYSTEMS

Secondary systems connect distribution transformers to customer service entrances. They can be extremely simple, like overhead service drop, and extremely complex, like a secondary network.

1.4.1 Secondary Mains and Service Drops

Customers are connected to distribution systems via service drops. In the US, service is typically 1ϕ 3-wire 120/240V, 3ϕ 4-wire 120/208V, or 3ϕ 4-wire 277/480V. Customers close to a distribution transformer are able to have service drops directly connected to transformer secondary connections. Other customers are reached by routing a secondary main for service drop connections. These two types of service connections are shown in Figure 1.19.

Systems utilizing secondary mains are characterized by a small number of large distribution transformers rather than a large number of small distribution transformers. This can be cost effective for areas with low load density and/or large lot size, but increases Ohmic losses and results in higher voltage drops. Increased line exposure tends to reduce reliability while fewer transformers tend to increase reliability.

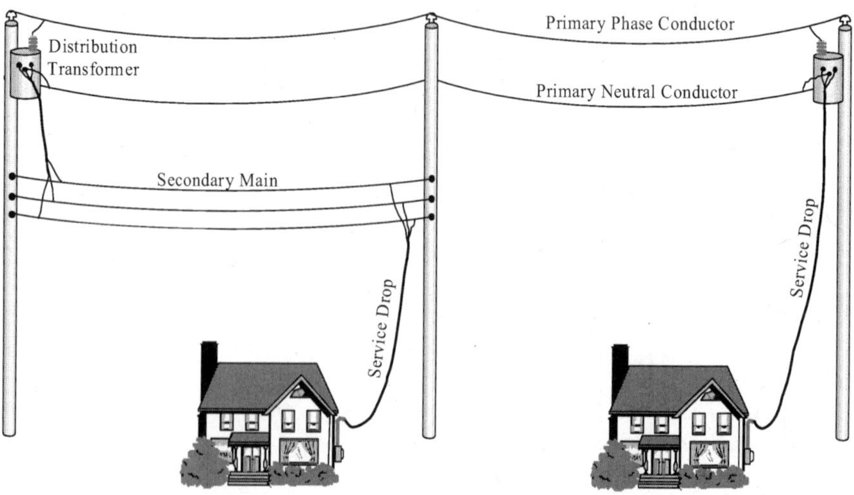

Figure 1.19. Customers near distribution transformers can be fed directly from a service drop (right). Other customers are reached by connecting service drops to secondary mains (left).

Many underground systems connect service drops directly to distribution transformers and do not use secondary mains. This forces distribution transformers to be located within several hundred feet of each customer, but eliminates the reliability concerns associated with T-splices that are required to connect underground service drops to underground secondary mains.

European-style systems use extensive secondary mains. To contrast, a typical American-style distribution transformer is 25 kVA and will have 1ϕ service drops to 4-7 customers. A European-style distribution transformer is 500 kVA and will have a 3ϕ secondary main with service drops to several hundred customers.[16] Confusingly, Europeans refer to distribution transformers as distribution substations and refer to secondary mains as low voltage networks. Extensive secondary mains make European-style distribution systems difficult to compare to American-style distribution systems, especially in the area of reliability.

1.4.2 Secondary Networks

The first AC secondary network was installed in the US in 1922.[17] Due to their exceptionally high reliability, most US cities followed suit and built secondary networks in their central business districts (although new secondary networks have not appeared for decades). These systems supply customers from a highly interconnected secondary grid that is routed under streets in duct banks, vaults, and manholes. The secondary grid is powered by at least two primary feeders that supply many transformers distributed throughout the network. A conceptual representation of a secondary network is shown in Figure 1.20.

Since a secondary network is supplied from many transformers in parallel, the loss of any transformer will not cause any noticeable disruption to customers. Most secondary networks can support the loss of an entire feeder without customer interruptions.

Secondary networks, like spot networks, require more sophisticated protection than radially operated systems. Each network transformer has a network protector on its secondary. This device uses sensitive reverse power relays to open an air circuit breaker when the secondary system is feeding a primary fault. Network protectors are also able to automatically close when the feeder is re-energized and able to supply load current.

Network protectors are not intended to protect against secondary faults. Faults occurring on a secondary mains persist until they burn clear (resulting in an open circuit). This process is reliable for 120/280V networks because electric arcs are not able to sustain themselves at this low voltage. Faults persist much more often on 277/480V networks.[18] To solve this problem, current limiters (small copper fuse links) are typically installed at all secondary network junctions for 277/480V systems.[19]

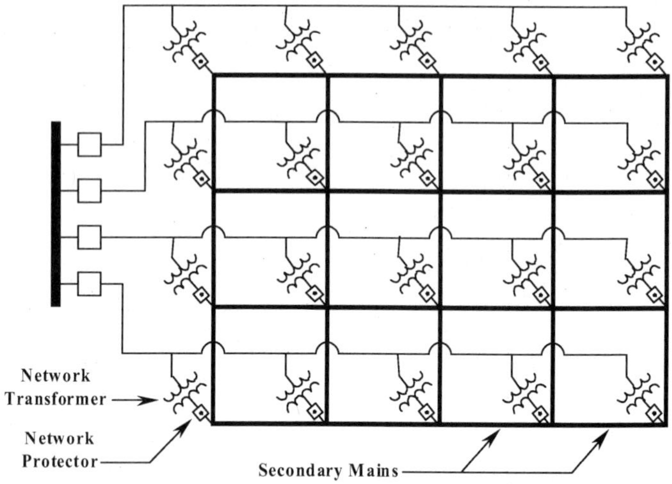

Figure 1.20. Secondary networks are supplied by a large number of network transformers. The loss of any transformer will not interrupt service to customers. Secondary faults are typically allowed to burn clear.

1.5 LOAD CHARACTERISTICS

Retail customers do not purchase electricity for electricity's sake. Rather, they purchase electricity to power electrical appliances. When appliances (referred to as *end use*) are turned on, electricity must be instantaneously supplied through the generation, transmission, and distribution systems. Instantaneous power is equal to the product of current and voltage, and RMS apparent power is equal to the product of RMS current and RMS voltage. This book assumes that readers are familiar with power system phasor analysis, but the following equations are supplied for reference.

$$S = I \times V \quad \text{Apparent Power (kVA)} \tag{1.5}$$

$$P = S \times \text{Cos}(\theta) \quad \text{Real Power (kW)} \tag{1.6}$$

$$Q = S \times \text{Sin}(\theta) \quad \text{Reactive Power (kVAR)} \tag{1.7}$$

$$\text{p.f.} = \text{Cos}(\theta) \quad \text{Power Factor} \tag{1.8}$$

$$\mathbf{I} = I \angle \theta_I \quad \text{Complex RMS Current (Amps)} \tag{1.9}$$

$$\mathbf{V} = V \angle \theta_V \quad \text{Complex RMS Voltage (kV)} \tag{1.10}$$

$$\theta = \theta_V - \theta_I \quad \text{Power Factor Angle (degrees)} \tag{1.11}$$

Loads are characterized by kVA and power factor. Unity power factor loads are purely resistive and do not draw reactive current. Examples include incandescent light bulbs and hot water heaters. Other types of equipment, like motors, utilize magnetic fields that cause current peaks to occur after voltage peaks. This is defined as a *lagging power factor* (as opposed to a leading power factor) and causes loads to draw more current than appliances with a high power factor.

1.5.1 Demand Factor and Load Coincidence

Individual customers are often characterized by their total amount of connected kVA. This number is easily computed by summing up the load of each electrical appliance. A customer's peak demand is typically much lower than this value. It is unlikely that all appliances will be simultaneously turned on; some electrical appliances are rarely used (e.g., backup devices) and certain combinations are not used at the same time (e.g., baseboard heaters and air conditioners). The ratio of peak kVA to total connected kVA is called *demand factor*.

$$\text{Demand Factor} = \frac{\text{Peak kVA}}{\text{Total Connected kVA}} \tag{1.12}$$

Most end use appliances are turned on and off randomly. Because of this, the probability of all customers simultaneously experiencing peak demand is small and decreases as the number of customers increases.[20] Because of this, distribution systems can be designed to supply less power than the sum of individual customer peak demands. The ratio of peak system demand to the sum of individual customer peak demands is called *coincidence factor*. The reciprocal of coincidence factor is *diversity factor*.

$$\text{Coincidence Factor} = \frac{\text{Peak System Demand}}{\sum \text{Individual Peak Demands}} \tag{1.13}$$

$$\text{Diversity Factor} = \frac{\sum \text{Individual Peak Demands}}{\text{Peak System Demand}} \tag{1.14}$$

Load coincidence and diversity are important to reliability because systems become less reliable as they become more heavily loaded. Heavily loaded equipment becomes hot and prone to failure. Heavily loaded systems are difficult to reconfigure to restore service to customers after a fault occurs. Increasing load diversity will reduce system peak load and result in higher reliability.

1.5.2 Load Curves

Load curves plot the expected demand of a customer over time. Each point on the curve is created by taking simultaneous demand measurements from many customers and dividing the sum by the number of customers sampled. No customer has actual load behavior resembling its load curve, but the concept is useful for modeling expected system loading without having to account for the random nature of individual appliances turning on and off.[21]

Load curves impact distribution system reliability by determining when and how long peak loading will occur. Since electrical equipment has thermal inertia, it will not heat up instantaneously when high currents flow through it. This allows equipment to have high emergency overload ratings if peak loading will not last very long. In contrast, systems that have near-constant loading levels cannot be overloaded without significant loss of life.

Since load curves vary by season and by day of the week, reliability characteristics of a system may change throughout the days of the year. Figure 1.21 shows the summer and winter daily load curves for residential homes in the southern US. Since summer electrical load is dominated by air conditioners and winter electric load is dominated by electric heating, the two load curves have different shapes. Figure 1.22 shows weekly load curves for a utility in the northern US. In general, loads tend to be greatest in the beginning of the week and lower on weekends. The most detailed load curves span all 365 days of the year and consist of 8760 hourly demand points.

Different types of customers have different load curve shapes. The most common breakdown of customer class is residential, commercial, and industrial (see Figure 1.23). Reliability needs for these customer classes vary greatly and different load curve shapes biases each towards different reliability characteristics. For example, residential loads peak over a three to four hour period. This allows small service transformers to be loaded up to 200%; by the time transformer temperatures reach critical levels, residential peak is over and transformers begin to cool down. In contrast, industrial loads can have peaks lasting eight hours or more. If transformers become overloaded to 200%, remedial actions must be taken long before the period of peak demand is over.

The average value of a load curve is equal to average power consumed. Average power consumed is useful for calculating the energy that a customer consumes and the energy that a customer would have consumed during a power interruption. Average demand is usually expressed as a percentage of peak demand, referred to as *load factor*.

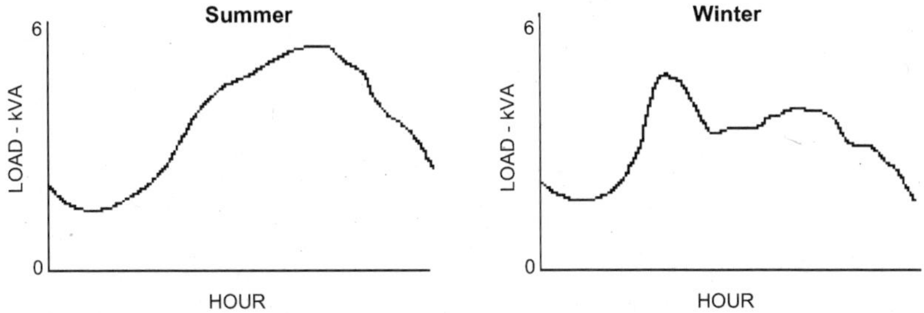

Figure 1.21. Electric load varies from hour to hour and season to season. These curves are for all-electric residential class (per customer, coincident load curves) from a utility system in northern Florida.

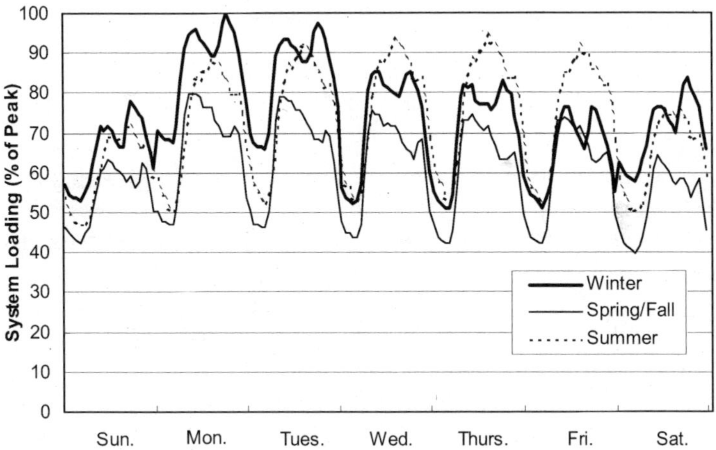

Figure 1.22. Electric load varies by day of the week. This figure shows a Midwestern utility's weekly load curves for different times of the year.

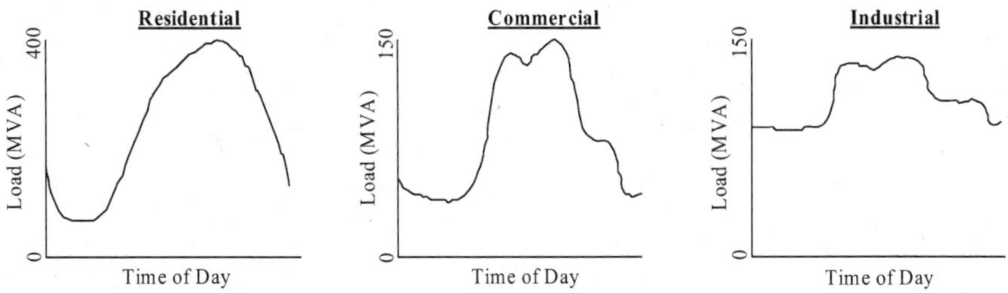

Figure 1.23. Different customer classes have different load curve shapes, particularly with regard to how demand varies with time. Here are summer peak day load curves for the three major classes of customer from the same utility system.

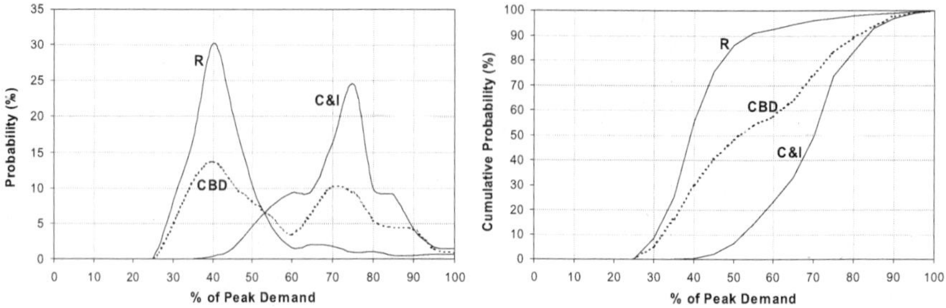

Figure 1.24. Substation load histograms (left) and load duration curves (right) for substations serving residential (R), commercial/industrial (C&I), and central business district (CBD) areas for a utility in the Midwestern US. Residential substations are lightly loaded a much higher percentage of the time when compared to commercial/industrial substations.

$$\text{Average Demand} = \frac{1}{t}\int_0^t \text{Demand}(t)\,dt \tag{1.15}$$

$$\text{Load Factor} = \frac{\text{Average Demand}}{\text{Peak Demand}} \tag{1.16}$$

Annual load curves are not always required when addressing reliability issues. Sometimes peak demand is sufficient and other times peak demand plus demand factor is sufficient. Another possibility is to utilize the percentage of time spent at each loading level (represented as a histogram). Doing so allows the reliability of each loading level to be weighted by its probability of occurrence. The integral of a loading histogram, called a *load duration curve*, represents the percentage of time that a system spends at or below each loading level. Sample load duration curves for various customer classes are shown in Figure 1.24.

1.5.3 Voltage Reduction and Reliability

Most end use loads can be characterized as *constant impedance* or *constant power*. Constant impedance loads, like incandescent lights and hot water heaters, consume less power if voltage is reduced. Since power is proportional to the square of voltage, lowering voltage can significantly reduce system demand. Some utilities utilize this phenomenon to improve reliability. If demand starts to approach the maximum amount of available generation, voltage is reduced with load tap changers and overall system demand is reduced. If a system consists of roughly 50% constant power and 50% constant impedance, a 5% reduction in voltage will result in approximately a 5% reduction in power since the constant impedance loads will draw 10% less power.

Constant power loads, like motors, use the same amount of power regardless of voltage. If voltage is reduced by a certain percentage, these loads will increase their current by an equal percentage to keep current times voltage constant. Increased current causes motors to heat up and may eventually result in equipment failure.

Since distribution system capacity is almost always limited by equipment ampacity, voltage reduction does not necessarily improve distribution reliability. When voltage is reduced, constant impedance loads draw proportionally less current and constant power loads draw proportionally more current (although many motors will operate more efficiently at lower voltages). If loads are split 50% constant impedance and 50% constant power, current does not appreciably change. If a higher percentage of loads are constant power, current will increase and distribution reliability can actually decrease. In addition, excessive low voltages can cause motors to overheat and lead to customer complaints.

1.6 DISTRIBUTION OPERATIONS

After distribution systems are planned, designed and built, they must be continually monitored, adjusted, expanded, maintained, and repaired. These activities play an important role in distribution reliability and are collectively referred to as distribution operations. Since distribution operations plays an important role in distribution reliability, a brief summary of operations, maintenance, and contingency response is provided in the following sections.

1.6.1 Dispatch Centers, Operators, and Crews

Real-time control and operation of distribution systems is performed by system operators located in dispatch centers. Each operator (sometimes called a *dispatcher*) is assigned a region to monitor and coordinate. During their shifts, operators will continuously monitor SCADA information such as feeder loading and device alarms such as breaker operations and equipment trouble (e.g., excessive temperature or internal pressure in power transformers). Customer interruption information, collected by a trouble call system, is also made available.

After a fault occurs, operators direct system reconfiguration efforts to restore as many customers as possible without violating equipment ratings. Certain devices are SCADA controlled and can be remotely operated. Crews are used to operate other devices.

Crews consist of workers specially trained to work on either overhead systems (linemen) or underground systems (cablemen). They are based out of regional service centers that serve as home bases for trucks and equipment.

Crews are responsible for locating faults, performing switching actions, repairing damaged equipment, performing routine maintenance, and constructing new facilities. There has been a long trend of using smaller crews and units of two to three people are becoming common.

A common scenario occurs after an operator receives trouble calls from customers with service interruptions. The operator first identifies the circuit associated with these customers and dispatches a crew to locate the fault. When located, the crew reports back and awaits further instructions. Typically, the operator directs the crew to isolate the fault by opening disconnect switches. The crew may also be instructed to close tie switches and restore additional customers before beginning repairs. After switching is accomplished, the crew repairs the damaged equipment and, when finished, returns the system to its pre-fault state.

Many utilities improve distribution operational efficiency with outage management systems. These software programs contain an electronic map of the distribution system and are used to coordinate system reconfiguration, alarms, trouble calls, and crew dispatch. A typical outage management system will automatically infer fault locations based on trouble calls, monitor crew locations and repair status, and update system topology as switches are opened and closed. Since outage management systems track the number of interrupted customers at each stage of system restoration and repair, they result in increased accuracy of customer reliability data when compared to manual methods based on crew reports.

1.6.2 Equipment Maintenance and Vegetation Management

Many types of distribution equipment require inspection, testing, and/or maintenance to ensure proper operation and minimize the probability of failures. Maintenance strategies can be broadly categorized as run-to-failure, periodic, condition based, and reliability centered.

Run-to-failure is the simplest maintenance strategy. After installation, equipment is not inspected or maintained until a failure occurs. This is cost effective for noncritical components with minimal maintenance requirements.

Periodic maintenance is the next simplest maintenance strategy. At specific time intervals, certain maintenance procedures are performed on equipment regardless of equipment condition or equipment criticality. Intervals are usually time-based, but other measures such as number of operations are also used.

Condition-based maintenance monitors equipment and only requires maintenance when signs of deterioration become evident. Condition assessment techniques are too numerous to list exhaustively, but include techniques such as visual inspection, acoustical inspection, infrared thermography, voltage withstand testing, partial discharge testing, and laboratory testing. For a more complete treatment, the reader is referred to Reference 22.

Reliability-centered maintenance (RCM) is a process used to determine maintenance requirements based on equipment condition, equipment criticality, and cost.[23] Effective applications of RCM result in maintenance schedules that maximize system reliability by maintaining components that are most likely to fail, have a high impact to customers when they fail, and can be maintained for a reasonable cost. This holistic approach makes intuitive sense, but is difficult to implement. RCM has been successfully applied to power plants, is beginning to be applied to substations and is in its infancy for feeder maintenance and vegetation management.

Vegetation management, often referred to as tree trimming, is the largest maintenance budget line item and the largest cause of line outages for many utilities. In the US alone, utilities spend more than $2 billion annually on this activity. Major factors impacting reliability include tree density (trees per mile), species, line clearance, pruning practices, and weather. To maintain reliability at acceptable levels, utilities will typically prune branches, use growth retardants, and remove danger trees on a two to five year cycle. To the detriment of reliability, tree trimming is often reduced due to budget cuts or public opposition (primarily for aesthetic reasons).

1.6.3 Contingency Response

Radial distribution systems are inherently insecure—failure of a major component always leads to customer interruptions. Reliable distribution systems minimize this impact by allowing faults to clear themselves, by minimizing the number of customers affected by protection device operations and by quickly restoring customers through system reconfiguration.[24] A clear understanding of this process is critical to understanding distribution reliability.

Fault on Main Feeder — for the system represented in Figure 1.25, consider a 3ϕ fault occurring on line section L5. Fault current immediately flows from subtransmission lines to the fault location, causing a voltage drop across transformer T1 and resulting in voltage sags for customers on L9-L12. Customers on L2 through L4 experience sags of increasing severity and the voltages on L5 and L6 are nearly zero.

Fault current causes breaker B1 to trip open. A reclosing relay then closes the breaker. If no fault current is detected, the fault has successfully cleared and all customers downstream of B1 experience a momentary interruption. If fault current persists, B1 trips and recloses again. After several reclosing attempts, B1 will lock out (remain open) and all customers downstream of B1 will experience a sustained interruption. After B1 sends an alarm to the system operator and customer trouble calls are received, the on-duty system operator dispatches a crew to locate the fault.

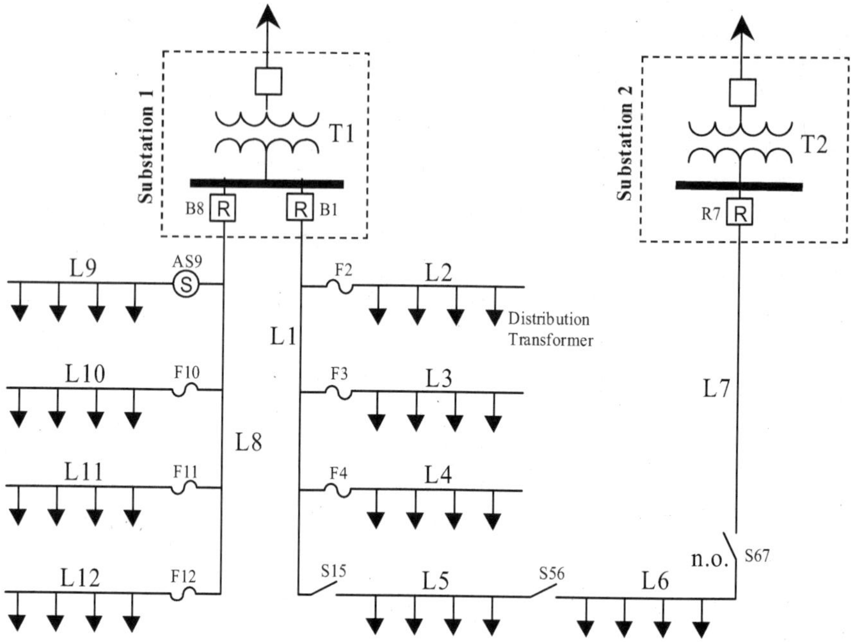

Figure 1.25. Example distribution system to illustrate system responses to various contingencies. T = Transformer, L = Switchable Line Section, S = Switch, B = Breaker with Reclosing Relay, F = Fuse, AS = Automatic Sectionalizer.

When the fault is located, the system operator instructs the crew to open switch S15. After this is done, the operator remotely closes B1, restoring power to customers on L2-L4. The crew estimates that repairs will take 5 hours to complete. Because of the long repair time, the operator instructs the crew to open S56 and close S67. Customers on L6 are restored and are now being supplied by Substation 2 rather than Substation 1. The first switching step is called *upstream restoration* and the second switching step is called *downstream restoration.* The entire switching sequence is called *system reconfiguration.* After the fault is repaired, the crew returns the system to its pre-fault state.

Fuse Saving — now consider a fault on L4. An instantaneous relay quickly opens B1 before the fuse element in F4 starts to melt. If the fault is cleared when B1 recloses, all customers downstream of B1 experience a momentary interruption. If not, a coordinated time-overcurrent relay allows F4 to blow before B1 trips again. Customers on L4 experience a sustained interruption and customers on the rest of the feeder experience a momentary interruption. Fuse saving is also referred to as *feeder selective relaying.*

Fuse Clearing — fuse saving schemes temporarily interrupt an entire feeder for all faults occurring on fused laterals. To reduce the high number of resulting customer momentary interruptions, utilities often block breaker instantaneous relays. Consider the same fault on L4. Instead of an instantaneous relay quickly opening B1, a coordinated time-overcurrent relay allows F4 to clear before B1 opens. L4 experiences a sustained interruption, and the rest of the feeder is prevented from experiencing an interruption.

Fuse clearing does not give faults on fused laterals a chance to clear themselves. This can significantly increase the total number of sustained interruptions and the total interruption time of customers supplied from these laterals. In addition, customers that are prevented from experiencing a momentary interruption will still see a voltage sag that can have a similar impact on sensitive loads.

Recloser/Fuse Coordination — consider a fault on L2 with fuse saving enabled. Since the fault is near the substation, the impedance of T1 limits its magnitude. T1 has a base rating of 20 MVA @ 12.47 kV and an impedance of 10%. This results in a fault current of:

$$\text{Fault Current} = 100{,}000\left(\frac{\text{Base MVA}}{\text{kV}_{\text{ll}} \cdot \sqrt{3} \cdot \%\text{Z}}\right) = 9260\ \text{A} \qquad (1.17)$$

The instantaneous relay on B1 will detect the fault in about 4 ms, but the breaker requires about 80 ms to physically open and clear the fault. Meanwhile, 9.26 kA of current will melt one of the slowest available fuses (200T) in about 40 ms. Since the breaker cannot beat the fuse and clear the fault, customers on L2 will experience a sustained interruption and all other customers on the feeder will still experience a momentary interruption.

The inability of utilities to coordinate the instantaneous operation of reclosers with fuses in areas of high fault currents leads many utilities to implement fuse clearing. A viable alternative is to block the instantaneous at the substation and install a recloser on the main feeder where fault currents are low enough to coordinate instantaneous recloser operation with downstream fuses. Another alternative is to use automatic sectionalizers.

Automatic Sectionalizers — consider a fault on L9 with a magnitude of 10 kA—too high to coordinate a fuse saving strategy. As an alternative, an automatic sectionalizer AS9 is used to protect the lateral. The automatic sectionalizer detects the fault and increments its counter to C = 1. B1 opens and recloses to allow the fault a chance to clear itself. If the fault persists, the AS9 counter will increment to C = 2. When its counter reaches a pre-set value, AS9 will automatically open the next time B1 opens. The fault is isolated, B1 closes a last time, and customers experience interruptions equivalent to a fuse saving scheme. Because sectionalizers use upstream devices to interrupt faults, they can be used wherever coordination issues become a problem.[25]

A distribution system fault impacts many different customers in many different ways. In general, the same fault results in voltage sags for some customers, momentary interruptions for other customers, and varying lengths of sustained interruptions for other customers depending on how the system is switched and how long it takes for the fault to be repaired—all critical to distribution system reliability. A thorough understanding of the examples presented in this section is recommended and will magnify the clarity and usefulness of future chapters.

1.7 STUDY QUESTIONS

1. What analyses and reliability measures are typically used for generation and transmission? How are generation and transmission different in this regard?
2. What are some of the pros and cons of a ring bus substation? Is this different for AIS versus GIS?
3. What percentage of customer interruption duration is typically due to problems on the distribution system? Why is this so high?
4. What are the typical insulation classes for primary distribution? What are some of the reliability issues that a utility should consider when choosing a nominal distribution voltage?
5. What are the main differences between typical US distribution systems and typical European distribution systems, and how might these differences impact reliability?
6. Why is it common for utilities to build new secondary spot networks, but no new secondary grid network have been built in decades?
7. Can voltage reduction schemes improve distribution reliability? Explain.
8. Why is it important to consider available fault current levels when deciding whether to block the instantaneous trip on reclosing relays?
9. From a reliability perspective, is the use of secondary mains desirable? Explain.
10. What are the primary maintenance strategies for distribution equipment? Why might a utility choose not to use reliability-centered maintenance?

REFERENCES

1. R. Billinton and R. N. Allan, *Reliability Assessment of Large Electric Power Systems,* Kluwer Academic Publishers, 1988.
2. A. R. Bergen, *Power Systems Analysis*, Prentice-Hall, 1986.
3. R. N. Allan, R. Billinton, A. M. Breipohl, and C. H. Grigg, "Bibliography on the Application of Probability Methods in Power System Reliability Evaluation," *IEEE Transactions on Power Systems,* Vol. 9, No. 1, Feb. 1994, pp. 41-49.
4. R. E. Brown and B. Howe, *Optimal Deployment of Reliability Investments*, E-Source Report PQ-6, April 2000.

5. R. Billinton and S. Jonnavitihula, "A Test System for Teaching Overall Power System Reliability Assessment," *IEEE Transactions on Power Systems,* Vol. 11, No. 4, Nov. 1996, pp. 1670-1676.
6. J. L. Blackburn, *Protective Relaying: Principles and Applications*, Marcel Dekker, 1987.
7. S. H. Horowitz and A. G. Phadke, *Power System Relaying*, Research Studies Press Ltd., 1992.
8. D. G. Fink and H. W. Beaty, *Standard Handbook for Electrical Engineers*, McGraw-Hill, 1987.
9. *Electric Utility Engineering Distribution Reference Book: Distribution Systems*, Westinghouse Electric Corporation, 1959.
10. E. B. Kurtz, T. M. Shoemaker, and J. E. Mack, *The Lineman's and Cableman's Handbook*, McGraw-Hill, 1997.
11. R. Thrash, G. Hudson, D. Cooper, and G. Sanders, *Overhead Conductor Manual*, Southwire Corporation, 1994.
12. T. Gönen, *Electric Power Distribution System Engineering*, McGraw-Hill, 1986.
13. T. P. Arnold and C. D. Mercier, *Power Cable Manual*, Southwire Company, 1997.
14. *National Electric Code (NEC),* National Fire Protection Agency, 1995.
15. J. J. Burke, *Power Distribution Engineering: Fundamentals and Applications*, Marcel Dekker, 1994.
16. E. Lakervi and E. J. Holmes, *Electricity Distribution Network Design*, Peter Peregrinus Ltd., 1995.
17. A. H. Kehoe, "Underground Alternating Current Network Distribution for Central Station Systems," *AIEE Transactions*, Vol. 43, June 1924, pp. 844-853.
18. D. Sutherland and D. S. MacCorkle, "Burn Off Characteristics of AC Low Voltage Network Cables," *AIEE Transactions,* Vol. 50, No. 3, Sept. 1931, pp. 831-844.
19. *Electrical Transmission and Distribution Reference Book (*formerly the *Westinghouse Electrical Transmission and Distribution Reference Book),* ABB Power T&D Company, Inc., 1997.
20. H. L. Willis, *Power Distribution Planning Reference Book*, Marcel Dekker, 1997.
21. H. L. Willis, *Spatial Electric Load Forecasting*, Marcel Dekker, 1996.
22. P. Gill, *Electrical Power Equipment Maintenance and Testing,* Marcel Dekker, 1998.
23. J. Moubray, *Reliability-Centered Maintenance*, John Moubray, 1991 (printed by Butterworth-Heinemann Ltd.).
24. L. L. Grigsby, ed., *The Electric Power Engineering Handbook*, CRC Press, 2000.
25. *Electrical Distribution System Protection*, Cooper Power Systems, 1990.

2
Reliability Metrics and Indices

Rigorous analytical treatment of distribution reliability requires well-defined units of measurements, referred to as metrics. Unfortunately, reliability vocabulary has not been used consistently across the industry, and standard definitions are just beginning to be adopted. This chapter begins by providing definitions for terms and indices commonly used in the areas of power quality and reliability. It continues by comparing engineering measures of reliability with economic measures, and concludes with a discussion on reliability targets.

2.1 POWER QUALITY, RELIABILITY, AND AVAILABILITY

"Power quality" is an ambiguous term that means many things to many people. From a customer perspective, a power quality problem might be defined as any electric supply condition that causes appliances to malfunction or prevents their use. From a utility perspective, a power quality problem might be perceived as noncompliance with various standards such as RMS voltage or harmonics. In fact, power is equal to the instantaneous product of current and voltage, and formulating a meaningful definition of power quality is difficult.

Power quality is better thought of as voltage quality. This is why the IEEE group focused on the subject is called the Working Group on Distribution Voltage Quality. The best a utility can do is to supply customers a perfect sinusoidal voltage source. Utilities have no control over the current drawn by end uses and should be generally unconcerned with current waveforms. Load current can affect voltage as it interacts with system impedances, but voltage is the ultimate measure of power quality.

In this book, power quality is defined as the absence of deviation from a perfect sinusoidal voltage source. Perfect power quality is a perfect sinusoid with constant frequency and amplitude. Less than perfect power quality occurs when a voltage waveform is distorted by transients or harmonics, changes its amplitude, or deviates in frequency.

According to this definition, customer interruptions are power quality concerns since they are a reduction in voltage magnitude to zero. Reliability is primarily concerned with customer interruptions and is, therefore, a subset of power quality. Although there is general agreement that power quality includes reliability, the boundary that separates the two is not well defined. Sustained interruptions (more than a few minutes) have always been categorized as a reliability issue, but many utilities have classified momentary interruptions (less than a few minutes) as a power quality issue. The reasons are (1) momentary interruptions are the result of intentional operating practices, (2) they did not generate a large number of customer complaints, and (3) they are difficult to measure. Today, momentary interruptions are an important customer issue and most distribution engineers consider them a reliability issue. Therefore, this book defines reliability as all aspects of customer interruptions, including momentary interruptions.

Availability is defined as the percentage of time a voltage source is uninterrupted. Its complement, unavailability, is the percentage of time a voltage source in interrupted. Since availability and unavailability deal strictly with interruptions, they are classified as a subset of reliability. The hierarchy of power quality, reliability, and availability is shown in Figure 2.1.

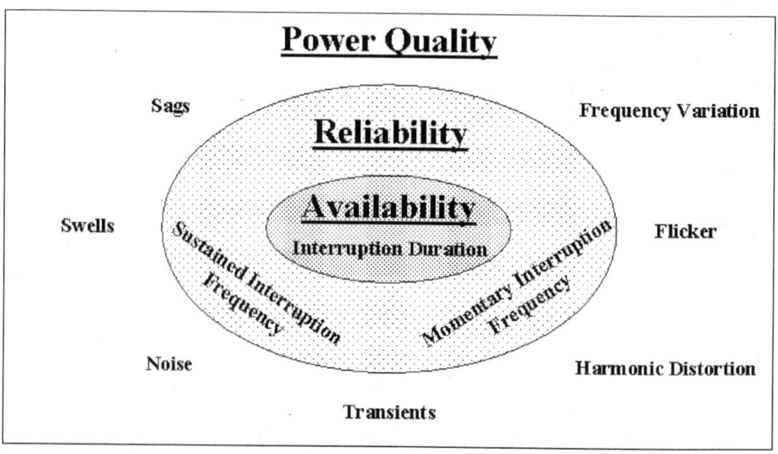

Figure 2.1. Availability is a subset of reliability and reliability is a subset of power quality. Power quality deals with any deviation from a perfect sinusoidal voltage source. Reliability deals with interruptions. Availability deals with the probability of being in an interrupted state.

2.1.1 Power Quality

Perfect power quality is characterized by a perfect sinusoidal voltage source without waveform distortion, variation in amplitude or variation in frequency. To attain near-perfect power quality, a utility could spend vast amounts of money and accommodate electrical equipment with high power quality needs. On the other hand, a utility could spend as little money as possible and require customers to compensate for the resulting power quality problems. Since neither extreme is desirable, utilities must find a balance between cost and the level of power quality provided to customers. Concerns arise when power quality levels do not meet equipment power quality needs.

Power quality concerns are becoming more frequent with the proliferation of sensitive electronic equipment and automated processes. A large part of this problem is due to the lack of communication between electric utilities and product manufacturers. Many manufacturers are not aware of common power quality problems and cannot design their equipment accordingly.

Power quality problems can be divided into many categories such as interruptions, sags, swells, transients, noise, flicker, harmonic distortion and frequency variation. Waveforms corresponding to these power quality problems are shown in Figure 2.2. Each of these categories is a complex topic in its own right, but can only be addressed briefly in this section. The reader is referred to chapter references for more detailed treatment.

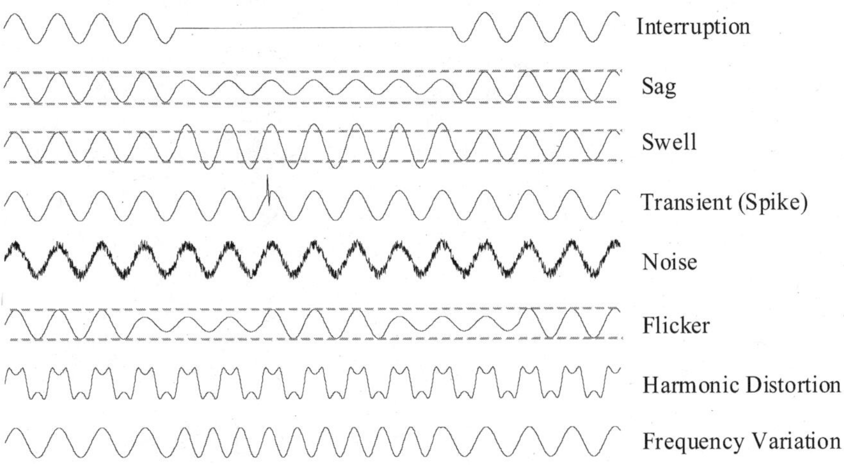

Figure 2.2. Common power quality problems. Perfect power quality corresponds to an undistorted sinusoidal voltage with constant amplitude and constant frequency. Most electrical appliances can accommodate slight disturbances in power quality, with some being much more sensitive than others. Utilities usually attempt to keep voltage quality within tolerances defined in industry standards such as ANSI, IEEE, and IEC.

Table 2.1. Acceptable steady state voltage ranges according to ANSI C84.1.

	Range A (Normal)		Range B (Infrequent)	
	Minimum	Maximum	Minimum	Maximum
Utilization Voltage	0.920	1.05	0.870	1.06
Service Voltage < 600 V	0.950	1.05	0.920	1.06
Service Voltage > 600 V	0.975	1.05	0.950	1.06

Overvoltages and Undervoltages — Steady state voltage magnitude is measured by taking the root-mean-squared (RMS) value of the voltage waveform. Electricity is sold to customers at a nominal RMS voltage, and deviations are referred to as overvoltages and undervoltages. Acceptable overvoltages and undervoltages are addressed by ANSI Standard C84.1[1] and the European Standard HD 472 S1.[2]

ANSI-C84.1 sets normal acceptable voltage levels (Range A) and infrequent acceptable voltage levels (Range B). Normal acceptable service voltages are within 5% of nominal voltage: 114 V to 126 V on a 120-V base. Under unusual circumstances such as extreme weather or temporary system contingencies, this range is extended to 110.4 V through 127.2 V. A summary of acceptable voltage ranges is shown in Table 2.1.

Interruptions — Interruptions are the loss of service voltage to a customer and can be momentary or sustained in nature. They are typically considered a reliability issue and will be discussed further in Section 2.1.2.

Sags — Voltage sags are temporary RMS reductions in voltage typically lasting from a half cycle to several seconds. They are a major power quality concern since they can cause sensitive electronic equipment to fail and motor contacts to drop out.[3] IEC documents use the term dip rather than sag.

Sags result from high currents, typically due to faults or starting motors, interacting with system impedances. The magnitude of a sag is described by either (1) the resulting per unit voltage, or (2) the per unit voltage decrease below nominal. An event that results in a voltage of 0.7 pu can be described as either a "sag to 0.7 pu" or a "sag of 0.3 pu."

The magnitude and duration of voltage sags are often compared against voltage tolerance envelopes. The most widely used curve, generated by the Computer Business Equipment Manufacturer's Association (CBEMA), describes the ability of mainframe computers to sustain voltage sags. This "CBEMA Curve" has been recently updated, renamed the "ITIC Curve" (for Information Technology Industry Council), and adopted by IEEE Std. 446-1995.[4] The ITIC Curve is shown in Figure 2.3.

Swells — Voltage swells are temporary RMS increases in voltage typically lasting from a half cycle to several seconds. Swells are commonly caused by the de-energizing of large loads or asymmetrical faults (a line to ground fault will cause a voltage rise in the other two phases). Swells can cause insulation breakdown in sensitive electronic equipment if voltage increases are high enough for a

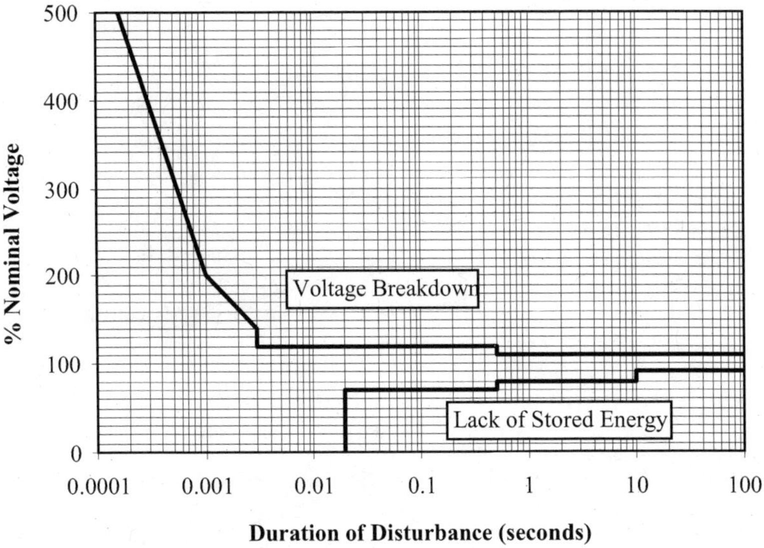

Figure 2.3. ITIC voltage tolerance envelope for Computer Equipment. If less than 20 ms, voltage sags do not typically result in computer failures. Longer durations are tolerable if voltage sags are less severe.

long enough period of time. Equipment tolerance to swells, like sags, is described by voltage tolerance envelopes like the ITIC Curve.

Transients — Voltage transients are sudden nonrecurring changes in voltage magnitude. An impulse transient, most commonly caused by lightning, is described by the time to reach its peak value and the time to decay to half of its peak value. For example, a 1.2 x 50-μs 500-kV transient rises to 500 kV in 1.2-μs and decays to 250 kV in 50 μs. Oscillatory transients, typically caused by capacitor switching and ferroresonance, alternate between positive and negative values at frequencies typically ranging between 1 kHz and 1 MHz. Both ANSI and IEC have standards and guides related to transients.[5-7] Transients are sometimes referred to as voltage surges or voltage spikes.

Noise — Noise can be broadly defined as unwanted voltage signals with broadband spectral content. Common causes include power electronics, control circuits, arcing equipment, loads with solid state rectifiers, and switched mode power supplies. Noise problems are often exacerbated by improper grounding.[8-9]

Flicker — Voltage flicker refers to low frequency variations in RMS voltage that cause visible changes in the brightness of incandescent lighting. These voltage variations are caused by the cycling of large loads such as refrigerators, air conditioners, elevators, arc furnaces, and spot welders. Tolerance levels, such as the flicker curve shown in Figure 2.4, have been generated by a number of empirical studies[10]. Determining whether customer service falls within these tolerance levels has been facilitated by the availability of flicker meters.[11]

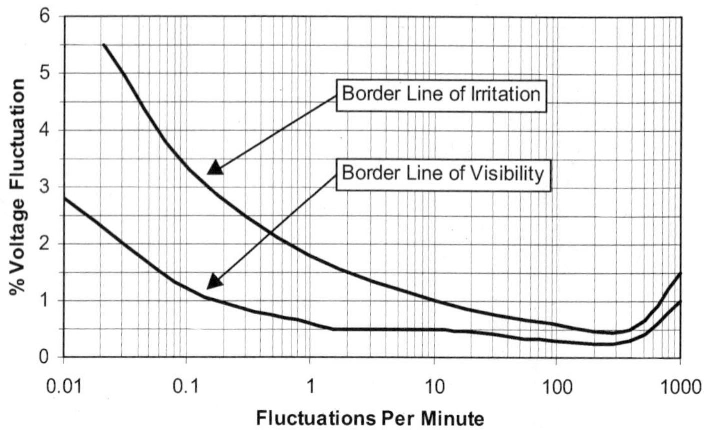

Figure 2.4. A flicker curve based on empirical studies of people observing 60-watt incandescent light bulbs. Flicker becomes perceptible at certain voltage fluctuation levels and frequencies. If fluctuation levels or frequency increase, flicker may become objectionable and irritate observers.

Harmonic Distortion — Periodic voltage waveforms can be described by a Fourier series consisting of a sinusoid with fundamental frequency (60 Hz in the US) and harmonic components with integer multiples of fundamental frequency.

$$V(t) = \sum_{h=1}^{\infty} \sqrt{2} V_h \mathrm{Cos}\left(\frac{2\pi \cdot h \cdot t}{f}\right)$$

t = time (s)

f = fundamental frequency (Hz) (2.1)

h = harmonic component

V_h = RMS Voltage of Harmonic Component h (kV)

A current waveform distorted by harmonics caused by a fluorescent light with a magnetic ballast is shown in Figure 2.5. The chart to the right shows the content of each harmonic and the graph to the left shows the corresponding wave shape. This current will cause harmonic voltage distortion as it interacts with system impedances. Voltages with high harmonic distortion can cause a variety of equipment problems and recommended harmonic limits are described in both US standards[12-13] and European standards.[14-15] Adherence to these standards is becoming more important with the proliferation nonlinear loads such as switched-mode power supplies and adjustable frequency motor drives.[16]

Frequencies that are not integer multiples of fundamental power system frequency are called *interharmonics*. They can appear as discrete frequencies or wideband spectrum and are primarily caused by static frequency converters, cycloconverters, induction motors, and arcing devices.[17]

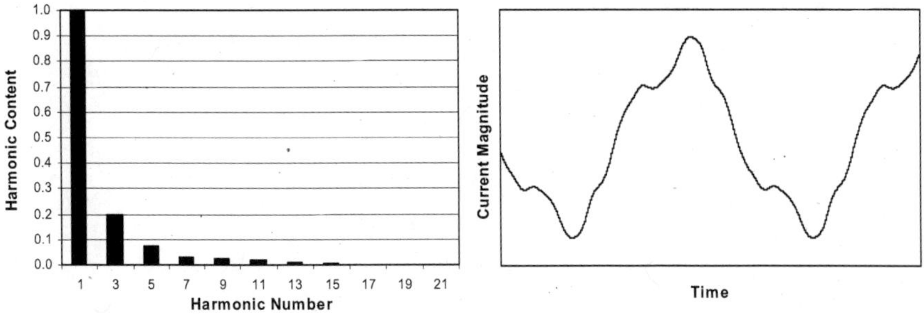

Figure 2.5. Harmonic content and current waveform of a fluorescent light with a magnetic ballast. The resulting severity of harmonic voltage distortion will depend upon system impedance.

The most common measure of harmonic content is total harmonic distortion (THD). THD is a measure related to the square of harmonic components, and is a good measure of the resistive heating potential of harmonic currents. The equation for THD is:

Total Harmonic Distortion:

$$\text{THD} = \frac{1}{V_1}\sqrt{\sum_{h=2}^{\infty} V_h^2} \qquad (2.2)$$

THD is a good measure of waveform distortion, but can be misleading since small currents can have very high levels of THD without being a power quality concern since they do not result in large voltage distortions. To account for current magnitude in addition to distortion, harmonic components can be normalized to peak current levels rather than the fundamental component. Doing so results in an index called Total Demand Distortion (TDD).

THD is not a good measure of heat generation in transformers since it does not accurately account for core losses. The K factor compensates by assuming that higher order harmonic currents contribute more to transformer heating than lower order harmonic currents. Since this phenomenon is generally more pronounced in large transformers, misapplication of K factor often leads to overspecifying small transformers.[22]

K Factor:

$$K = \left(\sum_{h=1}^{\infty} h^2 I_h^2\right) / \left(\sum_{h=1}^{\infty} I_h^2\right) \qquad (2.3)$$

THD and K Factor are good measures of heat generation, but are poor measures of dielectric stress, which is a strong function of peak voltage. The crest factor reflects dielectric stress by expressing the ratio of voltage amplitude and the RMS value of the waveform:

Crest Factor:

$$CF = \frac{\text{Peak Voltage}}{\text{RMS Voltage}} \tag{2.4}$$

A perfect sinusoid has a crest factor of 1.41. Harmonics can either increase this value or decrease this value. For example, a DC voltage has a crest factor of 1.00 and a half-wave rectified voltage source has a crest factor of 2.83.

DC Offset — Voltage has a DC offset if its average value is not equal to zero. This offset can be described as harmonic component h = 0 in a Fourier series, but is typically treated separately. Causes, concerns, and mitigation techniques for DC offset are different when compared to higher order harmonic components. DC offsets can be caused by geomagnetic disturbances and by half wave rectification that causes transformers to operate with saturated core flux.

Commutation Notches — Voltage notching occurs when power electronic switches temporarily cause short circuits between phases during commutation. Since notching is periodic, it can be categorized as harmonic distortion. In practice, notching involves high frequency components that cannot be adequately analyzed by most power quality equipment and must therefore be treated as a special case.

Frequency Variation — Voltage frequency is directly proportional to the rotational speed of synchronous generators. Frequency variations occur when generators adjust to changes in system loading. Large frequency variations are rare in highly interconnected systems and more commonly occur when isolated generators have control problems, cannot properly track large changes in load, or speed up after a system fault.

2.1.2 Reliability

Distribution reliability primarily relates to equipment outages and customer interruptions. In normal operating conditions, all equipment (except standby) is energized and all customers are energized. Scheduled and unscheduled events disrupt normal operating conditions and can lead to outages and interruptions. Several key definitions relating to distribution reliability include:

Contingency — an unexpected event such as a fault or an open circuit. Another term for a contingency is an unscheduled event.

Open Circuit — a point in a circuit that interrupts load current without causing fault current to flow. An example of an open circuit is the false tripping of a circuit breaker.

Fault — a short circuit. Faults are caused by dielectric breakdown of insulation systems and can be categorized as self-clearing, temporary, and permanent. A self-clearing fault will extinguish itself without any external intervention (e.g., a fault occurring on a secondary network that persists until it burns clear). A temporary fault is a short circuit that will clear if de-energized and then re-energized (e.g., an insulator flashover due to a lightning strike — after the circuit is de-energized, the fault path will de-ionize, restoring the insulator to full dielectric strength). A permanent fault is a short circuit that will persist until repaired by human intervention.

Outage — An outage occurs when a piece of equipment is de-energized. Outages can be either scheduled or unscheduled. Scheduled outages are known in advance (e.g., outages for periodic maintenance). Unscheduled outages result from contingencies.

Momentary Interruption — a momentary interruption occurs when a customer is de-energized for less than a few minutes. Most momentary interruptions result from reclosing or automated switching. Multiple reclosing operations result in multiple momentary interruptions (e.g., two recloser operations equals two momentary interruptions for downstream customers).

Momentary Interruption Event — a momentary interruption event consists of one or more momentary interruptions within a several minute time interval (e.g., if a recloser operates three times and then holds, downstream customers experience three momentary interruptions and one momentary event).

Sustained Interruption — a sustained interruption occurs when a customer is de-energized for more than a few minutes. Most sustained interruptions result from open circuits and faults.

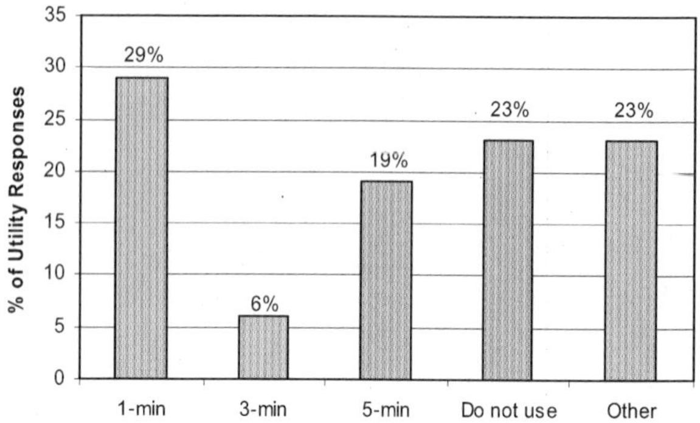

Figure 2.6. Utility maximum momentary interruption durations based on a 1995 survey. Even though the most common threshold is 1 minute, the IEEE has chosen to use 5 minutes in its standards.

Customers do not, in the strictest sense, experience power outages. Customers experience power interruptions. If power is restored within a few minutes, it is considered a momentary interruption. If not, it is considered a sustained interruption. The maximum duration of a momentary interruption varies from utility to utility, but is typically between one and five minutes. Results from a 1995 utility survey, summarized in Figure 2.6, indicate that 1 minute is the most commonly used value.[18] Despite this, the IEEE has chosen define the maximum momentary interruption value as 5 minutes in its 1366 standard.[19]

In the author's opinion, the IEEE momentary interruption definition based on five minutes is more appropriate than those based on one minute. From a customer perspective, there is little difference between a one-minute and a five-minute interruption. From a utility perspective, the difference is substantial since automated switching can usually be accomplished within five minutes but rarely within one minute. With a five-minute threshold, utilities can use automated switching to quickly restore customers before being charged with a sustained interruption — allowing automation to improve reliability frequency indices such as SAIFI (see Section 2.2.1). With a one-minute threshold, these customers will still experience a sustained interruption, concealing the effectiveness of automation and distorting average restoration time indices such as CAIDI.

2.1.3 Availability

Availability is the probability of something being energized. It is the most basic aspect of reliability and is typically measured in percent or per-unit. The complement of availability is unavailability.

Availability — the probability of being energized.
Unavailability — the probability of not being energized.

Table 2.2. Annual interruption times associated with different levels of availability. A developing nation may have "one nine" of availability while an internet data center may have "nine nines" of availability for their servers.

Availability (%)	"Nines"	Annual Interruption Time
90	1	36.5 days
99	2	3.7 days
99.9	3	8.8 hours
99.99	4	52.6 minutes
99.999	5	5.3 minutes
99.9999	6	31.5 seconds
99.99999	7	3.2 seconds
99.999999	8	0.3 seconds
99.9999999	9	1.9 cycles (60-Hz)

Unavailability can be computed directly from interruption duration information. If a customers experiences 9 hours of interrupted power in a year, unavailability is equal to 9 ÷ 8760 = 0.1% (there are 8760 hours in a year). Availability is equal to 100% - 0.1% = 99.9%.

With the growth of ultrasensitive loads, it has become common to describe high levels of reliability by the number of nines appearing at the left of availability values. Many manufacturing plants require "six nines," corresponding to an availability of 99.9999%. Internet data centers may demand reliability as high as nine nines for their servers—less than two cycles of interrupted power per year. Annual interruption times corresponding to various degrees of availability are shown in Table 2.2.

2.2 RELIABLITY INDICES

Reliability indices are statistical aggregations of reliability data for a well-defined set of loads, components, or customers. Most reliability indices are average values of a particular reliability characteristic for an entire system, operating region, substation service territory, or feeder. In the past decade, there has been a proliferation of reliability indices in the US and around the world. Comprehensive treatment is not practicable, but the following sections discuss the most important reliability indices used in the US.

The reliability index definitions provided, where applicable, follow the recently adopted IEEE standard 1366-2003.[19] This standard has not been universally adopted by US utilities, but is growing in popularity and provides a document to which individual utility practices can be compared.

2.2.1 Customer-Based Reliability Indices

The most widely used reliability indices are averages that weight each customer equally. Customer-based indices are popular with regulating authorities since a small residential customer has just as much importance as a large industrial customer. They have limitations, but are generally considered good aggregate measures of reliability and are often used as reliability benchmarks and improvement targets. Formulae for customer-based indices include (unless otherwise specified, interruptions refer to sustained interruptions):

System Average Interruption Frequency Index:

$$\text{SAIFI} = \frac{\text{Total Number of Customer Interruptions}}{\text{Total Number of Customers Served}} \quad \text{/yr} \qquad (2.5)$$

System Average Interruption Duration Index:

$$\text{SAIDI} = \frac{\sum \text{Customer Interruption Durations}}{\text{Total Number of Customers Served}} \quad \text{hr/yr} \qquad (2.6)$$

Customer Average Interruption Duration Index:

$$\text{CAIDI} = \frac{\sum \text{Customer Interruption Durations}}{\text{Total Number of Customer Interruptions}} \quad \text{hr} \qquad (2.7)$$

Average Service Availability Index:

$$\text{ASAI} = \frac{\text{Customer Hours Service Availability}}{\text{Customer Hours Service Demand}} \quad \text{pu} \qquad (2.8)$$

SAIFI is a measure of how many sustained interruptions an average customer will experience over the course of a year. For a fixed number of customers, the only way to improve SAIFI is to reduce the number of sustained interruptions experienced by customers.

SAIDI is a measure of how many interruption hours an average customer will experience over the course of a year. For a fixed number of customers, SAIDI can be improved by reducing the number of interruptions or by reducing the duration of these interruptions. Since both of these reflect reliability improvements, a reduction in SAIDI indicates an improvement in reliability.

CAIDI is a measure of how long an average interruption lasts, and is used as a measure of utility response time to system contingencies. CAIDI can be improved by reducing the length of interruptions, but can also be reduced by increasing the number of short interruptions. Consequently, a reduction in CAIDI does not necessarily reflect an improvement in reliability.

ASAI is the customer-weighted availability of the system and provides the same information as SAIDI. Higher ASAI values reflect higher levels of system reliability, with most US utilities having ASAI greater than 0.999.

Some less commonly used reliability indices are not based on the total number of customers served. The Customer Average Interruption Frequency Index (CAIFI) and the Customer Total Average Interruption Duration Index (CTAIDI) are based upon the number of customers that have experienced one or more interruptions in the relevant year. Formulae for these indices are:

Customer Average Interruption Frequency Index:

$$\text{CAIFI} = \frac{\text{Total Number of Customer Interruptions}}{\text{Customers Experiencing 1 or more Interruptions}} \quad \text{/yr} \qquad (2.9)$$

Customer Total Average Interruption Duration Index:

$$\text{CTAIDI} = \frac{\sum \text{Customer Interruption Durations}}{\text{Customers Experiencing 1 or more Interruptions}} \quad \text{hr/yr} \qquad (2.10)$$

On the surface, CAIFI seems similar to SAIFI. One large difference is that the lowest possible value for SAIFI is zero and the lowest possible value for CAIFI is one. Reducing the number of interruptions that a customer experiences from 2 to 1 will improve CAIFI. However, reducing the number of interruptions that this same customer experiences from 1 to 0 will make CAIFI worse. Improvements in CAIFI do not necessarily correspond to improvements in reliability. CTAIDI has the same difficulties as CAIFI since it can be improved by increasing the number of customers that experience a single interruption.

The increasing sensitivity of customer loads to brief disturbances has generated a need for indices related to momentary interruptions. Two momentary indices have become standard. One is based on the frequency of momentary interruptions and the other is based on the frequency of momentary events. Formulae for these indices are:

Momentary Average Interruption Frequency Index:

$$\text{MAIFI} = \frac{\text{Total Number of Customer Momentary Interruptions}}{\text{Total Number of Customers Served}} \quad \text{/yr} \qquad (2.11)$$

Momentary Event Average Interruption Frequency Index:

$$\text{MAIFI}_\text{E} = \frac{\text{Total Number of Customer Momentary Events}}{\text{Total Number of Customers Served}} \quad \text{/yr} \qquad (2.12)$$

These definitions are not followed in much of the literature. The acronym MAIFI is typically encountered and can refer to either Eq. 2.11 or Eq. 2.12. MAIFI is attractive to utilities because it can be easily computed from breaker and recloser counters. MAIFI_E is a better measure of customer satisfaction since multiple closely spaced momentary interruptions have much less impact than the same number of momentary interruptions spaced days or weeks apart.

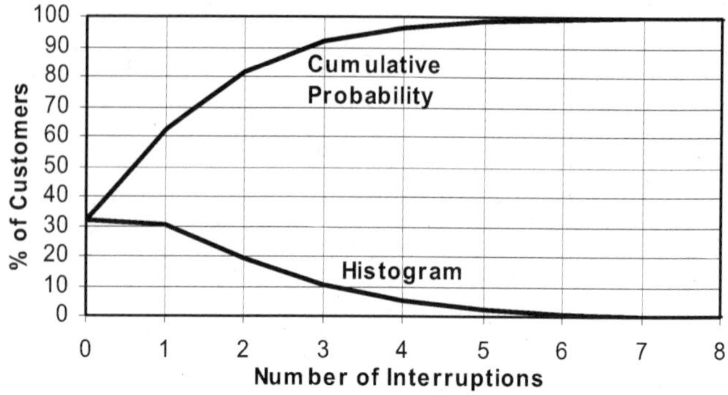

Figure 2.7. Probability distribution of interruption frequency for an area serving 27,000 customers in central Texas. About 32% of customers will not experience an interruption over the course of a year. About 99% of customers will experience 5 interruptions or less.

The previously discussed reliability indices reflect average system reliability. Such measures may not have a high correlation to customer satisfaction since few customers may actually experience average reliability. Ideally, each reliability measure could be plotted as a histogram to identify the percentage of customers receiving various levels of reliability.

Figure 2.7 shows the distribution of sustained interruptions for a 25 feeder area in central Texas. The histogram shows the percent of customers that experience each number of sustained interruptions. The cumulative probability curve shows the percent of customers that experience less than or equal to each number of interruptions. These types of curves are useful to examine variations in customer reliability and to identify the number of customers with relatively poor reliability.

There are two reliability indices that correspond to points on cumulative distribution functions such as the one shown in Figure 2.7. These indices have a subscript that defines a threshold level of reliability that a customer must exceed before being counted. One index considers sustained interruptions and another considers both momentary and sustained interruptions:

Customers Experiencing Multiple Interruptions:

$$\text{CEMI}_n = \frac{\text{Cust. Experiencing More Than n Interruptions}}{\text{Total Number of Customers Served}} \quad /\text{yr} \qquad (2.13)$$

Customers Experiencing Multiple Sustained and Momentary Interruptions:

$$\text{CEMSMI}_n = \frac{\text{Cust. Experiencing More Than n Combined Momentary and Sustained Interruptions}}{\text{Total Number of Customers Served}} \quad /\text{yr} \qquad (2.14)$$

Although few utilities are using CEMSMI, a growing number of utilities are using CEMI (both for internal and regulatory purposes). The rationale is simple. If customers tend to become dissatisfied after having more than "n" interruptions in a year, then $CEMI_n$ will indicate the percentage of customers who are potentially dissatisfied due to this factor.

2.2.2 Load-Based Reliability Indices

Two of the oldest distribution reliability indices weight customers based on connected kVA instead of weighting each customer equally. Because of this, the following are referred to as load-based indices:

Average System Interruption Frequency Index:

$$\text{ASIFI} = \frac{\text{Connected kVA Interrupted}}{\text{Total Connected kVA Served}} \quad /\text{yr} \qquad (2.15)$$

Average System Interruption Duration Index:

$$\text{ASIDI} = \frac{\text{Connected kVA Hours Interrupted}}{\text{Total Connected kVA Served}} \quad \text{hr/yr} \qquad (2.16)$$

The reason for load-based indices predating customer based indices is pragmatic. In the past, utilities knew the size of distribution transformers but did not know how many customers were connected to each transformer. When a protection device operated, interrupted transformer kVA was easily determined while interrupted customers required estimation. Today, customer information systems (CIS) associate customers with transformers and allow customer based indices to be easily computed.

From a utility perspective, ASIFI and ASIDI probably represent better measures of reliability than SAIFI and SAIDI. Larger kVA corresponds to higher revenue and should be considered when making investment decisions.

2.2.3 Power Quality Indices

Reliability is a subset of power quality, and many reliability decisions affect other areas of power quality. Because of this, familiarity with basic power quality indices is desirable. Many new power quality indices are being proposed,[20-21] but most are less appropriate for aggregate utility measures and are more appropriate for measuring power quality at specific customer locations. This section presents selected power quality indices that have applications similar to reliability indices.

Many sensitive loads cannot tell the difference between voltage sags and momentary interruptions. Because of this, interruption-based reliability indices are not able to reflect all customer concerns and voltage sag indices are becoming necessary. The *de facto* sag index standard is $SARFI_x$.[23]

System Average RMS Variation Frequency Index:

$$SARFI_X = \frac{\text{Total Customer Sags Below x\%}}{\text{Total Number of Customers Served}} \quad /yr \qquad (2.17)$$

Since the severity of voltage sags depends upon duration as well as magnitude, $SARFI_x$ may not contain enough information to be a useful index in all cases. One solution is to use an index based on the frequency of sags that violate voltage tolerance envelopes. For example, an index based on the ITIC curve (see Figure 2.3) could be called $SARFI_{ITIC}$. Another approach is to decompose $SARFI_x$ into an instantaneous, momentary, and temporary index based on the duration of the sag. Formulae for these duration dependent sag indices are:

System Instantaneous Average RMS Variation Frequency Index:

$$SIARFI_X = \frac{\text{Total Customer Sags Below x\% (0.5c - 30c)}}{\text{Total Number of Customers Served}} \quad /yr \qquad (2.18)$$

System Momentary Average RMS Variation Frequency Index:

$$SMARFI_X = \frac{\text{Total Customer Sags Below x\% (30c - 3s)}}{\text{Total Number of Customers Served}} \quad /yr \qquad (2.19)$$

System Temporary Average RMS Variation Frequency Index:

$$STARFI_X = \frac{\text{Total Customer Sags Below x\% (3s - 60s)}}{\text{Total Number of Customers Served}} \quad /yr \qquad (2.20)$$

In the past, voltage sags were not given broad attention since they were difficult for utilities to identify and quantify. In the future, electronic meters will be able to cost-effectively identify and record sags on a widespread basis. Customers will be able to identify sag problems, utilities will be able to record and track sag indices and regulators will be able to set sag index targets and integrate them into performance-based rates.

2.2.4 Interdependence of SAIDI, SAIFI, and MAIFI

All utilities track multiple reliability indices, and many have multiple goals associated with these indices. These multiple goals can be a result of multiple reliability indices (e.g., SAIFI, SAIDI, MAIFI), reliability indices for different levels of the system (e.g., SAIDI for the entire system, SAIDI for each operating area), or both.

Setting multiple performance targets for reliability is a straightforward and prudent way to track reliability over time. However, it greatly complicates the ability to prioritize reliability spending decisions when compared to using a single measure. If only one single reliability index is involved (e.g., SAIDI), each potential improvement project can be ranked in terms of its cost-to-benefit ratio (e.g., dollars per SAIDI minute). If multiple performance targets are involved, the reliability-spending problem becomes an optimization problem with multiple objectives and constraints. This type of problem is solvable with the use of computer optimization software, but it is questionable whether the added complexity is justified in terms of improved reliability spending.

To date, the most common way for utilities to prioritize distribution reliability spending decisions is based on SAIDI improvement. For each potential project, a cost-to-benefit ratio is computed based on the expected cost of the project and the expected reduction in SAIDI. The author has examined reliability improvement recommendations for thousands of distribution feeders based on dollars-per-SAIDI reduction and has found that the resulting recommendations do a good job of improving $MAIFI_E$, SAIFI, and SAIDI, even though the metric is based strictly on SAIDI.

Of course, every project that will improve SAIFI will also improve SAIDI. The question is whether the value of reliability improvement is proportional for short interruptions. Assume that the shortest possible sustained interruption is approximately one hour. If a two-hour interruption is twice as costly to a customer as a one-hour interruption, on average, then SAIDI improvement will be a good proxy for SAIFI improvement. On the other hand, if a two-hour interruption is valued about the same as a one-hour interruption to customers, SAIDI improvement will not be a good proxy for SAIFI improvement.

Research differs on this point, and one can easily dispute the validity of customer cost surveys in a variety of ways (See Section 2.3). However, a reasonable approach is to assume that typical residential customers will be willing to pay about 25 cents to avoid a one-minute interruption, and about two dollars to avoid a one-hour interruption. If these assumptions are valid, investments based on SAIDI will do a good job at reducing SAIFI.

The situation may be significantly different for commercial and industrial customers, where short interruptions could result in computer crashes and ruined batch processes. However, most utilities have approximately 90% residential customers. This means that SAIDI is a good single measure as long as all customers are to be given the same weight when making reliability investment decisions.

$MAIFI_E$ is a bit more complicated than SAIFI in its relationship to SAIDI. As discussed earlier, any effort to improve SAIFI will always improve SAIDI. This is not necessarily the case with $MAIFI_E$. For example, disabling the instantaneous trip on reclosing devices will reduce $MAIFI_E$, but actually make SAIDI get worse. Similarly, enabling the instantaneous trip on reclosing devices will improve SAIDI while worsening $MAIFI_E$.

There are two causes of momentary interruptions: reclosing and automated switching. The explicit goal of automated switching is to quickly restore customers so that what would have been a sustained interruption is converted into a momentary interruption. In this case, the momentary interruption is a deliberate and good thing. When a utility undertakes an aggressive distribution automation program, $MAIFI_E$ can be expected to increase, which is also a good thing.

Reclosing is used to allow temporary faults to clear themselves. For typical overhead distribution systems, 60% to 80% of faults will be temporary in nature. If a recloser interrupts a temporary fault and then recloses, the problem will often be solved automatically. This is desirable since the utility does not have to dispatch crews, and no repairs are necessary. With reclosing, fewer sustained interruptions are intentionally traded-off for more momentary interruptions.

In summary, SAIDI, SAIFI, and $MAIFI_E$ are interrelated. Experience has shown that reliability spending decisions based SAIDI alone do a good job of simultaneously improving SAIFI and the undesirable aspects of $MAIFI_E$.

2.2.5 Potential Problems with Standard Indices

Although the most commonly used indices do a reasonable job in tracking the reliability performance of utilities, they have the potential of allocating spending decisions that are not closely aligned with customer interests. This is especially true for utilities that are mature in their reliability improvement processes. Put simply, when there is a lot of "low hanging fruit," traditional reliability measures do a great job of prioritization. However, once this first round of investment is made, traditional reliability measures may present complications. The following describes potential problems with standard indices.

SAIDI and SAIFI – When making reliability investments, reductions in SAIDI and SAIFI are proportional to the number of affected customers. This means projects that affect many customers are preferred to those that affect few customers. In itself, this is not a problem. However, feeders with many customers typically have better-than-average reliability, and feeders with few customers typically have worse-than-average reliability. Therefore, reliability investments based on SAIDI and SAIFI can drive investments towards densely populated areas where reliability is already satisfactory.

CAIDI – Although popular with many utilities and regulators, CAIDI is problematic as a measure of reliability. In the authors opinion, this is because CAIDI does not mean what most think. Many view CAIDI as a measure of operational efficiency; when the utility responds more quickly after a fault, CAIDI will go down. This is true, but only part of the story. In fact, CAIDI is mathematically equal to SAIDI divided by SAIFI. Therefore, CAIDI will increase if SAIFI improves more quickly than SAIDI. That is, reliability could be improving in both frequency and duration, but CAIDI could be increasing.

Consider a utility embarking on reliability improvement initiatives. Typically, the most effective initial activities will focus on faults that occur frequently but are relatively quick and easy to fix (e.g., animal problems). When the quick and easy problems are solved, the remaining interruptions on the system will take longer to repair, causing CAIDI to increase. This situation is common and has caused frustration at many utilities; reliability is improving yet CAIDI is increasing. To avoid this problem, the author recommends against using CAIDI, preferring the use of SAIFI and SAIDI, which are mathematically equivalent.

$MAIFI_E$ – Like SAIFI and SAIDI, this index will also tend to drive investments towards densely populated areas where reliability is already satisfactory. To make matters worse, $MAIFI_E$ will discourage reliability investments in automated switching schemes because they will result in more momentary interruptions for customers. In addition, many utilities are only able to compute $MAIFI_E$ (or MAIFI) based on substation information, and do not have the ability to include the impact of line reclosers and automated switching actions.

The potential problems with SAIDI, SAIFI, and $MAIFI_E$ can be summarized as follows. Utilities are expected to provide adequate levels of reliability to customers, but these levels are not embedded in SAIDI, SAIFI, and $MAIFI_E$. Targets can be assigned to each of these indices, but they will be of no help. To reach a SAIDI target, the most cost effective solution will be biased towards densely populated areas that may already have good reliability.

Reliability targets can be set at a lower level (e.g., by substation or feeder), but the problem remains. For a feeder with a specific SAIDI target, it may be more cost effective to improve the reliability in areas of the feeder that already have good reliability rather than on parts of the feeder that have poor reliability but with relatively few customers.

$CEMI_n$ – Many utilities are beginning to understand the potential disconnect of SAIDI and SAIFI with customer satisfaction. As a result, there is a general industry trend towards the use of $CEMI_n$. Since $CEMI_n$ is based on customers that experience more than a threshold level of interruptions, it will never direct spending towards areas of the system that are already below this threshold.

However, $CEMI_n$ has problems of its own. First, $CEMI_n$ can only realistically be used in conjunction with other indices; it does not convey information on the duration of interruptions faced by customers. And when one has to include other indices, one will therefore also be subject to the limitation of those other indices.

Second, and more importantly, $CEMI_n$ has the following potential problems. $CEMI_n$ is strictly a frequency measure, and will assign zero value to projects designed to reduce the length of interruptions. Examples of reliability-improvement projects that will have no impact on $CEMI_n$ include: manual switches, ties between feeders, faulted circuit indicators, more aggressive customer restoration, and quicker repair times.

Even more disconcerting is the fact that $CEMI_n$ will assign zero value to most frequency-improvement projects that are targeted at the worst areas of the system. Consider a utility using $CEMI_3$, but has areas of the system where customers are experiencing ten interruptions per year. If the utility improves reliability such that these customers now only experience five interruptions per year, there is no improvement to $CEMI_3$. Any utility using $CEMI_n$ will heavily bias its spending towards customers near the $CEMI_n$ threshold. This may be result in spending decisions that are less aligned with customer interests than if $CEMI_n$ was not used at all.

In summary, SAIDI and SAIFI are generally good measures of reliability, but can potentially bias spending towards areas of the system that may already have adequate reliability. CAIDI is confusing since increasing CAIDI could be either good or bad. $MAIFI_E$ is also confusing since it can increase while customer reliability improves. $CEMI_n$ is problematic since it biases spending away from areas that have the worst levels of reliability.

2.2.6 A Potential New Index

Based on the potential shortcomings of standard indices, this section presents a new index that compensates for these shortcomings. It must be emphasized that this new index is not an industry standard. Rather, it is simply a possibility suggested by the author that addresses some of the limitations of standard indices. Goals for this new reliability measure include:

Goals for a Better Reliability Measure

- Easy to understand
- Avoids known problems with standard indices
- Aligned closely with customer satisfaction
- Useful for planning and budgeting
- Appropriate for performance management
- Useful for regulatory reporting
- Appropriate for performance-based regulation

As stated earlier, the author has found that SAIDI generally does a good job in driving investment decisions for a utility. The only potential problem is that SAIDI will sometimes encourage spending in areas that already have adequate reliability. To compensate for this shortcoming, a new index is proposed that is similar to SAIDI, but only considers customer interruption minutes when they exceed a threshold value for each individual customer. This new index is called the System Average Interruption Duration Exceeding Threshold, or SAIDET.

System Average Interruption Duration Exceeding Threshold:

$$SAIDET_X = \frac{\sum_{\text{Customers}} \text{Annual Interruption Duration Exceeding X}}{\text{Total Number of Customers Served}} \quad \text{min/yr} \qquad (2.21)$$

SAIDET can be viewed as the average violation experienced by customers. Consider the example shown in Table 2.3. There are five customers, each with a SAIDET threshold of 120 minutes. In the course of a year, each customer experiences an aggregate number of interruption minutes ranging from zero to nearly three hundred. The average of these numbers is equal to SAIDI. The number of interruption minutes exceeding the threshold value is then recorded for each customer. The average of these numbers is equal to SAIDET at the 120-minute threshold.

For customers exceeding the threshold, SAIDET will value projects in precisely the same way as SAIDI. For customers below the threshold, SAIDET will assign no value to projects. This is in close alignment with the regulatory concept of adequate service; if a customer is receiving adequate service, there is no incentive for additional improvement. It is also in close alignment with customer satisfaction; each customer will only contribute to SAIDET if reliability is worse that a threshold level, which acknowledges that customers do not expect perfect reliability.

SAIDET and other potential similar measures can be used in most cases to set reliability improvement targets and to direct reliability spending. However, some utilities and regulators will desire to supplement SAIDET in the way that SAIFI and MAIFI are used to supplement SAIDI. For this reason, two additional measures are suggested: SAIFET and MAIFET.

Table 2.3. Example of SAIDET calculation.

	Threshold (min)	Interruption Minutes	Minutes above Threshold
Customer A	120	150	30
Customer B	120	90	0
Customer C	120	295	175
Customer D	120	190	70
Customer E	120	0	0
Sum		750	275
		SAIDI = 150 min/yr	**SAIDET = 55 min/yr**

System Average Interruption Frequency Exceeding Threshold:

$$\mathrm{SAIFET_X} = \frac{\sum_{\text{Customers}} \text{Sustained Interruptions Exceeding X}}{\text{Total Number of Customers Served}} \ \text{/yr} \qquad (2.22)$$

Momentary Average Interruption Frequency Exceeding Threshold:

$$\mathrm{MAIFET_X} = \frac{\sum_{\text{Customers}} \text{Momentary Interruption Events Exceeding X}}{\text{Total Number of Customers Served}} \ \text{/yr} \qquad (2.23)$$

SAIFET can be viewed as the average sustained interruption violations experienced by customers. MAIFET can be viewed as the average momentary interruption violations experienced by customers.

These threshold indices are averages values, but are still closely aligned with customer satisfaction and the concept of adequate service. Consider a utility that defines adequate service for a specific customer as three sustained interruptions, six momentary interruption events, and three hundred minutes of total interruption time. This utility could use $\mathrm{SAIFET_3}$, $\mathrm{MAIFET_6}$, and $\mathrm{SAIDET_{300}}$. Only customers experiencing inadequate service on one or more of these measures would be eligible for reliability improvement dollars.

Of course, a utility could define adequate levels of service on a customer level without using threshold indices. However, it will be very difficult to prioritize spending without resorting to a simple index. $\mathrm{CEMI_n}$ and similar indices could be used to measure the percentage of customers not experiencing adequate service, but these types of measures can bias spending against areas with the worst reliability when used to prioritize spending.

Inevitably, there will be objections to the use of new indices such as SAIDET. The author has tried to anticipate some of these objections and offer responses, as indicated below.

Objection #1

"Customers care about multiple interruptions more than duration."

Reply to Objection #1

Customers that have frequent and long interruptions will be naturally addressed due to the duration aspect of SAIDET. Customers that have frequent and short interruptions generally have the ability to be restored after a fault occurs. Improvements for these customers will therefore tend to focus reducing the number of events. Improvements for surrounding customers on the same feeder will also tend to reduce the number of events. Based on these observations, SAIDET will naturally address customers that experience multiple interruptions.

Objection #2
"Customers care about very long outages rather than the total interruption time over a year."

Reply to Objection #2
Reducing very long outages will improve SAIDET. Since SAIDET is a system average, systematic issues associated with long interruptions will impact SAIDET from both a historical and predictive perspective. Based on these observations, SAIDET will naturally address customers that experience very long interruptions.

Objection #3
"A SAIDET threshold is not directly related to customer satisfaction. Customer do not have a good feel for the aggregate interruption duration over the past year."

Reply to Objection #3
The author is not convinced of this statement. It is true that customers can more easily recall the length of a specific interruption and the total number of interruptions. However, the correlation of customer satisfaction to actual SAIDET is probably stronger than SAIDI, SAIFI, MAIFI, or $CEMI_n$ alone.

Objection #4
"Nobody else is using this."

Reply to Objection #4
Many utilities are finding the need to supplement IEEE standard indices with their own internal measures. Clearly there is a need for additional indices, and measures like SAIDET should be considered along with others.

There is a particular aspect of SAIDET that many may initially find uncomfortable. It is possible to improve SAIDET while SAIDI actually gets worse. This could happen if the reliability of customers far below the threshold is allowed to move closer to the threshold level. Consider the example shown in Table 2.4. Two customers are experiencing reliability below the threshold, and three customers are experiencing reliability above the threshold. In this example, the reliability of all customers is moved closer to the threshold value. The result is a dramatic reduction in SAIDET, but a slight increase in SAIDI.

For the example in Table 2.4, an increase in SAIDI is not a bad thing. Resources that otherwise would have been used to maintain the reliability of customers experiencing reliability much better than the threshold levels have simply been shifted to customers experiencing reliability below threshold levels.

Table 2.4. Example of SAIDET calculation.

	Threshold (min)	Interruption Minutes		Minutes above Threshold	
		Before	After	Before	After
Customer A	120	150	130	30	10
Customer B	120	90	110	0	0
Customer C	120	295	250	175	130
Customer D	120	190	170	70	50
Customer E	120	25	115	0	0
Sum		750	775	275	190
SAIDI (min/yr)		**150**	**155**		
SAIDET (min/yr)				**55**	**38**

In realistic scenarios, it is unlikely that SAIDI is in danger of becoming worse unless threshold levels are set too high. Therefore, the threshold level used in SAIDET should correspond to adequate levels of service. If different areas are deemed to have different levels of adequate service (e.g., rural versus urban areas), different SAIDET thresholds should be used.

In summary, SAIDET is a suggested new reliability index that can effectively be used to drive reliability investment decisions within a utility. SAIDET is easy to understand, avoids known problems with standard indices, is aligned closely with customer satisfaction, is useful for planning and budgeting, is appropriate for performance management, is useful for regulatory reporting, and is appropriate for performance-based regulation. The only caution is that (1) the threshold value of SAIDET must correspond to an adequate level of service, and (2) it is not an industry standard.

2.2.7 Exclusion of Storms and Major Events

When electric utilities compute reliability indices, they often exclude interruptions caused by storms and other major events.[24] In the US, comparisons of reliability indices between utilities are difficult since more than 80% exclude major events but definitions of major events vary widely. Examples of major event definitions include:

- A severe storm, flood, or civil disturbance that requires three or more days to restore service.
- The following criteria are met: (1) the National Weather Service has issued a severe watch or warning for the area. (2) Extensive mechanical damage has been sustained. (3) More than 10% of customers are out of

service at some time during or immediately after the storm. (4) At least 1% of customers are out of service 24 hours after the beginning of the storm.

- An event resulting in more damaged sites than available crews to be dispatched to these sites.
- A certain percentage of customers in a specific dispatch area experience an interruption.

There are differing views about the appropriateness of excluding major events from reliability index calculations. From a customer cost perspective, it should not matter whether interruptions occur during mild or severe weather and reliability targets should be set to maximize societal welfare. From a utility perspective, distribution systems are not designed to withstand extreme weather such as earthquakes, floods, forest fires, hurricanes, ice storms, and tornadoes. Since doing so would require substantial rate increases, reliability measurements and improvement efforts should focus on nonstorm performance. In addition, customers tend to be more tolerant of interruptions during severe weather since the cause of the interruption is apparent and utility response is highly visible.

If a utility is financially penalized or rewarded based on reliability indices, major event exclusions may lead to gaming behavior. Consider a utility that categorizes storms based on three or more days to restore service to all customers. Now suppose that high winds result in widespread interruptions, but all customers can be restored within 2 days if crew overtime is utilized. If this utility excludes storms from index calculation, it may decline to use overtime and allow some customers to be interrupted for more than three days. Not only will the utility save money by not incurring overtime costs, but reliability indices will improve since the entire event will be excluded.

2.2.8 The IEEE 2.5 Beta Method

Since utilities have a wide range of approaches to storm exclusion for normal reliability index calculations, reliability index comparisons across utilities can be confusing. To address this concern, the IEEE has recommended a statistical-based approach to storm exclusion in Standard 1366. A utility is, of course, free to compute reliability indices using its traditional storm exclusion criteria, but the IEEE recommends that a utility also compute indices with its proposed approach for benchmarking purposes.

The IEEE storm exclusion process as detailed in Standard 1366 is called the "2.5 Beta Method." This is because the exclusion criterion is based on the log-normal parameter beta. [Note: Standard 1366 does not used the term "exclusion." Rather, it suggests that interruptions associated with an identified major event be classified as such and separately analyzed. Of course, this data is still effectively excluded from nonstorm reliability index calculations.]

The 2.5 Beta Method works by classifying each day as either a normal day or a major event day (MED). A major event day is a day in which the daily system SAIDI exceeds a threshold value, T_{MED}. Even though SAIDI is used to determine the major event days, all indices should be calculated based on removal of the identified days.

The threshold value of T_{MED} is computed at the end of each year for use in the following year. The process of computing T_{MED} is as follows:

Major Event Exclusion Using the 2.5 Beta Method

1. Collect the last five years of historical data including the daily SAIDI for each of these days.
2. Throw out all days that did not have any interruptions.
3. Take the natural logarithm for each remaining daily SAIDI value.
4. Compute the average value of these natural logarithms and set this equal to alpha (α).
5. Compute the standard deviation of these natural logarithms and set this equal to beta (β).
6. Compute the major event day threshold: $T_{MED} = \exp(\alpha + 2.5\ \beta)$.
7. Any day with a daily SAIDI greater than T_{MED} that occurs in the future year's reporting period is classified as a major event day and not included in normal reliability index calculations.

The 2.5 Beta method has been carefully formulated based on a balance of simplicity and actual results for a large body of historical utility data. Essentially, the IEEE found that utility SAIDI days, for the most part, follow a lognormal distribution. When utilities identified the days that they would consider major events based on other criteria, it was found that these days generally fall above the 2.5 Beta threshold point as described above.

Since the 2.5 Beta method was designed for simplicity, it encounters some problems that need discussing.

Events spanning multiple days. The 2.5 Beta method states that interruptions that span multiple days should have all of their associated interruption time allocated to the day in which the interruption started. This is a bookkeeping problem that some outage management systems may find difficult to accommodate. More importantly, an event contained within a single calendar day may qualify as a major event, but the same event split between two calendar days may not qualify as a major event. For example, consider an event that interrupts 200,000 customers for five hours, for a total of one million customer hours. If this event occurs within one calendar day, and T_{MED} corresponds to 800,000 customer hours, this day will qualify for exclusion. If the same event spans two days so that 500,000 customer interruption hours occur in each of two days, neither day will qualify for exclusion, even though the impact to customers is the same.

Zero days. Smaller distribution systems will sometimes have days with no customer interruptions. Since one cannot take the logarithm of zero, these "zero days" are difficult to account for in the 2.5 Beta method. After much deliberation, the IEEE decided that the best approach is to simply discard these days rather than complicate the methodology.

System size. Distribution systems in the US vary dramatically in size, from a few thousand customers to many millions of customers. The statistical nature of a distribution system depends in part on system size, since larger systems will have a larger number of events from which statistics can be computed, especially when considering a single day. In practice, this results in small utilities tending to qualify for more major event days than large utilities.

Even within a single utility, reliability indices will often be computed for various system sizes such as the entire system, an operating area, a substation, and a feeder. This creates a problem. Should major event days based on the entire system be used for reliability calculations for all subsystems? Should a separate T_{MED} be computed for each subsystem? If each subsystem has a separate T_{MED}, will the weighted average of reliability of all subregions be equal to the overall computed reliability? Consider a utility with five operating areas. One day, a storm interrupts 40% of customers in one operating area, which corresponds to 8% of all customers in the system. If T_{MED} is computed for each operating area, it is likely that this event will qualify for exclusion. If T_{MED} is computed for the system as a whole it is likely that this event will not qualify for exclusion.

Extreme events. Many distribution systems, at some point, experience very large and devastating events. A good example is the 2004 and 2005 hurricane season in Florida, where multiple storms pummeled Florida utilities and resulted in billions of dollars of damage. Of course, the 2.5 Beta method will identify these events for exclusion. However, the extreme nature of these events will dramatically increase T_{MED} for the next year, and make it much more difficult for a day to qualify as a major event day. Practically this could result in the following. In year 1 a utility has a storm that interrupts 10% of its customers for 10 hours, and this qualified for exclusion. In the same year, the utility has an extreme event that interrupts a large percentage of customers for weeks. In year 2 a utility has another storm that interrupts 10% of its customers for 10 hours. Because of the extreme event in the prior year, TMED has increased and this storm no longer qualifies for exclusion.

In summary, the 2.5 Beta method is a practical and consistent approach to identify days that are reliability outliers and exclude them from consideration for reliability index calculations. This method has been tested on data from a large number of utilities, has shown to acceptably identify days where standard operational procedures are stressed, and has growing acceptance in the industry.

2.2.9 Benchmarking

Benchmarking distribution reliability indices across utilities is difficult for a variety of reasons including geography, data gathering practices, index definitions and major event exclusions. Regardless, many US utilities report reliability indices and it is inevitable that comparisons will be made. Figure 2.8 shows US utility reliability index usage based on an IEEE survey of 64 utilities.[18]

Since so many utilities report SAIDI, SAIFI, CAIDI, it is possible to statistically examine historical values. Table 2.5 groups utility reliability index values into quartiles and shows the minimum values to qualify for each quartile. Each index is shown to have wide variation from top quartile to bottom quartile. Part of this variation is, no doubt, due to system design and equipment maintenance. This does not account for all of the variation, and one cannot necessarily assume that utilities with better indices are better managing their reliability.

Distribution systems across the country cover widely varying terrain and are exposed to widely varying weather. Some distribution systems are routed through dense vegetation while others are routed through open fields. Some areas of the country have high lightning activity while others areas experience little. Some distribution systems serve dense populations and are primarily underground while others serving rural populations are primarily overhead. Dense areas can have many feeder interconnections while sparse areas are typically radial. All of these factors can have a major impact on reliability, and these types of differences should always be taken into account when comparing reliability indices between different distribution systems.

Reliability trends based on annual IEEE reliability surveys is shown in Figure 2.9. Reliability indices have, on average, become gradually worse over recent years. Part of this trend may be due to better data collection, but part is likely due to actual worsening of reliability.

Benchmarking results from a 1999 EEI survey are show in Figure 2.10. The complete distributions of results are shown, and results are basically consistent with IEEE surveys.

Table 2.5. Typical reliability index values for US utilities based on 2005 IEEE survey data. Quartiles refer to the minimum value for a utility to fall in the top 25% of utilities (first quartile), next top 25% of utilities (second quartile), and so forth. Results are presented for indices calculated using both the 2.5 Beta major event exclusion methodology, and with all data included.

Quartile	SAIDI (min/yr)		SAIFI (/yr)		CAIDI (min)	
	2.5 Beta	All Data	2.5 Beta	All Data	2.5 Beta	All Data
1	98.3	128.4	1.09	1.25	82.5	101.4
2	144.6	216.5	1.39	1.63	104.9	134.1
3	191.8	366.0	1.63	2.13	127.4	179.8
4	410.4	2091.1	7.25	7.25	224.4	760.0

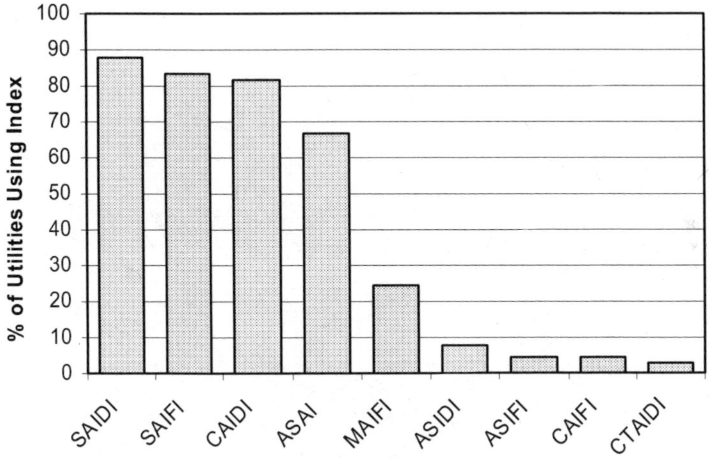

Figure 2.8. Usage of reliability indices by US utilities. Customer based indices (SAIDI, SAIFI, CAIDI, and ASAI) are the most common. MAIFI was not widely used a decade ago, but is quickly gaining popularity.

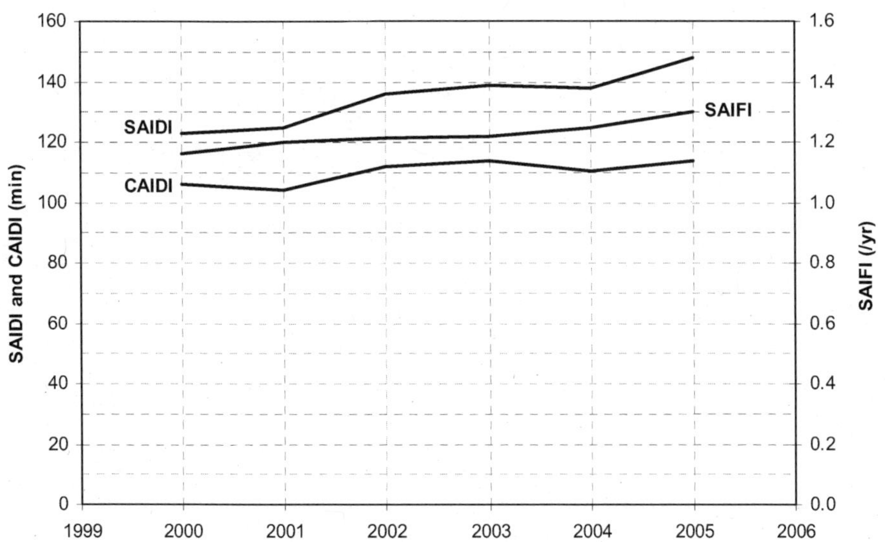

Figure 2.9. Reliability index trends (using 2.5 Beta) for US utilities based on 55 utilities reporting 2000 through 2005 results to the IEEE. Reliability indices have, on average, become gradually worse over recent years. Part of this trend may be due to better data collection, but part is likely due to actual worsening of reliability.

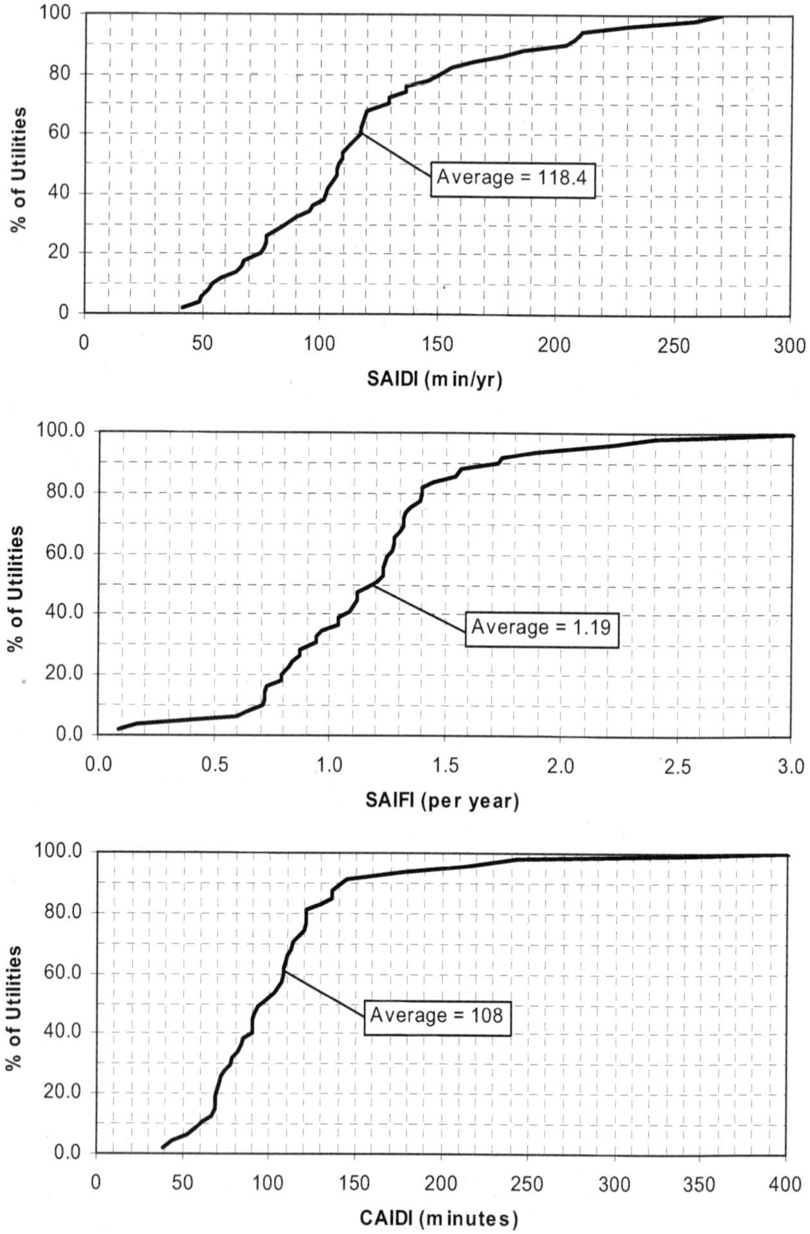

Figure 2.10. Reliability index values for US utilities based on 1999 survey data. In each graph, reliability indices are plotted from lowest to highest. Similar to Table 2.5, wide variability in reliability indices is seen.

In addition to geography, reliability indices can vary substantially based on utility data gathering practices. Many utilities compute reliability indices based on outage reports that are manually filled out by crews. Manual data collection tends to omit a significant percentage of interruptions and inaccurately accounts for customer restoration due to system reconfiguration. To overcome these data collection problems, many utilities are installing outage management systems (OMS) that automatically track customer interruption as they occur. When OMSs are installed, some utilities have found that SAIDI values have more than doubled. This does not imply that reliability is worse. Rather, reliability indices are now more accurate.

Reliability index definitions can also complicate comparisons between utilities. A good example is the definition of a sustained interruption. If a utility defines a sustained interruption based on 5 minutes, automatic switching will be effective in reducing SAIFI since most switching can be performed within this time frame. If another utility defines a sustained interruption based on 1 minute, automation cannot reduce SAIFI since most automated switching requires more than this amount of time.

Perhaps the greatest difficulty in comparing reliability indices is the exclusion of major events. Some utilities include all interruptions when computing indices and other have widely varying exclusion criteria. Some utilities will also exclude scheduled outages and bulk power events. In many cases, major events account for a majority of customer interruptions and exclusion will completely alter the reliability index characteristics of a distribution system.

Keeping in mind the difficulties involved, it is interesting to compare reliability indices between US systems and European systems. Results of a 1987 survey of seven countries are shown in Table 2.6.[25] This survey reports customer based reliability indices, and breaks down most countries into urban and rural components.

As expected, rural systems are less reliable than urban systems. This is due to both an increased number of interruptions (SAIFI) and an increase in average restoration time (CAIDI). Overall, the US has better reliability than Norway and lower reliability than the UK or the Netherlands. For rural and overall categories, reliability tends to increase as population density increases—feeders are shorter, crews can locate faults more quickly and more feeder interconnections are possible. In addition, the cost of improving and maintaining reliability can be distributed across a larger number of customers.

There is an increasing trend for regulatory bodies to impose reliability reporting requirements on electric utilities. Typically, these requirements will include reliability indices by region and a list of worst performing feeders. The more aggressive authorities financially penalize and/or reward utilities based on reported reliability (referred to as performance-based rates). This provides a financial incentive for utilities to maintain and/or improve reliability. A map of US state reporting requirements is shown in Figure 2.11.[26]

Table 2.6. Distribution reliability indices reported from utilities around the world (1987 data). The reliability of rural systems tends to get better as population density increases. The notable exception is Italy, which has poor reliability in both its urban and rural areas.

	SAIFI (/yr)	SAIDI (min/yr)	CAIDI (min)	ASAI (pu)	Density (people/mi^2)
Urban Systems					
Finland	0.8	33	41	0.99994	
Sweden	0.5	30	60	0.99994	
Denmark	0.3	7	20	0.99999	
Italy	2.5	120	48	0.99977	
Netherlands	0.3	15	58	0.99997	
Rural Systems					
Finland	5.0	390	78	0.99926	38.3
Sweden	1.5	180	120	0.99966	51.2
Denmark	1.2	54	45	0.99990	313.7
Italy	5.0	300	60	0.99943	496.9
Netherlands	0.4	34	79	0.99994	975.3
Overall					
Norway	2.0	300	150	0.99943	34.6
United States	1.3	120	90	0.99940	73.2
United Kingdom	0.7	67	92	0.99987	653.4
Netherlands	0.4	27	73	0.99995	975.3

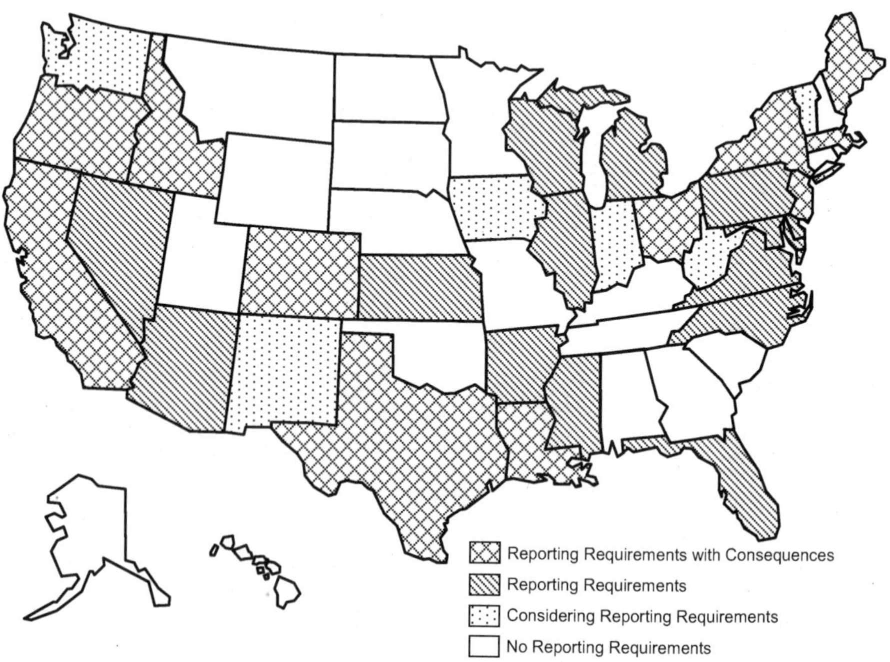

Figure 2.11. State reliability reporting requirements as of December 2006.

2.2.10 Reliability Data Collection

At present, most states in the US have distribution reliability reporting requirements for investor-owned utilities. Since reliability reporting has become a major regulatory issue, there has been increasing interest in the area of the reliability reporting process and its impact on the regulatory usefulness of reliability reports.

There is substantial complexity with regards to utility outage reporting process and, in addition, there is a wide range of practices employed by utilities. Obviously, a large variation in data collection and reporting practices can lead to a great disparity of data quality. It is desirable not only to know to what extent reporting practices vary, but also to know how many utilities conform to best practices in data reporting *vis-a-vis* some other lower standard. To investigate this issue, the author has conducted a benchmarking survey to ascertain where the industry stands today in many aspects of outage reporting.

This section presents the survey results of twenty-two large investor-owned utilities in the US, all with reliability reporting obligations. Based on these surveys, comparisons are made for eight aspects of outage reporting. An overall comparison is also provided, and the section ends with a discussion on industry trends.

Utilities for the survey were selected from among all US electric utilities with operating companies in states with regulatory reporting requirements. From these, the author contacted as many large companies as possible and conducted interviews with respect to reliability reporting systems and processes. In all, 22 interviews were performed representing approximately 55% of all retail electricity sold in states with regulatory reporting requirements.

When conducting these interviews, the following issues were specifically investigated due to their potential implications in reliability reporting: event capture, start time, customer count, step restoration, restoration time, verification of events, and the computation of indices. In addition, the ability of utility systems and processes to handle escalated events (e.g., major storms) was also discussed. Based on these interviews, descriptions of these aspects and a general range of utility practices are now provided.

Event Capture: Event capture refers to events that involve customer interruptions being reflected in reliability indices. A perfect system would have voltage sensors at each customer, and automatically create an event in a database whenever an interruption is detected. Practically, such a system is not yet cost-effective and event capture is handled in more pragmatic ways.

For utilities with regulatory reporting requirements, by far the most common method of capturing an event is through a combination of call centers and SCADA. When a customer dials the call center to report an interruption, the service representative either enters the event into a tracking system or an automated call handling system performs this function without human intervention. In either case, both the customer and the time of the call are logged.

Some of the major pieces of distribution equipment are connected to a communications network referred to as SCADA (supervisory control and data acquisition). For example, if a feeder circuit breaker is connected to SCADA and trips open, an event message is automatically generated and relayed to the dispatcher. Some utilities have outage events automatically generated from SCADA events, but most require the dispatcher to manually create an event and transcribe data from the SCADA log.

Historically, many utilities relied on paper forms, commonly called *outage tickets*, filled out by crews to capture events. In these cases, it is common for tickets to be not filled out, lost, or not turned in. This results in reliability indices not reflecting these events. Today, most of the utilities surveyed capture events in electronic databases near the time of occurrence (within 24 hours).

Most utilities intentionally omit certain events that result in customer interruptions. For example, if customer-owned equipment has led to the interruption, it is typically not included in reliability event data. Many utilities also omit service-level interruptions; many omit interruptions due to scheduled events such as maintenance; and many omit bulk power events such as transmission outages or intentional load shedding.

Historically, many utilities only included events that caused a feeder main circuit breaker to operate. This would result in a large number of events occurring on fused laterals to be omitted from reliability calculations. Although this still occurs, most of the utilities surveyed capture all events resulting in a customer call to the service center.

Start Time: Start time refers to the time that customers are interrupted. Accurate start times are typically available if SCADA information is available, but a large percentage of events are not detectable through SCADA; the time of the first customer call is commonly used. Since interrupted customers do not call the utility immediately, using a call center time stamp results in a bias that tends to make SAIDI smaller. A perfect system would have voltage sensors at each customer and automatically create a time stamp whenever an interruption is detected. Practically, such a system is not cost-effective.

Start times can also be impacted by their method of entry into the outage database. If entry automatically occurs from both the call center and from SCADA, errors will be extremely rare. If times must be transcribed by dispatchers, errors become more likely. Individual events will be incorrect, but will errors will tend to average out and therefore have little impact on computed reliability indices.

Perhaps the worst systems are those that rely solely on paper outage tickets. Consider a scenario where a service representative fields a customer call. He communicates start time information to a dispatcher by telephone. The dispatcher then relays this information to a crew by radio. The crew fills out a paper ticket with a start time, and this start time is later entered into a spreadsheet for reliability index calculations. Clearly, this scenario provides many opportunities for start time errors (one of the utilities surveyed has such a process).

Customer Count: Obtaining the correct number of customers interrupted by an outage event is of critical importance for calculating customer-based reliability indices such as SAIFI and SAIDI. Since only a fraction of interrupted customers will call the utility, customer count must typically be inferred.

The most sophisticated systems use real-time models that are able to dynamically track the distribution transformers that are out of service due to protection device operations and/or switching device operations. They are also linked to a utility's customer information system (CIS) so that the specific customers associated with each transformer are known and are synchronized with billing information.

Many systems that utilities have developed in-house utilize hierarchical device data instead of actual connectivity data. In these systems, each customer is associated with an upstream transformer, each transformer is associated with an upstream device such as a fuse, and so on until the main feeder circuit breaker is reached. Such models are able to generate reasonably accurate customer counts if the system is in its normal operating state, but may result in errors if the system topology has been temporarily reconfigured due to maintenance, a prior outage event, and so forth.

Whether a utility system uses a connectivity model or a hierarchical model, interrupted customers are typically inferred through a rule base. For example, if a single customer calls, the system may assume that this is the only interrupted customer. If two customers fed from the same transformer call in, the system may infer that this transformer is out. If two transformers on the same lateral are inferred to be out, the system may deduce that the entire lateral is out. This type of logic is straight forward, but specific rules and thresholds vary by utility.

In the not-so-distant past, many utilities did not have databases showing the transformer to which each customer was connected. Consequently, interrupted customer counts were not easily deduced and were typically inferred by transformer size (one of the utilities surveyed still uses a similarly flawed process).

Step Restoration: Step restoration (also referred to as partial restoration) refers to the restoration of some customers before the outage is repaired and the system is returned to its normal operating condition. For example, many feeders are broken into switchable sections by the use of normally-closed switches. This allows a main feeder fault to be isolated. When a fault is isolated, all customers upstream of the switchable section can be restored by closing the main feeder breaker. Customers downstream of the switchable section may also be able to be restored by closing tie switches and temporarily transferring customers to another feeder.

The most sophisticated outage reporting systems dynamically track system topology and are able to compute each restoration step and the impact to each customer. Less sophisticated systems are able to infer the impact of step restoration based on a base connectivity model. Other systems require crews or dispatchers to manually estimate the impact of step restoration, and many systems to not handle step restoration at all (although all surveyed utilities consider step restoration in some form).

Underground residential distribution (URD) loops present a specific problem for step restoration. These systems are commonly configured so that customers can be sequentially restored until the faulted section is located. This process can easily result in 10 or more restoration steps that are in close temporal proximity. It would be overly burdensome to record each of these steps, and could actually prolong the time required for partial restoration. Neither of these issues is a problem when reliability reporting practices are kept constant from year to year. But changes in how URD restoration is handled could result in a change in reported indices (e.g., always using the time of the first step rather than the time of the last step).

Automated switching is another aspect that can pose problems for reliability reporting systems. It is becoming more common to have automation schemes that will automatically restore power to a set of customers after an interruption is detected. Perhaps the simplest example is an automated transfer switch that will transfer a single customer to an alternate feeder if voltage is lost on the primary feeder. Another example is an automatic rollover scheme where an entire feeder section is automatically transferred to another feeder.

Most reliability reporting systems are not able to automatically account for automated switching and rely on dispatcher intervention to account for these automated events. For example, if a lot of customers on a feeder call in to report an interruption, many systems will infer that the feeder circuit breaker is open and that the entire feeder is interrupted. If part of the feeder has no associated customer calls, it is up to the dispatcher to infer that automation has occurred and to modify the outage report accordingly. The quality of such a process is a strong function of the dispatcher's familiarity with the system topology, use of automation, and typical trouble call patterns.

Automation, in a sense, is special since it has the ability to impact SAIFI as well as SAIDI and CAIDI. Although every aspect of reliability reporting can impact SAIDI and CAIDI, only event capture, customer count, and automation can have a major impact on SAIFI. Automation does this by quickly restoring customers before a time threshold is reached. If customers are restored before this threshold (typically 5 minutes), it is not supposed to be treated as a sustained interruption and is not supposed to impact SAIFI. Utilities are increasing their use automation and it is becoming increasingly important to consider these factors when computing reliability indices.

Restoration Time: The point at which the last customer associated with an outage is restored is referred to as restoration time. Commonly, a single troubleman will be responsible for restoring an outage and it is at the complete discretion of this person to (1) look at the time when the last customer is restored, (2) remember this time, and (3) correctly record this time or communicate it to a dispatcher via a radio. The precision and accuracy of restoration times, therefore, are highly dependent on training and awareness that this task is important.

Another issue with restoration times occurs when outages are referred from troublemen to construction crews. When this happens, it is the job of the construction crew to record the restoration time, but construction trucks may not be fitted with mobile data terminals, and construction crews may not have the same training and awareness of the issue as troublemen.

Although precise and accurate restoration times are desirable for a host of reasons, they are generally not required for reported reliability indices to be useful. Unbiased guesses will not significantly impact reliability indices, and biased entries will not impact the ability to detect year-to-year trends as long as the biases are constant from year to year.

Verification of Events: Most utilities have some process for verifying outage event data before it is used to report reliability data. Ideally, each event is examined close to the time that the event occurs. This allows dispatchers, troublemen, and construction crews to be contacted if an outage event is missing data or appears unusual in some way. Less aggressive processes sample a portion of outage events based on criteria ranging from computer-generated exception reports, outages impacting more than a threshold number of customers, all feeder-level outages, and random sample. Based on the surveys, event verification is generally considered an important part of the reliability reporting process.

Computing Indices: After raw outage data is entered into an electronic database, reliability indices must be computed. The best systems are able to do this automatically through integrated functions. In such cases, reliability indices can be tracked daily. More commonly, however, is to either have a separate application that queries the original data, or to create a new database from the original data. In either case, there is a wide range of sophistication from web-based interfaces to simple spreadsheets.

It should be emphasized that computing reliability indices from an outage database is nontrivial. Since there is an immense amount of data, the process must be automated and errors in the automation process can lead to errors in index calculations. Further, outage databases typically include events that should not be used for reliability indices such as customer-level events, nonoutage construction jobs, and excluded storm events. The way these exclusions are handled can also impact reliability index calculations.

The benefit of using a reliability analyst to generate reliability indices is in flexibility. Typically, this person is utilized for a host of issues related to reliability and familiarity with the database. The disadvantage is in continuity when this person is replaced. If a new person alters the way that reliability indices are generated, new indices may not be directly comparable to past indices. Therefore, automated index calculations are preferred since consistency is guaranteed when different people are used to compute reliability indices.

Escalated Events: Utility processes typically change when many outages are occurring within close temporal proximity (such as a major storm). For example, it is not uncommon for crews from other utilities to be used during major storms, and these crews may not be familiar with the reliability reporting processes of the utility they are aiding. The capacity of an outage management system can also be overloaded in certain situations, requiring the system to be de-activated and resulting in the potential loss of outage event data.

For some utilities, escalated event handling is not a reliability reporting issue since most escalated events are excluded from reported reliability indices. However, most utilities enter "storm mode" well before the exclusion criteria is reached. In such cases, escalated event handling becomes critical since years with many escalated events may have data that is not comparable to years with few escalated events.

For example consider a utility that captures 100% of events during normal mode but only 80% of events in "storm mode." If storms account for 50% of events in a year, 10% of all events are not captured. If the next year storms account for only 25% of events, only 5% of all events are not recorded.

The best utilities have systems that perform the same for all events that will be included in reported reliability indices (other events are excluded). The next best utilities utilize systems that perform nearly the same during storms, except that borrowed crews may not be able to exactly follow the same processes (e.g., no computers in trucks). In the worst situations, a utility is not supposed to exclude any storm event data when computing reliability indices, but captures little data during actual storms (no utility in the survey falls under this category).

It is now of interest to compare the reliability data collection systems of the surveyed utilities using a consistent evaluation methodology. Towards this end, reliability data collection assessments are now made by assigning each utility with a score of best, good, fair, poor, or worst for each aspect. This range spans all utility practices rather than just the range of the benchmarked utilities.

A summary of the scoring criteria is provided below describing how each of the reliability reporting aspects is evaluated. When considering utility scores, the reader should keep in mind that these surveys were performed in 2003, and utilities across the country continue to implement new reliability data collection systems and processes.

Event Capture

Best: Automatically captured from both call system and SCADA (i.e., an electronic link between SCADA and the outage management system).

Good: Required to be captured before a troubleman can be dispatched.

Fair: Manually captured when crew is dispatched.

Poor: Captured from paper tickets collected daily.

Worst: Captured from paper tickets collected monthly.

<u>**Start Time**</u>

Best: Automatically generated from both call center and SCADA and does not require manual transcription.

Good: Automatically generated from both call center and SCADA but requires manual transcription from customer service representative and/or dispatcher.

Fair: Automatically generated from SCADA, but events from customer calls are relayed to dispatcher via telephone.

Poor: No SCADA on main feeder circuit breakers, requiring feeder-level events to be reported through call centers.

Worst: Taken from paper ticket.

<u>**Customer Count**</u>

Best: Based on dynamic connectivity model with direct access to customer information system and affected device inference logic.

Good: Based on device hierarchy that is periodically updated in terms of system topology changes, new customer connections, etc., and uses affected device inference logic.

Fair: Based on customer call patterns with no affected device inference logic.

Poor: Inferred from transformer power ratings.

Worst: Guess by troubleman on trouble ticket.

<u>**Step Restoration**</u>

Best: All system topology changes are communicated in real-time to dispatcher, who updates a dynamic connectivity model that is able to precisely track all restoration. Automation schemes are embedded into the systems model and are automatically handled.

Good: Step restoration is tracked at the customer level, but based on a normal-state connectivity model rather than the dynamic state of the system.

Fair: Step restoration is recorded as a percentage of customers restored, with individual customers impact not explicitly tracked.

Poor: Step restoration is never reflected in event records.

Worst: Step restoration is haphazardly reflected in event records; many are recorded properly, many are recorded improperly, and many are not recorded.

<u>**Restoration Time**</u>

Best: Radio communication is required for all feeder-level switching actions and mobile data terminals in trucks are used for all lower level events.

Good: Radio communication is required for all switching events, but there are no mobile data terminals in trucks.

Fair: Radio communication is used for feeder-level switching, but lower level events are recorded on paper tickets.

Poor: Times are recorded on paper tickets that are turned in daily.

Worst: Times are recorded on paper tickets that are turned in monthly.

Verification of Events

Best: All events are examined daily.
Good: Most events are examined daily.
Fair: Exception report logic generates a list of suspicious events, which are then reviewed.
Poor: Events are reviewed on an ad-hoc basis at the time that reliability reports are generated.
Worst: Events are not reviewed.

Computing Indices

Best: Indices calculated automatically from dispatch system database.
Good: Indices calculated automatically by a separate system that periodically extracts outage data from the outage management system.
Fair: Indices calculated from standardized query searches that accesses either the outage management system directly or an extraction used specifically for reliability reporting.
Poor: Indices calculated from a reliability analyst using spreadsheets populated from outage database queries.
Worst: Indices calculated directly from paper tickets.

Escalated Events

Best: All events that will be included in reported reliability indices are captured during escalated events. Any escalated event that may alter the use of systems and processes is likely to be excluded from reported indices.
Good: Systems and processes are able to handle escalated events, but reportable events may require the utilization of contract or borrowed crews.
Fair: Systems are significantly altered during storms, resulting in a substantial change in how outage data is collected (e.g., changing from central to regional dispatch).
Poor: Dispatch system capacity is exceeded during reportable events. In these cases, the system must be turned off and outage data must be collected manually.
Worst: No exclusions are allowed, and outage data is not captured during storms.

Each utility was ranked Best, Good, Fair, Poor, or Worst for each of the eight reporting practices, using the criteria described in previous section. A summary of results for the 22 utilities surveyed is shown in Table 2.7. It should be emphasized that all comparison utilities are major investor-owned utilities that will generally have more sophisticated systems and processes than smaller utilities. As such, this benchmarking analysis is comparing utilities that can be expected to score well on comparison attributes.

In general these larger companies are using good reliability reporting practices. The majority of utilities fell in the Good category for each reporting practice except Verification of Events, where there was a wider variance in the results. More utilities were ranked Best for event verification than for any other practice, but an equal number of utilities were ranked Fair or Poor in that category. Uniformly high scores were given for Event Capture, Start Time and Customer Count, a reflection of the greater use of automated systems and good system topology models. The weakest reporting practice involved Escalated Events with more Fair and Poor ratings than any other category. This indicates that even the better utilities suffer in data collection during the periods when a large percentage of their outage events occur.

In general, utility reliability reporting practices have been improving because of greater attention to the reporting processes and systems. In particular, more utilities are using integrated Outage Management Systems (OMS). These improvements not only have increased the accuracy of outage reporting but also have enabled improved customer communications during power interruptions. Most enhanced capabilities arise from automated data interfaces between the corporate GIS, OMS, SCADA and Call Centers, resulting in greater data accuracy, reporting productivity, more timely information for dispatchers and customers, and quicker outage restoration.

Table 2.7. Reporting practices survey results

Category	Best	Good	Fair	Poor	Worst
Event Capture	3	17	1	1	0
Start Time	2	18	1	1	0
Customer Count	3	15	3	1	0
Step Restoration	4	11	4	2	1
Restoration Time	5	12	3	2	0
Verification of Events	8	6	6	2	0
Computing Indices	2	12	4	4	0
Escalated Events	1	11	9	1	0

2.3 CUSTOMER COST OF RELIABILITY

When a customer experiences an interruption, there is an amount of money that the customer is willing to pay to have avoided the event. This amount is referred to as the customer cost of reliability. In the US alone, power interruptions results in more than $13 billion in lost production, not including scrapped product or damage to equipment.[27]

The customer cost of reliability becomes important when a utility wishes to balance their costs with customer costs. Consider Figure 2.12, which represents a utility serving strictly industrial customers with an average size of 500-kW. As SAIDI increases, these industrial customers incur higher losses due to lost production. To reduce SAIDI decreases, the utility must spend increasing amounts of money on system reinforcements and maintenance. The sum of these two curves is the total societal cost of reliability, and minimizing total cost maximizes social welfare. In this case, the optimal level of reliability corresponds to a SAIDI of 105 minutes. This point is where the marginal customer cost of reliability is equal to the marginal utility cost of reliability.

Designing and operating a distribution system to minimize total societal cost is referred to as value-based planning. Value-based planning is the goal of publicly owned utilities (e.g., public utility districts, municipals, cooperatives) and has been required by law in certain countries such as Norway.[28] Much recent research has focused on minimizing the total cost of reliability,[29-34] and good knowledge of customer costs is vital for these methods to be practical.

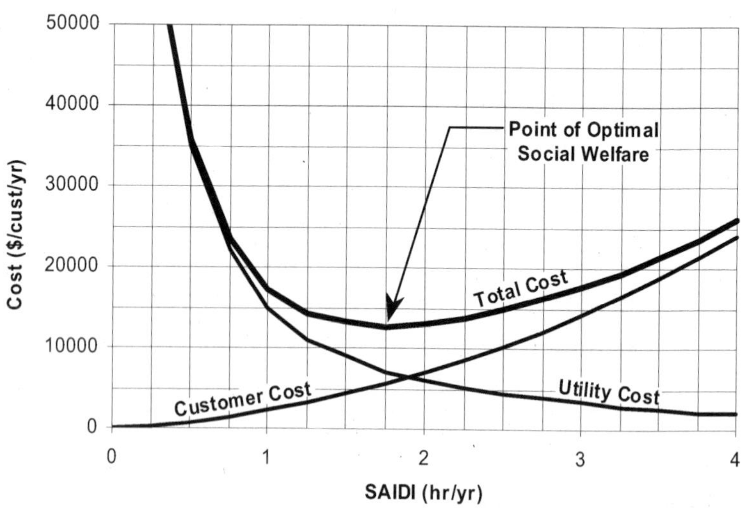

Figure 2.12. The total societal cost of reliability is the sum of utility cost plus customer cost. Societal welfare is maximized when the total cost of reliability is minimized. It is acceptable for publicly owned utilities to consider total cost minimization, but it is more appropriate for privately owned utilities to satisfy regulatory expectations for the least possible cost.

Investor owned utilities attempt to maximize profits rather than maximize social welfare, which is presumably accomplished by satisfying all regulatory expectations for the least possible cost. Customer cost information becomes less important because a customer dollar is no longer equal to a utility dollar. Regardless, a good feel for customer costs can help investor owed utilities to better understand their customers and offer value added reliability services that can increase profits and increase customer satisfaction.

2.3.1 Customer Cost Surveys

The customer cost of reliability is typically estimated based on surveys. These surveys capture information about tangible costs, opportunity costs, and intangible costs. Tangible costs include items such as computer crashes, ruined processes, scrapped product, spoiled food, overtime pay, and the cost of going out to eat. Opportunity costs includes lost production and lost sales. Intangible costs include inconveniences such as water pump failures, difficulties in getting ready for work, impact on leisure time, and needing to reset digital clocks.

The cost of an interruption varies widely from customer to customer and from country to country. Other important factors include duration, time of year, day of the week, time of day, and whether advanced warning is provided. Customers will also be impacted less if they have back-up generators, uninterruptible power supplies and other on-site interruption mitigation equipment. Good customer surveys attempt to capture as much of this information as possible, but the quantity and interaction of parameters makes complete models difficult to achieve.

Estimates of customer cost are well documented by a host of surveys.[35-46] An interesting observation is the widely varying costs associated with different industries. Results from a University of Saskatchewan survey are shown in Table 2.8.[47] Costs of a typical one-hour interruption, normalized to peak load, are provided for a variety of commercial and industrial customers and shown to vary from virtually zero cost to more than $276 per kW. On average, industrial customers incur about $8.40/kW for a 1-hr interruption and commercial customers incur about $19.38/kW for a 1-hr interruption. Based on these results, large customers with high costs can easily incur millions of dollars per interruption hour.

The cost of an interruption is highly dependent on its duration. Short interruptions can result in computer crashes, ruined processes, and broken equipment. Longer interruptions result in lost production and ruined inventory. For specific customers, curves tend to be highly nonlinear. A semiconductor factory may incur a high initial cost due to a ruined process and a small time-dependent cost due to lost production. A plastic extrusion facility may incur small costs for short

Table 2.8. One-hour interruption costs for industrial and commercial customers. Results are based on a University of Saskatchewan survey and are presented in 2001 dollars.

Industrial	**$/kW$_{peak}$**	**Commercial**	**$/kW$_{peak}$**
Logging	2.11	Food and Drug	18.52
Forestry	0.00	Clothing Stores	18.92
Mining	3.00	Household Furniture	39.88
Crude Petroleum	276.01	Automotive	42.39
Quarry and Sand	5.33	General Merchandise	30.10
Services to Mining	2.13	Other Retail	5.95
Food Industries	20.46	Vending and Direct	0.00
Beverage Industries	1.55	Accommodations	1.32
Rubber Products	1.80	Food Service	19.90
Plastic Products	2.91	Entertainment	23.81
Leather Products	1.37	Personal Services	0.39
Primary Textiles	17.29	Other Services	3.51
Textile Products	8.93		
Clothing	8.68		
Wood Industries	2.93		
Furniture	23.20		
Paper Products	7.52		
Printing and Publishing	6.01		
Primary Metal	3.54		
Fabricated Metal	8.41		
Machinery	7.70		
Transportation	42.96		
Electrical Products	8.78		
Nonmetal Minerals	9.59		
Refined Petroleum	0.00		
Chemical Products	4.65		
Other Manufacturing	15.31		
Total Industrial	**8.40**	**Total Commercial**	**19.38**

interruptions, but incur an extremely high cost if the interruption is long enough for plastic to solidify within the extrusion equipment. A refrigeration warehouse may not incur any cost for short interruptions. At a certain point, food will begin to spoil and severe economic losses will occur. After all of the food is spoiled, additional interruption time will not harm this particular customer much more. Cost functions reflecting these three cases are shown in Figure 2.13.

Average customer cost curves tend to be linear and can be modeled as an initial cost plus a first-order time dependent cost. For planning purposes, it is useful to group results into a few basic customer classes: commercial, industrial, and residential. Since larger customers will have a higher cost of reliability, results are normalized to the peak kW load of each customer. Reliability cost curves for typical US customers are shown in Figure 2.14.[48]

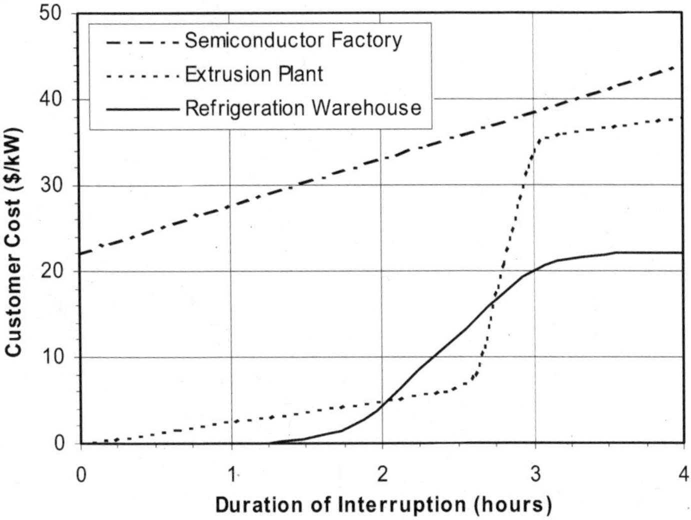

Figure 2.13. Interruption cost curves for several customer types. The semiconductor factory has high initial costs and linear time dependent costs. The extrusion plant and refrigeration warehouse have low initial costs and highly nonlinear time related costs.

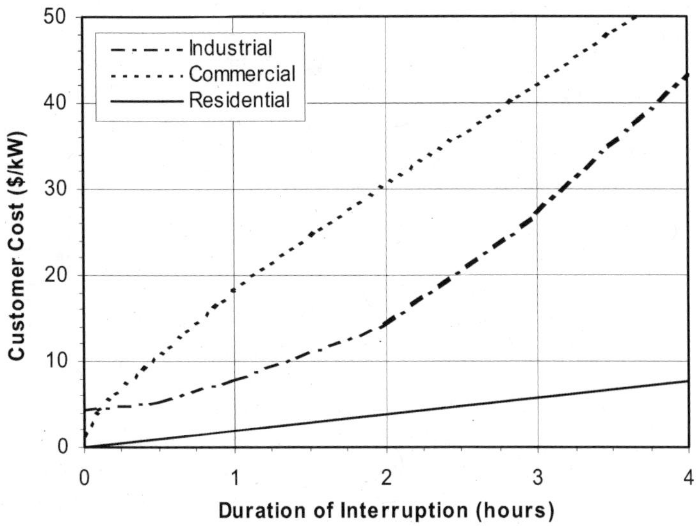

Figure 2.14. Average interruption costs for industrial, commercial, and residential customers. Average costs can be approximated with an initial cost and a first-order time related cost. Results are based on several North American surveys and are presented in 2001 dollars.

2.3.2 Problems with Surveys

Difficulties in applying the results of customer cost surveys to utility decision making are well known.[49] Customers incur real costs when an interruption occurs, but accurately capturing these costs is elusive. It is common to receive customer survey results that express a high cost of reliability, but customers will typically not be willing to pay for reliability improvements unless the associated payback period is much shorter than would be required for other purchasing decisions.

An example of the disconnect between survey costs and willingness to pay is illustrated by a paper plant in the Northeast that had been experiencing more than a dozen momentary interruptions per year due to lightning. According to this customer, each momentary interruption resulted in economic losses exceeding $100,000. A pilot study funded by the local utility successfully eliminated all momentary interruptions by installing $1 million in lightning protection equipment. When asked if the customer wanted to pay for similar solutions on other locations, the customer declined even though the presumed payback period was less than one year.

A computer chip manufacturer in the Midwest displayed similar behavior. This factory was experiencing 8 to 10 momentary interruptions per year and claimed that each event resulted in economic losses of $500,000. The utility investigated the problem, identified a $560,000 solution, and offered to split the cost of the solution with the factory. Even though the apparent payback period was several months, the factory declined the offer.

Similar difficulties occur with commercial districts. One utility survey indicated that restaurants were experiencing large costs due to sustained interruptions. The utility proactively offered to improve reliability for a price much less than the surveyed cost of reliability, but not a single customer was willing to pay for the solution.

There are several reasons why survey results do not typically reflect willingness to pay. The first is customer perception that utilities are obligated to address reliability problems for no additional cost. Utilities are regulated monopolies with an obligation to serve, and this obligation includes adequate reliability. Other customers understand that higher reliability requires higher rates, but that many reliability decisions are based on value-based planning. These customers may overstate their costs in the hope of directing more utility resources towards their reliability problems. Last, many customers prefer to have neighbors pay for reliability improvements and reap reliability benefits without having to pay (referred to as the free rider problem).

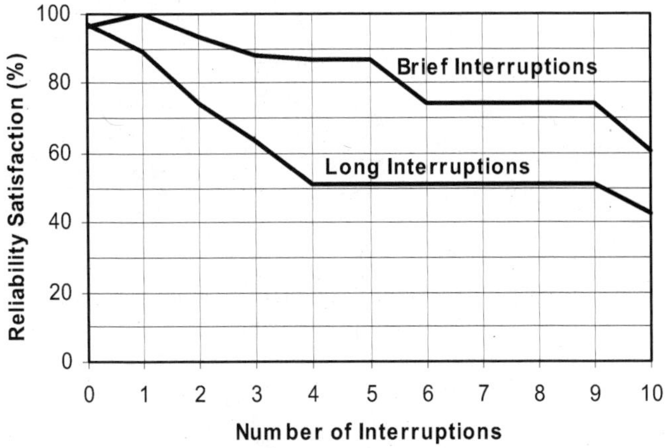

Figure 2.15. Satisfaction results from a survey of 1200 customers. Customers have a much higher tolerance for brief interruptions than long interruptions. Satisfaction drops quickly if long interruptions occur more than once per year, but remains high for up to five brief interruptions per year.

Customer cost surveys for residential customers are even more concerning. Since most residential customers do not experience any tangible economic losses when an interruption occurs, cost surveys are based on subjective issues. When asked, most residential customers respond that they are not willing to pay anything to improve reliability. When surveyed, typical residential customers state that interruptions incur costs of $5 to $10 per hour.

Customer satisfaction surveys have similar problems and are not highly correlated to reliability. In fact, some utilities have even experienced increased customer satisfaction with decreased customer reliability. Reliability problems give the utility an opportunity to interact with the customer and provide education on reliability issues, improvement efforts, and associated costs.

There is a point at which poor reliability does begin to affect customer satisfaction. One survey of 1200 customers shows that customers are generally tolerant if brief interruptions are kept to less than six per year and long interruptions do not occur more than once per year[50] (customers were not provided with definitions of "brief" and "long"). These results, shown in Figure 2.15, are difficult to generalize since satisfaction is a strong function of customer perception and expectation.

2.3.3 Problems with Minimizing Societal Cost

It costs money for a utility to improve the reliability of its distribution system. There is also a cost that customers incur when they experience interruptions in their electrical supply. In theory, it is better for society to improve reliability if the money saved by customers due to improved reliability exceeds the cost to the utility to achieve the reliability improvement. Since the goal of minimizing societal cost seems altruistic, it is often discussed as a possible driver of utility expenditures.

Minimizing societal cost, however, is a narrow view in the sense that it does not address the issue of cost allocation. Consider an industrial factory that experiences large economic costs whenever its electrical supply is interrupted. Now suppose that a perfect economic analysis shows that the factory will save a bit more than one million dollars if the electric utility spends a bit less than one million dollars. From a societal perspective, it appears that the utility should spend the money.

However, if revenues are unchanged, this situation amounts to a transfer of wealth from the owners of the utility to the owners of the factory. If the utility is not compensated in some way, the stock price of the utility will drop, owners of the stock will become less wealthy, and the ability of the utility to attract capital will weaken.

Problem 1: Without compensation, minimizing societal cost transfers wealth from utility owners to utility customers.

Now assume that "prudent expenses" are allowed in the utility rate base so that investors can attain their expected return on investment. Investors are happy, but the utility will encounter other problems. Consider again the factory on which the utility has now spent one million dollars in reliability improvements. Now consider a competitor's factory that happens to be located on a part of the system that is expensive to improve reliability. Due to the higher cost to improve reliability, societal cost is minimized by not improving reliability. This leaves the second factory in a situation where it is paying the same rate as its competitor, but receives a smaller amount of utility investment. In effect, the second factory is subsidizing the profits of the first factory.

Problem 2: With an inflexible rate structure, minimizing societal cost creates cross-subsidies from areas with a high cost to improve reliability to areas with a low cost to improve reliability.

In the same way, minimizing societal cost can bias investment towards customers with a high cost of poor reliability. Consider an affluent neighborhood where most residential houses have home computers and have efficient heat pumps. Now consider a poor neighborhood where most residential houses do not have home computers, and have baseboard heating. The total electrical demand for each neighborhood is the same, but the cost of poor reliability for the affluent neighborhood is higher due to the presence of computers and other expensive electronics. By minimizing societal cost, the utility will be required to spend more on the affluent neighborhood, even though all residential rates may be the same.

Problem 3: With an inflexible rate structure, minimizing societal cost creates cross-subsidies from areas with a low cost of poor reliability to areas with a high cost of poor reliability.

Now consider the factory again. It may well be the case that the factory can save itself one million dollars in reliability improvement by spending far less than one million dollars. This could be done through on-site emergency generation, uninterruptible power supplies, load desensitization, and so forth. However, the factory owners would prefer that the utility fix the problem, even though the societal cost is higher. The more confident that the factory owners are that they can get the utility to pay for the improvement, the less likely they are to explore customer-funded alternatives.

Problem 4: When a utility makes decisions by minimizing societal cost, it leads under investment in customer-funded reliability improvement projects, which defeats the purpose of minimizing societal cost.

The first four problems assume that the customer cost of poor reliability is perfectly known. In reality, the cost of poor reliability is very customer specific, and difficult to ascertain. As a result, customer survey data is of questionable value. In all cases of which the author is aware, surveyed industrial customers overstate their cost of poor reliability. This is evidenced by presenting a solution by which the customer will pay for the reliability improvement for an amount that, according to the survey, results in a positive NPV investment. In virtually all cases the "customer pays" solution is rejected as too expensive.

A related problem is gaming. If a customer knows that utility investments are based on survey results, gaming behavior can result in overstating the cost of poor reliability to increase the level of utility reliability investment.

Problem 5: Surveys almost always overstate a customer's cost of poor reliability when compared to the customer's willingness to pay for reliability improvements. Further, customer gaming behavior can lead to intentionally overstated responses.

The only practical way to minimize societal cost is to set reasonable reliability standards and make customers pay for all reliability above these minimum standards. The difficulty in having customers pay for reliability improvements is related to free-riding. That is, reliability improvements paid for by one customer often benefits other customers. Consider a transmission line that serves ten identical factories. When the transmission line experiences an outage, all ten factories are equally affected, and reliability improvements on the transmission line equally benefit all factories. Now assume that all ten of the factories will benefit if $1 million dollars is spent to improve the reliability of the transmission line. Divided equally, each factory will spend $100,000 but will realize $130,000 in savings. One of the clever factory managers computes that if only nine factories share the cost, each will pay $111,111 but will still receive more in benefits. Therefore, this manager refuses to pay for the project in hopes that the remaining nine will still make a "rational economic decision." All of the other managers realize now that if only eight factories share the cost, each will pay $125,000. At this point, most of the plant managers will simply refuse to pay for the project unless everyone pays his fair share. In the real world, it is difficult to measure the reliability benefit of each customer and the issue of "fair share" is difficult to solve. The end result is typically a perception of free riders, a refusal to pay for free riders, and an underinvestment in reliability.

Of course, one scenario is to allow a utility to allocate reliability improvement costs to all customers that benefit. This, however, amounts to rate-base design, which is a function of regulatory authorities, not utilities.

In the end, minimizing total societal cost is simply not a practical approach. Such analyses can provide helpful guidelines, sanity checks, and insights into many reliability issues, but should not drive investment decisions. A much better approach is to set reliability targets based on minimum acceptable levels of reliability.

2.4 RELIABILITY TARGETS

Historically, rates that electric utilities charge are based on the cost of generating and delivering electricity. In return for fulfilling their obligation to serve customers in an exclusive service territory, utilities are guaranteed a reasonable return on their investments. Reliability is achieved implicitly through conservative design and maintenance standards, and reliability targets are not needed.

Recent regulatory changes have put immense pressure on electric utilities to reduce costs. Since the National Electric Policy Act (NEPA) was passed in 1992, virtually every major utility has undergone massive downsizing and has drastically reduced spending by deferring capital projects, reducing in-house expertise, and increasing maintenance intervals. As a direct consequence, the reliability on these systems is starting to deteriorate.

Regulatory agencies are well aware that regulatory changes might have a negative impact on system reliability. In a perfect free market, this would not be a concern. Customers would simply select an electricity provider based on a balance between price and reliability. In reality, customers are connected to a unique distribution system that largely determines system reliability. These customers are captive, and cannot switch distribution systems if reliability becomes unacceptable. For this reason, regulators are searching for proper ways to define, measure, and enforce reliability targets.

In a perfect world, reliability targets and incentives would provide utilities with price signals that simulate a free market. If the marginal benefit of improved customer reliability is greater than the marginal utility cost, proper reliability targets will provide incentives for utilities to make reliability improvements. The concept is simple in theory and difficult in application. As such, most regulators are taking a step-by-step approach. In order of increasing complexity, some common strategies are:

- Reliability Reporting
- Worst Performing Circuit Programs
- Reliability Index Targets
- Performance-based Rates
- Reliability Guarantees

Reliability reporting and worst performing circuit programs do not require reliability targets. They simply increase the transparency of utility reliability and encourage reactive treatment of reliability concerns. Reliability index targets, performance-based rates, and reliability guarantees utilize reliability targets and encourages proactive treatment of reliability issues.

2.4.1 Reliability Index Targets

Reliability index targets are set internally or by state regulators and are based on historical index values and desired reliability trends. Once set, utilities act in good faith to achieve and maintain these targets. Targets are typically based on maintaining or improving recent SAIDI and SAIFI values.

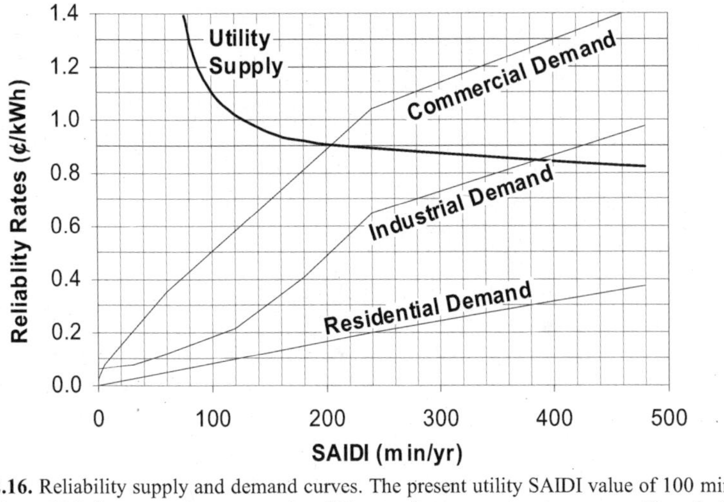

Figure 2.16. Reliability supply and demand curves. The present utility SAIDI value of 100 minutes per year is too high from a market perspective. If free to choose, most customers would prefer lower rates and lower reliability.

If reliability targets are set near existing levels, there is an implicit assumption that existing reliability levels are good for the utility and its customers. From a free market perspective, the most efficient target occurs where the reliability supply curve of a utility intersects the reliability demand curve of its customers. Both of these curves can be estimated to compare present reliability targets with efficient reliability targets.

Figure 2.16 shows the reliability supply curve of a large investor owned utility in the Midwestern US. This utility has a SAIDI of 100 minutes and a corresponding reliability cost of 1.1¢/kWh. Existing cost is based upon FERC Form 1 data (50% of distribution rate base plus 100% of distribution O&M cost). The remainder of the reliability cost curve is generated by the marginal cost of reliability projects submitted by district engineers.[51] The same figure shows customer demand curves for typical commercial, industrial and residential customers based on interruption cost surveys and translated into equivalent rate increases.

If the reliability supply and demand curves of Figure 2.16 are approximately correct, reliability is too high. From a market perspective, customers would prefer to pay lower rates and experience SAIDI values much higher than 100 minutes. A purely commercial system should have a SAIDI target of 200 minutes, a purely industrial system should have a SAIDI target of 380 minutes and a purely residential system should have a SAIDI target of about 15 hours per year. These SAIDI targets include interruptions caused by major events.

These reliability targets will seem high to many people. In fact, they are conservative since customer cost surveys typically overestimate customer willingness to pay. A small percentage of customers require high levels of reliability, but most do not. The needs of relatively few vocal customers often drive reliability targets rather than the needs of average customers.

This example, though based on actual data, is approximate and intended for demonstrative purposes. Specific reliability targets for specific utilities should be carefully considered on a case by case basis, but allowing reliability to become worse should always be considered a viable option and potentially appropriate.

2.4.2 Performance-Based Rates

In their most general form, performance-based rates (PBRs) are regulatory statutes that reward utilities for good reliability and penalize them for poor reliability. Regulators use PBRs under rate cap situations to counteract the tendency of utilities to cut costs and allow reliability to suffer. Performance is usually based on average customer interruption measures such as SAIDI and SAIFI.

A common method of implementing a PBR is to have a "dead zone" without bonuses or penalties. If reliability is worse than the upper dead zone threshold, a penalty is assessed. Penalties increase as performance worsens and are capped when a maximum penalty is reached. Rewards for good reliability can be implemented in a similar manner. If reliability is better than the lower dead zone threshold, a bonus is given. The bonus grows as reliability improves and is capped at a maximum value. Bonuses are far less common than penalties since regulatory agencies do not have sources of revenue. A graph of a PBR based on SAIDI is shown in Figure 2.17.

Regulatory agencies can simulate bonuses and rewards by annually adjusting rates based on reliability performance. Rates are increased if reliability targets are exceeded and are decreased if reliability targets are not met. Rate adjustments can be phased in over several years to lessen financial volatility and spread out the impact of particularly good and/or bad years.

Under PBRs, a utility will attempt to minimize the sum of the reliability costs and PBR costs. Reliability is increased if one dollar in improvements saves more than one dollar in penalties. Conversely, reliability is decreased if one dollar in reduced spending results in less than one dollar in additional penalties. The optimal solution occurs when the marginal cost of increasing reliability is equal to the marginal increase in performance penalties.

Consider a utility in the Pacific Northwest that has generated a reliability-cost curve for their service territory. Assume that this utility is subjected to a PBR with penalties starting at SAIDI = 90 min/yr and capping at $100 million when SAIDI = 180 min/yr (see Figure 2.18). Since there is no economic incentive to improve SAIDI higher than 90 min/yr, reliability will gravitate towards this value. If penalties increase at a slower rate than cost decreases, the utility will spend nothing on reliability and simply pay penalties to minimize total cost.

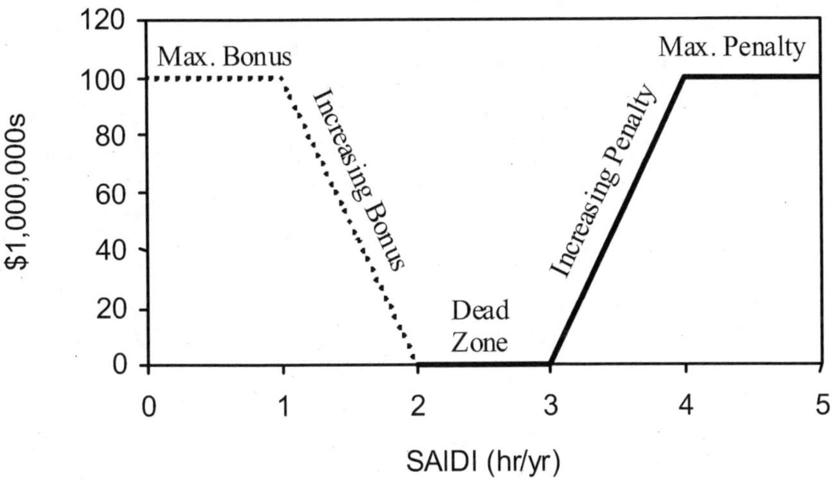

Figure 2.17. A typical performance-based rate structure will have a dead zone (no penalties or rewards) and a penalty zone that increases until a certain point level of reliability is reached. Some performance-based rate structures will also have a bonus zone to reward good reliability.

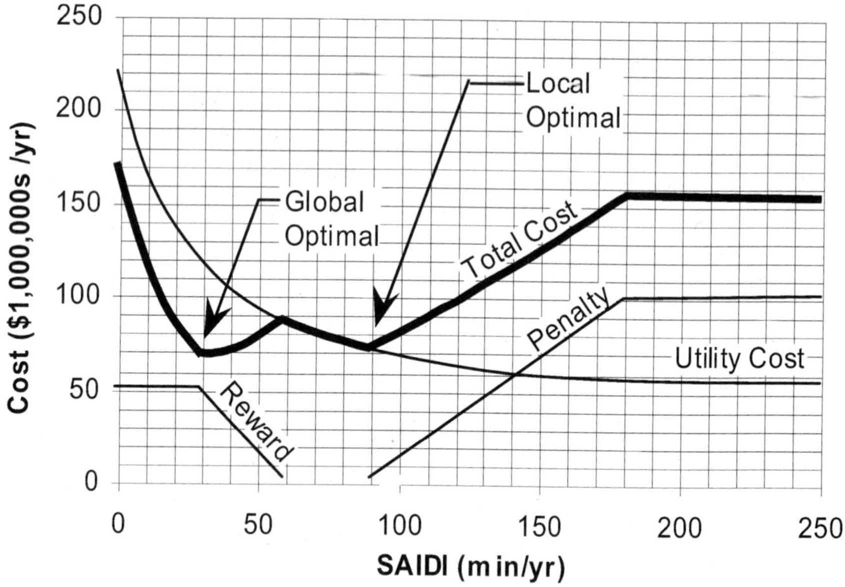

Figure 2.18. Total cost is equal to utility cost plus performance-based rate cost. To minimize total cost, this utility should set a SAIDI target of 30 minutes per year. This target may be difficult to achieve if the utility is at 90 minutes per year and small changes will increase total cost.

Now consider PBR rewards starting at SAIDI = 60 min/yr and capping at $50 million when SAIDI = 30 min/yr. If rewards increase faster than utility cost, a new locally optimal target occurs. The total cost associated with this target may be lower than the target associated with penalties. If a utility finds itself with locally optimal reliability, it may be difficult to target globally optimal reliability since small changes will increase total cost. In Figure 2.18, this will occur if a utility has a SAIDI of 90 min/yr. Even if the utility is able to improve SAIDI by 10 min/yr, it will have to increase its costs for 6 years until the optimal SAIDI of 30 min/yr is reached, and will not recover these costs for many more years.

PBRs have many difficulties. First, the possibility of multiple optimal targets can send ambiguous price signals and does not accurately reflect free markets or societal cost. Second, they only address average levels of reliability and can result in a small number of customers experiencing reliability levels much worse than average. Last, PBRs subject utilities to new financial risk. Since reliability will vary naturally from year to year, aggressive PBRs can expose utilities to potential financial distress, increasing investor risk and depressing market capitalization.[52]

2.4.3 Reliability Guarantees

Reliability guarantees are the simplest method of allowing customer choice to influence reliability targets. Each customer chooses a reliability contract. Expensive contracts guarantee high levels of reliability, basic contracts guarantee modest levels of reliability and the cheapest contracts do not provide guarantees. Customers experiencing reliability below guaranteed levels receive rebate checks or credits on their energy bill.

Reliability guarantees allow neighboring customers to select different contracts associated with different levels of reliability. If low reliability is experienced in an area where many customers are signed up for high reliability, rebate costs will be high and the utility will spend money to improve reliability. If high reliability is experienced in an area where many customers are signed up for low reliability, the utility will reduce spending in this area. Reliability guarantees create a free-market-like scenario with customer choices influencing utility reliability targets. Customers usually react positively to choice, and many utilities are beginning to offer reliability guarantees to major accounts.

Widespread implementation of reliability guarantees to customers is problematic due to free rider behavior and gaming behavior. Most consumers are aware that they receive electricity via the same wires as their neighbors and will experience the same reliability. This may lead customers to choose less expensive reliability guarantees and hope to benefit from neighbors selecting expensive guarantees. Other customers may research reliability levels and choose guarantees based on expected payouts minus expected premiums. Even if customers have low reliability needs, they may choose expensive guarantees, opportunistically receive payments and switch back to inexpensive plans after reliability improves (requiring longer-term contracts can help to restrict this type of behavior).

Gaming of guarantees can be prevented if utilities are able to change the reliability of customers nearly as fast as they are able to switch plans. This is possible in areas with multiple feeders and a sufficient number of tie points. The reliability of each feeder can be altered by changing its configuration. Feeders can be made more reliable by reducing their service territory and can be made less reliable by increasing their service territory. Figure 2.19 shows a 16 feeder system with uniform feeder service territories (left) and this same system reconfigured to reflect the customer distribution of reliability guarantees.

Except in dense urban areas, many new switches and feeder ties will need to be installed on distribution systems before system reconfiguration can flexibly differentiate customer reliability. Adding these switches will tend increase cost and increase SAIDI. This may not be desirable if most customers are not willing to pay for improved reliability.

For areas where reliability guarantees are desirable, utilities must identify the feeder configuration that minimizes expected payouts to customers. After this is done, reliability targets must be set for each individual feeder. Utilities must set higher targets for feeders where reliability can be improved for lower costs than resulting reductions in payouts. Similarly, utilities must set lower targets for areas where current expenses can be lowered more than consequential increases in payouts.

Reliability guarantees may soon become widespread. They are easily implemented and allow customer choice to influence reliability targets. They also provide an opportunity for customers to choose, if they wish, lower reliability rather than higher reliability (something regulators and utilities are hesitant to endorse).

Widespread reliability guarantees have other implications. First, utilities must allocate capital for payouts and ensure that these reserves do not erode overall financial performance. Second, utilities will have to hedge against the risk of extraordinary events by indemnifying themselves with insurance policies. Last, and perhaps most importantly, utilities will have to build core competencies around reliability planning, reliability engineering, and reliability centered maintenance. They must begin to make decisions based on their impact to financial statements rather than engineering standards. They must discard old mindsets based on capacity planning and guaranteed rates-of-return, and integrate new processes based on customer choice and profit maximization. Reliability will no longer be a by-product of design and operation, but a viable commodity with price and reliability targets governed by the laws of supply and demand.

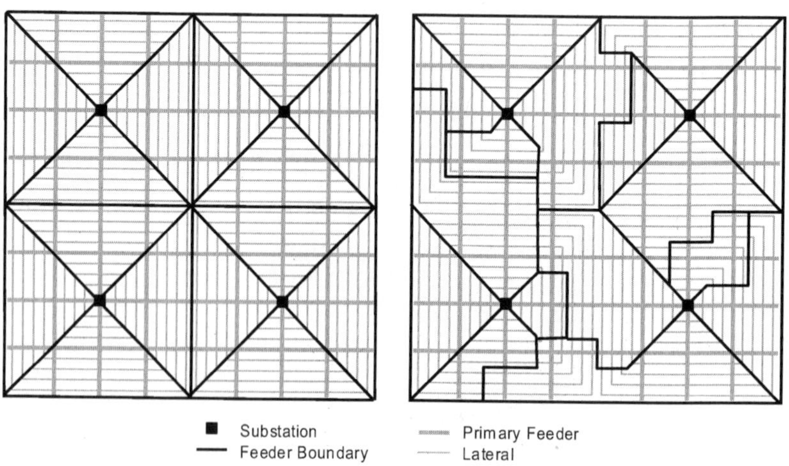

Figure 2.19. Reliability guarantees are much more attractive if a utility is able to quickly reconfigure feeder service territories. The left drawing represents 16 feeders serving equivalent service territories. These service territories can be reconfigured to improve reliability to customers with expensive reliability guarantees and degrade reliability to customers with inexpensive reliability guarantees (right).

2.5 HISTORY OF RELIABILITY INDICES

The interested reader may be interested in the story behind reliability indices and reliability reporting. Today, reliability indices have a strong regulatory flavor to them. This was not always the case. This section describes the history of reliability indices, reliability data collection, industry standards, and outage management systems to explain how this transition occurred.

Pre-1970s

Electric utilities, since their beginnings in the late 1800s, have always had to respond to faults on their distribution system. After a fault, a utility generally strove to ensure that (1) protection systems were able to safely clear faults, and (2) faults could be efficiently located and fixed. Customer-level reliability was rarely an explicit consideration.

Immediately after a fault occurred, certain customers lost power and some would call the utility. Each call would result in somebody sticking a pin in a large paper map mounted on the wall. The locations of the pins would allow system operators to infer the potential location of the fault and direct crews to the appropriate location. This method tended to break down during adverse weather when many faults occurred in a short period of time.

For each outage event, a crew would typically fill out a paper "outage ticket" describing the outage and related repair efforts. These outage tickets were collected at various field offices and used for a variety of purposes. It was common for a large percentage of outage tickets to be lost or otherwise missing.

Through the mid-1940s, there was no standard approach to addressing distribution reliability. An early step in this direction was a joint study in 1947 by the Edison Electric Institute (EEI) and the American Institute of Electrical Engineers (AIEE) on the different types of construction and protection for distribution circuits (the AIEE is now the IEEE). This study included fault rates for the major causes of faults on overhead distribution lines (e.g., wind, trees, lightning, equipment failure, human error). Results of this study were summarized by Westinghouse in 1959 in the definitive and comprehensive *Electric Utility Engineering Reference Book: Distribution Systems*. This book makes no mention of distribution system reliability, reliability indices, or outage reporting systems.

After the Northeast Blackout of 1965 (a transmission reliability issue), system reliability assessment started to gain more attention. This was true for generation, transmission, substations, and distribution. One of the earliest works related to distribution was a 1969 paper discussing the ability to design distribution systems to a specified level of reliability.[53] One of the reliability metrics used by this paper includes annual interruption rates as seen by the average customer, similar to the index SAIFI which is used today.

1970s

In the early 1970s, it became increasingly common for large utilities to compile paper outage tickets in an attempt to compute the reliability characteristics of their system. The most commonly used reliability indices included SAIFI, SAIDI, CAIDI, ASIFI, and ASIDI.

In the mid-1970s, workstation-based computers became available in the marketplace. Some utilities saw potential advantages in using these computers to aid in the distribution outage management process. The primary goal was typically to replacing the "pushpin" method of outage tracking with a computerized method, which would be particularly helpful during adverse weather. This often resulted in a custom software application being developed, referred to as an outage management system (OMS). These OMSs were designed to group incoming customer calls, infer the location of likely faults, and track the status of identified outages. They initially were not designed to collect reliability data, compute reliability indices, or be used for regulatory purposes. Most utilities still used paper outage tickets for reliability data.

1980s

More large utilities developed custom (sometimes called "in-house") OMSs in the 1980s, but the vast majority of utilities continued to use paper-based systems. In 1981, the industry funded an EPRI program to develop a computer program to predict distribution reliability indices.[54] The first textbook to discuss reliability indices was published in 1988,[55] and defines SAIFI, SAIDI, and CAIDI.

In the mid-1980s, many utilities began large workforce reductions. Up until this point, distribution reliability was not a significant regulatory issue. With utility workforce reductions and other spending cutbacks, state regulators began to worry about the possibility of reliability getting worse. Regulators were now beginning to ask questions about reliability, and made it known that it was not acceptable for reduced spending to result in worse reliability. Some state regulators began to ask for annual reliability performance reports (even though the systems and processes that formed the basis of these reports were never designed for this purpose).

In the late 1980s the Power Engineering Society of the IEEE formed a working group under the Distribution Subcommittee on Performance Records for Optimizing System Design. This group initially focused on design issues, but realized that the calculation of reliability indices varied widely by utility. This group therefore compiled a report in 1989 defining reliability indices commonly in use by US utilities.[56] The group was later renamed the Working Group on Distribution Reliability (WGDR). There was no reliability index associated with momentary interruptions at this time.

In the late 1980s, commercial vendors began to develop computerized OMSs. The first was Westinghouse (later ABB), which began the development in 1988 (a software program called CADOPS). Other vendors were soon to follow.

1990s

In the early 1990s, most electric utilities still did not have a computerized OMS, and relied on paper systems to collect reliability data. Due in large part to participation in the WGDR, calculation of reliability indices by utilities became increasingly popular. In 1998, the WGDR published a trial use guide on distribution reliability indices,[57] which was the precursor to the current standard.[19] This guide was noteworthy in two respects. First, it included indices related to momentary interruptions. Second, it presented the results of a reliability index survey showing the reliability associated with different "quartiles" of performance. At this point, nearly all US utilities began computing annual reliability indices in an attempt to determine how the reliability of their systems compares with the overall industry. Many of the calculations were still based on paper systems.

As reliability index benchmarking became more common, it became apparent that many factors can limit the ability to compare the reported indices of different utilities. This includes issues such as excluded outages (e.g., transmission, substations, planned, major events), and major event definition. To investigate these issues, the WGDR conducted a survey on reliability measurement practices and published the results in 1999.[58] This investigation did not examine issues related to data quality.

As more utilities began implementing computerized OMSs, it became apparent that reliability indices often, but not always, saw a dramatic worsening (even though the customer experience remained essentially unchanged). Major issues included data completeness, data quality, and the ability of the system to track the true duration and number of customers associated with sequential restoration steps. To investigate these issues, the WGDR in 1998 sent surveys to 161 US utilities and received responses from 71 (results published in 2003).[59] Key results from this survey include:

Results from 1998 IEEE Survey

- 64% are required to report reliability to state regulators.
- 62% consider step restoration when computing indices.
- There are wide variations in what events are included when computing reliability indices. For example, about half of all responding utilities do not include outages that are downstream of line reclosers and fuses.
- Only 18% compute momentary interruption indices.
- 70% exclude major events when computing indices.

In the late 1990s, utilities and software vendors began to give increased attention to the ability of an OMS to compute "true" reliability indices as seen by the customer so that the reliability of different utilities could be compared on a fair basis. Many of the custom OMS applications had inherent limitation that made it cost-prohibitive to more accurately track and compute "true" reliability.

Up until this point, an OMS was an expensive proposition that involved workstation-based hardware and either custom-developed software or expensive commercial software. Because of the high price, computerized OMSs were rare except for large utilities.

Towards the end of the 1990s, personal computers (PCs) became inexpensive and commonly used for a variety of business functions. Software vendors began creating affordable PC-based OMSs. These systems offer much greater flexibility for interfacing with many other corporate IT systems, and eliminate ongoing support requirements for antiquated hardware and software. These systems also pay close attention to reliability data collection and the ability to accurately compute reliability indices. PC-based systems were quickly adopted by many utilities without an OMS. This transition to PC-based systems was motivated by several performance benefits: enhanced customer call responsiveness, accelerated system restoration, and improved outage reporting capabilities. The performance improvements are made possible because of the effective IT integration that can be implemented with other utility systems such as: customer call centers, geographic information systems, SCADA, workforce management, and system operator interfaces. Some utilities with custom OMS applications also converted, but many did not because their existing system was "still working well."

2000s

In 2001, the IEEE trial use guide on reliability indices was balloted and become a full standard.[19] Several key features of this standard included (1) the distinction between momentary interruptions and momentary interruption events; (2) the five-minute threshold between momentary and sustained interruptions; and (3) the "2.5 Beta" method for major event identification.

By the early 2000s, reliability reporting to state regulatory agencies had become common. As of August 2001, twenty six states had reliability reporting requirements, and eleven of these had associated penalties and/or rewards based on reliability performance. At this point, reliability index calculation had become as much of a regulatory issue as a benchmarking issue, and companies increasingly began to examine the ability of their systems and processes to serve in this regulatory context.

Many vendors were now offering low-cost OMS deployment, and almost all but the very smallest utilities without an OMS in the 1990s had purchased one. Many others converted from older systems to newer systems. All of these new OMSs were able to handle reliability data collection and calculation in manner that allowed computed reliability to be a very good approximation of true reliability as experienced by customers.

Today

Today, most utilities have robust systems and processes related to reliability data collection and reporting. This typically includes a PC-based OMS that is interfaced with the customer call center, customer information system, SCADA system, and geographic information system. Reliability indices are computed from data provided from the OMS system, and this data is generally complete and auditable. Reliability benchmarking problems of the past, at least as they relate to data completeness and data quality, have largely gone away. However, even with all the standardization, there still remain differences between utilities such as: what is allowed to be excluded from reliability indices, the categorization of causes, weather normalization, customer density, geography, and system topology. These factors make it difficult to make direct benchmark comparisons. This said, the increased use of reliability indices for regulatory purposes has been a key driver in providing a degree of uniformity in reliability reporting.

In order to properly understand distribution reliability indices, it is important to understand the history involved. This section has provided history as it related to US utilities. Other areas of the world have different historical experiences that have shaped their current state of reliability indices, and the reader should note that the US experience may differ from other countries. A timeline summary of reliability index history is provided in Table 2.9.

Table 2.9. Timeline summary of reliability index history.

Time	Event
1947	Survey on distribution failure causes and failure rates
Early 1970s	Utilities begin to compute indices from paper outage tickets
Late 1970s	Some large utilities develop custom outage management systems (OMS)
mid 1980s	Regulators start to take an increased interest in distribution reliability
1988	First commercial OMS begins to be developed
1989	IEEE defines commonly used reliability indices
1998	IEEE P1366 published (including reliability index benchmark data); there are still wide variations in OMS usage from newer PC-based to older workstation-based to none at all
2001	Reliability indices become an IEEE standard
Early 2000s	Mass adoption of PC-based OMS

2.6 STUDY QUESTIONS

1. How are power quality and reliability related, and how power quality and reliability issues typically addressed within a utility?
2. Discuss similarities, differences, and interrelationships with regards to sustained interruptions, momentary interruptions, and voltage sags.
3. Why are customer-based indices most commonly used, even though the interruption of a large customer is typically more important than the interruption of a small customer?
4. How can the selection of the momentary interruption threshold impact reliability indices and reliability investment decisions?
5. Although SAIDI is generally considered a good measure of reliability, what are some of the potential limitations of SAIDI when used to drive investment decisions?
6. How does the IEEE recommended method for storm exclusion differ from traditional approaches? Why does the IEEE recommend this approach? What are some potential problems with this approach?
7. Describe the factors that should be considered when comparing the historical reliability performance of separate utilities.
8. Explain why it may be a good or bad idea for a utility with below average reliability performance to set a goal of being in the top 25%.
9. How can reliability data collection and reporting processes result in higher and/or lower reported reliability indices?
10. What are some of the arguments against the use of customer cost surveys to drive utility investment decisions?

REFERENCES

1. ANSI Std. C84.1-1977, *Voltage Ratings for Electric Power Systems and Equipment (60Hz).*
2. HD 472 S1, *Nominal Voltages for Low Voltage Public Electricity Systems.*
3. M. E. Baran, J. Maclaga, A. W. Kelley, and K. Craven, "Effects of Power Disturbances on Computer Systems," *IEEE Transactions on Power Delivery,* Vol. 13, No. 4, Oct. 1998, pp. 1309-1315.
4. IEEE Std. 446-1995, *IEEE Recommended Practice for Emergency and Standby Power Systems for Industrial and Commercial Applications.*
5. IEC Publication 801-4, *Electromagnetic Compatibility for Industrial Process Measurement and Control Equipment, Part 4: Electrical Fast Transient/Burst Requirements.*
6. ANSI/IEEE Std. C37.90.1-1989, *IEEE Standard Surge Withstand Capability Tests for Protective Relays and Relay Systems.*
7. ANSI/IEEE Std. C62.41.1991, *IEEE Recommended Practice on Surge Voltages in Low Voltage AC Power Circuits.*
8. ANSI/IEEE Std. 1100-1992, *IEEE Recommended Practice for Powering and Grounding Sensitive Electronic Equipment.*
9. ANSI/IEEE Std. 518-1982, *IEEE Guide for the Installation of Electrical Equipment to Minimize Electrical Noise Inputs to Controllers from External Sources.*

10. *Electrical Transmission and Distribution Reference Book (*formerly the *Westinghouse Electrical Transmission and Distribution Reference Book),* ABB Power T&D Company, Inc., 1997.
11. IEC Publication 868/EN 60868, *Flickermeter; Functional and Design Specification.*
12. ANSI/IEEE Std. 519-1992, *IEEE Recommended Practice and Requirements for Harmonic Control in Electric Power Systems.*
13. ANSI/IEEE Std. C57.110-1986, *IEEE Recommended Practices for Establishing Transformer Capacity When Supplying Nonsinusoidal Load Currents.*
14. IEC Publication 555-2, *Disturbances in Supply Systems Caused by Household Appliances and Similar Electric Equipment, Part 2: Harmonics.*
15. IEC Publication 1000-4-7, *General Guide on Harmonics and Interharmonic Measurements and Instrumentation for Power Supply Systems and Equipment Connected Thereto.*
16. I. M. Nejdawi, A. E. Emanuel, D. J. Pileggi, M. J. Corridori, and R. D. Archambeault, "Harmonics Trend in NE USA: A Preliminary Survey," *IEEE Transactions on Power Delivery,* Vol. 14, No. 4, Oct. 1999, pp. 1488-1494.
17. R. C. Dugan, M. F. McGranaghan, and H. W. Beaty, *Electrical Power Systems Quality,* McGraw-Hill, 1996.
18. IEEE Working Group on System Design, "A Survey of Distribution Reliability Measurement Practices in the U.S.," *IEEE Transactions on Power Delivery,* Vol. 14, No. 1, Jan. 1999, pp. 250-257.
19. IEEE Standard 1366, *Guide for Electric Power Distribution Reliability Indices*, 2003.
20. S. H. Jaramillo, G. T. Heydt, and E. O'Neill-Carrillo, "Power Quality Indices for Aperiodic Voltages and Currents," *IEEE Transactions on Power Delivery,* Vol. 15, No. 2, April 2000, pp. 784-790.
21. D. D. Sabin, D. L. Brooks, and A. Sundaram, "Indices for Assessing Harmonic Distortion from Power Quality Measurements: Definitions and Benchmark Data," *IEEE Transactions on Power Delivery,* Vol. 14, No. 2, April 1999, pp. 489-496.
22. G. T. Heydt and W. T. Jewell, "Pitfalls of Power Quality Indices," *IEEE Transactions on Power Delivery,* Vol. 13, No. 2, April 1998, pp. 570-578.
23. D. L. Brooks, R. C. Dugan, M. Waclawiak, and A. Sundaram, "Indices for Assessing Utility Distribution System RMS Variation Performance," *IEEE Transactions on Power Delivery,* Vol. 13, No. 1, Jan. 1998, pp. 254-259.
24. R. E. Brown, S. Gupta, R. D. Christie, S. S. Venkata, and R. D. Fletcher, "Distribution System Reliability: Momentary Interruptions and Storms," *IEEE Transactions on Power Delivery,* Vol. 12, No. 4, October 1997, pp. 1569-1575.
25. H. V. Freund und ihre Kosten, Elektrizitätswirtschaft Jg. (90) 1991, pp. 913-915.
26. C. A. Warren and M. J. Adams, "Reliability on the Regulatory Horizon," *IEEE Transmission and Distribution Conference and Exposition,* Atlanta, GA, Oct. 2001.
27. J. M. Clemmensen, "A Systems Approach to Power Quality," *IEEE Spectrum*, June 1993.
28. G. H. Kjølle and A. T. Holen, "Delivery Point Interruption Costs: Probabilistic Modeling, Time Correlations, and Calculation Methods," *Proceedings of the Power Systems Computation Conference,* PSCC 1996, Dresden, Germany.
29. R. E. Brown, S. Gupta, R. D. Christie, S. S. Venkata, and R. D. Fletcher, "Automated Primary Distribution System Design: Reliability and Cost Optimization," *IEEE Transactions on Power Delivery*, Vol. 12, No. 2, April 1997, pp. 1017-1022.
30. R. E. Brown, "The Impact of Heuristic Initialization on Distribution System Reliability Optimization," *International Journal of Engineering Intelligent Systems for Electrical Engineering and Communications*, Vol. 8, No. 1, March 2000, pp. 45-52.
31. W. F. Horton, S. Goldberg, and R. A. Hartwell, "A Cost/Benefit Analysis in Feeder Reliability Studies," *IEEE Transactions on Power Delivery,* Vol. 4, No. 1, Jan. 1989, pp. 446-452.
32. G. Kjølle, L. Rolfseng, and E. Dahl, "The Economic Aspect of Reliability in Distribution System Planning," *IEEE Transactions on Power Delivery,* Vol. 5, No. 2, April 1990, pp. 1153-1157.
33. R. E. Brown, S. S. Venkata, and R. D. Christie, "Hybrid Reliability Optimization Methods for Electric Power Distribution Systems," *International Conference on Intelligent Systems Applications to Power Systems*, Seoul, Korea, IEEE, July 1997.

34. R. E. Brown, S. S. Gupta, R. D. Christie, and S. S. Venkata, "A Genetic Algorithm for Reliable Distribution System Design," *International Conference on Intelligent Systems Applications to Power Systems*, Orlando, FL, January 1996, pp. 29-33.
35. G. Tollefson, R. Billinton, and G. Wacker, "Comprehensive Bibliography on Reliability Worth and Electric Service Consumer Interruption Costs 1980-1990," *IEEE Transactions on Power Systems,* Vol. 6, No. 4, Nov. 1991, pp. 1508-1514.
36. R. Billinton, G. Wacker, and E. Wojczynski, "Comprehensive Bibliography on Electrical Service Interruption Costs," *IEEE Transactions on Power Apparatus and Systems,* Vol. PAS-102, No. 6, June 1983, pp. 1831-1837.
37. R. Billinton, J. Oteng-Adji, and R. Ghajar, "Comparison of Two Alternate Methods to Establish an Interrupted Energy Assessment Rate," *IEEE Transactions on Power Systems,* Vol. 2, No. 3, Aug. 1987, pp. 751-757.
38. S. Burns and G. Gross, "Value of Service Reliability," *IEEE Transactions on Power Systems,* Vol. 5, No. 3, Aug. 1990, pp. 825-834.
39. M. Lehtonen, B. Lemström, R. Stilling-Peterson, K. Kølbæk-Jensen, F. Vilhjálmsson, A. T. Holen, and L. Livéus, "Electric Supply Outage Costs in the Nordic Countries," CIRED 1995, Brüssel, Belgium.
40. P. Wang and R. Billinton, "Time Sequential Distribution System Reliability Worth Analysis Considering Time Varying Load and Cost Models," *IEEE Transactions on Power Delivery,* Vol. 14, No. 3, July 1999, pp. 1046-1051.
41. R. Billinton and M. Pandey, "Reliability Worth Assessment in a Developing Country – Residential Survey Results," *IEEE Transactions on Power Systems,* Vol. 14, No. 4, Nov. 1999, pp. 1226-1231.
42. L. Goel and R. Billinton, "Evaluation of Interrupted Energy Assessment Rates in Distribution Systems," *IEEE Transactions on Power Delivery,* Vol. 6, No. 4, Oct. 1991, pp. 1876-1882.
43. L. Goel and R. Billinton, "Determination of Reliability Worth for Distribution System Planning," *IEEE Transactions on Power Delivery,* Vol. 9, No. 3, July 1994, pp. 1577-1583.
44. L. Goel and R. Billinton, "Utilization of Interrupted Energy Assessment Rates to Evaluate Reliability Worth in Electrical Power Systems," *IEEE Transactions on Power Systems,* Vol. 8, No. 3, Aug. 1993, pp. 929-936.
45. A. Makinen J. Partanen, and E. Lakervi, "A Practical Approach for Estimating of Future Outage Costs in Power Distribution Networks," *IEEE Transactions on Power Delivery,* Vol. 5, No. 1, Jan. 1990, pp. 311-316.
46. R. Billinton and M. Pandey, "Reliability Worth Assessment in a Developing Country – Commercial and Industrial Survey Results," *IEEE Transactions on Power Systems,* Vol. 14, No. 4, Nov. 1999, pp. 1232-1237.
47. G. Tollefson, R. Billinton, G. Wacker, E. Chan, and J. Aweya, "A Canadian Customer Survey to Assess Power System Reliability Worth," *IEEE Transactions on Power Systems,* Vol. 9, No. 1, Feb. 1994, pp. 443-450.
48. L. L. Grigsby, ed., *The Electric Power Engineering Handbook*, CRC Press, 2000.
49. A. P. Sanghvi, "Measurement and Application of Customer Interruption Costs/Value of Service for Cost-Benefit Reliability Evaluation: Some Commonly Raised Issues," *IEEE Transactions on Power Systems,* Vol. 5, No. 4, Nov. 1990, pp. 1333-1344.
50. H. Schrayshuen, "Case Study: Niagara Mohawk' Approach to Setting Reliability Targets," *Electric Distribution Reliability Planning: Course Handbook,* Chicago, IL, Sept. 2000.
51. R. E. Brown and M. M. Marshall, "Budget Constrained Planning to Optimize Power System Reliability," *IEEE Transactions on Power Systems*, Vol. 15, No. 2, May 2000, pp. 887-892.
52. R. E. Brown and J. J. Burke, "Managing the Risk of Performance Based Rates," *IEEE Transactions on Power Systems*, Vol. 15, No. 2, May 2000, pp. 893-898.
53. R. L. Capra, M.W. Gangel, and S.V. Lyon, "Underground Distribution System Design for Reliability," *IEEE Transactions on Power Apparatus and Systems*, Vol. PAS-88, No. 6, June 1969, pp. 834-842.
54. EPRI Report EL-2018, Development of Distribution Reliability and Risk Analysis Models, Aug. 1981. (Report prepared by Westinghouse Corp.).

55. R. Billinton and R. Allan, *Reliability Assessment of Large Electric Power Systems*, Kluwer Academic Publishers, 1988.
56. "Definitions of Customer and Load Reliability Indices for Evaluating Electric Power System Performance," IEEE Working Group on Performance Records for Optimizing System Design, A 75 588-4.
57. *IEEE Guide for Electric Power Distribution Reliability Indices*, IEEE 1366-2003, 2003.
58. Trial C. A. Warren, R. Ammon, and G. Welch, "A Survey of Distribution Reliability Measurement Practices in the U.S.," *IEEE Transactions on Power Delivery*, Vol. 14, No. 1, Jan. 1999, pp. 250-257.
59. C. A. Warren, D. J. Pearson, and M. T. Sheehan, "A Nationwide Survey of Recorded Information Used for Calculating Distribution Reliability Indices," *IEEE Transactions on Power Delivery*, Vol. 18, No. 2, April 2003, pp. 449-453.

3
Interruption Causes

Customer interruptions are caused by a wide range of phenomena including equipment failure, animals, trees, severe weather, and human error. These causes are at the root of distribution reliability, and understanding them allows abstract topics like reliability modeling and computer optimization to be viewed from a practical perspective. In addition, identifying and addressing physical root causes is often the most cost effective way to address reliability problems.

3.1 EQUIPMENT FAILURES

Each piece of equipment on a distribution system has a probability of failing. When first installed, a piece of equipment can fail due to poor manufacturing, damage during shipping, or improper installation. Healthy equipment can fail due to extreme currents, extreme voltages, mischievous animals, severe weather, and many other causes. Sometimes equipment will fail spontaneously for reasons such as chronological age, thermal age, state of chemical decomposition, state of contamination, and state of mechanical wear. Comprehensive treatment of equipment failures is beyond the scope of this book, but the following sections present the most common modes of failure for equipment that is most critical to distribution system reliability.

3.1.1 Transformers

Transformers impact distribution system reliability in two related ways: failures and overloads. Catastrophic transformer failures within substations can result in interruptions to thousands of customers. When this happens, other transformers are often called upon to pick up the interrupted load. If there is not enough spare transformer capacity, a decision must be made whether or not to overload in-service transformers and accept the resulting loss-of-life. Accepting loss-of-life will improve reliability for the moment, but will increase the probability that the overloaded transformers will fail at a future date. Understanding these issues requires a basic knowledge of transformer ratings and thermal aging.

Transformer ratings are based on the expected life of winding insulation at a specified temperature. Ratings typically assume an ambient temperature of 30°C, an average winding temperature rise of either 55°C or 65°C, and an additional hot spot rise of 10°C or 15°C. A summary of transformer design temperatures is shown in Table 3.1. Older 55°C rise transformers are shown to have a hot spot design temperature of 95°C and newer 65°C rise transformers are shown to have a hot spot design temperature of 110°C.

The life of a transformer is often defined as the time required for the mechanical strength of the insulation material to lose 50% of its mechanical strength (many other definitions are also possible). Loss of mechanical strength occurs when insulation polymers break down due to heat. The rate of breakdown increases exponentially with temperature, allowing the expected life of insulation to be expressed by the Arrhenius theory of electrolytic dissociation:[1]

$$\text{life} = 10^{(K_1/(273+°C))+K_2} \text{ (hours)} \tag{3.1}$$

Constants for this equation have been experimentally determined for both power transformers and distribution transformers and are documented in standard transformer loading guides[2]. A summary of these values is shown in Table 3.2. This table also shows the required hot spot temperature rise above normal limits that will cause the rate of thermal aging to double.

Table 3.1. Temperatures used for transformer ratings. Many older transformers have 55°C rise insulation, but most new transformers have 65°C rise insulation. Higher insulation ratings allow transformers to operate at a higher temperature and, therefore, serve higher loads.

Temperature	55°C Rise Insulation	65°C Rise Insulation
Ambient Temperature	30°C	30°C
Average Winding Temperature Rise	+ 55°C	+ 65°C
Average Winding Temperature	85°C	95°C
Additional Hot Spot Temperature Rise	+ 10°C	+ 15°C
Hot Spot Temperature	95°C	110°C

Table 3.2. Transformer aging constants for Equation 3.1, determined by accelerated aging tests that measure the time required for transformer insulation to lose 50% of its initial strength.

Description	K_1	K_2	Rise that Doubles Aging Rate
Power Transformer (55°C Insulation)	6972.15	-14.133	95°C + 5.9°C
Power Transformer (65°C Insulation)	6972.15	-13.391	110°C + 6.4°C
Distribution Transformer (55°C Insulation)	6328.80	-11.968	95°C + 6.6°C
Distribution Transformer (65°C Insulation)	6328.80	-11.269	110°C + 7.1°C

Plots of life versus hot spot temperature are shown in Figure 3.1. These curves can be used to determine the expected life of a transformer and to estimate the loss of life that will occur during an overload. If run constantly at hot spot design temperatures (95°C for 55°C insulation and 110°C for 65°C), power transformers have an expected insulation half-life of about 7.2 years and distribution transformers have an expected insulation half-life of about 20 years. Transformers are not normally loaded constantly at their nameplate rating and are re-rated by utilities based on weekly load curves to result in an acceptable lifetime (e.g., 30 years).

Transformer temperatures do not instantaneously increase when overloads are applied. The substantial thermal capacity of cases, windings, cores, and oil can result in thermal time constants of many hours. This allows all transformers to be overloaded for a brief period of time without any loss-of-life if initial temperatures are below normal ratings. Loss-of-life will occur if temperature is allowed to rise above normal ratings, and many utilities will accept a certain amount of loss-of-life during emergency situations.

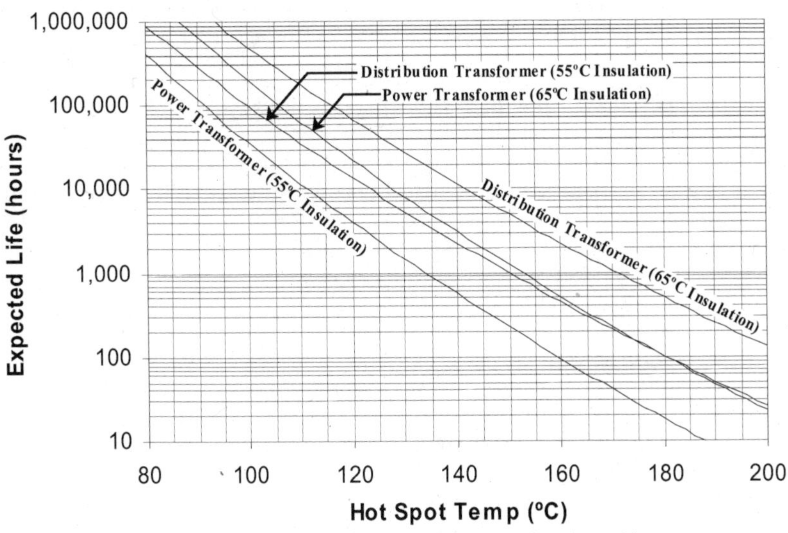

Figure 3.1. Expected insulation half-life of transformers as a function of hot spot temperature. Typical transformers are designed for hot spot temperatures of 95°C (55°C rise insulation) or 110°C (65°C rise insulation). Transformer life reduces exponentially with hot spot temperature.

Extreme overloads can result in catastrophic transformer failure. The temperature of the top oil should never exceed 100°C for power transformers with 55°C insulation or 110°C for those with 65°C insulation. The consequence of exceeding these limits could be oil overflow, excessive pressure, or tank rupture. If winding hot spot temperatures exceed 140°C when moisture is present, free bubbles may form and result in internal faults. Due to these considerations, the peak short-duration loading of power transformers less than 100 MVA should never exceed 200% of nameplate rating.[3]

Loading has a significant influence on transformer life and is a critical aspect of distribution reliability, but most catastrophic transformer failures are due to faults that occur downstream of the transformer secondary bushings. The resulting fault current passes through the transformer and shakes the windings with a mechanical force proportional to the square of the fault current magnitude. If thermal aging has caused insulation to become sufficiently brittle, a crack will form and an internal transformer fault will result. Low impedance transformers tend to fail more often than high impedance transformers since they will experience more severe fault currents. Autotransformers generally have very low impedances and tend to fail more often than multiple-winding transformers.

Extensive research has been done in the area of transformer condition assessment. The goal is to determine the health of the transformer and identify incipient problems before they lead to catastrophic failure. Simple methods include visual and audio inspection. More complex methods include thermal load tests, power factor tests, high potential tests and dissolved gas analysis, and other electrical, mechanical, and chemical techniques.[4] Some of these tests can be performed continuously, referred to as condition monitoring, and can automatically notify operators if monitored values exceed warning thresholds.

Transformers have many accessories that can also fail. Failures associated with pumps, fans, and blocked radiators reduce the ability of transformers to dissipate heat but do not cause outages directly. Other failures, like cracked insulators and broken seals, may result in outages but can be fixed quickly and inexpensively. Oil-filled load tap changers have historically been prone to failure and can substantially reduce the reliability of a transformer.[5] Manufacturers have addressed this problem and new devices using vacuum technology have succeeded in reducing load tap changer failure rates.

3.1.2 Underground Cable

A major reliability concern pertaining to underground cables is electrochemical and water treeing. Treeing occurs when moisture penetration in the presence of an electric field reduces the dielectric strength of cable insulation. When moisture invades extruded dielectrics such as cross-linked polyethylene (XLPE) or ethylene-propylene rubber (EPR), breakdown patterns resembling a tree reduce the voltage withstand capability of the cable. When insulation strength is degraded sufficiently, voltage transients caused by lightning or switching can result in dielectric breakdown. The severity of treeing is strongly correlated with thermal age since moisture absorption occurs more rapidly at high temperatures.[6]

Water treeing has been a widespread and costly problem for utilities with aging XLPE cable[7-9] (EPR cables have generally not encountered these problems). To address utility concerns, cable manufacturers have developed both jacketed cable and tree retardant cable (TR-XLPE, see Figure 3.2). Cable jackets protect insulation from moisture ingress and protect concentric neutral conductors from corrosion. Tree retardant insulation slows the development of water trees after moisture is present. Utilities can also install surge protection devices on riser poles to limit the magnitude of voltage transients seen by old cable.

Treeing is largely attributed to bad manufacturing. Insulation impurities and insulation voids accelerate moisture absorption and can substantially reduce the life expectancy of cable. To minimize the probability of installing inferior cable, utilities are encouraged to test all reels before acceptance. Some common cable testing methods include:

- **Steady State Voltage Withstand Test** — a destructive test that applies a constant AC or DC voltage (about 3 times nominal) to see if dielectric breakdown occurs.
- **Impulse Voltage Withstand Test** — a destructive test that applies a voltage transient about equal to the BIL rating of the cable to see if dielectric breakdown occurs.
- **Partial Discharge Test** — a small high frequency signal is injected at one end of the cable. Signal reflections detected by sensors indicate the number and location of partial discharge points.

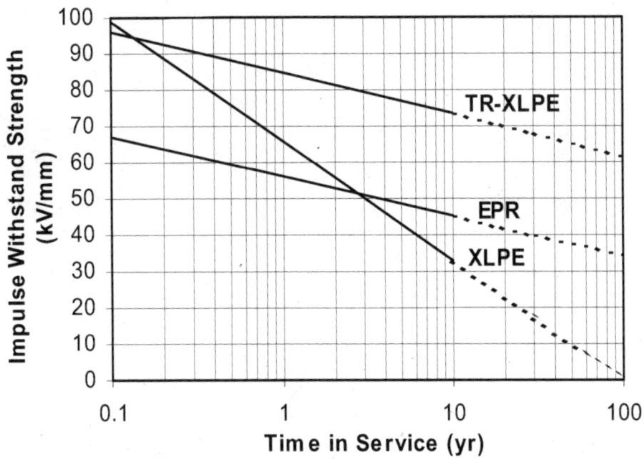

Figure 3.2. Insulation breakdown versus age for 35-kV cable at Houston Lighting and Power. Tree retardant cable retains much more strength than EPR or XLPE cable.[10] Actual results are shown up to 10 years and are extrapolated to 100 years.

- **Power Factor Test** — the power factor of cable impedance is measured and compared to cables in known states of deterioration.
- **Dielectric Spectroscopy** — measuring power factor over a range of frequencies (e.g., 0.1-Hz to 20-kHz). Graphs of power factor versus frequency allow cable samples to be sorted by relative health.
- **Degree of Polymerization** — analyzing physical insulation samples in a lab for polymeric breakdown. The average polymer length, referred to as the degree of polymerization (DP), is directly related to the strength of the insulation. DP testing is suitable for celluloid insulation such as oil impregnated paper, but is less applicable for extruded dielectrics.
- **Indentor Test** — using special cable indentors to test the hardness of the cable insulation.

Unfortunately, most mature testing techniques require cable sections to be de-energized before examination. This can cause significant inconveniences if customers have to be interrupted, alternate cables overloaded or traffic lanes shut down. Sometimes this inconvenience is so great that utilities forgo testing and install new cable whenever old cable is de-energized. Online partial discharge testing is still in its infancy, but has already shown promising results.[11]

An alternative to cable replacement is cable rejuvenation. In this process, special elbows are placed on the ends of cable sections. A special fluid is injected into one end and allowed to slowly migrate though the cable over a period of weeks, filling voids and restoring dielectric strength. Cable rejuvenation can add decades of life to old cable and may be a cost-effective solution in areas where cable removal and replacement are expensive.

For most utilities, a majority of cable system failures occur at splices, terminations, and joints rather than at the cable itself. In addition to water treeing, cable accessory failures are caused by incipient mechanical stresses due to water reacting with aluminum and liberating hydrogen gas.[12] Water ingress in cable accessories is largely due to poor workmanship, and can be mitigated by proper training and the use of waterproof heat shrink covers.

In extreme situations, currents can cause cable insulation to melt or chemically break down. Equation 3.2 represents the time that a short circuit must be cleared before insulation damage begins (see Figures 3.3 and 3.4).

$$t_{max} = C_1 \log_{10}\left(\frac{T_m + C_2}{T_i + C_2}\right) \cdot \left(\frac{A_{kcmil}}{I_{sc}}\right)^2 \tag{3.2}$$

t_{max} = maximum short circuit duration before insulation damage (seconds)
T_i = Initial °C operating temperature (typically 90°C for thermoset insulation)
T_m = Maximum °C short circuit temperature (typically 250°C for thermoset insulation)
C_1 = Conductor Constant (29700 for Cu, 12500 for Al)
C_2 = Conductor Constant (234 for Cu, 228 for Al)
A_{cmil} = Conductor area in kcmil
I_{sc} = Short circuit current magnitude in amps

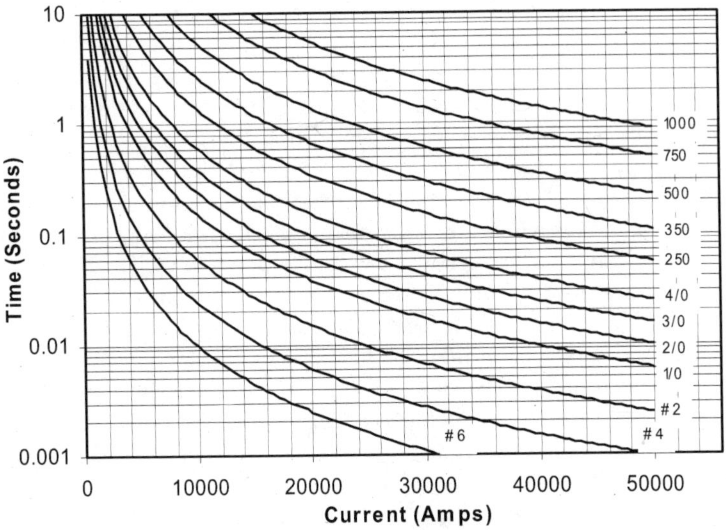

Figure 3.3. Damage curves for XLPE and EPR cable with aluminum conductor. If the fault is not cleared in time, the thermoset material will begin to break down, reducing the dielectric strength of the insulation.

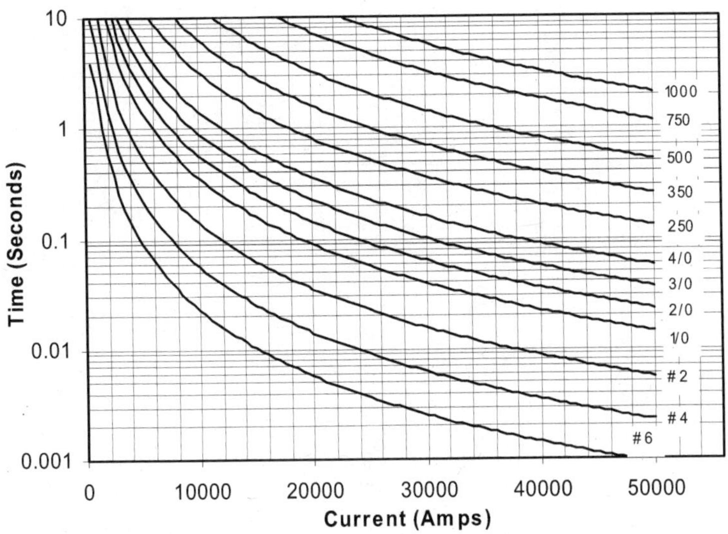

Figure 3.4. Damage curves for XLPE and EPR cable with copper conductor. If the fault is not cleared in time, the thermoset material will begin to break down, reducing the dielectric strength of the insulation.

3.1.3 Overhead Lines

Due to high exposure, most overhead line damage is caused by external factors such as vegetation, animals, and severe weather. Bare conductor is able to withstand much higher temperatures than insulated conductors and damage due to high currents is less of a concern. Regardless, high currents do impact the reliability of overhead lines in several ways.[13] High currents will cause lines to sag, reducing ground clearance and increasing the probability of phase conductors swinging into contact. Higher currents can cause conductors to anneal, reducing tensile strength and increasing the probability of a break occurring. Fault currents, if not cleared fast enough, can cause conductors to melt and burn down.

The normal current rating of overhead lines is often limited by ground clearances. As temperature increases, conductors will elongate based on their coefficients of thermal expansion. This expansion will cause lower sags, increase the likelihood of phase conductor contact, and may result in unsafe clearances. Due to thermal inertia, conductor sag will not occur instantaneously. Typical lines have thermal time constants between 5 and 20 minutes, allowing temporary overloading without sag concerns. Detailed treatment of sag calculations and emergency line ratings are beyond the scope of this book and the reader is referred to Reference 14.

If not cleared fast enough, short circuit currents can cause lines to melt and fall to the ground. The maximum short circuit current depends upon many factors including clearing time, conductor resistance, thermal capacity, initial temperature, and fusing temperature. Assuming that no heat can be dissipated to the environment, short circuit levels are governed by Sverak's ampacity equation:[15]

$$I_{sc} = 5.0671 \cdot A_{kcmil} \cdot \sqrt{\frac{TCAP}{\lambda_m \rho_{20} \alpha_{20} t} \ln\left|\frac{K_0 + T_m}{K_0 + T_i}\right|} \tag{3.3}$$

I_{sc} = Short circuit current magnitude in amps
t = maximum short circuit duration before conductor begins to fuse (seconds)
A_{kcmil} = Conductor area in kcmil
T_i = Initial °C conductor temperature
T_m = Maximum °C temperature before conductor begins to fuse
TCAP = Conductor thermal capacity (J/cm³/°C)
λ_m = magnetic effect (≈1)
ρ_{20} = conductor resistivity at 20°C (μΩ/cm)
α_{20} = thermal coefficient of resistivity at 20°C (1/°C)
K_0 = thermal coefficient of conductance at 0°C (°C)

Conductor damage curves for various sizes of aluminum and copper wire are shown in Figure 3.5 and Figure 3.6, respectively. Associated parameters for various types of conductor material are shown in Table 3.3.

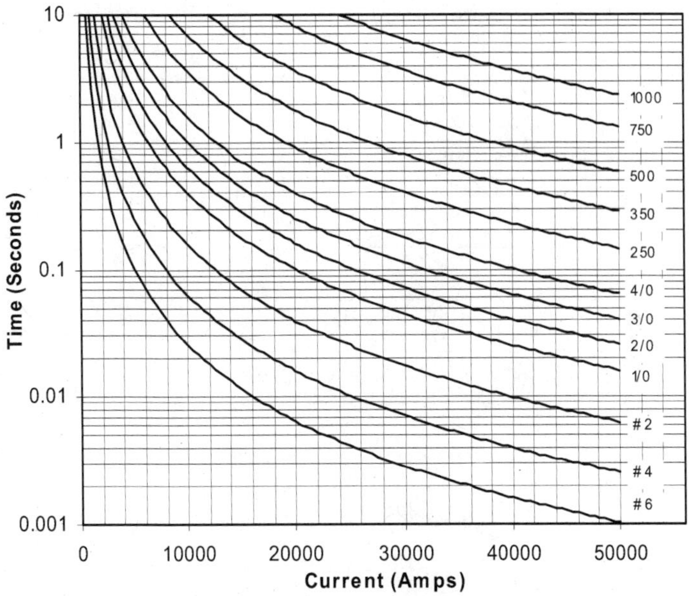

Figure 3.5. Damage curves for bare aluminum wire with a pre-fault temperature of 90°C. The steel core of an ASCR conductor will be damaged much later than the aluminum and may prevent conductor burndown.

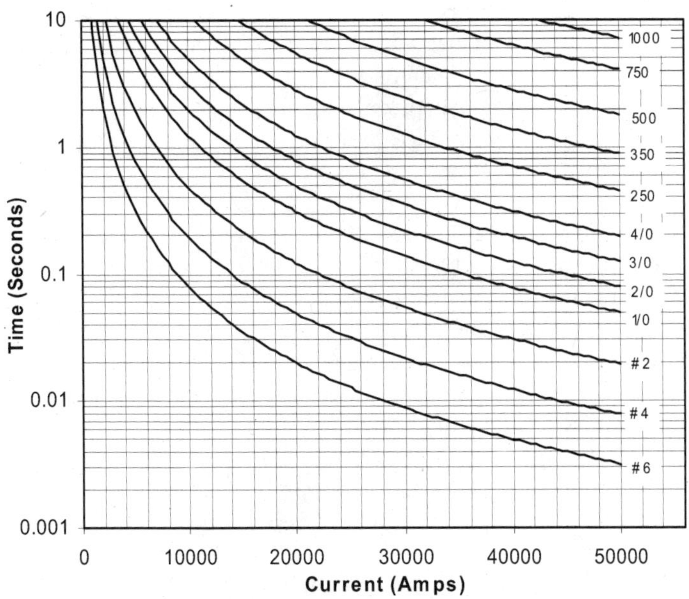

Figure 3.6. Damage curves for hard drawn bare copper wire with a pre-fault temperature of 90°C. If the fault is not cleared in time, the wire will burndown.

Table 3.3. Conductor properties required to compute short circuit damage characteristics. The ability of conductors to withstand high short circuit currents depends upon thermal capacity, fusing temperature, resistivity, and the sensitivity of resistivity to temperature.

Description	Conductivity (%)	α_{20} (1/°C)	K_0 (°C)	Fusing Temp. (°C)	ρ_{20} μΩ-cm	TCAP (J/cm^3/°C)
Standard Annealed Cu	100.0	0.00393	234	1083	1.7241	3.422
Hard Drawn Cu Wire	97.0	0.00381	242	1084	1.7774	3.422
Copper Clad Steel Wire	40.0	0.00378	245	1084/ 1300	4.397	3.846
Copper Clad Steel Wire	30.0	0.00378	245	1084/ 1300	5.862	3.846
Commercial EC Al Wire	61.0	0.00403	228	657	2.862	2.556
Al Alloy Wire 5005	53.5	0.00353	263	660	3.2226	2.598
Al Alloy Wire 6201	52.5	0.00347	268	660	3.2840	2.598
Al Clad Steel Wire	20.3	0.00360	258	660/ 1300	8.4805	2.670
Zn Coated Steel Wire	8.5	0.00320	293	419/ 1300	20.1	3.931
Stainless Steel No. 304	2.4	0.00130	749	1400	72.0	4.032

Many reliability problems on overhead systems are associated with auxiliary components rather than the actual wires. This includes energized equipment such as hot clamps, splices, switches, cutouts, arresters, capacitor banks, and voltage regulators. It also includes nonenergized equipment such as poles and crossarms. Potential reliability problems related to these devices can be identified by various inspection methods including:

- **Visual Inspection** — Feeders can be visually inspected by following their routes on foot, bicycle, car, or truck. Doing so can identify advanced stages of damage to crossarms, hardware, and conductors.
- **Radio Frequency Interference** — Damaged overhead equipment such as cracked insulators, loose connections or broken conductor strands can emit radio frequency noise. After a radio receiver picks up noise, further investigations can be made to identify the source.
- **Infrared Inspection** — Nonferrous line hardware in good condition will operate cooler than the conductor.[16] Unreliable or loose connections tend to operate hotter and can be identified by infrared inspection equipment.
- **Aerial Inspection** — Visual and infrared inspections can be effectively performed from a helicopter or small airplane. This allows many miles of circuit inspection to be performed quickly, facilitates the inspection

of inaccessible areas, and can identify problems that will not be noticed from a ground inspection.

- **Switch and Cutout Testing** — Switches that have not been recently operated are subject to many reliability problems. They could be rusted shut, fused shut, frozen shut or have warped or broken components. These problems will not result in a failure, but will prevent the switch from being used after a contingency occurs. Operational testing can prevent these situations, but is costly and may result in customer interruptions.
- **Fuse Testing** — Distribution protection systems are coordinated based on specific fuse sizes. After a fuse blows, crews may replace it with a different size and cause coordination problems. Worse, the fuse may be replaced with a piece of conductor if no fuse is available or if inrush currents are causing replacement fuses to blow. To verify fuse sizes without causing customer interruptions, it is usually necessary to jumper around and de-energize the fuse cutout. If an overhead conductor breaks and falls to the ground, it will obviously cause an open circuit, but may or may not cause fault current to flow. If the broken conductor touches or arcs to another phase conductor or the neutral conductor when it is falling, a "low impedance" fault occurs and the fault current will be limited by circuit impedance. Low impedance faults typically have contact resistances of less than 2Ω. If the wire falls to the ground without contacting or arcing to another conductor, the contact impedance will limit current to low levels and a "high impedance" fault occurs. Tests have shown that faults to surfaces such as asphalt, grass, and gravel are typically less than 50 amperes while faults to reinforced concrete can approach 200 amperes.[17]

3.1.4 Poles

Poles are the critical structural element for overhead distribution systems. In terms of pure reliability, most pole-related issues are based on contributing factors such as wind and vehicular accidents. Even so, it is important to have a basic understanding of the design criteria, reliability characteristics, and failure modes associated with poles.

The basic pole-related reliability and safety concern is the pole falling over due to wind loading. It is impossible to give comprehensive treatment of this subject here, but a basic overview is now provided.

Wind force on a pole is due to the following four components: (1) wind blowing on the pole itself, (2) wind blowing on equipment attached to the pole, (3) wind blowing on conductors attached to the pole, and (4) wind blowing on third-party attachments such as telephone, television, and fiber optic cables.

Wind pressure on an object is proportional to the square of wind velocity. The pressure is also related to the aerodynamic shape of the object. Wind pressures related to cylindrical and flat surfaces are as follows:

$$p = 0.025 \cdot v^2 \quad ; \text{cylindrical surfaces} \tag{3.4}$$

$$p = 0.004 \cdot v^2 \quad ; \text{flat surfaces} \tag{3.5}$$

p = Wind pressure in lb/in^2
v = Wind velocity in mph

When wind blows on a pole, the pressure exerts a moment at the location of ultimate failure. For distribution poles, this location is typically at the ground line. Therefore, the moment of each force is computed by multiplying the wind pressure by the projected wind exposure area by the attachment height above ground. The sum total of all of the moments can then be compared to the strength of the pole at the groundline.

The strength of a pole with a circular cross section at a specific location is proportional to the cube of the circumference at this location. It is also proportional to material strength. Consider a wood pole with a groundline circumference of c (in) and wood fiber strength of w (lb/in^2). The maximum moment for this pole at the groundline (M_{max}) is the following:

$$M_{max} = \frac{w \cdot c^3}{3790} \quad ; \text{Maximum moment (ft-lb)} \tag{3.6}$$

The maximum moment is the expected breaking point of the pole. Typically utilities will only allow pole loadings lower than this maximum value to account for variations in materials, uncertainty in assumptions, and safety considerations. The most common approach for distribution is to allow 50% loading for most poles, and to allow 25% loading for poles with special safety considerations (e.g., railway and freeway crossings).

The fiber strength of a pole is a function of both tree species and the treatment that is used to dry the tree. Fiber strength values used in the standard ANSI O5-1 are show in Table 3.4. Most of the values in this table represent the mean value of a large sample. Fiber strength can vary widely for each pole and a typical coefficient of variation for fiber strength is 20%.

Wind on the pole. Consider a pole with a base diameter of b (inches), a top diameter of t (inches), and a height-above-ground of h (in feet). The projected wind exposure is a trapezoid with an area equal to the following:

$$A = \frac{1}{2} b \cdot t \cdot h \cdot 12 \quad ; \text{Projected area of a pole (in}^2\text{)} \tag{3.7}$$

Table 3.4. Fiber strength of poles based on species and treatment type. This table is based on ANSI-O5-1, and most values correspond to the mean fiber strength of a large sample.

Treatment Type	Species	Fiber Strength (psi)
Air Seasoned	Cedar, northern white	4000
	Cedar, western red	6000
	Pine, ponderosa	6000
	Pine, jack	6600
	Pine, lodgepole	6600
	Pine, red	6600
	Pine, Scots	7800
	Cedar, Alaska yellow	7400
	Douglass fir, interior	8000
Boulton Drying	Douglass fir, coastal	8000
	Larch, western	8400
Steam Conditioning	Pine, southern	8000
Kiln Drying	Cedar, western red	6000
	Douglass fir, interior	8000
	Douglass fir, coastal	8000
	Larch, western	8400
	Pine, jack	6600
	Pine, lodgepole	6600
	Pine, ponderosa	6000
	Pine, radiata	6600
	Pine, red	6600
	Pine, Scots	7800
	Pine, southern	8000

The moment of this pole at the groundling is computed based on the height of the center-of-mass location (h_{cm}) for the wind-exposed portion of the pole. This is computed as follows:

$$h_{cm} = \frac{h(2b+t)}{3(b+t)} \text{ ; Center of mass location (ft above ground)} \tag{3.8}$$

The wind moment can easily be calculated as the wind pressure times the projected area times the center-of-mass height. For typical designs, the percentage of total force due to wind on the pole is small, often in the range of 10% to 15%. This is not the case when designing for extreme wind conditions, such as tornados and hurricanes. In these cases, the force due to wind on the pole will often be half or more of the total force.

Wind on conductors. When wind blows on a conductor span, the force is transferred to the poles at each end of the span. Since about half of the force is transferred to each pole, conductor force on a pole is typically based on the average span length. For example, if a pole has its upstream span at one hundred feet long and its downstream span at two hundred feet long, an average span length of one hundred fifty feet can be used for calculations. For a conductor with a diameter of d (in), an average span length of s (ft), a mounting height of h (ft), and a wind velocity of v (mph), the applied moment of the conductor to the pole (M_c) is:

$$M_c = h \cdot d \cdot s \cdot 12 \cdot \left(0.025 \cdot v^2\right) \quad \text{; moment of conductor (ft-lb)} \qquad (3.9)$$

Typically there will be at least one conductor and one neutral attached to a pole. Many times there will be multiple circuits involving six phase conductors and a neutral. In any case, the moment arm of each conductor can be calculated separately and added together to compute the total applied moment to the pole.

Wind on attachments. The impact of wind blowing on third-party attachments is calculated in an identical manner as wind on conductors. It is listed here separately since new pole installations may not know the number and size of future attachments. When designing a new distribution pole, a utility will typically contact local telephone and cable companies and ask whether they intend to attach to the pole. If so, the loads of the third party attachments can be considered directly in the design. If not, the utility must decide whether to allow for potential, but not-yet identified attachments.

When a third party places new attachments on existing poles, there is the potential that this pole will no longer meet wind loading criteria (utility and/or regulatory). If so, actions should be taken to ensure that wind loading criteria is met. The process is simple as long as a utility knows when and where attachments occur. In reality, attachments sometimes occur without utility knowledge and utilities should perform periodic attachment audits. Typically these audits are performed for billing purposes. It is a small incremental effort to perform loading calculations as a part of these audits.

Ice Loading. Certain geographic regions are prone to ice buildup on conductors (see Section 3.3.6). The NESC requires that areas in the heavy loading district account for 0.5 inches of radial ice buildup on conductors, which increases the effective diameter by 0.5 inches. The medium loading district requires a 0.25 inch radial buildup of ice, and the light loading district assumes no ice buildup.

Pole-Mounted Equipment. Equipment that is mounted on poles will increase the wind loading on the pole. Typically, this is accounted for by computing the projected wind profile of the equipment, multiplying this by the design wind pressure, and multiplying this by the center-height to determine the moment.

When a pole fails due to a wind load, the failure will occur where the material stress is highest. For a distribution pole, this will typically occur at the groundline. For very tall poles or for poles with high taper, this failure may occur above the groundline. For a single load point and a pole with constant taper, failure will occur where the diameter of the pole is 150% that of the location of the load point. All calculations can then be referenced to this location.

The above calculations assume that the pole material will fail. This will only happen if the pole does not lean over or pop out of the ground (i.e., foundation failure). Many utilities set poles in typical soil to a depth of 10% of the height of the pole plus 2 feet. For example, a 50 foot pole would be set 7 feet deep. This is often modified based on soil condition. Setting depth will be less for rocky soil and more for sandy soil. These rules are generally acceptable for wood poles of standard distribution sizes, but may be insufficient for very strong poles. A very strong pole must be set deeper to take full advantage of its strength.

The governing safety standard for distribution pole strength is the National Electrical Safety Code (NESC). This document is intended to provide minimum design criteria to ensure public safety. It is not intended to be a design manual, nor is it intended to address issues other than public safety. A pole meeting the NESC requirements can be considered safe, but may or may not be desirable from an economic or reliability perspective.

The NESC defines three different grades of safety requirements depending upon the public safety issues related to a particular installation. These are termed Grade B, Grade C, and Grade N, with Grade B being the highest requirement. In general, the NESC requires distribution structures to meet Grade C construction except when crossing railroad tracks or limited-access highways (these require Grade B construction).

According to the NESC, a structure must be able to withstand loading due to combined ice buildup and wind. The NESC also requires structures taller than 60 feet above the ground to be designed to withstand extreme wind speeds (but with no ice buildup since icing does not occur during extreme winds). Most distribution structures do not fall into this category.

The NESC specifies the required strength of structures based on grade of construction and loading conditions. Criteria based on combined wind and ice loading is specified in Section 250B. Criteria based on extreme wind loading are specified in Section 250C. Each of these sections uses an "overload" factor, which is essentially a safety factor that results in added strength. The NESC also allows for deterioration of structure strength, resulting in a specified strength for "initial installation" and a separate specified strength for "at replacement." Engineered materials like concrete are assumed to not lose strength over time, and therefore have a single specified strength.

As previously mentioned, the force that wind exerts on a pole is proportional to the square of wind speed. This means that 120-mph winds exert four times more force than 60-mph winds. Using this relationship, relative strengths for different design criteria can be computed (these do not appear in the NESC). In this case, Grade C construction with light combined ice and wind loading is assumed to have a relative strength of 1.

Table 3.5. Relative strengths of various NESC design criteria. This example assumes light loading district for 250B and 250C calculations and 145 mph winds for 250C calculations.

	250B Grade C	250B Grade B	250C (145 mph) Grade B&C
Wood at Installation			
Wind Speed (mph)	60	60	145
Overload Factor	2.67	4.00	1.33
Relative Strength	1.00	1.50	2.91
Wood at Replacement			
Wind Speed (mph)	60	60	145
Overload Factor	1.33	2.67	1.00
Relative Strength	0.50	1.00	2.19
Concrete			
Wind	60	60	145
Overload Factor	2.20	2.50	1.00
Relative Strength	0.82	0.94	2.19

Table 3.5 shows the relative strength requirements of different NESC scenarios assuming a light loading district for Grade B and C calculations and winds of 145 mph for extreme wind calculations. For railway and freeway crossings, 250B Grade B applies and pole strength must be 50% greater (relative strength of 1.5). For extreme wind, a wind speed of 145 mph is assumed, which is the worst case in the US. For structures taller than sixty feet in height, 250C applies and poles must be nearly three times as strong (relative strength of 2.91).

Wood poles will naturally degrade in strength over time due to wood deterioration and other factors. The NESC accounts for this deterioration by specifying the overload factor to be used to determine when pole replacement is required. For example, the 250B Grade C overload factor is 2.67 for initial installation, but is 1.33 at replacement. This implies that a fully loaded wood pole can lose 50% of its initial strength before replacement is required. At minimum replacement strength, Grade B construction is twice as strong as Grade C construction, and 250C construction is more than four times as strong.

To prevent deterioration, new wood poles are typically treated with decay-resistant substances. Older poles were typically treated with coal-tar creosote. Popular treatments today include pentachlorophenol ("penta") and chromated copper arsenate ("CCA"). Both creosote and penta poles will deteriorate after time, and require periodic inspections and supplemental preservative treatments to prevent excessive loss of strength. CCA poles have not generally shown signs of decay, but must still be specified assuming decay will occur (utilities may choose to periodically inspect CCA poles for reasons other than decay).

When wood poles decay over time, most problems typically occur near ground level where trapped moisture causes rotting. Additional damage can occur due to singular events such as automobile collisions and fires. A simple way to test a wood pole is to (1) visually inspect the pole for signs of decay, and (2) strike the pole near its base with a hammer and listen for suspicious hollow sounds. A more complete inspection will excavate around the pole and take core samples. Less common inspection methods include sonograms, indentation testing, and electrical resistance testing.

If a pole shows excessive signs of rot, it should be replaced. It the rot is less severe, it may be possible to take remedial actions. This will include removing all existing rot, fumigating the pole, and possibly filling internal cavities with a filler paste. If the pole has lost strength, it can be reinforced with an adjacent wood stub, a steel brace, or a fiberglass wrap.

In part due to deterioration considerations, some utilities are beginning to use non-wood poles for distribution. The following is a summary of the most viable candidates.

Laminated Wood. These poles are created by gluing together metric lumber into a tapered pole with a square cross section. These poles are lighter that an equivalent round pole, have less variability in strength, and can be climbed with the same equipment as round wood poles. However, these poles tend to be expensive, and have a less desirable shape for wind loading calculations.

Cast Concrete. This type of pole is made by pouring concrete into a tapered form with a square cross section. Pre-stressed steel strands are also typically included to increase the strength of the pole. Some people prefer the look of concrete poles over wood, and concrete is essentially maintenance free. These poles are relatively inexpensive when compared to spun concrete. Cast concrete poles should be pre-drilled since they are very difficult to drill in the field.

Spun Concrete. These poles are similar in characteristics to cast concrete, but are circular in cross section and have a hollow interior. They are manufactured in a circular mold that is spun at a high rate so that the centrifugal force compresses the concrete against the inner wall of the mold. For a given strength (compensating for shape factor), spun concrete will weigh less than cast concrete, but will be more expensive. Spun concrete poles have the additional advantage of being round, thereby being less affected by wind. Spun concrete poles should be pre-drilled since they are very difficult to drill in the field.

Steel. Round steel poles are common for many applications such as transmission towers, street lights, traffic signals, and so forth. Steel has an excellent strength-to-weight ratio and can be used for very strong structures that can still be installed with bucket trucks. Drawbacks to steel include high price, climbability, low BIL, and susceptibility to corrosion.

Steel Hybrid. This refers to a round steel pole mounted on a concrete foundation. This approach has several advantages. First, the concrete foundation is not susceptible to corrosion. Second, the pole structure is in two pieces, thereby reducing the likelihood that a crane will be required for installation. Steel hybrid still has the disadvantages of having a low BIL, being difficult to climb, and being expensive. Also, the steel pole, although completely above ground, is still susceptible to corrosion.

Lattice. A lattice structure has exceptional strength-to-weight characteristics and wind characteristics. They are often used for transmission structures, but are less common for distribution due to high cost and the requirement of a foundation. However, lattice distribution towers are available in both steel and, more recently, composite materials. The composite lattices towers are exceptionally light and can be climbed, but are approximately twice as wide as an equivalent wood pole due to mechanical considerations.

Ductile Iron. Ductile iron pipe has been used for many years for water pipes, and recently has been adapted for distribution poles. The characteristics of these poles are similar to those of steel, including cost. Potential advantages are that ductile iron produces a natural corrosion barrier similar to weathering steel, there are no seams required in the manufacturing process, and the manufacturing process can produce poles at a rapid rate.

Aluminum. Although aluminum is sometimes used for light-duty applications such as street lights and flag poles, it is rare for use in transmission and distribution poles. This is due to the relatively poor strength-to-weight properties of aluminum when compared to steel.

Composite. Composite poles are made by injecting an epoxy resin into a matrix of reinforcing fibers such as fiberglass, carbon fiber, and Kevlar. The result is exceptional strength-to-weight ratio, no susceptibility to corrosion, and high BIL. Manufacturers also claim that new technologies prevent deterioration due to high sun exposure. The use of composite poles is becoming more common in areas subject to woodpecker and insect damage and in areas not assessable by digger/derrick trucks (such as back lots). Some composite poles are constructed through extrusion, giving them a non-tapered cross section. This increases wind loading and somewhat reduces visual appeal.

Distribution poles are becoming a major consideration for many US utilities due to their quantity and age. Periodic pole inspections are expensive, and pole replacement is even more expensive. Up until now, pole deterioration has been more of a safety and compliance issue than a reliability issue. Moving forward, it will be increasingly important to consider poles in the context of distribution reliability.

3.1.5 Circuit Breakers

Circuit breakers are complicated devices that can fail in many different ways. They can spontaneously fail due to an internal fault, spontaneously open when they should not, fail to open when they should, fail to close when they should, and so forth. Table 3.6 lists the most common circuit breaker failure modes and their relative frequencies of occurrence.[18]

Table 3.6. Typical failure modes of circuit breakers. The most common failures occur when circuit breakers open when they should not (false tripping). The next most common failures are due to spontaneous internal faults.

Failure Mode	% of Failures
Opened when it should not	42
Failed while in service (not opening or closing)	32
Failed while opening	9
Damaged while successfully opening	7
Failed to close when it should	5
Damaged while closing	2
Failed during testing or maintenance	1
Damage discovered during testing or maintenance	1
Other	1
Total	**100**

A circuit breaker opening when it should not is referred to as false tripping. False tripping is typically associated with miscoordinated protection devices or problems with relays and associated equipment. False trips can be reduced by testing protection device coordination, relay settings, CT/PT ratios, and control wiring.

Circuit breakers can fail to open or close due to faulty control wiring, uncharged actuators or by simply being stuck. The probability of these types of operational failures occurring can be reduced through periodic exercising and testing all circuit breakers.

Circuit breakers can also experience internal faults while neither opening nor closing. These faults are due to dielectric breakdowns similar to those caused in transformers. Specific tests that check for dielectric strength will vary slightly based on the insulating medium of the circuit breaker (air, oil, SF_6 or vacuum).

In addition to insulation aging, circuit breakers are subject to contact erosion. Each time the contacts are used to interrupt current, a small amount of contact material is vaporized. The amount of contact erosion associated with each opening or closing event can be computed as follows.[19]

$$C = (V \cdot I \cdot t)/H \tag{3.10}$$

C = Amount of contact erosion (grams)
V = Voltage drop across contacts (volts)
I = Current across contacts (amps)
t = Duration of current (sec)
H = Heat of vaporization of contact material (Joules per gram)

Continuous monitoring can be used to estimate circuit breaker condition including the cumulative loss of contact material. Typical monitored values are current, opening time, closing time, contact speed, contact bounce, and recharge time. After excessive values or negative trends are identified, circuit breaker maintenance or replacement can be performed before failures occur.

3.1.6 Surge Arresters

Surge arresters come in two basic forms: silicon carbide and metal oxide varistors (MOVs). Silicon carbide is the older of the two technologies and requires an air gap to avoid excessive currents during normal operation. When voltage exceeds a certain threshold, the air gap arcs over and voltage is clamped across the arrester. MOVs have a highly nonlinear resistance (as a function of voltage), do not conduct excessive currents during normal operation, and generally do not require an air gap. When voltage exceeds a certain threshold, the resistance of the MOV drops sharply and clamps voltage across the device (similar to a Zener diode).

Silicon carbide arresters tend to fail more often than MOVs due to the presence of air gaps. Moisture entering an air gap can cause corrosion and reduce the voltage withstand strength of the gap. Further, thermal expansion of water vapor under heat can result in damaging mechanical stress to the interior of the arrester and can result in catastrophic failure during normal or overvoltage situations. Other failure modes include bad or aged blocks and direct lightning strikes.

There are four other major failure modes associated with surge arresters: puncture, thermal runaway, cracking under tension, and cracking under compression.[20] Cracking and puncture are caused by a localization of the current, which causes concentrated heating leading to nonuniform thermal expansion and thermal stresses. Puncture is most likely in varistor disks with low geometrical aspect ratio, when the current density has intermediate values. Cracking dominates at higher current densities and for disks with high aspect ratio. Puncture and cracking do not occur when the current is small, because the time evolution of the nonuniform heating is slow enough for the temperature distribution to flatten. For low and very high current densities, the most likely failure mode is thermal runaway—the surge arrester simply is not able to handle the energy levels flowing through it.

3.1.7 Insulators and Bushings

Insulators and bushings are made from three basic materials: glass, porcelain, and polymeric. Glass and porcelain are the oldest technologies, but polymeric materials are gaining popularity due to their increased strength and reduced brittleness.

Insulator and bushing failures are associated with dielectric breakdown. A dielectric breakdown in an insulator allows current to arc across the device. A dielectric breakdown in a bushing allows current to arc from the internal conductor to the outside of the device. Sometimes these currents will be small or self-extinguishing. At other times these currents lead to a low impedance arc, resulting in a short circuit and a potentially catastrophic failure of the insulator.

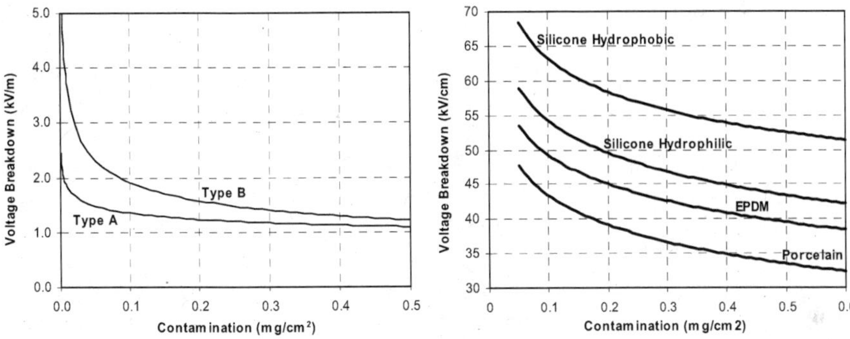

Figure 3.7. The left figure shows the decrease in voltage withstand with increasing contamination for two types of polymer insulator design used by a Florida utility. The right figure shows laboratory test results for different weather shed material types under fully wetted conditions. Contamination is measured in equivalent salt deposit density (ESDD).

Insulators and bushings can lose dielectric strength when exposed to contamination such as sea salt, fertilizers, industrial pollution, desert sand, vehicular deposits, road salt, and salt fog. Many field and laboratory studies have been performed and show that performance of contaminated conductors is generally good, but dielectric strength gradually decreases with contamination and highly contaminated insulators under wet conditions may be likely to flash over.[21-24]

Figure 3.7 shows the degradation of dielectric strength with salt contamination for various types of insulator designs and weather shed materials.[25] Voltage withstand capability drops dramatically with contamination, but can vary substantially for different insulator designs. This can correspond to a much higher probability of flashover for certain designs. Other important factors include uniformity of contamination, ratio of soluble to nonsoluble contaminants, wetting agents, altitude, and mounting position.

Regular washing of insulators will reduce the probability of flashover by keeping contamination density low and, consequently, dielectric strength high. The frequency of washing will depend on a variety of factors including the accumulation rate of contamination, the impact of contamination on particular insulator designs, cost, and overall system impact.

3.2 ANIMALS

Animals are one the largest causes of customer interruptions for nearly every electric utility. Problems and mitigation techniques are as varied as the animals involved. This section describes reliability concerns and common reliability improvement strategies for the following classes of animals: squirrels, mice, rats, gophers, birds, snakes, fire ants, and large animals.

3.2.1 Squirrels

Squirrels are a reliability concern for all overhead distribution systems near wooded areas. Squirrels will not typically climb utility poles, but will leap onto them from nearby trees and cause faults by bridging grounded equipment with phase conductors. Of the more than 365 species throughout the world, reliability concerns lie primarily with gray squirrels and red squirrels (see Table 3.7).

The gray squirrel, inhabiting most of the Northern Hemisphere, is the most common type of tree squirrel. Typical gray squirrels are fifteen inches long, weigh about one pound, and have bushy tails about as long as the combined head and body length. They have a typical diet of nuts, seeds, and fruit.

North American red squirrels (also known as pine squirrels and chickarees) are smaller then the gray (eight to ten inches in typical length), have fur that ranges from red to black, and have bushy tails about the same length as their body. Some subspecies can have red or black ear tufts and most have a white belly. The red squirrel is found in coniferous forests, where it feeds on the seeds and cones of the pines, fir, and spruce trees.

Squirrels are extremely agile, persistent and committed to travel routes that may include utility poles. Because of this, attempts to keep squirrels off of overhead utility equipment are usually futile and reliability improvement efforts should focus on equipment protection. The most common technique is to install plastic animal guards on bushings and insulators to prevent squirrels from simultaneously touching the tank and phase conductors (see Figure 3.8).

Table 3.7. Common types of squirrels and their typical maximum length.

Squirrel Type	Location	Length
Eastern Gray	Eastern US, southern Canada	20"
Western Gray	West Coast of US	22"
Fox	Eastern US	29"
North American Red	Alaska, Canada, North US, and West US	12"

Figure 3.8. The figure to the left shows a squirrel causing a fault by bridging the phase voltage to the grounded transformer tank. The middle figure shows the bushing fitted with a plastic animal guard to prevent the squirrel from causing a fault. The right figure shows the bushing fitted with an electrostatic guard that deters an approaching squirrel with a harmless but noticeable electric shock.

3.2.2 Mice, Rats and Gophers

Mice, rats, and gophers are rodents that cause faults by gnawing through the insulation of underground cable. Rats and mice are the most common cause of animal related outages on underground equipment, and gophers are third (snakes are the second most common cause).

Besides chewing through insulation, mice and rats also create reliability problems by attracting snakes. To prevent these rodents from gaining access to underground equipment, all equipment cabinets and duct openings should be tightly sealed. Some utilities have successfully used ultrasonic devices to ward off mice and rats.

When gophers dig tunnels, they chew their way through any obstacle in their path, including electric cables and plastic conduit. Gophers are most common in the Midwestern US where they are often the most common cause of underground equipment failure. Effective gopher control is difficult to achieve, with methods including traps, poisoned bait, poison gas, and ultrasonic techniques.

3.2.3 Birds

Birds are the most common cause of animal faults on transmission systems, subtransmission systems and air insulated substations. Different types of birds cause different types of problems, but can generally be classified as nesting birds, roosting birds, raptors, and woodpeckers.

Nesting birds commonly build their homes on lattice towers, poles, and in substations. Nesting materials can cause faults, and bird excrement can contaminate insulators. Removing nests sites can be difficult since the instinctual nature of certain birds will cause them to rebuild nests in the same location. To further complicate matters, nesting birds can attract predators (e.g., raccoons, snakes, cats) that can be a worse reliability problem than the birds themselves.

Roosting birds use electrical equipment to rest on or to search for prey. They cause faults by bridging conductors with their wings and contaminating insulators with their excrement. A small number of roosting birds is typically not a reliability concern, but the instinctual nature of certain birds can result in many thousands of birds roosting in a single location. To prevent birds from roosting, anti-roosting devices can be placed on attractive perches (see Figure 3.9). For locations that cater to thousands of roosting birds, more extreme deterrent methods such as pyrotechnics and falconry may be required.

Some of the larger roosting birds have a tendency to empty their bowels when they take off (referred to as streaming). In addition to contamination, streaming can result in faults if it bridges phase-to-phase or phase-to-ground potentials.

Table 3.8. Typical wingspans for large birds. Long wingspans can bridge conductors and result in a fault. Many large birds have threatened or endangered status, requiring utilities to take measures to reduce the number of electrocutions regardless of reliability concerns.

Bird	Typical Wingspan
Golden Eagle	8 ft
Bald Eagle	6 ft
Vulture	6 ft
Gray Owl	5 ft
Osprey	5 ft
Blue Heron	6 ft
Hawk	4 ft

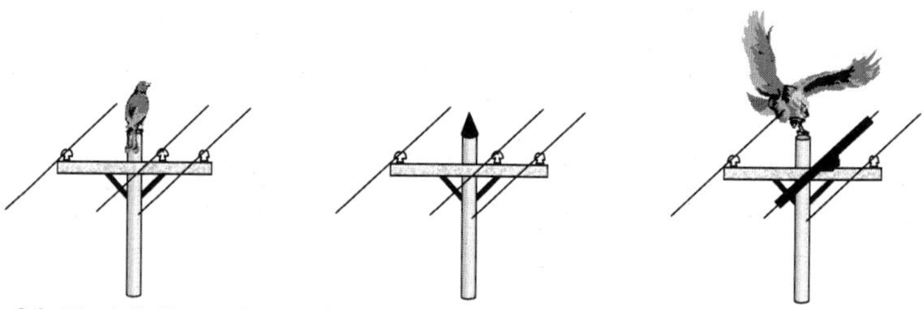

Figure 3.9. The left figure shows a bird perched on a distribution pole. This behavior can be deterred through the use of anti-roosting devices like the cone shown in the center figure. Phase-to-phase wing contact can be avoided by protecting the center phase with an insulated covering like the one shown on the right.

Raptors are birds of prey such as eagles, hawks, ospreys, owls, and vultures. In addition to the types of reliability problem associated with roosting and nesting birds, raptors have large wingspans that can bridge phase conductors during takeoff and landing. In addition to causing a fault, this can kill the raptor—a legal problem since most raptors are protected by federal statute. Typical raptor wingspans are shown in Table 3.8.

Woodpeckers peck holes in wood with their beaks as they search for insects or build nests. This does not harm trees (the bark regenerates), but can cause devastating damage to utility poles. Woodpecker reliability problems can be mitigated by using steel or concrete poles, by placing new poles next to damaged poles, and by using woodpecker repellent. Utilities addressing woodpecker problems should be aware that certain types are considered threatened or endangered and are protected by federal statute.

3.2.4 Snakes

Snakes are major reliability concerns in both substations and underground systems. They can squeeze through very small openings, can climb almost anything, and have the length to easily span phase conductors (see Table 3.9 for the lengths of different types of snakes). Snakes cause more substation outages than any other animal except birds and more underground outages than any other animal except mice.

Table 3.9. Common snakes and their typical maximum length. The length of a snake is an important reliability consideration since many faults occur when snakes bridge phase conductors.

Name	Location	Venomous	Length
Royal Python	Southeast Asia	No	30 ft
Anaconda	South America	No	25 ft
King Cobra	South Asia	Yes	18 ft
Black Mamba	Sub-Saharan Africa	Yes	14 ft
Bushmaster	Central America	Yes	12 ft
Boa Constrictor	Latin America	No	9 ft
Rattlesnake	Across the US	Yes	8 ft
Black	Northeastern US	No	8 ft
Water Moccasin	Southern US	Yes	6 ft
Racer	Eastern US	No	6 ft
Indian Cobra	South Asia	Yes	5 ft
King	Eastern US	No	5 ft
Copperhead	Eastern US	Yes	4 ft
Rosy Boa	Southwestern US	No	3 ft
Eastern coral	Southeastern US	Yes	2.5 ft
Garter	North America	No	2.0 ft
Rubber Boa	Western US	No	1.5 ft

Snakes are cold-blooded animals focused on body temperature regulation and obtaining food. During cold weather, snakes may seek out warm electrical cabinets. While hunting, snakes may seek out rats and mice in underground distribution systems and birds and nests in substations. Snakes problems can usually be mitigated if electrical cabinets are well sealed and if food supplies are removed. Another common solution is to install specialized snake fences around substations and critical pieces of equipment. Some snake fences use a low-voltage electric shock to increase their effectiveness.

3.2.5 Fire Ants

Fire ants, originally from South America, are small but have a severe sting. They have proliferated across much of the Southern US and it is becoming common in these regions for fire ants to build nests in pad-mounted equipment. Nesting materials can cause short circuits, the ants can eat away at conductor insulation, and their presence can make equipment maintenance a challenge.

There are actually two types of fire ant: black and red. The black fire ant, accidentally imported from South America into Mobile, Alabama, was first reported in 1918. Its distribution is still restricted to parts of Mississippi and Alabama. The red fire ant was imported around the 1930s and has spread across most states in the Southeast.[26] This species has become very abundant, displacing many native ant species. Fire ants have the potential of spreading widely across the South, the Southeast, and the Pacific Coast (see Figure 3.10).

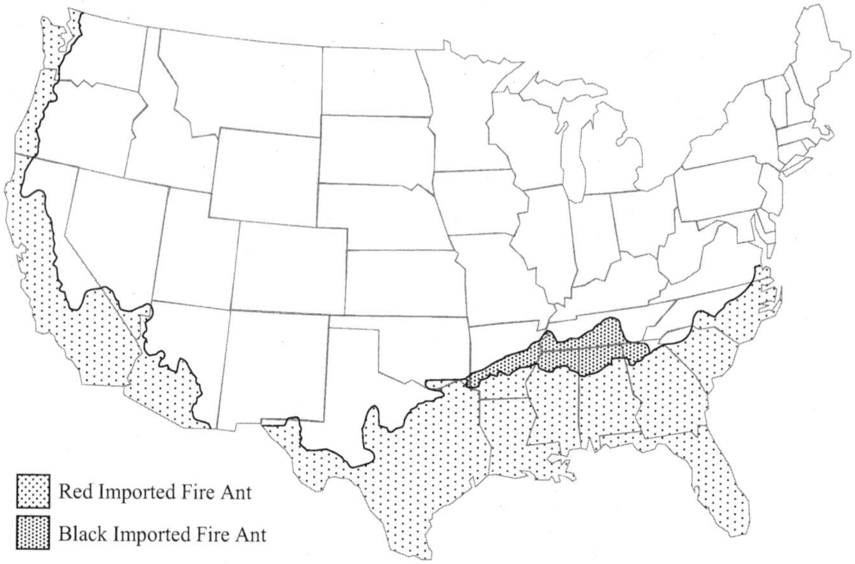

Figure 3.10. Likely areas of future fire ant infestation. Fire ants were accidentally introduced into the US from South America in the early 1900s. Since then, they have spread widely across the southern US, displacing native ant species and causing reliability problems by building nests in electrical equipment.

Controlling fire ants can be difficult due to their sheer number. In multiple queen colonies, there may be more than 200 ant mounds and 40 million fire ants per acre. Control methods generally consist of using poisoned bait for broadcast application and insecticides for individual mound treatment.

3.2.6 Large Animals

Large animals typically cause reliability problems through contact with guy wires and poles. Since they are large and strong, they can do physical damage and cause an immediate outage or make the system more prone to outages in the future. Most large animal problems are due to cattle, horses, bison, and bears.

Cattle cause reliability problems by rubbing on guy wires. Since they are habit-forming animals, they will tend to keep rubbing on the same guy wires and will eventually cause poles to lean. It is difficult to deter this behavior, and mitigation strategies should focus on barricading or removing guy wires.

Horses can also cause reliability problems by rubbing on guy wires, but more commonly collide with those that they cannot see. Even if these collisions do not cause reliability problems, they can kill or injure the horse and create liability concerns. Collisions can be avoided through barricades or by making the guy wires more visible.

Bison are large and powerful and like to rub their heads on objects. If this object is a utility pole, it may be pushed completely over and cause a major outage. There is usually no practicable way to deter this behavior and mitigation efforts must resort to barricades or pole removal.

Bears are becoming more of a distribution reliability concern as the US population expands into previously unpopulated areas. Most reliability problems are associated with wooden utility poles, since both brown and black bears can destroy wooden poles by using them as scratching posts. In addition, brown bears can climb wooden utility poles and cause faults by contacting live conductors. Bear related problems can be mitigated by placing fences around wooden poles or by replacing wooden poles with steel or concrete.

3.3 SEVERE WEATHER

Severe weather can take many forms and is the most frequent cause of customer interruptions for many utilities. During normal weather, equipment failures are independent events—the failure of one device is completely independent of another device and multiple overlapping failures are rare. During severe weather, many equipment failures can occur at the same time. This puts a strain on utility resources, can lead to long restoration times for many interrupted customers, and can challenge data collection systems. This section describes severe weather conditions that have the most impact on distribution reliability including wind, lightning, icing, extreme heat, and earthquakes.

3.3.1 Extreme Wind and Extreme Wind Ratings

The terms "extreme wind" and "wind storm" can either refer to linear winds that blow down trees and utility poles, or to circular winds such as tornadoes and cyclones. The severity of wind storms is a function of sustained wind speed, gust speed, wind direction, and the length of the storm. Severity is also sensitive to vegetation management and the time elapsed since the last wind storm. Since a wind storm will tend to blow over all of the weak trees, a similar storm occurring a few months later may have little impact. On the other hand, a wind storm may kill a tree without blowing it over. This dead tree will lose strength and become more susceptible to future wind storms.

A US map showing wind speeds for the worst expected storm in 50 years is shown in Figure 3.11. The highest winds (as measured by 3-second gusts) occur in special geographic areas that funnel winds, have high temperature gradients, or are exposed to hurricanes. The probability of equipment failure increases rapidly with increasing wind speed since the pressure exerted on trees and poles is proportional to the square of wind speed.

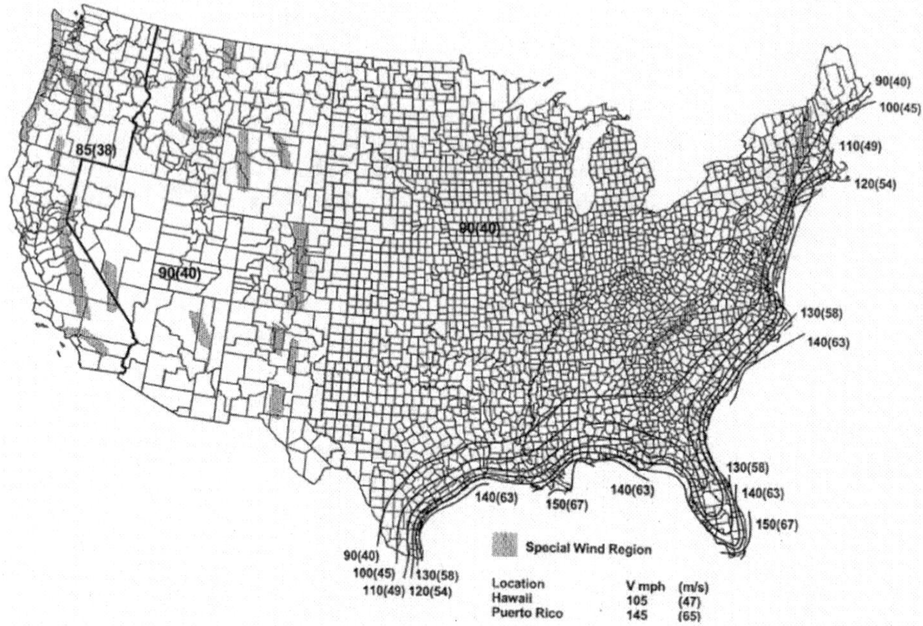

Figure 3.11. Map of 50-year strongest expected 3-second wind gust. The most severe wind storms occur in the southeast US during hurricane season (late summer).

The governing safety standard for distribution pole strength, including the ability to withstand extreme wind, is the National Electrical Safety Code, or NESC. According to the NESC, a structure must be able to withstand loading due to combined ice buildup and wind (the ice is both heavy and provides more area for the wind to affect; see Section 3.3.6). The NESC also requires certain structures to be designed to withstand extreme wind speeds. The extreme wind speed criteria of the NESC changed in 2002, and are now based on 3-second gust speeds. It is important to note that only structures taller than sixty feet must meet this extreme wind criterion. Most distribution structures do not fall into this category.

The NESC specifies the required strength of structures based on grade of construction and loading conditions. Criteria based on combined wind and ice loading is specified in Section 250B. Criteria based on extreme wind loading are specified in Section 250C. Each of these sections uses an "overload" factor, which is essentially a safety factor that results in added strength. The NESC also allows for deterioration of structure strength, resulting in a specified strength for "initial installation" and a separate specified strength for "at replacement." Engineered materials like concrete are assumed to not lose strength over time, and therefore only have a single specified strength.

The NESC also describes how to design a structure to extreme wind criteria. However, since most distribution structures are not required to meet these criteria, the wind loading of distribution poles is most commonly linked to grade such as "a Grade B pole loaded to 80%" or a "Grade C pole loaded to 95%." When considering extreme wind, it is beneficial to move away from this type of description and towards the concept of extreme wind ratings.

Grade C and Grade B construction can easily be viewed as having an equivalent extreme wind rating. This is done by considering two things: overload factors and wind forces. In the NESC, Grade B requires an overload factor of 4, Grade C requires an overload factor of 2.67, and extreme wind designs (for wood poles) require an overload factor of 1.33.

When an overload factor is reduced, the usable strength of the pole in increases accordingly. For example, if a Grade C pole with an overload factor of 2.67 is re-assigned a new overload factor of 1.33, the pole can be subject to twice as much force without violating the new overload factor. Since the wind force on a pole is proportional to the square of wind speed, this corresponds to a new rated wind speed of the old rated wind speed times the square root of the allowed force increase. The conversion equation from base wind rating (W_{base}) to extreme wind rating (W_{ew}) is the following:

$$W_{ew} = W_{base}\sqrt{\frac{\text{Base Overload Factor}}{\text{Extreme Wind Overload Factor}}} \tag{3.11}$$

Consider Grade B and Grade C construction in a light combined wind-and-ice loading district, where the base wind speed is 60 mph. The extreme wind ratings for 100% loaded structures are computed as follows:

$$\text{Grade C:} \quad 60\sqrt{\frac{2.67}{1.33}} = 85 \text{ mph} \tag{3.12}$$

$$\text{Grade B:} \quad 60\sqrt{\frac{4}{1.33}} = 104 \text{ mph} \tag{3.13}$$

A fully-loaded Grade C pole has an equivalent extreme wind rating of 85 mph and a fully-loaded Grade B pole has an equivalent extreme wind rating of 104 mph. If poles are not loaded to 100%, the extreme wind rating is calculated as follows:

$$W_{ew} \text{ for Grade C: } \quad \frac{85}{\sqrt{\text{Grade C Loading}}} \text{ mph} \tag{3.14}$$

$$W_{ew} \text{ for Grade B: } \quad \frac{104}{\sqrt{\text{Grade B Loading}}} \text{ mph} \tag{3.15}$$

Using the above equations, conversion from "percent loaded" to "extreme wind rating" is straightforward. Consider the following examples for Grade B poles loaded to 50%, 100%, and 150%:

$$\text{Grade B (50\% loaded): } \quad \frac{104}{\sqrt{0.5}} = 147 \text{ mph} \tag{3.16}$$

$$\text{Grade B (100\% loaded): } \quad \frac{104}{\sqrt{1.0}} = 104 \text{ mph} \tag{3.17}$$

$$\text{Grade B (150\% loaded): } \quad \frac{104}{\sqrt{1.5}} = 85 \text{ mph} \tag{3.18}$$

The extreme wind rating of a structure can be interpreted as the maximum extreme wind speed that could be applied to a structure with that structure still meeting the extreme wind requirements of NESC 250C.

It must be emphasized that the concept of extreme wind ratings is not explicitly addressed in the NESC, and most distribution structures are not required to meet the extreme wind requirements of NESC 250C. However, assigning an extreme wind rating to a distribution structure provides helpful information about the ability of the structure to withstand extreme wind while using calculations compatible with NESC Section 250C.

3.3.2 Tornados

Tornadoes are concentrated circular winds that can reach speeds of devastating magnitude. A tornado begins when a strong thunderstorm develops a rotating mass of air a few miles up in the atmosphere, becoming a supercell. This mass of rotating air is called a mesocyclone. As rainfall in the storm increases, a downdraft is created that drags the mesocyclone towards the ground with it, resulting in a tornado.

Tornado intensity is typically measured using the Fujita scale, which assigns each tornado a rating based on rotational wind speed. The least severe rating is F0 with increasing severity designated by F1, F2, etc. Very few tornadoes ever exceed the rating of F5. The tornado density for US states is shown in Figure 3.12 and the destructive potential of tornadoes with different ratings is shown below.

F0 (0-72 mph) — These tornadoes can break off tree branches and push over shallow-rooted trees into overhead distribution lines.

F1 (73-112 mph) — The lower limit is the beginning of hurricane wind speed. This significant destructive force can push over some trees, utility poles and towers.

F2 (113-157 mph) — These wind speeds can uproot large trees and snap wooden utility poles. Light object missiles can be generated and cause damage to substation equipment.

F3 (158-206 mph) — Trees and utility poles have little chance of survival. Weakly constructed control buildings can be blown over.

F4 (207-260 mph) — Substations buildings are leveled. Equipment with weak foundations are blown over and large missiles weighing thousands of pounds are generated.

F5 (261-318 mph) — Strong frame structures are lifted off foundations and carried considerable distances to disintegrate. Power transformers fly through the air in excess of 100 meters. Steel reinforced concrete structures can be badly damaged.

Figure 3.13 shows an image of a tornado about to strike an overhead distribution system. Since distribution systems are typically not designed to withstand three second gusts greater than 85 mph, any F2 or stronger tornado will likely destroy any overhead distribution system that is encounters.

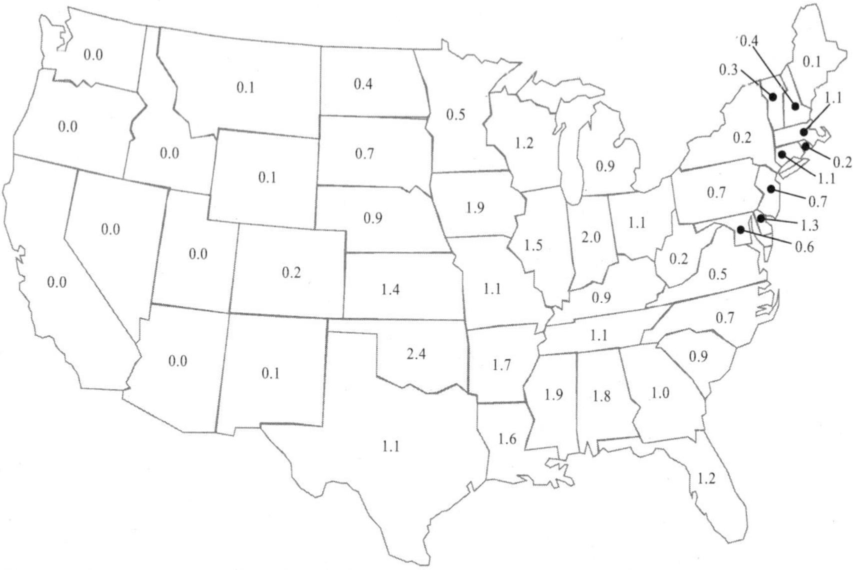

Figure 3.12. Average annual number of strong to violent tornadoes per 10,000 square miles (greater than 113 mph, or F2-F5 on the Fujita scale of tornado ratings). Oklahoma has a higher tornado density than any other state.

Figure 3.13. This image shows a tornado about to strike a distribution line. Overhead distribution structures are typically not designed to withstand the forces associated with strong tornadoes.

3.3.3 Hurricanes

In the 2004 and 2005 hurricane seasons, the following hurricanes made landfall and caused widespread damage to distribution systems: Charlie, Frances, Ivan, Jeanne, Dennis, Emily, Katrina, Rita, Wilma. As a result of these storms (and the memory of other prior hurricanes such as Andrew, Isabel, and Fran), hurricanes have become an important topic for electric power distribution reliability.

A tropical cyclone is a low-pressure system that develops over tropical waters. A hurricane is the name for a tropical cyclone that occurs in the Atlantic Ocean. Tropical cyclones with maximum sustained surface winds of less than 39 mph are called *tropical depressions*. Once the tropical cyclone reaches winds of at least 39 mph it is called a tropical storm and assigned a name. If winds reach 74 mph, the tropical cyclone is called a *hurricane*.

A hurricane forms when a mass of warm moist air over the ocean begins to rise. When the moist air reaches higher and cooler altitudes, water vapor condenses, releasing heat and causing the air to rise further. The rising air creates low surface pressure that causes surrounding air to flow into the area of low pressure. This inflowing air then rises and the cycle repeats. The Coriolis effect of the Earth's rotation causes the incoming surface winds to rotate counter clockwise in the Northern Hemisphere. If high altitude wind speeds are not similar at all altitudes, the resulting "wind shear" causes the tropical cyclone to lose organization and weaken.

A hurricane is typically assigned a "category" of one through five based on its maximum 1-minute wind speeds according to the Saffir-Simpson Hurricane Scale. The minimum and maximum sustained wind speeds corresponding to each hurricane category are shown in Table 3.10. Since the extreme wind rating of distribution structures is based on a three second gust, it is useful to also think of hurricane categories in terms of gust speeds. A typical hurricane will have 3-second gusts that are about 25% faster than 1-minute sustained wind speeds (this can vary). Using this 25% gust adder, the minimum and maximum expected 3-second gust speeds corresponding to each hurricane category are also shown in Table 3.10.

Hurricanes cause damage to distribution systems in a variety of ways. Many utilities report that a majority of damage is due to entire trees blowing over into power lines, which results in broken conductors, broken crossarms, broken insulators, broken poles, and leaning poles. Other hurricanes, such as Wilma in 2005, caused damage primarily by blowing over poles. Damage can also result from flying tree branches, sheet metal, and a variety of other debris. After a hurricane, utilities also typically report wind-related damage to riser shields and streetlights.

Figure 3.10 shows images of distribution system damage caused by hurricanes. This emphasized the range of damage that hurricanes can do including overhead system damage, underground system damage, and flooding.

Table 3.10. Saffir-Simpson hurricane categories and corresponding 1-minute sustained wind speeds (mph). Corresponding 3-second gusts are shown assuming a 25% increase.

Category	1-min sustained(mph)		3-sec gust (mph)	
	Min	Max	Min	Max
1	74	95	93	119
2	96	110	120	138
3	111	130	139	163
4	131	155	164	194
5	156	180	195	225

Figure 3.14. This figure shows images of hurricane damage to distribution system: (upper left) overhead line damage in a coastal area; (upper right) substation flooding; (lower left) a padmount transformer damaged from storm surge; (lower right) a concrete pole broken during a hurricane.

When a hurricane approaches land, it blows a wall of water onto shore called a *storm surge*. A storm surge tends to pick up a large amount of sand and debris. The sand can bury and contaminate pad-mounted equipment, and the debris can damage and dislodge pad-mounted equipment. When the storm surge recedes, it can carry away sand and dirt, leaving formerly underground cables, vaults, and manholes exposed.

When a storm surge floods coastal areas, salt water immerses all of the pad-mounted and sub-surface electrical equipment in the storm surge area. When the storm surge recedes, a salt residue can be left on insulators, bushings, and other components. This contamination can result in an immediate failure when the equipment is energized, or can result in a future failure when the contamination is exposed to moisture.

With a hurricane comes an extensive amount of rain and the potential for flooding. This causes water-immersion problems similar to a storm surge but somewhat less severe since the flooding is with fresh water instead of salt water. Typically live-front equipment performs worst when flooded, dead-front equipment is preferable to live-front equipment, and only sub-surface submersible equipment can be considered immune from hurricane damage.

Even if distribution equipment survives a hurricane, it may be damaged during the cleanup effort. Typically a hurricane will result in piles of debris that can easily cover pad-mounted equipment. When bulldozers come through the area, nonvisible electrical equipment will incur severe damage if struck.

Figure 3.15 illustrates several issues related to hurricane debris. The left image shows a corrugated steel roof that detached and flew into power lines, acted as a sail, and caused strong concrete poles to blow down. The right image shows a pile of debris that may be covering undamaged pad-mounted equipment. When bulldozers clear this pile, the pad-mounted equipment is vulnerable to damage (some utilities scout debris piles and mark buried utility equipment with flags).

Figure 3.15. Debris is a major hurricane concern. The left image shows a corrugated steel roof that flew into power lines. The right image shows a pile of debris that may be covering undamaged pad-mounted equipment. When bulldozers clear this pile, the pad-mounted equipment is vulnerable to damage.

3.3.4 Swinging, Galloping, and Aeolian Vibration

In addition to blowing over trees and poles, wind induces several types of conductor motion that can be reliability concerns: swinging, galloping, and aeolian vibration.

Swinging refers to pendulum motions of conductor spans. If the swing amplitude is high enough, phase conductors can contact each other, referred to as a "blowout," and cause a fault. Blowouts can be reduced by increasing phase spacing and increasing span tension.

Galloping occurs when ice accumulates on conductors, creates an airfoil, and allows wind to exert a vertical force on conductor spans. Galloping exerts extreme forces on conductors and towers and can have amplitudes approaching that of conductor sag.[27] Galloping can be mitigated by increasing conductor tension, installing dampers or placing de-tuning devices on conductor spans.[28] Wire manufacturers also offer specially shaped conductors that are less prone to galloping.

Wind creates vortices as it blows across conductors. If the frequency of these vortices coincides with the natural frequency of the conductor, a low amplitude oscillation (less than one conductor diameter) occurs. This motion, referred to as aeolian vibration, does not result in an immediate outage, but causes conductor fatigue at clamps that will ultimately result in an outage.

The natural frequency of a conductor span is a function of weight, tension and length. Natural frequency increases as tension increases, and decreases as length and weight increase. Typical spans have natural frequencies of 0.1 Hz to 1.0 Hz, and the natural frequency of any span can be calculated as follows:

$$f_n = \frac{n}{2L}\sqrt{\frac{T \cdot 32.2}{W}} \tag{3.19}$$

f_n = Natural frequency (Hz)
n = Nodal number
L = Span length (ft)
T = Tension (lb)
W = Conductor weight (lb/ft)

The frequency at which wind vortices exert alternating pressure on conductor spans is a function of wind speed and conductor diameter. Frequency increases with increasing wind speed and decreases with increasing conductor diameter according to the following formula:

$$f_w = \frac{3.26 \cdot s}{d} \tag{3.20}$$

f_w = Frequency of wind force (Hz)
s = Wind speed (mph)
d = Conductor diameter (inches)

A 5-mph wind blowing across a 500-kcmil conductor with a diameter of 0.8 inches will exert force with a frequency of about 20 Hz. At these frequencies, resonance will usually occur at multiples of the fundamental span mode. Aeolian vibration can be mitigated by reducing conductor tension or by installing conductor dampers.

3.3.5 Lightning Storms

A lightning strike occurs when the voltage generated between a cloud and the ground exceeds the dielectric strength of the air. This results in a massive current stroke that usually exceeds 30,000 amps. To make matters worse, most strokes consist of multiple discharges within a fraction of a second. Lightning activity varies greatly throughout the US. It is the most important reliability concern for certain areas and a very small concern for others. A map showing the expected number of days that thunderstorms will occur for various parts of the US is shown in Figure 3.16 (another measure of lightning activity that is gaining popularity is flash density).

Lightning can affect power systems through direct strikes (the stroke contacts the power system) or through indirect strikes (the stroke contacts something in close proximity and induces a traveling voltage wave on the power system). The lightning strike causes a flashover that causes a short circuit (most of the fault energy comes from the power system—not the lightning). The number of expected direct strikes on a distribution line in open ground with no nearby trees or buildings can be calculated with the following equation:[29]

$$N = \frac{\left(28 \cdot h^{0.6} + w\right) \cdot k^{1.25}}{25000} \tag{3.21}$$

N = Lightning strikes per mi/yr
h = Pole height (m)
w = Crossarm width (m)
k = Keraunic level (thunderstorm days per year)

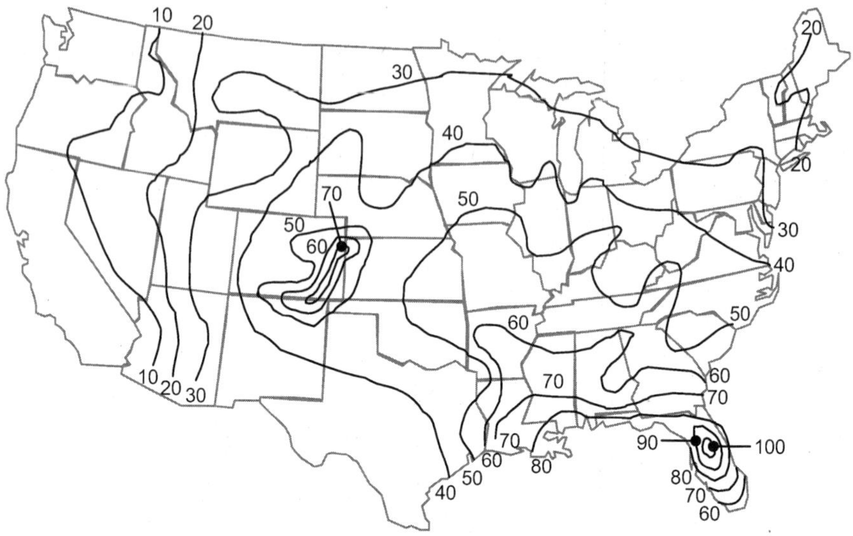

Figure 3.16. Isokeraunic map of the US in thunderstorm days per year. In the Southeast, lightning storms are the most important reliability issue and can easily occur once or twice per week. In contrast, lightning storms are rare on the West Coast and reliability priorities lie elsewhere.

Unfortunately, it is almost impossible to protect a distribution system (with standard BIL levels) from direct lightning strikes. Consider a small 10-kA strike that contacts a line and splits into two 5-kA traveling waves. If the line has a characteristic impedance of 300Ω (a typical value), each traveling wave will generate 1.5 MV and almost certainly cause a flashover. To protect against direct strikes, utilities must commit to using shield wire, surge arresters on every pole and transmission class BILs. Even these extreme measures may prove to be ineffective.

Luckily, most lightning strikes will not strike distribution equipment directly. Rather, they will strike a nearby object such as a tree or a building. As current flows from the cloud to this object, a magnetic field with a high rate-of-change will induce a voltage in nearby conductors that can cause similar, but less severe conditions when compared to direct strikes. The reader is referred elsewhere for calculations,[30] but induced voltage magnitudes are much lower than for direct strikes, and flashovers can be prevented by the careful application of shield wire and arresters.

3.3.6 Ice Storms

Ice storms occur when supercooled rain freezes on contact with tree branches and overhead conductors and forms a layer of ice. This generally occurs when a winter warm front passes through an area after the ground-level temperature falls below freezing. Ice buildup on conductors places a heavy physical load on the conductors and support structures and increases the cross-sectional area that is exposed to the wind. Combined ice and wind can cause conductors to gallop (see Section 3.3.4), and ice can break off and cause a conductor to jump into the phase wires located above it. Ice accumulation in trees can cause limbs to break off and fall into conductors, and can cause entire trunks to fall over into power lines.

Figure 3.17 shows the National Electric Safety Code recommended ice loading districts based on expected ice accumulation and wind speed.[31] Ice loading typically assumes an ice density of 57-lb/ft^3 and can be computed as follows:

$$W_i = 1.244 \cdot T \cdot (D + T) \tag{3.22}$$

W_i = Ice load (lb/ft)
T = Radial thickness of ice (in)
D = Conductor diameter (in)

The wind loading is equal to wind pressure multiplied by the conductor diameter plus ice thickness:

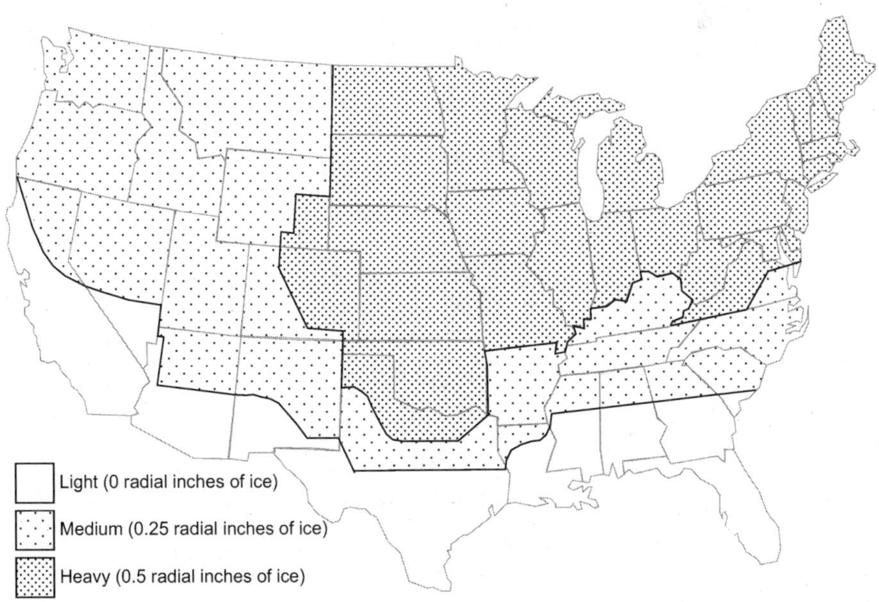

Figure 3.17. Ice loading districts in the US for ice accumulation on surfaces.[31] Ice accumulation on overhead lines increases physical loading and allows wind to exert more force. Ice accumulation can also cause trees and branches to fall into overhead conductors.

$$W_w = \frac{V^2(D+2T)}{4800} \tag{3.23}$$

W_w = Wind load (lb/ft)
V = Wind speed (mi/hr)

Since ice load and wind load are not in the same direction, the total conductor load must be computed as the vector sum of the two values:

$$W = \sqrt{(W_c + W_i)^2 + W_w^2} \tag{3.24}$$

W = Total conductor load (lb/ft)
W_c = Bare conductor weight (lb/ft)

Overhead distribution systems should be designed so that conductor and structure strength can accommodate expected icing and wind conditions. This book is not intended to be a structural design reference, but distribution engineers should be aware of the potential impact that icing, wind, and design strength can have on system reliability.

Figure 3.18 shows ice buildup on overhead distribution wires, telephone cables, and cable television cables. If ice loading becomes severe, widespread damage can occur to the distribution system.

Figure 3.18. Ice buildup on trees, conductors, and third-party attachments. This weight of this ice can break tree branches, tree trunks, crossarms, conductors, and poles. The ice also increases susceptibility to wind damage due to increased surface area.

3.3.7 Heat Storms

Heat storms are extended periods of exceedingly hot weather. This hot weather causes electricity demand to skyrocket due to air conditioning loads. At the same time, conductors and transformers cannot carry as much electricity since they cannot transfer heat as effectively to their surroundings. This combination of heavy loading and equipment de-rating can cause equipment overloads and loss-of-life. Overloading can lead to equipment failure, increase the load on remaining equipment and potentially lead to widespread outages. In a worse scenario, the maximum power transfer capabilities of the system can be approached resulting in a voltage collapse condition.

Most modern distribution systems that experience peak loads during the summer are designed based on weather assumptions. Some utilities design for average annual maximum temperature, while others design for the hottest day expected over five or ten years. It is not generally prudent to design a distribution system to a 100-year heat storm. Doing so is expensive and the event is not likely to occur within the useful lifetime of the equipment.[32] Table 3.11 shows typical daily maximum temperatures that will occur in most major US cities during summer peak.

Maximum temperature is only one of four weather factors that significantly impact electric load. The other three are humidity, solar illumination, and the number of consecutive extreme days. Humidity has a substantial impact on heat storm load since (1) air conditioners must use energy for dehumidification, and (2) humid air has a higher thermal capacity than dry air and requires more energy to cool. Solar illumination increases building air conditioning loads through radiant heating. Consecutive extreme days further increases loads since (1) the thermal inertia of buildings will cause them to slowly increase in temperature over several days, and (2) many people will not utilize air conditioning until it has been uncomfortably hot for several days.

The negative reliability impact of heat storms is difficult to mitigate after a system has been designed since weather cannot be controlled. The most effective method to avoid heat storm problems is to perform periodic load forecasts (five to ten years in the future) and ensure that enough system capacity is installed to meet forecasted demand.

Even if this is done, reliability during heat storms will worsen since the ability of the system to be reconfigured after a contingency is diminished. During lightly loaded conditions, an interrupted section of feeder can be easily transferred to an adjacent feeder without overloading wires or substation power transformers. During a heat storm, many post-contingency transfers may be disallowed because they would result in overloaded wire, overloaded cable, overloaded transformers or unacceptable voltage drops. If operators can choose to overload equipment, accelerated loss-of-life will occur and the risk of additional equipment failure increases (see Sections 3.1.1, 3.1.2 and 3.1.3).

Table 3.11. Typical daily maximum temperatures during summer peak. Values are computed by taking the average of daily maximums for July and August from 1961-1990.

City	°F	City	°F	City	°F
BIRMINGHAM, AL	89.5	HONOLULU,HI	88.1	ROSWELL, NM	93.3
HUNTSVILLE, AL	88.9	BOISE, ID	89.2	ALBANY, NY	82.7
MOBILE, AL	90.9	LEWISTON, ID	88.7	BUFFALO, NY	79.1
MONTGOMERY, AL	90.8	CHICAGO,IL	82.8	ROCHESTER, NY	79.4
ANCHORAGE, AK	64.1	PEORIA, IL	84.4	SYRACUSE, NY	80.4
BARROW, AK	43.7	ROCKFORD, IL	82.6	ASHEVILLE, NC	82.6
COLD BAY,AK	55.5	SPRINGFIELD, IL	85.6	CAPE HATTERAS, NC	84.7
FAIRBANKS, AK	69.3	EVANSVILLE, IN	88.2	CHARLOTTE, NC	88.3
HOMER, AK	60.5	FORT WAYNE, IN	83.4	GREENSBORO,NC	86.2
JUNEAU, AK	63.3	INDIANAPOLIS, IN	84.6	RALEIGH, NC	87.4
NOME, AK	56.9	SOUTH BEND, IN	81.8	WILMINGTON, NC	88.1
ST. PAUL ISLAND, AK	50.3	DES MOINES, IA	85.5	BISMARCK, ND	83.6
VALDEZ, AK	61.6	DUBUQUE,IA	81.2	FARGO, ND	82.4
FLAGSTAFF, AZ	80.6	SIOUX CITY, IA	85.0	AKRON, OH	81.4
PHOENIX, AZ	104.8	DODGE CITY, KS	91.9	CLEVELAND, OH	81.5
TUCSON, AZ	98.1	GOODLAND, KS	88.6	COLUMBUS, OH	82.9
WINSLOW, AZ	92.0	TOPEKA, KS	88.4	DAYTON, OH	84.0
YUMA, AZ	106.0	WICHITA, KS	91.8	TOLEDO, OH	82.4
FORT SMITH, AR	92.7	JACKSON, KY	83.6	OKLAHOMA CITY, OK	93.0
LITTLE ROCK, AR	91.9	LEXINGTON, KY	85.4	TULSA, OK	93.1
BAKERSFIELD, CA	97.6	LOUISVILLE, KY	86.4	EUGENE, OR	81.8
BISHOP, CA	96.1	BATON ROUGE, LA	91.3	PENDLETON, OR	87.0
EUREKA, CA.	62.2	NEW ORLEANS, LA	90.4	PORTLAND, OR	80.1
FRESNO, CA	97.7	SHREVEPORT, LA	93.1	SALEM, OR	81.8
LONG BEACH, CA	83.4	CARIBOU, ME	75.1	ALLENTOWN, PA	83.4
LOS ANGELES AP, CA	76.0	PORTLAND, ME	78.1	ERIE, PA.	79.2
LOS ANGELES C.O., CA	84.3	BALTIMORE, MD	86.3	PHILADELPHIA, PA	85.4
REDDING, CA	97.0	BOSTON, MA	80.8	PITTSBURGH, PA	81.7
SACRAMENTO, CA	92.7	WORCESTER, MA	78.3	PROVIDENCE, RI	81.4
SAN DIEGO, CA	77.0	DETROIT, MI	82.3	CHARLESTON,SC	89.6
SAN FRANCISCO, CA	72.0	FLINT, MI	80.4	COLUMBIA, SC	90.9
SANTA BARBARA, CA	74.7	GRAND RAPIDS, MI	81.7	GREENVILLE, SC	87.5
SANTA MARIA, CA	73.7	LANSING, MI	81.5	SIOUX FALLS, SD	84.8
STOCKTON, CA	93.7	MARQUETTE, MI	75.0	CHATTANOOGA, TN	88.6
ALAMOSA, CO	80.6	DULUTH, MN	75.5	KNOXVILLE, TN	86.9
COLORADO SPRINGS, CO	82.9	MINNEAPOLIS, MN	82.4	MEMPHIS, TN	91.6
DENVER, CO	87.0	ROCHESTER, MN	80.3	NASHVILLE, TN	89.0
GRAND JUNCTION, CO	92.1	SAINT CLOUD, MN	81.0	OAK RIDGE,TN	86.4
PUEBLO, CO	91.4	JACKSON, MS	92.2	AMARILLO, TX	90.4
BRIDGEPORT, CT	81.3	COLUMBIA, MO	87.7	AUSTIN, TX	95.3
HARTFORD, CT	83.9	KANSAS CITY, MO	87.6	DALLAS, TX	96.4
WILMINGTON, DE	84.9	ST. LOUIS, MO	88.3	EL PASO, TX	94.8
WASHINGTON, D.C.	87.7	SPRINGFIELD, MO	89.1	HOUSTON, TX	92.6
DAYTONA BEACH, FL	89.5	BILLINGS, MT	85.7	SAN ANTONIO, TX	95.2
FORT MYERS, FL	91.2	GREAT FALLS, MT	82.5	WACO, TX	97.0
GAINESVILLE, FL	90.4	HELENA, MT	84.1	SALT LAKE CITY, UT	90.8
JACKSONVILLE, FL	91.1	KALISPELL, MT	79.8	BURLINGTON, VT	79.6
KEY WEST, FL	89.2	MISSOULA, MT	82.8	RICHMOND, VA	87.8
MIAMI, FL	89.0	LINCOLN, NE	88.4	ROANOKE, VA	85.9
ORLANDO, FL	91.5	NORFOLK, NE	85.3	OLYMPIA, WA	76.8
PENSACOLA, FL	89.6	NORTH PLATTE, NE	86.9	SEATTLE, WA	75.2
TALLAHASSEE, FL	91.2	OMAHA, NE	86.6	SPOKANE, WA	82.8
TAMPA, FL	90.2	LAS VEGAS, NV	104.6	YAKIMA, WA	86.2
PALM BEACH, FL	90.0	RENO, NV	90.8	HUNTINGTON, WV	83.7
ATHENS, GA	88.9	CONCORD, NH	81.1	GREEN BAY, WI	79.0
ATLANTA, GA	87.6	MT. WASHINGTON, NH	52.7	MADISON, WI	81.0
AUGUSTA,GA	91.0	ATLANTIC CITY, NJ	83.9	MILWAUKEE, WI	78.9
COLUMBUS, GA	91.5	NEWARK, NJ	86.2	CASPER, WY	86.7

3.3.8 Earthquakes

On January 17th 1995 at 5:46 in the morning, an earthquake measuring 7.2 on the Richter scale hit Kobe, the fifth largest city in Japan. This earthquake destroyed more than 10,000 distribution poles, interrupted service to more than one million people for two and a half days and required \$2.3 billion to repair all damaged electrical facilities.[33] Severe earthquakes are not common, but can decimate distribution systems, cause safety hazards and result in widespread customer interruptions.

Earthquakes are caused when a sudden rupture occurs along a pre-existing geological fault. This rupture sends vibratory shock waves through the ground, causing both horizontal and vertical ground motion. If the frequency of earthquake vibrations matches an oscillatory mode of distribution structures, significant forces, motion, and damage can result. To make matters worse, rigid connections (like those found in substations) will transmit forces to adjacent equipment.

In general, it is not cost effective to design primary distribution systems to withstand earthquakes. This is not the case for substations, which should be structurally engineered in accordance to adopted state seismic codes.[34-36] These codes provide design guidelines based on seismic activity that can be expected in any given area (see Figure 3.19). Further information can be found in IEEE Standard 693, *Seismic Design for Substations* (1997).

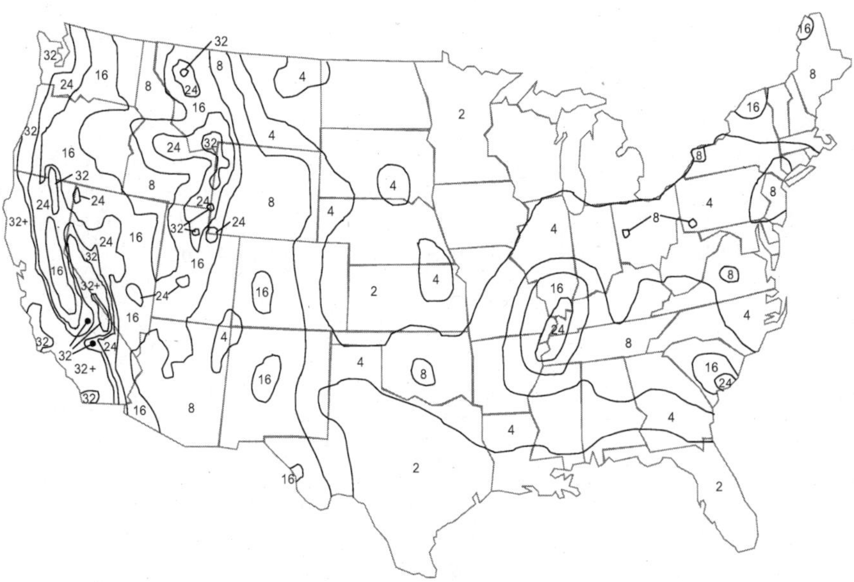

Figure 3.19. Seismic map of the US. Numbers indicate peak seismic acceleration (%g) with a 10% probability of occurring in 50 years. The West Coast (especially California) is most susceptible to strong earthquakes.

3.3.9 Fires

Brush and forest fires can cause major damage to distribution systems. When exposed to mild fires, overhead lines cannot effectively dissipate heat and must be de-rated. When exposed to severe fires, overhead lines may start to anneal and lose mechanical strength. In the worst of situations, lines become too weak to support themselves and break.

Fires are also a threat to wooden poles. If a pole catches on fire, it will lose mechanical strength and be susceptible to falling over. If the fire reaches the top of the pole, transformers and hardware will heat up and be susceptible to loss-of-life or catastrophic failure.

3.4 TREES

Trees are one of the top three causes of customer interruptions for most utilities (animals and lightning being the other two). Several modes of failure associated with trees include:

- Mechanical damage to overhead conductors when struck by a falling trunk or branch.
- Faults caused by animals that use trees as a gateway to electrical poles.
- Faults caused when growing branches push two conductors together.
- Faults caused when wind blows branches into conductors, resulting in two wires contacting each other.
- Faults caused when a branch falls across two conductors.

There are many misconceptions about distribution system reliability and trees, especially in the area of faults. The following sections address branch-induced faults in some detail, discuss the characteristics of common North American trees, and present some basic principles concerning tree trimming.

3.4.1 Faults Caused by Trees

A freshly cut tree branch will not cause a fault if it is placed across two distribution phase conductors. The resistance of the branch typically limits current flow to several amps. Similarly, a desiccated branch will not cause a fault if it is placed across two distribution phase conductors because its resistance is even higher. Regardless, faults often but not always occur after a branch falls across two conductors. Understanding the physics responsible for determining whether a fault occurs or not is important and can help when making tree-related distribution reliability decisions.

When a branch breaks off of a tree and bridges two conductors, a fault does not occur immediately because a moist tree branch has a substantial resistance. Even though a fault has not immediately occurred, the following two processes initiate: (1) a small current begins to flow and starts to dry out the wood fibers, and (2) the electric field initiates carbonization of cellulose near each conductor (see the left drawing in Figure 3.20).

Desiccation steadily increases the resistance of the wood and reduces current flow. At the same time, the carbon path continues to grow from each side (see the middle drawing in Figure 3.20). If the resistance of the branch becomes too great before the carbon paths meet, no fault occurs. If the carbon paths meet before resistance becomes too high, the low impedance path between the conductors results in a phase-to-phase fault (see the right drawing in Figure 3.20).

The most important factor in determining whether branch contact will result in a fault is voltage gradient. The high voltage gradients associated with 35-kV class distribution systems (10-ft crossarm construction) result in faults for nearly 100% of branch contacts. The moderate voltage gradients associated with 15-kV class distribution systems (8-ft crossarm construction) result in faults for approximately 80% of branch contacts. Branch contacts on 5-kV class distribution systems almost never result in faults. Large diameter branches pose a greater risk due to their lower resistance, and live branches pose a greater risk since they begin with a higher moisture content. Branches brushing against a single-phase conductor typically *do not* result in system faults.[37]

There are several strategies to prevent faults due to tree branch contact: line construction, insulated conductor, and tree trimming. Line construction can reduce faults by utilizing lower distribution voltages, by avoiding compact designs with high phase-to-phase voltage gradients, and by utilizing designs that are less likely to result in sustained phase-to-phase branch contact (e.g., vertical construction). Insulated wire can be effective, but faults tend to result in conductor burndown since they will not motor (move themselves along the conductor) like faults on bare conductor. Tree trimming is fundamental to overhead distribution reliability and is discussed in detail in Section 3.4.3.

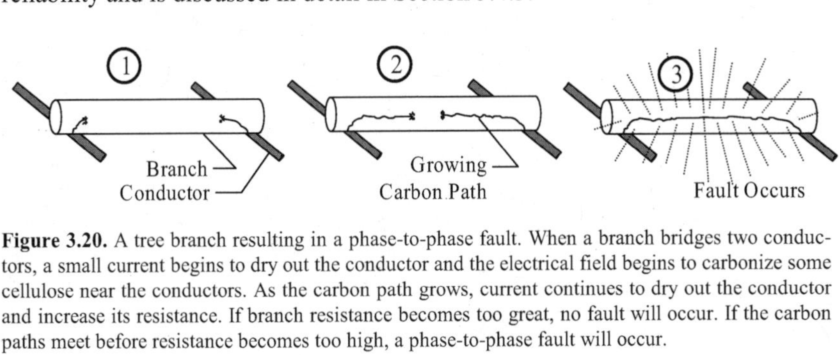

Figure 3.20. A tree branch resulting in a phase-to-phase fault. When a branch bridges two conductors, a small current begins to dry out the conductor and the electrical field begins to carbonize some cellulose near the conductors. As the carbon path grows, current continues to dry out the conductor and increase its resistance. If branch resistance becomes too great, no fault will occur. If the carbon paths meet before resistance becomes too high, a phase-to-phase fault will occur.

3.4.2 Tree Characteristics

A basic understanding of tree characteristics is helpful for ensuring reliability during distribution planning, engineering, and operations. For a given tree species, it is helpful to know its vertical growth rate, its sucker growth rate, the strength of its wood, and its ultimate height.[38] Typical values for the most common trees found in the US are shown in Table 3.12. It should be noted that the growth values in this table are averages and can vary dramatically based on a number of external factors including precipitation, temperature, sunlight, soil condition, competition, wind, and fire.[39-40] The following is a short description of each of these growth factors.

<u>**Factors Affecting Tree Growth**</u>

- **Precipitation** — The amount of water that a tree receives is one of the most significant factors determining its growth. Too little water affects many physiological processes, results in leaf wilting, and can slow or stop growth. Too much water results in poor soil aeration, also resulting in retarded growth.
- **Temperature** — Temperature is also critical to the growth rate of trees. Low temperatures can damage tree cells through dehydration and freezing. High temperatures above a critical level slow growth by inhibiting photosynthesis.
- **Sunlight** — Tree growth requires sunlight so that photosynthesis can convert oxygen and water into sugar (and CO_2 as a waste product). Insufficient sunlight limits the amount of sugar that can be produced and can result from increased cloud cover or from the shade of nearby trees.
- **Soil Condition** — Trees rely on soil for most of their nutrition. Soil not containing proper nutrients will retard tree growth and may ultimately result in tree death.
- **Competition** — Dense tree populations retard growth due to competition for root space and competition for sunlight.
- **Wind** — High winds exert strong forces on tree branches. These forces create a torque that may uproot the tree or damage root systems. Damaged root systems will limit the amount of water and nutrients that a tree can utilize, slowing growth and possibly resulting in death.
- **Fire** — Periodic fires are healthy for forested areas. They keep combustible fuel levels low, reduce the probability of an intense fire capable of killing mature trees, and decompose organic matter into its mineral components, accelerating growth for several seasons. Certain types of trees, such as lodgepole pines and jack pines, have cones that will only open and spread their seeds when they have been exposed to the heat of a fire.

Table 3.12. Tree characteristics. Annual growth rates are averages, and actual rates will vary considerably depending on soil conditions and rainfall. Strength is the tensile breaking point of green wood, and does not consider factors such as deformities or weak points.

Tree	Vertical Growth (inches/yr)	Sucker Growth (inches/yr)	Mature Height (feet)	Strength (lb/in^2)
Ash	18	36	80	9600
Basswood	18	27	75	5000
Beech	12	30	60	8600
Birch	21	52	50	6400
Black Cherry	14	24	60	8000
Black Walnut	20	40	80	9500
Box-elder	26	72	50	5000
Cedar	15	15	50	7000
Cottonwood	52	80	85	5300
Douglas Fir	18	21	125	7600
Elm (American)	26	60	85	7200
Elm (Chinese)	40	72	65	5200
Eucalyptus	24	50	125	11200
Ficus	36	60	50	5800
Hackberry	18	30	60	6500
Hickory	14	21	65	11000
Locust (Black)	18	80	80	13800
Locust (Honey)	22	33	80	10200
Madrona	8	24	50	7600
Magnolia	20	42	90	7400
Maple (Big Leaf)	30	72	60	7400
Maple (Norway)	15	35	50	9400
Maple (Red)	18	42	75	7700
Maple (Silver)	25	65	65	5800
Maple (Sugar)	18	40	75	9400
Oak (Black and Red)	18	30	85	6900
Oak (Live)	30	45	70	11900
Oak (Pin)	24	36	100	8300
Oak (Water)	30	45	75	8900
Oak (White)	9	18	75	8300
Oak (Willow)	24	40	50	7400
Palm (Coconut)	60	---	90	---
Palm (Queen)	36	---	65	---
Palm (Royal)	48	---	80	---
Palm (Washington)	36	---	70	---
Pine	36	48	80	7300
Poplar (Lombardi)	45	72	60	5000
Red Alder	36	84	120	6500
Sassafras	24	36	50	6000
Sweet Gum	12	20	90	6800
Sycamore	34	72	100	6500
Tulip Tree	30	52	100	5400
Willow (Black)	40	70	50	3800
Willow (Weeping)	48	72	50	3800

3.4.3 Tree Trimming

Tree trimming, periodically pruning vegetation adjacent to power lines to ensure safe and reliable clearances, is a critical utility activity for a variety of reasons. The first is reliability, which has already been discussed in some detail. The second is cost—tree trimming is one of the largest operational costs for most distribution systems, with US utilities spending more than $2 billion annually. The third is customer relations. Many customers feel passionately about trees and can have extremely negative responses to the prospect of tree trimming, the sight of tree trimming crews or the aesthetic results of tree trimming.

Periodic tree trimming is required to maintain a safe and reliable distribution system. Most distribution systems are trimmed on two to six year cycles, with longer cycles becoming more common as utilities attempt to reduce operating budgets. Several trimming styles are shown in Figure 3.21. Some utilities have even eliminated trimming on lateral branches and are only trimming main trunks. Tree trimming should always be performed by a trained crew to ensure safety, maintain tree health and direct re-growth away from conductor location.

From a tree trimming perspective, removed trees are the best, slow growing trees are next best and fast growing trees are the worst. Where fast growing trees present a safety or reliability problem, removal or replacement with a slower growing species should be considered (growth can also be slowed by coating pruned locations with special retardants). Removal should also be considered for trees having major damage such as splits in the main trunk or the loss of more than 50% of the crown.

Tree trimming is an area well suited for a technique referred to as reliability-centered maintenance (RCM).[41] Instead of trimming each location periodically, each location is assigned a score based on the benefit that tree trimming will have on customer reliability. Locations are then ranked based on the ratio of tree trimming benefit to tree trimming cost and locations with the highest scores are trimmed first.

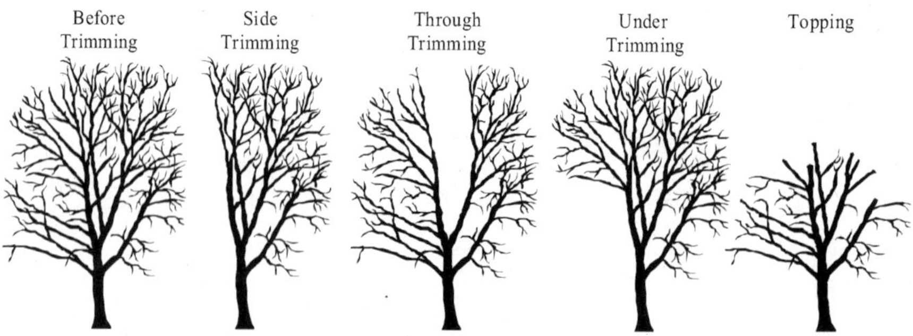

Figure 3.21. Various types of tree trimming. Side trimming is preferable from a reliability perspective, but other types of trimming may be required due to wire location or aesthetic needs.

3.5 HUMAN FACTORS

In addition to equipment, animals, vegetation and weather, humans are directly responsible for many customer interruptions. Sometimes these interruptions are intentional (e.g., scheduled outages, service disconnections, vandalism) and sometimes these interruptions are unintentional (e.g., operational errors, traffic accidents, dig ins). There are too many ways for humans to cause interruptions for the subject to be treated exhaustively, but the following sections provide an overview of the most common occurrences.

3.5.1 Scheduled Interruptions

It is sometimes necessary to interrupt customer service when performing work on radial distribution systems. Since this work is scheduled in advance, customers can be notified as to the time and expected duration of the interruption. Advance knowledge greatly reduces the economic impact and negative perception that interruptions have on customers.

Certain types of distribution maintenance require equipment to be de-energized and grounded. During maintenance, all customers downstream of the maintenance location will experience interruptions unless they can be fed from an alternate path. Even if the system can be reconfigured to restore certain customers, short interruptions may be necessary since many switches can only be switched while de-energized. Even if all switches are capable of making and breaking load, operating the system as a temporary network may be unacceptable and customer interruptions may be required.

The situation is similar for feeder cutovers. Before load can be transferred from an original feeder to an existing or new adjacent feeder, it may need to be de-energized. If load break switches are not available, both feeders need to be interrupted at the feeder breaker until the cutover is complete.

Feeder expansions also require scheduled interruptions. Since the expansion location must be de-energized before a feeder extension can be connected, all customers downstream of this point (on the original feeder) may need to be interrupted. The use of live-line construction techniques can avoid these types of interruptions.

3.5.2 Human Errors

Mistakes by utility workers often result in customer interruptions. Examples are many and varied, but can generally be classified into switching errors, direct faults, and indirect faults.

Switching errors occur when operators or crews reconfigure the system in a manner that unintentionally interrupts customers. This usually occurs when a wrong circuit breaker or SCADA controlled switch is opened. Customers can also be interrupted when the system is being reconfigured to mitigate losses or low voltage conditions. Interruptions can also occur when mistakes are made during relay testing, resulting in undesired breaker tripping.

When working near energized equipment, crews mistakes can cause faults resulting in customer interruptions. A good example is tree trimming—accidentally dropping a heavy branch into distribution lines can easily result in customer interruptions. Substation work is another example of an activity where faults can easily occur due the close proximity of equipment and the large amount exposed buswork (not to mention the risk of digging into buried cable).

Mistakes made while equipment is de-energized can manifest itself as a fault when the equipment is energized. A common example is forgetting to remove grounding connections. Other examples include damaging de-energized equipment, leaving tools inside of equipment, and making improper phase connections.

3.5.3 Vehicular Accidents

Vehicular accidents are a major reliability problem along congested thoroughfares. A speeding automobile striking a wooden distribution pole will usually cause the pole to lean, causing unsafe line clearances and requiring repair. These collisions may also result in a fault by causing conductors to swing together, sagging lines into the ground or snapping the pole entirely. The situation is similar for pad-mounted equipment. The impact of automobile collisions can be reduced by using concrete barriers or crash-resistant poles, but these techniques may negatively impact driver safety and result in increased litigation.

Cranes are another common cause of faults and interruptions. If an operator is not careful, the crane cable can easily be swung into overhead lines as the boom is rotated. If the crane is unloaded or lightly loaded, the cable may simply cause a fault to occur. If the crane is heavily loaded, conductors can be damaged and poles can be knocked over.

Although less common, low-flying aircraft can cause interruptions by flying into overhead wires. These types of events typically involve small aircraft used for agricultural purposes. To help prevent these life-threatening accidents, overhead lines can be fitted with special markers, such as colored spheres, to increase their visibility to pilots. Nighttime collisions can be reduced by utilizing lighted markers.

3.5.4 Dig Ins

A dig in occurs when excavation equipment cuts through one or more underground cables. They cause a substantial number of interruptions and are usually associated with construction sites. To prevent dig ins from occurring, utilities should encourage the public to have cable routes identified before initiating site excavation. In extreme cases where high reliability is required, utilities can place cable in concrete-encased duct banks (although a determined backhoe operator still may not be deterred).

3.5.5 Mischief and Vandalism

A common form of vandalism occurs when people shoot at utility equipment for gunshot practice. Perhaps the biggest problem is ceramic bushings and insulators, which shatter upon impact as a reward for good aim. Some utilities have found that interruptions decrease after switching to polymeric bushings and insulators. Another common problem occurs when bullets puncture transformer radiator fans. The resulting oil leaks can cause overheating, internal faults, and environmental concerns.

Another reliability concern arises when thieves steal conductor wire to sell as scrap metal. Removal of ground wire, counterpoise and substation grounding may not directly result in interruptions, but may cause safety concerns and compromise protection schemes. More sophisticated criminals may choose to steal phase conductors, resulting in open circuits and customer interruptions.

Terrorism is a reliability concern in certain second-world and third-world nations and is of growing concern in more developed nations. Typical terrorist attacks involve the use of explosives targeted at towers, poles, or substations. Most successful attacks result in customer interruptions since the usual intent of the terrorists is precisely to cause these interruptions.

To end on a less malicious note, interruptions can be caused by leisure activities such as remote controlled aircraft, kites with metallic strings and metallic helium balloons. Phase-to-phase contact is not necessarily required for a fault to occur. Conductive devices placed between phases act as a floating electrode and can cause a flashover if voltage withstand strength is sufficiently reduced.

3.6 MOST COMMON CAUSES

This chapter has presented a large number of potential interruption causes. It is now desirable to place this information in perspective by examining the most common causes of interruptions for typical utilities.

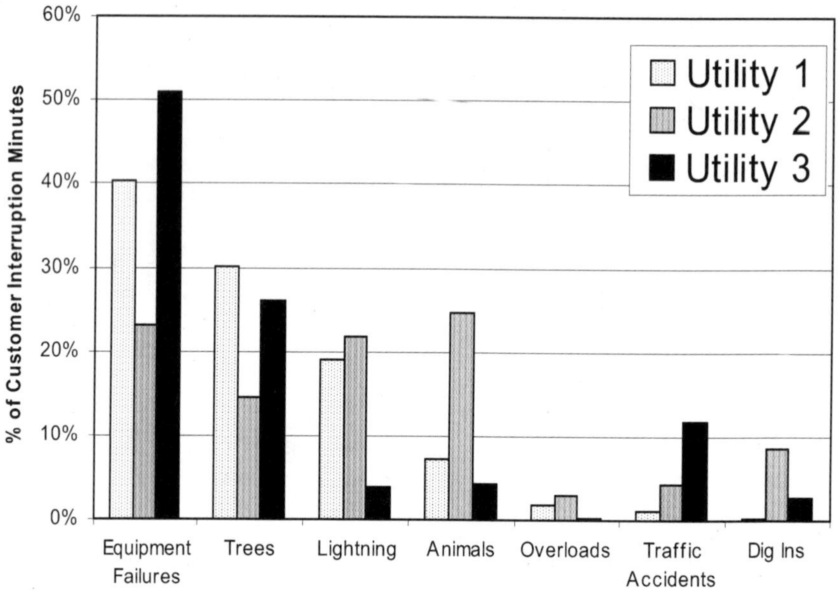

Figure 3.22. Major causes of interruptions for three US utilities. Equipment failures are almost always a top cause of customer interruptions and can be mitigated by purchasing quality equipment and ensuring proper installation. Other top causes include trees, lightning, and animals.

Figure 3.22 shows a breakdown of interruption causes for three major US utilities. Numbers are based on customer interruption minutes and, therefore, are dependent upon both the frequency and duration of interruptions (storm related interruptions are not included). Although substantial variation is seen from utility to utility, most interruptions can generally be attributed to equipment failures, trees, lightning, and animals. Utility 3 is primarily an urban utility in a low keraunic area, and has a correspondingly high value for traffic accidents and a correspondingly low value for lightning.

Equipment failures and trees are almost always the major causes of interruptions on distribution systems. As such, there are several important strategies for utilities striving for high reliability. First and foremost, utilities should purchase quality equipment and install it properly. Cheap products can have substantially higher failure rates and make high levels of reliability impossible to achieve. Even if quality equipment is purchased, poor installation can result in higher than expected failure rates and lower than expected reliability. Second, utilities should implement a sound vegetation management program. Although vegetation management budgets are tempting targets for cost reduction, insufficient tree trimming can lead to low clearances, a proliferation of dead and weakened trees and ultimately, a substantial reduction in system reliability.

Lightning and animals are common interruption causes, but specifics of these problems tend to vary. Lightning should be the primary reliability concern for utilities in Florida, but may not be a concern for utilities on the West Coast. Most utilities have a substantial number of animal related outages, but the types of animals involved, and therefore mitigation strategies, can vary substantially. In any case, utilities should invest the resources required to obtain data about interruption causes on their system. After common causes are identified, mitigation efforts can be studied and selected so that maximum reliability is obtained for the lowest possible cost.

3.7 STUDY QUESTIONS

1. Does overloading cause transformers to fail? Explain.
2. What are some of the problems utilities have historically experienced with underground cable? What can utilities do to help avoid these problems?
3. Why might a utility consider using nonwood distribution poles? What are some pros and cons of several non-wood alternatives?
4. List some of the ways that birds can cause reliability problems. How can these problems be mitigated?
5. List some of the ways that wind can cause reliability problems. How can these problems be mitigated?
6. Explain why a well-grounded shield wire may or may not be effective in protective a distribution system against lightning strikes.
7. Explain what happens when a tree branch blows into a phase conductor and becomes entangled.
8. In an area with many trees, will reliability be different for compact construction (2-ft phase separation) when compared to crossarm construction (4-ft phase separation)? Explain.
9. How can the use of live-line techniques improve reliability? Will these improvements be reflected in reliability indices?
10. What is typically the top cause of customer interruptions?

REFERENCES

1. Robert C. Weast, Ph.D., and Melvin J. Astle, Ph.D., *Handbook of Chemistry and Physics, 63rd Edition,* CRC Press, 1982.
2. ANSI/IEEE Std. C57.91-1995. *IEEE Guide for Loading Mineral-Oil-Immersed Transformers.*
3. D. G. Fink and H. W. Beaty, *Standard Handbook for Electrical Engineers*, McGraw Hill, 1987.
4. M. Darveniza, T. K. Saha, D. J. T. Hill, and T. T. Le, "Investigations into Effective Methods for Assessing the Condition of Insulation in Aged Power Transformers," *IEEE Transactions on Power Delivery,* Vol. 13, No. 4, Oct. 1998, pp. 1214-1223.
5. H. L. Willis, *Power Distribution Planning Reference Book*, Marcel Dekker, Inc., 1997.

6. S. V. Nikolajevic, "The Behavior of Water in XLPE and EPR Cables and Its Influence on the Electrical Characteristics of Insulation," *IEEE Transactions on Power Delivery,* Vol. 14, No. 1, Jan. 1999, pp. 39-45.
7. M. D. Walton, B. S. Bernstein, W. A. Thue, and J. T Smith III, "Dielectric Breakdown Strength Correlation with Time-to-Failure During Wet Aging of XLPE-Insulated Cables," *IEEE Transactions on Power Delivery,* Vol. 14, No. 3, July 1999, pp. 750-755.
8. B. S. Bernstein, W. A. Thue, M. D. Walton, and J. T. Smith III, "Accelerated Aging of Extruded Dielectric Power Cables, Part II: Life Testing of 15 kV XLPE-Insulated Cables," *IEEE Transactions on Power Delivery,* Vol. 7, No. 2, April 1992, pp. 603-608.
9. M. D. Walton, B. S. Bernstein, J. T. Smith III, W. A. Thue, and J. H. Groeger, "Accelerated Cable Life Testing of EPR-Insulated Medium Voltage Distribution Cables," *IEEE Transactions on Power Delivery,* Vol. 9, No. 3, July 1994, pp. 1195-1208.
10. C. Katz and M. Walker, "Evaluation of Service Aged 35 kV TR-XLPE URD Cables," *IEEE Transactions on Power Delivery,* Vol. 13, No. 1, Jan. 1998, pp. 1-6.
11. N. Ahmed and N. Srinivas, "Experience Gained with On-Line Partial Discharge Testing in Power Cable System," *IEEE 2001 Transmission and Distribution Conference and Exposition,* Atlanta, GA, Oct. 2001.
12. H. N. Nagamani and Channakeshava, "Investigations of the Failure Modes of XLPE Cables and Joints," *IEEE Transactions on Power Delivery,* Vol. 13, No. 3, July 1998, pp. 706-711.
13. F. Jakl and A. Jakl, "Effect of Elevated Temperatures on Mechanical Properties of Overhead Conductors Under Steady State and Short Circuit Conditions," *IEEE Transactions on Power Delivery,* Vol. 15, No. 1, Jan. 2000, pp. 242-246.
14. L. L. Grigsby, ed., *The Electric Power Engineering Handbook*, CRC Press, 2000.
15. *IEEE Guide for Safety in AC Substation Grounding (ANSI/IEEE Std 80-1986),* Institute of Electrical and Electronics Engineers, Inc., 1986.
16. G. J. Clark, "Summary Report on the Effects of High Operating Temperatures on Conductors and Hardware Behavior," Ontario Hydro Research Division, Report No. 77-177-H, April 25, 1977.
17. J. J. Burke, *Power Distribution Engineering: Fundamentals and Applications,* Marcel Dekker, 1994.
18. IEEE Std 493-1990, *IEEE Recommended Practice for the Design of Reliable Industrial and Commercial Power Systems,* IEEE, 1991.
19. R. D. Garzon, *High Voltage Circuit Breakers: Design and Applications,* Marcel Dekker, Inc., 1997.
20. M. Bartkowiak, M. G. Comber, and G. D. Mahan, "Failure Modes and Energy Absorption Capability of ZnO Varistors," *IEEE Transactions on Power Delivery,* Vol. 14, No. 1, Jan. 1999, pp. 152-162.
21. R. S. Gorur and B. S. Bernstein, "Field and Laboratory Aging of Polymeric Distribution Cable Terminations: Part 1—Field Aging," *IEEE Transactions on Power Delivery,* Vol. 13, No. 2, April 1998, pp. 316-322.
22. R. S. Gorur and B. S. Bernstein, "Field and Laboratory Aging of Polymeric Distribution Cable Terminations: Part 2—Laboratory Aging," *IEEE Transactions on Power Delivery,* Vol. 13, No. 2, April 1998, pp. 323-329.
23. N. H. Ahmed and N. N. Srinivas, "Contamination Performance of Silicone Rubber Cable Termination," *IEEE Transactions on Power Delivery,* Vol. 12, No. 4, Oct. 1997, pp. 1672-1677.
24. M. A. R. M. Fernando and S. M. Gubanski, "Performance of Nonceramic Insulators Under Tropical Conditions," *IEEE Transactions on Power Delivery,* Vol. 15, No. 1, Jan. 2000, pp. 355-360.
25. J. Montesinos, R. S. Gorur, and J. Burnham, "Estimation of Flashover Probability of Aged Nonceramic Insulators in Service," *IEEE Transactions on Power Delivery,* Vol. 15, No. 2, April 2000, pp. 820-826.

26. C. S. Lofgren, "The Economic Importance and Control of Imported Fire Ants in the United States," *Economic Impact and Control of Social Insects*, Praeger Publishers, New York, 1985, pp. 227-256.
27. J. L. Lilien and D. G. Havard, "Galloping Data Base on single and Bundle Conductors: Prediction of Maximum Amplitudes," *IEEE Transactions on Power Delivery,* Vol. 15, No. 2, April 2000, pp. 670-674
28. R. Keutgen and J-L Lilien, "A New Damper to Solve Galloping on Bundled Lines: Theoretical Background, Laboratory Results and Field Results," *IEEE Transactions on Power Delivery,* Vol. 13, No. 1, Jan. 1998, pp. 260-265.
29. A. J. Erioksson, "The Incidence of Lightning Strikes to Power Lines," *IEEE Transactions on Power Delivery,* Vol. 2, No. 3, July 1987, pp. 859-870.
30. L. L. Grigsby, ed., *The Electric Power Engineering Handbook*, CRC Press, 2000.
31. *National Electric Safety Code*, U.S. National Bureau of Standards, 1993.
32. H. L. Willis, *Spatial Electric Load Forecasting*, Marcel Dekker, Inc.,, 1996.
33. M. Kobayashi, "What We Must Learn From 1995 The Great Kobe Earthquake: Considerations on the Distribution Power Poles," *IEEE Transactions on Power Delivery,* Vol. 12, No. 3, July 1997, pp. 1368-1377.
34. *Uniform Building Code,* International Conference of Building Officials, 1997.
35. *Standard Building Code*, Southern Building Code Congress International, 1997.
36. *The Boca National Building Code,* Building Officials and Code Administrators International, 1996.
37. E. Hoffmann, H. V. Rheinbaben, and B. Stösser, "Trees in Electrical Contact with High Voltage Lines," *Proceedings of the 30th Session, International Conference on Large High Voltage Electric Systems*, Paris, France, CIGRE, 1984, Vol. 2, pp. 22.3/1-6.
38. B. F. Wilson, *The Growing Tree*, University of Massachusetts Press, 1970.
39. T. T. Kozlowski, *Tree Growth and Environmental Stresses*, University of Washington Press, 1979.
40. D. B. Botkin, *Forest Dynamics*, Oxford University Press, Inc., 1993.
41. H. L. Willis, G. V. Welch,and R. R. Schrieber, *Aging Power Delivery Infrastructures*, Marcel Dekker, Inc., 2001.

4
Component Modeling

A distribution system consists of thousands of components such as transformers, circuit breakers, overhead lines, underground cable, fuse cutouts, and sectionalizing switches. These components are building blocks that can be assembled in a myriad of ways to create a wide variety of distribution system designs, each with its own unique characteristics.

From a reliability perspective, nearly all of the information needed to create a distribution system model is contained in distribution component information—a highly desirable feature. Given a palette of components, systems can be constructed from scratch by choosing components and connecting them together. Once a system model has been created, modifications can be easily made by adding components, removing components, and modifying component characteristics.

Needless to say, component models are critical to distribution system reliability. A component model should be as simple as possible, but needs to capture all of the features critical to system reliability. This chapter presents the reliability parameters typically assigned to components, discusses how these parameters can be modeled, and provides some guidelines for assigning default component reliability data.

4.1 COMPONENT RELIABILITY PARAMETERS

Each distribution system component can be described by a set of reliability parameters. Simple reliability models are based on component failure rates and component repair times, but sophisticated models make use of many other reliability parameters. A detailed description of some of the most common component reliability parameters is now provided.

Permanent Short Circuit Failure Rate (λ_P) — λ_P describes the number of times per year that a component can expect to experience a permanent short circuit. This type of failure causes fault current to flow, requires the protection system to operate, and requires a crew to be dispatched for the fault to be repaired.

Temporary Short Circuit Failure Rate (λ_T) — λ_T describes the number of times per year that a component can expect to experience a temporary short circuit. This type of failure causes fault current to flow, but will clear itself if the circuit is de-energized (allowing the arc to de-ionize) and then re-energized.

Open Circuit Failure Rate (λ_{OC}) — λ_{OC} describes the number of times per year that a component will interrupt the flow of current without causing fault current to flow. An example of a component causing an open circuit is when a circuit breaker false trips.

Mean Time To Repair (MTTR) — MTTR represents the expected time it will take for a failure to be repaired (measured from the time that the failure occurs). A single MTTR is typically used for each component, but separate values can be used for different failure modes.

Mean Time To Switch (MTTS) — MTTS represents the expected time it will take for a sectionalizing switch to operate after a fault occurs on the system. For manual switches, this is the time that is takes for a crew to be dispatched and drive to the switch location. For an automated switch, the MTTS will be much shorter.

Probability of Operational Failure (POF) — POF is the conditional probability that a device will not operate if it is supposed to operate. For example, if an automated switch fails to function properly 5 times out of every 100 attempted operations, it has a POF of 5%. This reliability parameter is typically associated with switching devices and protection devices.

Scheduled Maintenance Frequency (λ_M) — λ_M represents the frequency of scheduled maintenance for a piece of equipment. For example, a maintenance frequency of 2 per year means that the equipment is maintained every 6 months.

Mean Time To Maintain (MTTM) — MTTM represents the average amount of time that it takes to perform scheduled maintenance on a piece of equipment.

All of the above-mentioned reliability parameters are important, but component failure rates have historically received the most attention. This is because failure rates have unique characteristics and are essential for all types of reliability analyses. The next section looks at failure rates in more detail and explains how electrical component failure rates tend to vary with time.

4.2 FAILURE RATES AND BATHTUB CURVES

It is typical to model component reliability parameters by a single scalar value. For example, a power transformer might be modeled with a failure rate of 0.03 per year. These scalar values, though useful, might not tell the entire story. Perhaps the most obvious example is the observation that the failure rates of certain components tend to vary with age.[1-2]

It might seem reasonable to conclude that new equipment fails less than old equipment. When dealing with complex components, this is usually not the case. In fact, newly installed electrical equipment has a relatively high failure rate due to the possibility that the equipment has manufacturing flaws, was damaged during shipping, was damaged during installation, or was installed incorrectly. This period of high failure rate is referred to as the infant mortality period or the equipment break-in period.

If a piece of equipment survives its break-in period, it is likely that there are no manufacturing defects, that the equipment is properly installed, and that the equipment is being used within its design specifications. It now enters a period referred to as its useful life, characterized by a nearly constant failure rate that can be accurately modeled by a single scalar number.

As the useful life of a piece of equipment comes to an end, the previously constant failure rate will start to increase as the component starts to wear out. That is why this time is referred to as the wearout period of the equipment. During the wearout period, the failure rate of a component tends to increase exponentially until the component fails. Upon failure, the component should be replaced.

A graph that is commonly used to represent how a component's failure rate changes with time is the bathtub curve.[3] The bathtub curve begins with a high failure rate (infant mortality), lowers to a constant failure rate (useful life), and then increases again (wearout). Another name for the bathtub curve is the bathtub hazard function. The use of the term "hazard rate" is common in the field of reliability assessment and is equivalent to the failure rate of the component.

Hazard Rate (Failure Rate) — The hazard rate of a component at time t is the probability of a component failing at time t if the component is still functioning at time t.

A more detailed curve used to represent a component's hazard function is the sawtooth bathtub curve. Instead of using a constant failure rate in the useful life period, this curve uses an increasing failure rate. The increase is attributed to normal wear, and can be mitigated by periodic maintenance. This is analogous to changing the oil in an automobile. If done regularly, the reliability of the car will not degrade substantially. If changing the oil is neglected, reliability will quickly degrade and the probability of a failure occurring increases accordingly. If performing maintenance on the component reduces the failure rate to the same level each time, it is referred to as perfect maintenance. A bathtub function and a sawtooth bathtub function are shown in Figure 4.1.

In the real world, maintenance is rarely perfect. After each maintenance effort, component reliability will usually be a bit worse than the last time maintenance was performed. If performing maintenance on the component reduces the failure rate to a slightly higher level each time, it is referred to as imperfect maintenance. A further complication is that failure rates after maintenance can often increase temporarily. This phenomenon, similar to infant mortality, is due to the possibility of maintenance crews causing damage, making errors during re-assembly, leaving tools inside equipment, and so forth. If the maintained equipment survives for a short period of time, the maintenance was probably performed properly and failure rates are decreased accordingly. A detailed maintenance interval hazard function is shown in Figure 4.2.

A standard bathtub curve is an approximation of a sawtooth bathtub curve. It models the useful life as the average useful life of the sawtooth curve. This approximation is sufficient for most reliability models, but a full sawtooth curve must be used if decisions about maintenance are to be made.

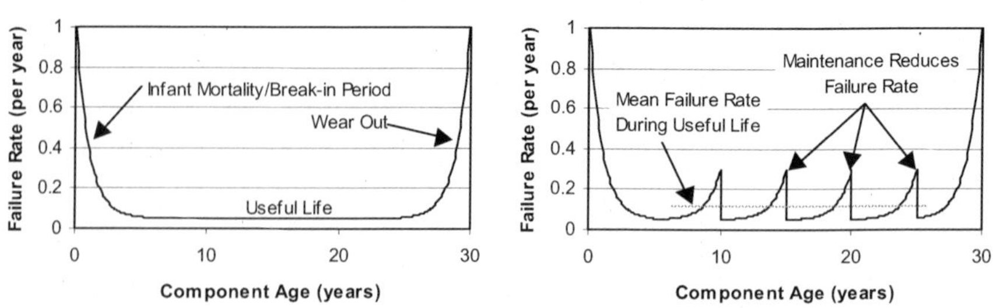

Figure 4.1. A standard bathtub curve (left) and a sawtooth bathtub curve (right). The standard bathtub curve is characteristic of the failure rates of many electrical components that are prone to shipping damage and installation errors. The standard bathtub curve is usually an approximation of the sawtooth bathtub curve, which models the increasing failure rate of a component between maintenance and shows a reliability improvement after maintenance has been performed.

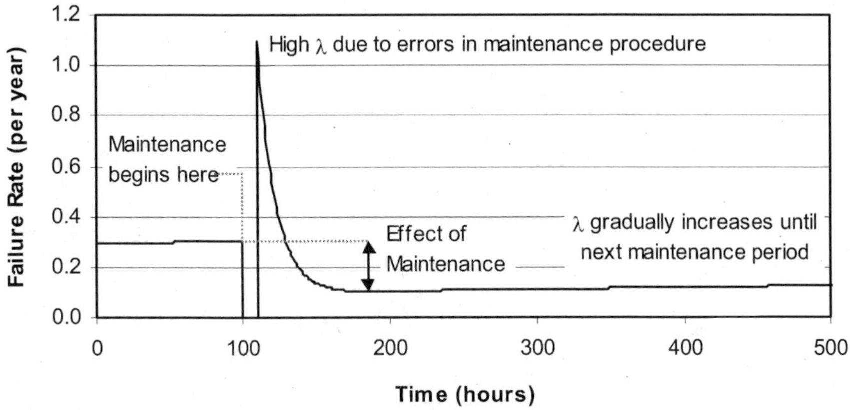

Figure 4.2. A hazard function showing the detailed behavior of equipment reliability during maintenance. When maintenance is performed at hour 100, the failure rate (λ) is relatively high. λ is reduced to zero during maintenance, and then spikes to a very high level due to the possibility of mistakes happening during maintenance. λ quickly decreases to a level lower than pre-maintenance, and then gradually rises until it is time for the next maintenance activity.

4.3 PROBABILITY DISTRIBUTION FUNCTIONS

Many parameters in the field of reliability vary from component to component or from situation to situation. To illustrate, consider the time that it takes to repair a component after it fails. The MTTR of a component represents the expected repair time—the average repair time of the component considering many failures. After each individual failure, the actual repair time may be much lower than the expected value or much higher than the expected value. Because the actual repair time varies and cannot be known beforehand, it is referred to as a random variable. Random variables are represented by probability distribution functions.[4-6]

Probability distribution functions are mathematical equations allowing a large amount of information, characteristics, and behavior to be described by a small number of parameters. Those averse to equations should not skip this section. A deep understanding of each equation is not mandatory, but the reader should, at a minimum, become (1) familiar with the vocabulary that is introduced and (2) have a conceptual understanding of probability distribution functions and related topics.

A probability distribution function has an associated density function, f(x), that represents the likelihood that a random variable x will be a particular value. For example, the probability density function for a random coin toss is:

f(heads) = 0.5
f(tails) = 0.5

There are two important characteristics associated with each probability density function. First, each function always returns a value between (and including) zero and one. A value of zero indicates no probability of occurrence and a value of one indicates certain occurrence. The next characteristic is that the integral of the function over all possible outcomes must equal unity because each random event will have precisely one actual outcome.

$$f(x) \in [0,1] \quad ; f = \text{probability density function} \tag{4.1}$$

$$\int_{-\infty}^{\infty} f(x)\,dx = 1 \tag{4.2}$$

A close cousin to the probability density function is the cumulative distribution function, F(x). The cumulative distribution function is the integral of the probability density function, and reflects the probability that f(x) will be equal to or less than x.

$$F(x) = \int_{-\infty}^{x} f(y)\,dy \quad ; F = \text{cumulative distribution function} \tag{4.3}$$

To illustrate the concept of probability density functions and cumulative distribution functions, consider a random toss of two six-sided dice. There are 36 possible combinations resulting in 12 possible scores. Rolling double ones will only happen once every 36 times, making f(2) = 1/36 = 0.028. Since two is the lowest possible score, F(2) = f(2) = 0.028. Rolling the number 3 will happen twice as often: f(3) = 0.056. The cumulative distribution function at this point will be the sum of all previous density function values: F(3) = f(2) + f(3) = 0.083. This computation continues until the cumulative distribution function is equal to unity. The probability density function and the cumulative distribution function for this example are shown in Figure 4.3. Commonly used probability density functions and their corresponding cumulative distribution functions are shown in Figure 4.4.

Notice that the probability density function shown in Figure 4.3 is a triangle with a peak at the value seven. This high point is called the *mode* of the density function. It is possible for a density function to have two peaks (bimodal), three peaks (trimodal), or more (polymodal).

A function closely related to both the probability density function and the cumulative distribution function is the hazard function, $\lambda(x)$. The hazard function is typically used when cumulative distribution and probability density functions describe the behavior of component failures. The hazard function is equal to the probability of a component failing if it has not already failed. Since the density function is the probability of a component failing, and the cumulative distribution function is the probability that it has already failed, the hazard rate can be mathematically characterized as follows:

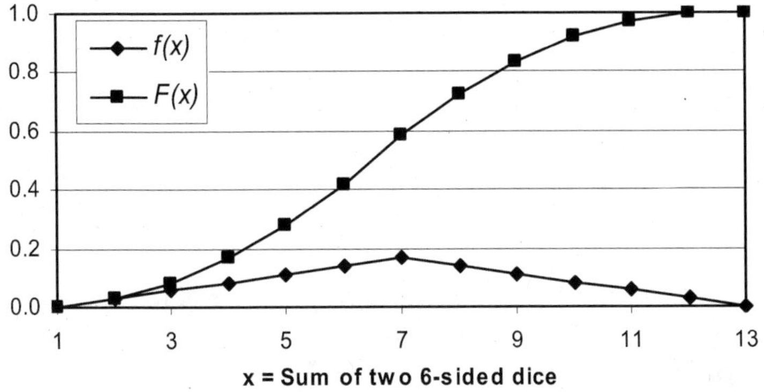

Figure 4.3 Probability density function, f(x), and the cumulative distribution function, F(x) for a random toss of two six-sided dice. The probability density function can assume any value between zero and one. The cumulative distribution function is equal to the integral of the density function, begins with a value of zero, ends with a value of 1, and is monotonically increasing.

$$\lambda(x) = \frac{f(x)}{1 - F(x)} \quad ; \lambda = \text{Hazard Function} \tag{4.4}$$

As has been previously mentioned, hazard functions are identical to failure rate functions. Bathtub curves are a good example of hazard functions—they represent how the failure rate of a piece of equipment changes as the equipment ages.

Another value used commonly when probability density and cumulative distribution functions are used to characterize failure rates is the survivor function, R(x). The survivor function is the probability that a component will survive until a certain point. Mathematically, the survivor function is equal to the arithmetic complement of the cumulative failure distribution function, which is the probability that a component has failed before a specific time.

$$R(x) = 1 - F(x) \quad ; R = \text{Survivor Function} \tag{4.5}$$

Probability distribution functions can be, and often are, characterized by statistical measures such as the expected value (sometimes called the *mean value*), variance, and standard deviation. The expected value is simply the geometric mean of the function. Since the integral of all density functions is equal to unity, this is equal to the integral of first-order moments:

$$\text{Expected Value} = \bar{x} = \int_{-\infty}^{\infty} x \cdot f(x) dx \tag{4.6}$$

The variance is a measure of how the function varies about the mean. A small variance indicates that a value close to the mean is likely, and a large variance indicates that a value close to the mean is unlikely:

$$\text{Variance} = \int_{-\infty}^{\infty} \left[f(x) - \bar{x}\right]^2 dx \tag{4.7}$$

The standard deviation is the square root of the variance. It is convenient because it has the same units as the density function and the expected value. To compare standard deviations, it is sometimes helpful to use the units of "percent of mean." To achieve this, divide the standard deviation by its associated expected value and multiply by 100.

$$\text{Standard Deviation} = \sqrt{\text{Variance}} \tag{4.8}$$

$$\text{Standard Deviation (\% of mean)} = 100 \times \frac{\sqrt{\text{Variance}}}{\text{Expected Value}} \tag{4.9}$$

Now that the basic theory and vocabulary for probability distribution functions have been discussed, some common and useful distribution functions will be presented. After this, methods for fitting the density functions to historical data will be addressed.

4.3.1 Normal Distribution

Perhaps the most well known distribution of all is the normal distribution—the proverbial "bell curve" that is mathematically characterized by its expected value and standard deviation. Formulae corresponding to the normal distribution are:

$$f(x) = \frac{1}{\sigma\sqrt{2\pi}} \exp\left[-\frac{(x-\mu)^2}{2\sigma^2}\right]; -\infty \le x \le \infty$$

$$\text{; Normal Distribution} \tag{4.10}$$

$$\text{Expected Value} = \mu$$

$$\text{Variance} = \sigma^2$$

Use of the normal distribution is so common that its parameters, μ and σ, are often times misrepresented to be means and standard distributions for functions other than the normal distribution. This is not always true—μ and σ correspond to mean and standard deviation for the normal distribution, but not necessarily for other distributions.

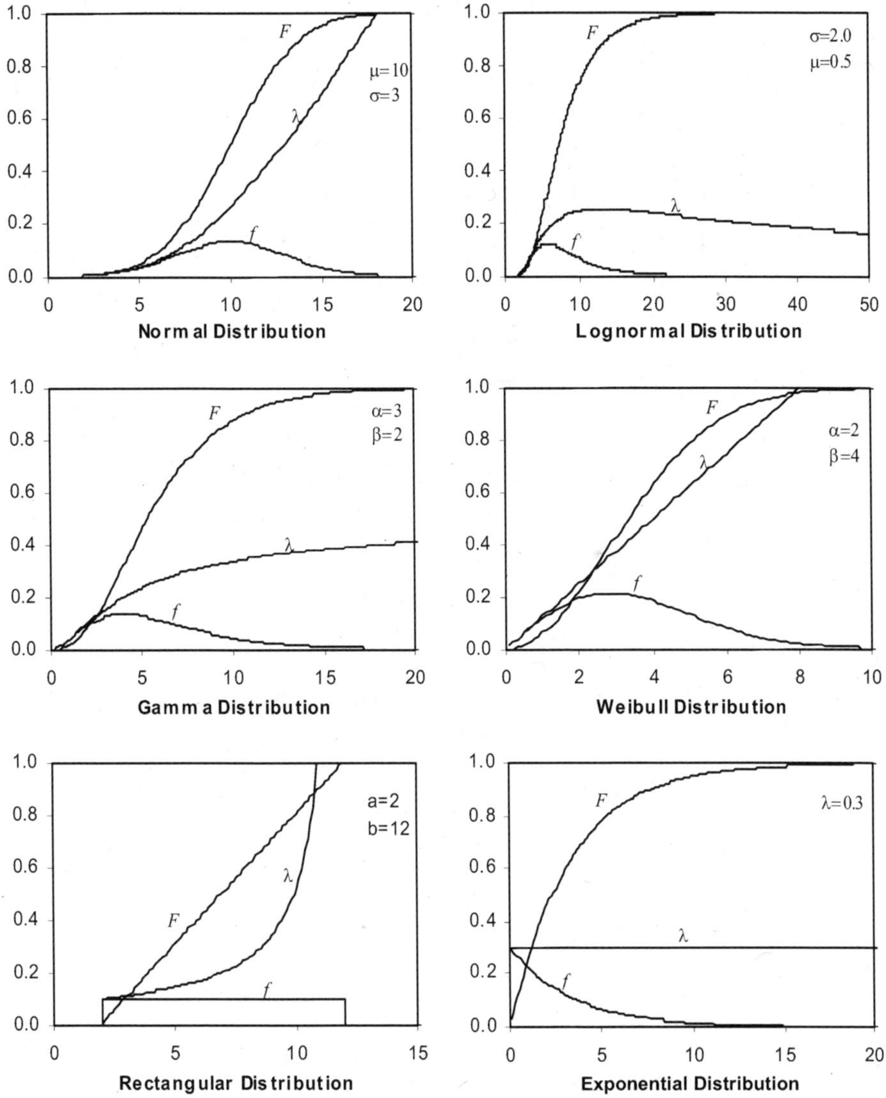

Figure 4.4. Shown above are six probability distribution functions that have wide application in reliability assessment. If a failure rate is being modeled, the hazard function (λ) should represent the desired characteristics of the failure rate. Exponential distributions are often used to model failure rates since they are characterized by a constant hazard function. Other distributions tend to be more characteristic of repair times and switching times.

The normal distribution is convenient because its parameters can be easily computed from experimental data. For any data set, computing the mean and standard deviation produces the μ and σ of the normal distribution. Though convenient, using the normal distribution is rarely appropriate in the area of distribution reliability. Random variables characterizing component reliability parameters are typically constrained to be positive while the normal distribution allows for negative values. Distribution random variable values are often times skewed in shape while the normal distribution is always symmetrical. For random variables with nonnormal characteristics, other distribution functions are more appropriate to apply.[7]

4.3.2 Lognormal Distribution

The lognormal distribution, as one may guess, is a close relative to the normal distribution. The primary differences are that the lognormal distribution uses the natural logarithm of the independent variable, and the independent variable is constrained to be nonnegative. Formulae for the lognormal distribution are:

$$f(x) = \frac{1}{x\sigma\sqrt{2\pi}} \exp\left[-\frac{(\ln x - \mu)^2}{2\sigma^2}\right] \quad ; \ x \geq 0$$

$$\text{Expected Value} = \exp\left[\mu + \frac{\sigma^2}{2}\right] \qquad ; \text{Lognormal Distribution} \qquad (4.11)$$

$$\text{Variance} = \exp\left[2\mu + 2\sigma^2\right] - \exp\left[2\mu + \sigma^2\right]$$

The lognormal distribution has a skewed shape—its peak will always be to the left of its mean. Depending upon its parameters, this skew can be slight or severe. Density functions of this shape are useful for characterizing a host of reliability parameters such as repair times and switching times.

4.3.3 Exponential Distribution

The exponential distribution is the most common distribution function used in the field of reliability analysis. This is because it is characterized by a constant hazard function, which is representative of electrical components during their useful life. A further advantage of the exponential distribution is that it is fully characterized by a single parameter, λ. Formulae related to the exponential distribution are:

$$f(x) = \lambda \exp[-\lambda x]\ ; \quad x \geq 0$$
$$F(x) = 1 - \exp[-\lambda x]$$
$$\lambda(x) = \lambda \qquad ;\ \text{Exponential Distribution} \qquad (4.12)$$
$$\text{Expected Value} = \frac{1}{\lambda}$$
$$\text{Variance} = \frac{1}{\lambda^2}$$

Notice that the exponential distribution has closed-form expressions for its probability distribution function (F) and its hazard function (λ). This is in contrast to the normal distribution and the lognormal distribution, and is a computational advantage.

If a component's failure rate is constant (characteristic of the exponential distribution) and the component is repaired quickly after each failure, the component can be characterized by a Poisson Process. This allows the probability of multiple failures occurring in a given time period to be easily computed. The Poisson Process is described in more detail in later sections.

4.3.4 Rectangular Distribution

The rectangular distribution has a uniform value over a lower bound, a, and upper bound, b. This distribution is mathematically simple and all of its important functions have closed-form solutions. Formulae are:

$$f(x) = \frac{1}{b-a}\ ; \quad a \leq x \leq b$$
$$F(x) = \frac{x-a}{b-a}$$
$$\lambda(x) = \frac{1}{b-x} \qquad ;\ \text{Rectangular Distribution} \qquad (4.13)$$
$$\text{Expected Value} = \frac{a+b}{2}$$
$$\text{Variance} = \frac{(b-a)^2}{12}$$

The rectangular distribution is useful for modeling random variables that have a well-defined upper and lower limit. Perhaps the most common example is the random number generator used by most computers. These algorithms are equally likely to produce any number within the specified boundaries. To check whether such an algorithm is behaving properly, a large number of random numbers can be generated. The mean and variance are then computed and checked against the theoretical values for a rectangular distribution (to be complete, random number generator testing must also include a trend analysis).

4.3.5 Gamma Distribution

Before the gamma distribution can be presented, the gamma function, $\Gamma(x)$, must be defined. This function is closely related to the factorial function, $x!$, but allows nonintegers to serve as an input. The gamma function is:

$$\Gamma(x) = \int_0^{\infty} t^{x-1} e^{-t} \qquad \text{; Gamma Function} \qquad (4.14)$$
$$\Gamma(x) = (x-1)(x-2)\ldots(2) \quad ; \; x = [2,3,\ldots]$$

While $\Gamma(x)$ is a bit complicated, it allows the gamma distribution function to be represented in a compact form. The gamma distribution function is versatile because it can assume various shapes to fit disparate data sets. The gamma distribution is a function of two parameters: a scale parameter α and a shape parameter β. It is described by the following formulae:

$$f(x) = \frac{x^{\beta-1}}{\alpha^{\beta}\Gamma(\beta)} \exp\left[-\frac{x}{\alpha}\right] \quad ; \quad x \geq 0$$
$$\text{Expected Value} = \alpha\beta \qquad \text{; Gamma Distribution} \qquad (4.15)$$
$$\text{Variance} = \alpha^2\beta$$

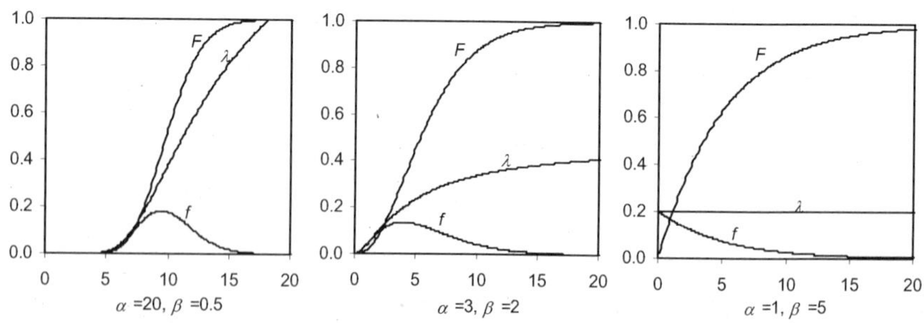

Figure 4.5. Three different gamma distributions. By varying the scale parameter, α, and the shape parameter, β, a wide variety of distribution shapes can be modeled. The left graph represents a bell shaped density curve, the middle graph shows a density curve that is skewed to the left, and the right graph shows a density curve that is exponentially decaying.

Three example gamma distributions are shown in Figure 4.5. Of particular interest are the different density curves and hazard curves that result from different parameter selection, with low betas resulting in steep hazard functions and high betas resulting in near-constant hazard functions.

4.3.6 Weibull Distribution

The Weibull distribution is similar to the gamma distribution in two ways. First, it is a function of two parameters: a scale parameter α and a shape parameter β. Second, it is able to assume various shapes to fit varying data sets. It has the additional advantages of having a closed-form cumulative distribution function and a closed-form hazard function. Formulae for the Weibull distribution are:

$$f(x)=\frac{\beta x^{\beta-1}}{\alpha^{\beta}}\exp\left[-\left(\frac{x}{\alpha}\right)^{\beta}\right] \quad ; \quad x\geq 0$$

$$F(x)=1-\exp\left[-\left(\frac{x}{\alpha}\right)^{\beta}\right]$$

$$\lambda(x)=\frac{\beta x^{\beta-1}}{\alpha^{\beta}} \qquad ;\ \text{Weibull Distribution} \qquad (4.16)$$

$$\text{Expected Value}=\alpha\Gamma\left(\frac{1}{\beta}+1\right)$$

$$\text{Variance}=\alpha^2\left[\Gamma\left(\frac{2}{\beta}+1\right)-\Gamma^2\left(\frac{1}{\beta}+1\right)\right]$$

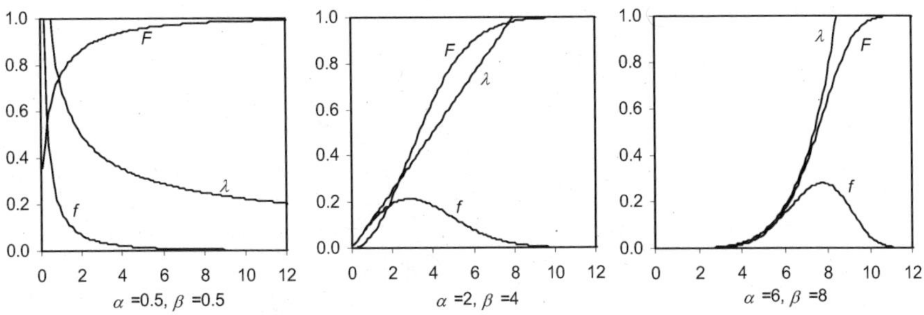

Figure 4.6. Thee different Weibull distributions. By varying the scale parameter, α, and the shape parameter, β, a wide variety of distribution shapes can be modeled. The left graph represents an exponentially decaying density curve, the middle graph shows a density curve that is skewed to the left, and the right graph shows a density curve that is skewed to the right.

Three examples of Weibull distributions are shown in Figure 4.6. Notice that $\beta < 1$ corresponds to a decreasing hazard function and a $\beta > 1$ corresponds to an increasing hazard function. $\beta = 1$ (not shown) corresponds to a constant hazard function equivalent to an exponential distribution with $\lambda = 1/\alpha$.

4.3.7 Raleigh Distribution

A special case of the Weibull distribution is the Raleigh distribution. The Raleigh distribution is a function of the single parameter k and occurs when $\beta = 2$ and k is defined as $2\alpha^{-2}$. Formulae for the Raleigh distribution are:

$$f(x) = kx \cdot \exp\left[\frac{-kx^2}{2}\right] \quad ; \quad x \geq 0$$

$$F(x) = \exp\left[\frac{-kx^2}{2}\right]$$

$$\lambda(x) = kx \qquad ; \text{Raleigh Distribution} \qquad (4.17)$$

$$\text{Expected Value} = \sqrt{\frac{\pi}{2k}}$$

$$\text{Variance} = \frac{2}{k}\left(1 - \frac{\pi^2}{4}\right)$$

The advantage of the Raleigh distribution, like the exponential distribution, is that it is represented by a single parameter. The Raleigh distribution is characterized by a linearly increasing hazard function, and has a density curve shape characteristic of repair times, switching times, and a host of other phenomena.

4.4 FITTING CURVES TO MEASURED DATA

Probability distribution curves are usually used to represent empirical data. In this way, the information associated with thousands of data points can be reasonably modeled with one or two parameters. This section addresses the question of determining which distribution function to use and what parameter values to use. It first discusses straightforward ways of computing distribution parameters, and then discusses "goodness of fit" tests to identify the curve that best represents the measured data. It also discusses a graphical method for determining Weibull parameters and a perturbation method for "fine tuning" parameters.

4.4.1 Direct Parameter Extraction

The easiest method of fitting a set of data to a distribution function is to directly compute distribution parameters from statistical results. For most distribution functions with one or two parameters, a reasonable curve fit can be found by using the mean and variance of the data set. If direct parameter extraction is used, the mean and variance of the density function will exactly match the mean and variance of the data set.

Direct parameter extraction formulae for six density functions are now provided. These equations are based on a data set with a pre-computed mean and variance.

Statistics from Data Set

Size of Data Set = n

$$\text{Data Set} = [x_1, x_2, \ldots, x_n]$$

$$\text{Mean Value} = \bar{x} = \frac{1}{n}\sum_{i=1}^{n} x_i \tag{4.18}$$

$$\text{Variance} = \frac{1}{n}\sum_{i=1}^{n} (x_i - \bar{x})^2 \tag{4.19}$$

Normal Distribution

$$\mu = \bar{x} \tag{4.20}$$

$$\sigma = \sqrt{\text{variance}} \tag{4.21}$$

Lognormal Distribution

$$\sigma = \sqrt{\ln\left(\text{variance} + e^{2\ln\bar{x}}\right) - 2\ln\bar{x}} \tag{4.22}$$

$$\mu = \ln\bar{x} - \frac{1}{2}\sigma^2 \tag{4.23}$$

Exponential Distribution

$$\lambda = \frac{1}{\bar{x}} \tag{4.24}$$

Rectangular Distribution

$$a = x - \sqrt{3 \cdot \text{variance}} \tag{4.25}$$

$$b = x + \sqrt{3 \cdot \text{variance}} \tag{4.26}$$

Gamma Distribution

$$\alpha = \frac{\text{variance}}{\overline{x}} \tag{4.27}$$

$$\beta = \frac{\overline{x}^2}{\text{variance}} \tag{4.28}$$

Raleigh Distribution

$$k = \frac{\pi}{2\overline{x}^2} \tag{4.29}$$

Figure 4.7 shows an example of how direct parameter extraction can be applied to a practical problem. A utility has gathered data on how long it has historically taken its crews to perform maintenance on circuit breakers. A histogram of the data is plotted and it appears that a lognormal distribution will provide a good fit. The mean and variance of the data set are computed, and the lognormal parameters μ and σ are computed using Eq. 4.22 and Eq. 4.23.

4.4.2 Weibull Parameter Extraction

Due to the presence of the Gamma function in the mean and variance equations, there is no easy formula to directly compute Weibull parameters from mean values and variances. Instead, a transformation of variables is performed on the cumulative distribution function and parameters are computed graphically or by linear regression.[8] Consider the distribution function in its normal form:

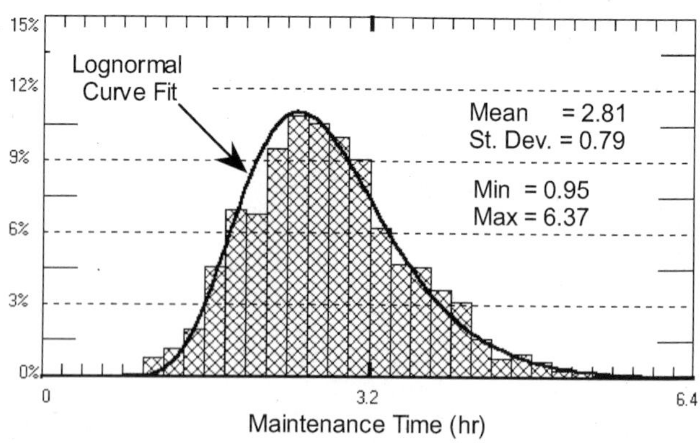

Figure 4.7. Histogram of data and the associated lognormal curve fit. The lognormal distribution parameters were extracted directly from the mean and variance of the histogram data. Notice that the shape of the curve is slightly skewed causing the mean to be located to the right of the peak.

$$F(x) = 1 - \exp\left[-\left(\frac{x}{\alpha}\right)^{\beta}\right] \quad ; \text{Weibull Distribution} \tag{4.30}$$

For the parameters α and β to be extracted, it is convenient to rearrange this equation into the following form:

$$\ln\ln\frac{1}{1-F(x)} = \beta\ln x - \beta\ln\alpha \tag{4.31}$$

According to this equation, ln ln [1/(1-F(x)] plots as a straight line against ln(x) with a slope of β and a y-intersection of -β·ln(α). Once plotted, the Weibull parameters can be computed as follows:

$$\beta = \text{Slope of Linear Fit} \tag{4.32}$$

$$\alpha = \exp\left(-\frac{\text{y - intercept}}{\beta}\right) \tag{4.33}$$

An example illustrates this process. Utility outage reports are used to obtain the time that crews have historically taken to repair faults on overhead lines. The data ranges between 0 and 30 hours, but 90% of the data points range between 0 and 9 hours. The data points are grouped into bins of one hour, and are placed on a Weibull plot. Each data point is represented as an "x" on the left graph of Figure 4.8. Linear regression is then used to identify the line that minimizes the least squared error. This line is shown on the same graph and has a slope of 1.07 and a y-intercept of minus 1.08.

Once the slope and intercept of the line have been determined, α and β can be easily computed from Eq. 4.32 and Eq. 4.33. In this case, β = 1.07 and α = 2.74. Using these parameters, the Weibull density function can be compared to the density of each data bin (if a bin contains x data points, and there are n total samples, then the bin has a density of x/n). This is shown in the right graph of Figure 4.8.

After comparing the Weibull density function to the data bin densities, it is immediately apparent that the shapes of the two curves are different. The Weibull density function looks like a decreasing exponential, and the data bins look like a skewed bell curve. The mismatch occurs because the shape of the histogram is largely dependent upon the first two data points (bin_{0-1} and bin_{1-2}). In order to create the best linear fit in the Weibull plot, these two data points are largely ignored so that the errors for the remaining 27 data points can be minimized. A good linear fit for Eq. 4.31 does not necessarily result in a representative probability density function.

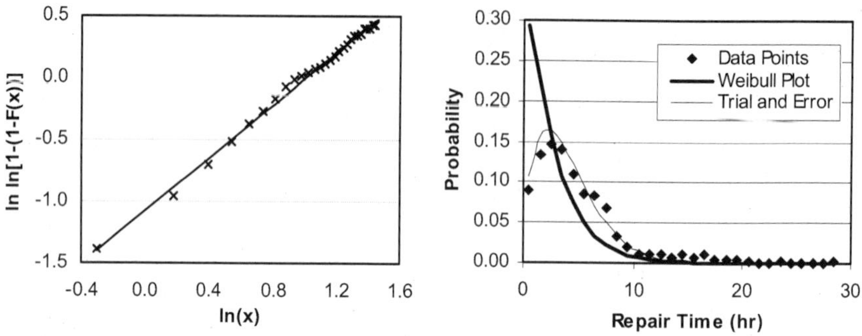

Figure 4.8 A Weibull plot of data points and the line resulting from linear regression (left) and the density function that result from this direct parameter extraction. This curve fit does not match the shape of the data points, and a much more representative Weibull curve was quickly found by trial and error.

The Weibull density function is quite capable of representing the shape of the data bin densities. Through trial and error of approximately 10 parameter combination, the parameters $\beta = 1.5$ and $\alpha = 4.5$ are shown to accurately represent the data points (see Figure 4.8). This example should caution the reader to visually check modeled curves with data curves to insure that the shape of the model is representative of the data.

4.4.3 Chi Squared Criterion

When modeling reliability data, it is desirable to have a "goodness of fit" metric that indicates how well a model matches the data that it is suppose to represent. The most popular metric is called the *chi squared* criterion, or χ^2. It is based on density functions and data bin densities, and is computed using the following formulae:

$$\chi^2 = \sum_{\text{bins}} \frac{(\text{Observed Freq. in Bin - Expected Freq. in Bin})^2}{\text{Expected Freq. in Bin}}$$

$$\text{Observed Freq. in Bin} = \frac{\text{\# of Samples in Bin}}{\text{Total \# of Samples}} \qquad (4.34)$$

$$\text{Expected Freq. in Bin} = \int_a^b f(x)dx = F(b) - F(a)$$

$$\text{Bin} = a \le x < b$$

The chi squared criterion can be used to help identify distribution function parameters, and can be used to compare the fit of various distribution functions to a single data set. For example, assume that the best curve fit is desired for a certain data set. Parameters for a lognormal distribution, a rectangular distribution, a Weibull distribution, and a Gamma distribution can be found that minimize the chi squared error in each case. The distribution function that is ultimately selected will be the one with the smallest error.

4.4.4 Minimizing Chi Squared Error

If a distribution function is being fitted to a data set, it is usually desirable to find the distribution function parameters that minimize the chi squared error. There are a host of methods to do this, but a straightforward way is to start with a reasonable guess and then perturb the parameters as long as the error continues to decrease. This algorithm, referred to as a hill climbing, is summarized in Figure 4.9.

Hill climbing algorithms guarantee that the parameters are locally optimal. This means that any small changes to parameters will always cause the error to increase. Local optimality does not preclude the possibility of better parameter combinations. In fact, the globally optimal solution may be much better than certain locally optimal solutions. For this reason, it is good to start with reasonable parameters that generally represent the shape of the data set. This can be done by trial and error or by direct parameter extraction. Ambitious users can explore more complicated optimization methods such as nonlinear programming, genetic algorithms, simulated annealing, and others.

1. Initialize parameters to achieve a reasonable fit
2. Set Δ equal to the precision desired for each parameter
3. Compute χ^2
4. Start with the first parameter, p
5. Compute χ^2_{test}, based on $p_{test} = p + \Delta$
6. If $\chi^2_{test} < \chi^2$ then set $p = p_{test}$, set $\chi^2 = \chi^2_{test}$, and go to step 9
7. Compute χ^2_{test}, based on $p_{test} = p - \Delta$
8. If $\chi^2_{test} < \chi^2$ then set $p = p_{test}$, set $\chi^2 = \chi^2_{test}$
9. Repeat steps 5-9 for all parameters
10. Has χ^2 changed its value since step 4? If so, go to step 4
11. End

Figure 4.9 A hill climbing algorithm to minimize the chi squared error of a curve fit. It guarantees local optimality, but does not ensure global optimality. For this reason, the algorithm should be initialized with parameters that already represent the general shape of the data set.

4.5 COMPONENT RELIABILITY DATA

Component reliability data is one of the most important aspects of distribution system reliability assessment. Without good data, the answers provided by complicated analyses and sophisticated computer programs are baseless. Good information is still hard to come by, but most utilities are recognizing the importance of component reliability data and are undertaking collection efforts by both manual and automated methods.

This section presents a large amount of component reliability information based on historical utility data, manufacturer test data, professional organizations such as the IEEE and Cigré, and technical publications from journals and conference proceedings. Most of the reliability parameters listed in this section are assigned three numbers: a low value, a typical value, and a high value. The low and high numbers are the lowest and highest values found in published literature. The typical number is a reasonable generic value for a typical US distribution system, but may not be representative of any specific system. Component reliability data can vary widely from system to system and, if possible, should be calibrated to specific historical data.

A final word of caution: all of the numbers listed in the following tables (low, typical, and high) represent average system values. Low and high values are not the typical low and high values for specific components. To clarify, consider the reliability of overhead lines (primary trunk) as seen in Table 4.1. The high failure rate listed is 0.3 failures per mile per year. This represents the average failure rate for overhead lines on a system with worse than average overhead line reliability. Specific line sections on this system may have failure rates much higher or much lower, but on average, the failure rate is 0.3 failures per mile per year.

4.5.1 Overhead Distribution

Overhead distribution equipment refers to pole-mounted devices that are operated radially and have voltage ratings between 5 kV and 35 kV. Because this type of equipment is directly exposed to weather, vegetation, and animals, it typically has failure rates that are higher than corresponding underground equipment. To help compensate for this, overhead equipment failures are relatively easy to locate and fix, making overhead MTTR values lower than corresponding underground equipment. Failure rates of overhead distribution equipment are, in general, very system specific due to their dependence on geography, weather, animals, and other factors.[9] The typical reliability values for overhead distribution equipment shown in Table 4.1 can serve as a good guide for comparing system designs, but caution should be exercised if they are to be used for predicting actual system performance.

Table 4.1. Reliability of overhead distribution components.[10-21]

Description	λ_P (per year)			MTTR (hours)		
	Low	Typical	High	Low	Typical	High
Overhead Lines						
Primary Trunk	0.020*	0.100*	0.300*	2.0	4.0	8.0
Lateral Tap	0.020*	0.160*	0.300*	2.0	4.0	8.0
Secondary & Service Drop	0.020*	0.088*	0.030*	1.5	2.5	8.0
Pole Mounted Transformer	0.004	0.010	0.015	3.0	5.0	10.0
Disconnect Switch	0.004	0.014	0.140	1.5	4.0	12.0
Fuse Cutout	0.004	0.009	0.030	0.5	2.0	4.0
Line Recloser	0.005	0.015	0.030	3.0	4.0	10.0
Shunt Capacitor	0.011	0.020	0.085	0.5	1.5	4.0
Voltage Regulator	0.010	0.029	0.100	1.0	4.0	10.0

*Failure rates for overhead lines are per circuit mile.

An important reliability parameter for the reliability assessment of overhead distribution systems is the temporary short circuit failure rate of overhead conductors (λ_T). Unfortunately, available data is scarce. A reasonable way to deal with this lack of data is to assume that between 50% and 80% of all faults are temporary in nature if they are allowed to clear themselves.[15-16] Assuming this, λ_T can be calculated by multiplying λ_P by a momentary failure rate factor that will typically vary between 1 and 4.

4.5.2 Underground Distribution

Although the majority of distribution circuit-miles are overhead, underground distribution system are steadily rising in popularity. This increase can be largely attributed to two factors: aesthetics and reliability. This book is interested in the reliability implications of underground equipment.

Underground equipment is sheltered from vegetation and weather, and usually has lower failure rates than associated overhead equipment. Faults, however, can be difficult to locate and can be time-consuming to repair. These characteristics are reflected in the reliability values shown in Table 4.2.

The reliability of underground equipment has received a fair amount of attention in recent years. As a result, more data is available than for many other types of components. This data can be used to track the failure rate of components as they age. Figure 4.10 shows the results of a typical study. Results are shown for cable sections, cable joints, and pad-mounted transformers, and are broken down by voltage class. The failure rates of all components increase with age, but the rate of increase is small for certain components and large for other components.

Table 4.2 Reliability of underground distribution components.[17-22]

Description	λP (per year)			MTTR (hours)		
	Low	Typical	High	Low	Typical	High
Underground Cable						
Primary Cable	0.003*	0.070*	0.587*	1.5	10.0	30
Secondary Cable	0.005*	0.100*	0.150*	1.5	10.0	30
Elbows Connectors	6.0e-5	0.0006	0.001	1.0	4.5	8.0
Cable Splices and Joints	6.0e-5	0.030	0.159	0.5	2.5	8.0
Padmount Transformers	0.001	0.010	0.050	4.0	6.5	7.8
Padmount Switches	0.001	0.003	0.005	0.8	2.5	5.0

*Failure rates for underground cable are per circuit mile.

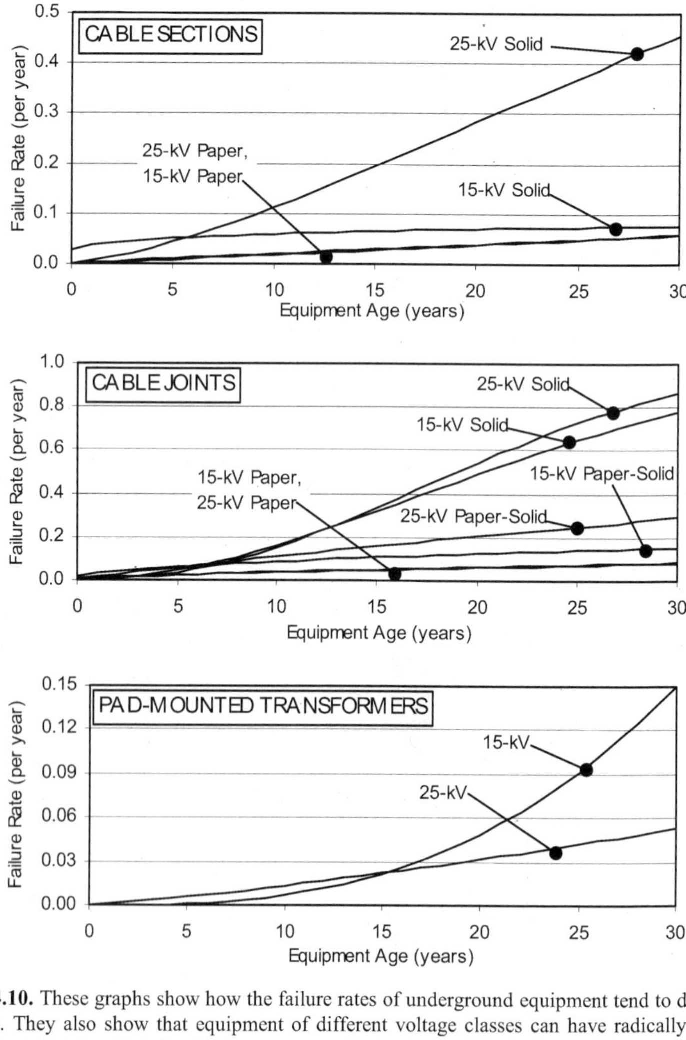

Figure 4.10. These graphs show how the failure rates of underground equipment tend to deteriorate with age. They also show that equipment of different voltage classes can have radically different failure characteristics. Data for these graphs came from a large utility in the northeastern US.[23]

4.5.3 Air Insulated Substations

Most utility and industrial substations utilize air to electrically insulate live conductors and, therefore, are referred to as air insulated substations (AIS). Since live conductors are exposed to open air, these substations experience reliability problems associated with weather, birds, and terrestrial animals. In addition, these substations contain power transformers, which play a critical role in distribution system reliability. Typical reliability parameters for AIS equipment are shown in Table 4.3.

The mean time to repair of power transformers can vary widely from utility to utility and from substation to substation. Since power transformer repair time can have a substantial impact on customer reliability, care should be taken so that this parameter is as accurate as possible. Key factors to be considered are the availability of on-site spares, the availability of off-site spares, the availability of mobile substations, and the actual repair time of failed units.

Air insulated substations have large footprints, are subject to exposure-related reliability problems, and are visually unappealing. An ugly AIS can be concealed by surrounding it with an architectural wall. Aesthetics and reliability can be simultaneously addressed by placing the entire AIS in a building. Both of these strategies are expensive, can actually increase the repair time of substation equipment, and do not address footprint issues. For these reasons, many utilities choose to utilize gas insulated substations where compactness, aesthetics, and/or reliability are major concerns.

4.5.4 Gas Insulated Substations

Gas insulated substations (GIS) enclose all conductors inside pipes filled with sulfur hexafluoride gas (SF_6). Doing so protects conductors from many failure modes and allows the footprint of buswork and switchgear to be reduced by as much as 90%. Although more expensive to purchase, GIS has been successfully used to eliminate reliability problems associated with AIS installations in harsh environments such as chemical plants and seacoasts.

Table 4.3. Reliability of AIS components.[24-31]

Description	λ_P (per year)			**MTTR** (hours)		
	Low	**Typical**	**High**	**Low**	**Typical**	**High**
Power Transformers						
Less than 25 MVA	0.015	0.04	0.070	15	40	85
Bigger than 25 MVA	0.010	0.03	0.060	15	70	170
Circuit Breakers	0.001	0.01	0.030	2.5	12	80
Disconnect Switches	0.004	0.01	0.160	1.5	4	12
Instrument Transformers	0.000	0.01	0.003	4	4	24
Air Insulated Busbar	0.001	0.01	0.038	2	4	36

Table 4.4. Reliability of GIS components.[32-34]

Description	λP (per year)			MTTR (hours)		
	Low	**Typical**	**High**	**Low**	**Typical**	**High**
GIS Bay (before 1985)	0.0006	0.002	0.015	12	90	240
GIS Bay (after 1985)	0.0004	0.001	0.009	12	90	240
Bushing		13.2%				
Metallic Earthed Component		1.6%				
Gas Insulation		7.0%				
Circuit Breaker		8.5%				
Main Contact Assembly		7.7%				
Solid Insulation		13.0%				
Gas Sealing		11.2%				
Control and Protection		9.1%				
Motors/Pumps		19.8%				
Interlocking Elements		1.6%				
Gas Density Supervision		4.2%				
Unknown		3.0%				

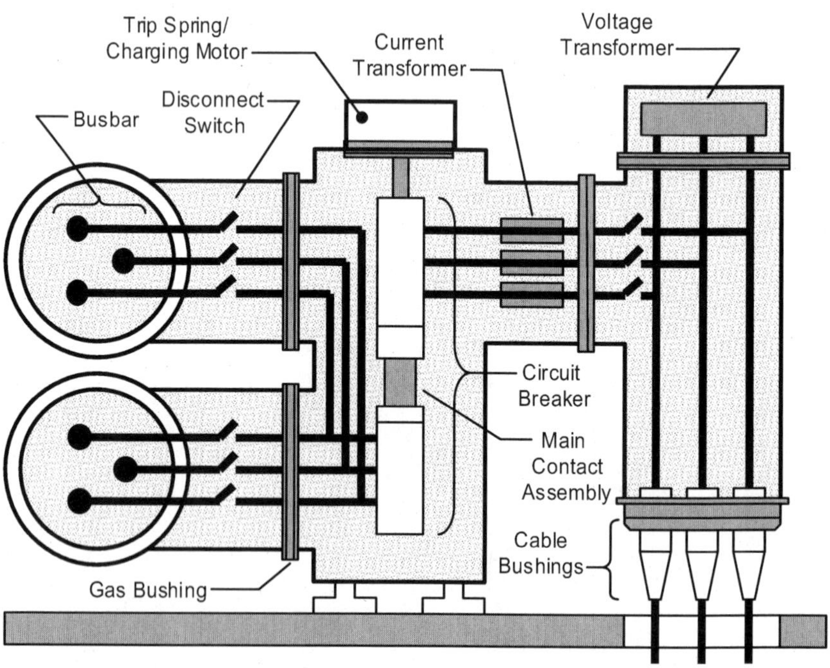

Figure 4.11. A typical GIS bay consisting of (from left to right) two busbar units, a circuit breaker unit, and a cable end unit. GIS offers increased reliability over AIS since conductors are enclosed in hermetically sealed containers.

The reliability of gas insulated substations is commonly reported in terms of bay failures. A GIS bay consists of a circuit breaker, connections from the circuit breaker to buswork connections from the circuit breaker to line or cable bushings, and auxiliary equipment such as current transformers, voltage transformers, protection, and control (see Figure 4.11). Typical GIS bay failure rates are shown in Table 4.4. These bay failures can be attributed to a variety of phenomena such as circuit breaker failures and gas leaks. Table 4.4 shows a breakdown of the most common bay failure causes and their associated frequencies of occurrence.

Although GIS has been in widespread use for more than 30 years, many utilities still consider it new technology and question its reliability. For this reason, several reliability studies have examined the reliability of old versus new GIS installations.[33-34] Results show that older GIS installations are more reliable than comparable AIS configurations and that newer GIS installations fail even less often. Repair times are typically higher for GIS when compared to AIS, but the drastically reduced failure rate improves overall availability and may allow for simpler configurations to achieve comparable performance.

4.5.5 Industrial Equipment

Although the focus of this book is on utility distribution systems, many of the concepts and techniques are applicable to industrial systems (transmission reliability parameters are available in References 35-36). Reliability parameters for typical industrial components are provided in Table 4.5. These values, based on IEEE surveys,[37] show that industrial equipment failure rates tend to vary less widely than utility failure rates, but repair times tend to vary more widely. If an industrial plant keeps an inventory of spare electrical parts, repair time can be very short. In contrast, repair time can be weeks if parts need to be ordered and/or service appointments must be scheduled. Hence, the low versus high MTTR for drawout breakers varies by more than a factor of two hundred (high repair times are most likely associated with redundant components that do not result in prolonged process interruptions should they fail).

Table 4.5. Reliability of industrial components.[37]

Description	λ_P (per year)			**MTTR** (hours)		
	Low	Typical	High	Low	Typical	High
Liquid Filled Transformers	0.0053	0.0060	0.0073	39	300	1000
Molded Circuit Breakers	0.0030	0.0052	0.0176	1.0	5.8	10.6
Drawout Breakers	0.0023	0.0030	0.0036	1.0	7.6	232
Disconnect Switches	0.0020	0.0061	0.0100	1.0	2.8	10.6
Switchgear Bus	0.0008[1]	0.0030[1]	0.0192[1]	17	28	550
Cable (not buried)	0.0014[2]	0.0100[2]	0.0492[2]	5.3	7.0	457
Cable (buried)	0.0034[2]	0.0050[2]	0.0062[2]	15	35	97
Cable Terminations	0.0003	0.0010	0.0042	1.0	2.8	10.6

[1]Failure rates for switchgear bus are per circuit foot.
[2]Failure rates for cable are per 1000 circuit feet.

4.5.6 Summary

This section has provided a large amount of component reliability data based on a wide variety of sources. It has also attempted to (1) illustrate the wide variations seen in published data, (2) emphasize that failure rates and repair times are system specific, and (3) encourage the practice of deriving reliability parameters from historical data of the specific system being modeled. That said, the typical values provided are appropriate to use as a first step in the absence of specific data. They are also appropriate to be used when comparing the relative reliability of various system configurations. It is generally acceptable to use generic data to make value judgments such as "the SAIDI of Configuration A is better than the SAIDI of Configuration B." To make absolute judgments such as "the SAIDI of Configuration A is X hours per year," component reliability data must typically be calibrated to historical data, a process discussed in later chapters.

4.6 STUDY QUESTIONS

1. What are three important types of equipment failure rates?
2. What are the three basic parts of a bathtub hazard function?
3. Why is a normal distribution function typically not appropriate to use for reliability analyses?
4. Name two probability distribution functions that are described by a single parameter. What is each of these distribution functions commonly used to describe?
5. Why are Weibull functions popular for modeling equipment failure rates? What are some potential concerns when using Weibull functions?
6. What is meant by saying that a failure rate is exponentially distributed?
7. Describe some of the reliability differences between overhead distribution equipment and underground distribution equipment.
8. Describe some of the reliability differences between utility distribution equipment and industrial distribution equipment.
9. What factors should a utility consider when deciding whether a substation should be air insulated or gas insulated?
10. Is it appropriate to use generic industry reliability data when performing reliability analyses? Explain.

REFERENCES

1. R. E. Barlow and F. Proschan, *Statistical Theory of Reliability and Life Testing: Probability Models*, Holt, Rinehart and Winston, Inc., 1975.

2. N. R. Mann, R. E. Schafer and N. D. Singpurwalla, *Methods for Statistical Analysis of Reliability and Life Data*, John Wiley and Sons, 1977.
3. J. Endrenyi, *Reliability in Electric Power Systems*, John Wiley & Sons Ltd., 1978.
4. W. H. Beyer, *Standard Mathematical Tables*, 26th Edition, CRC Press, Inc., 1981.
5. R. Ramakumar, *Engineering Reliability: Fundamentals and Applications*, Prentice-Hall, Inc., 1993.
6. T. Gönen, *Electric Power Distribution System Engineering*, McGraw Hill, 1986.
7. S. Asgarpoor and M. J. Mathine, "Reliability Evaluation of Distribution Systems with Non-Exponential Down Times," *IEEE/PES 1996 Winter Meeting,* Baltimore, MD, IEEE, 1996.
8. R. Billinton and R. N. Allan, *Reliability Evaluation of Engineering Systems: Concepts and Techniques*, 2nd Edition, Plenum Press, 1992.
9. R. E. Brown and J. R. Ochoa, "Distribution System Reliability: Default Data and Model Validation," *IEEE Transactions on Power Systems*, Vol. 13, No. 2, May 1998, pp. 704-709.
10. W. F. Horton, S. Goldberg, and R. A. Hartwell, "A Cost/Benefit Analysis in Feeder Reliability Studies," *IEEE Transactions on Power Delivery*, Vol. 4, No. 1, Jan. 1989, pp. 446-452.
11. P. F. Albrecht and H. E. Campbell, "Reliability Analysis of Distribution Equipment Reliability Data," Presented to Edison Electric Institute Transmission and Distribution Committee, New Orleans, LA, Jan. 1972.
12. D. A. Duckett and C. M. McDonough, "A Guide for Transformer Replacement Based on Reliability and Economics," *42nd Annual Power Distribution Conference, Austin*, TX, Oct. 1989.
13. R. J. Butrym, "Distribution Reliability Considerations at Wisconsin Electric," *IEEE PES Winter Meeting*, New York, NY, January, 1978.
14. R. Brown, S. Gupta, S.S. Venkata, R.D. Christie, and R. Fletcher, 'Distribution System Reliability Assessment Using Hierarchical Markov Modeling, *IEEE Transactions on Power Delivery*, Vol. 11, No. 4, Oct., 1996, pp. 1929-1934.
15. R. Brown, S. Gupta, S.S. Venkata, R.D. Christie, and R. Fletcher, "Distribution System Reliability Assessment: Momentary Interruptions and Storms," *IEEE Transactions on Power Delivery,* Vol. 12, No. 4, Oct 1997, pp. 1569-1575.
16. *Electrical Distribution System Protection (3rd Edition)*, Cooper Power Systems, 1990.
17. G. Kjolle and K. Sand, "RELRAD - An Analytical Approach for Distribution System Reliability Assessment," *IEEE Transactions on Power Delivery*, Vol. 7, No. 2, April 1992, pp. 809-814.
18. P. Ghose, "Distribution System Reliability," *1978 IEEE PES Winter Power Meeting*, New York, NY, Jan. 1978.
19. R. N. Allen, R. Billinton, I. Sjarief, L. Goel, and K. S. So, "A Reliability Test System for Educational Purposes—Basic Distribution System Data and Results," *IEEE Transactions on Power Systems*, Vol. 6, No. 2, May 1991, pp. 813-820.
20. C. A. Volkman, S. Goldberg, and W. F. Horton, "A Probabilistic Approach to Distribution System Reliability Assessment," *Third International Conference on Probabilistic Methods Applied to Electric Power Systems,* London, UK, July 1991.
21. M. Dussart, "Statistics on MV Network Incidents and the Average Amount of Time of Power Cuts," *CIRED 13: International Conference on Electricity Distribution*, Brussels, 1995.
22. J. J. Burke, *Power Distribution Engineering: Fundamentals and Applications,* Marcel Dekker, Inc., 1994.
23. R. M. Bucci, R. V. Rebbapragada, A. J. McElroy, E. A. Chebli, and S. Driller, "Failure Prediction of Underground Distribution Feeder Cables," *IEEE Transactions on Power Delivery,* Vol. 9, No. 4, Oct. 1994, pp. 1943-1955.
24. H. L. Willis, *Power Distribution Planning Reference Book*, Marcel Dekker, Inc., 1997.
25. A. J. L. Janssen, W. Degen, C. R. Heising, H. Bruvik, E. Colombo, W. Lanz, P. Fletcher, and G. Sanchia, *A Summary of the Final Results and Conclusions of the Second International Enquiry of the Reliability of High Voltage Circuit Breakers,* Cigré Working Group 13.06, Report 13-202.

26. W. J. Lannes, III, "Cost-Effectiveness Analysis of Substation Arrangements," *IEEE Transactions on Reliability,* Vol. R-31, No. 4, Oct. 1982, pp. 404-409.
27. P. F. Albrecht, "Substation Reliability Evaluation," *Fourth Annual Reliability Engineering Conference for the Electric Power Industry,* New York, NY, June 1977.
28. P. Kulik, H. Röhsler, K. Reichert, and H-G Schütte, *Reliability in the Design of Substations Influence of the Location within the Grid and the Kind of Connections to the Grid,* Cigré Study Committee 23, Report 23-01.
29. M. S. Grover and R. Billinton, "A Computerized Approach to Substation and Switching Station Reliability Evaluation," *IEEE Transactions on Power Apparatus and Systems,* Vol. PAS-93, No. 5, Sept/Oct 1974, pp. 1488-1497.
30. C. R. Heising, R. J. Ringlee, and H. O. Simmons, "A Look at Substation Reliability," *Proceedings of the American Power Conference,* Vol. 30, 1968, pp. 1252-1261.
31. P. Save, "Substation Reliability—Practical Application and System Approach," *IEEE Transactions on Power Systems,* Vol. 10, No. 1, Feb. 1995, pp. 380-386.
32. D. Karlsson, H-E Olovsson, L. Walliin, and C-E Sölver, "Reliability and Life Cycle Cost Estimates of 400-kV Substation Layouts," *IEEE Transactions on Power Delivery,* Vol. 12, No. 4, Oct. 1997, pp. 1486-1492.
33. D. Kopejtkova, T. Molony, S. Kobayashi, and I. M. Welch, *A Twenty-Five-Year Review of Experience with SF_6 Gas Insulated Substations (GIS),* Cigré Working Group 23.10, Report 23-101.
34. C. N. Cheung, F. Heil, S. Kobayashi, D. Kopejtkova, T. Molony, P. O'Connell, B. Skyberg, J. P. Taillebois, and I. Welch, *Report on the Second International Survey on High Voltage Gas Insulated Substations (GIS) Service Experience,* Cigré Working Group 23.02, Feb. 2000.
35. *Forced Outage Performance of Transmission Equipment for the Period January 1, 1988 to December 31, 1992*, Oct. 1993.
36. D. O. Koval, "Transmission Equipment Reliability Data from Canadian Electrical Association," *IEEE Transactions on Industry Applications*, Vol. 32, No. 6, Nov./Dec. 1996.
37. *IEEE Recommended Practice for the Design of Reliable Industrial and Commercial Power Systems*, IEEE Std 493-1990.

5
System Modeling

It is invaluable to mathematically model the behavior of a system. For distribution systems, this is exemplified by the power flow, which computes steady-state voltages and currents on a power system and allows engineers to ensure that a proposed design will meet important criteria such as equipment loading, voltage drops, and system losses. A power flow model can also be used to calculate the impact of system expansions and system upgrades before they are actually constructed. In essence, a power flow predicts the electrical behavior of the modeled system. This is such a powerful concept that power flow models have become fundamental to the distribution planning process.

In the same manner that a power flow model can predict the steady state electrical properties of a distribution system, a reliability model can predict the outage and interruption characteristics of a distribution system. As reliability becomes more important to utilities and electricity consumers, predictive reliability models will become just as important as (or even more important than) power flow models. Examples of planning functions enabled by reliability models include:

- Design new systems to meet reliability targets
- Identify reliability problems on existing systems
- Test the effectiveness of reliability improvement projects
- Determine the reliability impact of system expansion
- Design systems that can offer different levels of reliability
- Design systems that are best suited for performance-based rates

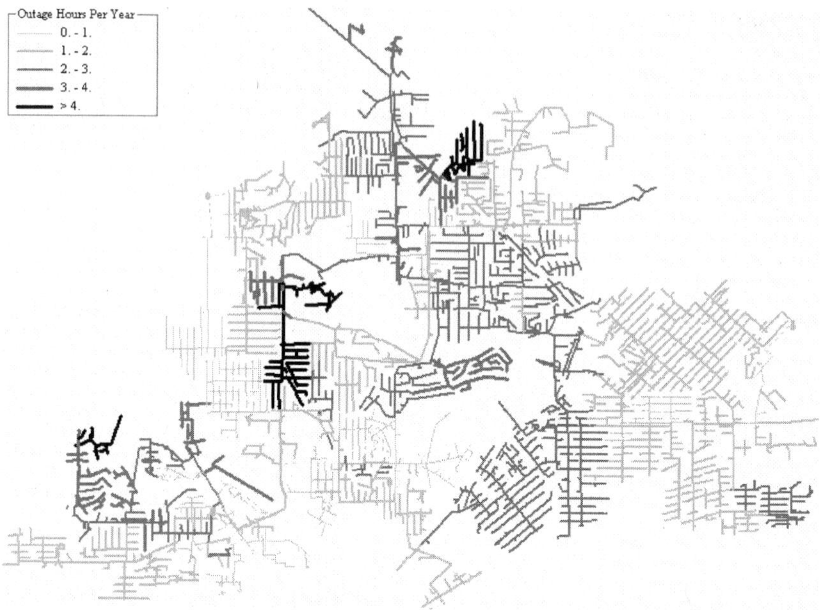

Figure 5.1. Results of a distribution system reliability assessment for a model containing approximately twenty five feeders, ten thousand components, twenty seven thousand customers, and two hundred fifty MVA of peak load. Shading is based on computed outage hours per year, with dark areas having more expected outage time than light areas.

This chapter addresses predictive reliability models for power distribution systems. These are models capable of predicting the reliability of each customer based on system topology, system operation, and component reliability behavior. At a minimum, reliability assessment results include the number of annual interruptions and the number of interruption minutes that each customer can expect in a typical year. More sophisticated models are able to predict behavior associated with momentary interruptions and voltage sags. Figure 5.1 shows the output of a modern predictive reliability assessment application.

Distribution system reliability assessment is a relative newcomer in the field of power system reliability. Despite this late start, it is quickly becoming one of the most important subjects in the industry. This chapter begins by introducing some fundamental concepts associated with system modeling. It continues by presenting basic analytical reliability modeling techniques and concludes by presenting basic Monte Carlo techniques.

5.1 SYSTEM EVENTS AND SYSTEM STATES

A system consists of a group of components that interact to achieve a specific purpose. For distribution systems, this purpose is to supply customers with reliable power within acceptable voltage tolerances. All first-world distribution systems are able to do this in normal conditions—when all components are operating properly, the system is configured in its usual manner, and loading levels are within design limits. This condition is referred to as the *normal operating state* of the distribution system.

> **Normal Operating State** – the state of the distribution system when switch positions are in their usual positions, no protection devices have tripped, all components are operating properly, and loading levels are within design limits.

The normal operating state is the preferred state of a distribution system. If a distribution system is in its normal operating state, all demand is being adequately supplied and reliability is perfect. The key to system reliability is to identify the events that will cause the system to leave this normal operating state and to quantify the reliability impact that these states have on customers.

Events can by divided into unscheduled events and scheduled events. Unscheduled events are brought about by chance occurrences called contingencies (e.g., a component fault or a breaker false tripping). Scheduled events are usually due to maintenance or system expansion.

> **Contingency** – a chance event (such as a fault) that causes the distribution system to leave its normal operating state.

> **Scheduled Event** – a planned activity (such as maintenance) that causes the distribution system to leave its normal operating state.

It is useful to view a system in terms of a state space. Since a state space is defined as a set of all possible system situations, the probability of the system being somewhere in the state space is unity. For a distribution system, the state space will consist of the normal operating state and all other states. As a simple example, consider a system that is either in its normal operating state, N, or in an outage state, A. If the probability of being in an outage state is P(A), then the probability of being in the normal operating state is 1-P(A) since the sum of all possible states must be equal to unity. This example can be extended to include the possibility of being in a maintenance state, B. If the probability of being in a maintenance state is P(B), then the probability of being in the normal operating state is 1-P(A)-P(B).

Figure 5.2 represents the previous example in a Venn diagram (a graphical representation of a state space in which events are drawn as areas within a state space). Notice that in the right diagram, Area A and Area B do not overlap. Since the two states do not overlap, they cannot simultaneously occur at the same time and are referred to as being mutually exclusive.

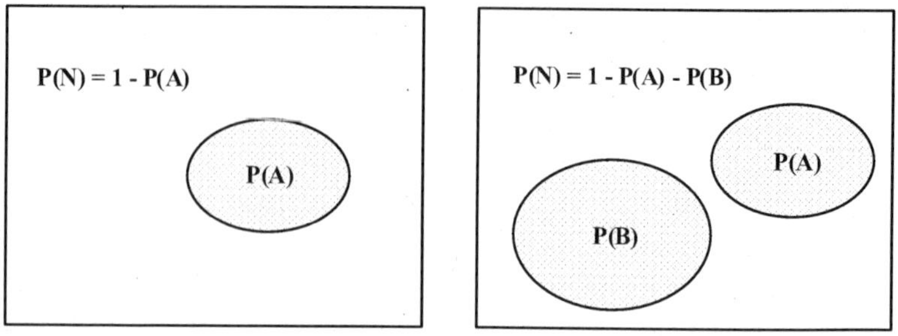

Figure 5.2. Two examples of Venn diagrams. The area inside the box of a Venn diagram reflects all possible outcomes and areas within the box represent specific outcomes. The left diagram represents a distribution system that has a probability of being in an outage state of P(A). The right diagram adds a probability of being in a maintenance state, P(B).

Mutually Exclusive – when states (or events) cannot occur simultaneously.

Mutual exclusiveness is a critical assumption made by many system models to simplify calculations and to reduce computing time. For some systems, simultaneous events are crucial to consider. In these situations, assuming that events are mutually exclusive will lead to inaccurate results. For example, a system may be designed to handle any outage without impacting customers, and may be able to maintain any component without impacting customers. It cannot, however, handle a fault that occurs when certain components are being maintained. To accurately compute reliability in this situation, a reliability model must account for maintenance events and fault events that occur at the same time.

Two or more events occurring simultaneously is referred to as an event intersection. The intersection of State A and State B can also be described by the logical expression of "State A and State B." The situation described by being in "State A and/or State B" is referred to as the union of the two states. Intersection and union are mathematically represented as follows:

Intersection of A and B:	$A \cap B$	(A and B)
Union of A and B:	$A \cup B$	(A and/or B)

Intersections and unions are easily visualized in Venn diagrams. Figure 5.3 shows the overlapping area of State A and State B. The entire shaded area is the union of the two states and the darker shaded area in the middle is the intersection of the two areas.

Computing the probability of being in the normal operating state is a bit more complicated if events are not mutually exclusive. Care must be taken so that intersection areas are not counted more than once. Consider Figure 5.3. The sum of P(A) and P(B) is greater than the union of the two areas. This excess is equal to the intersection of the two areas, and must be added back when computing P(N). This formula is shown in the figure.

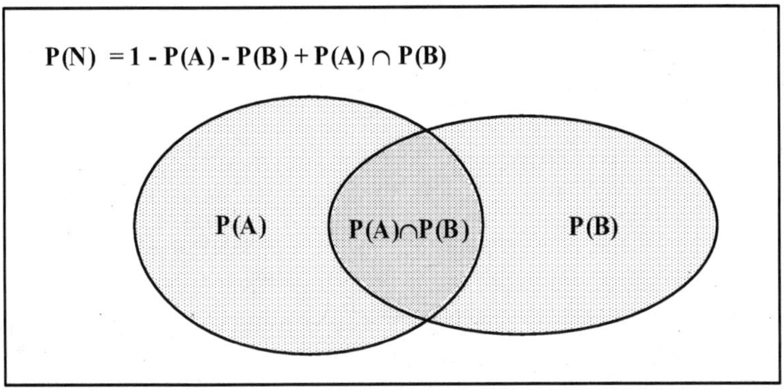

Figure 5.3. Event unions and event intersections. The total shaded area is the union of P(A) and P(B), P(A)∪P(B), and represents the probability of being in a maintenance state and/or being in an outage state. The heavily shaded area in the middle is the intersection of P(A) and P(B), P(A)∩P(B), and represents the probability of being in a maintenance state and being in an outage state.

5.2 EVENT INDEPENDENCE

Another important concept relating to system modeling is event independence. Two events are independent if the occurrence of one event does not affect the probability of the other event occurring:

> **Independent Events** – events in which the occurrence of one does not affect the probability of the other.

A good example of independent events is rolling two dice. The result of the second die is in no manner affected by the result of the first die. In contrast, system reliability is not independent of weather. If there happens to be a thunderstorm in the area, the chance of a fault occurring on overhead lines is greater than if there is not a thunderstorm in the area.

If two events are independent, computing the probability of their intersection is straightforward, and is simply equal to the product of the probability of the two events.

$$P(A \cap B) = P(A) \cdot P(B) \text{ ; true if A and B are independent} \tag{5.1}$$

Similarly, the probability of the intersection of N independent events is equal to the product of the probability of each event. If events are not independent, then the occurrence of one event will affect the probability of other events occurring. If the probability of one event depends upon the outcome of another event, it is said to have a conditional probability of occurrence.

Conditional Probability of Occurrence – the probability of an event occurring assuming that another event (or combination of events) has occurred.

A good example of conditional probability is a circuit breaker getting stuck during an attempted operation. If the circuit is not attempting to open or close, the probability of getting stuck is obviously zero. For a probability of operational failure, $P(F_{operational})$ to exist, the breaker must be attempting to operate, $B_{operating}$. Mathematically, this is stated as "the probability of operational failure given the breaker is operating" and is denoted $P(F_{operational} \mid B_{operating})$. The mathematics of conditional probabilities are beyond the scope of this book, but the reader should be aware that Equation 5.1 is not true if P(A) depends (i.e. is conditional) on the occurrence of B.

5.3 NETWORK MODELING

Network modeling translates a physical network into a reliability network based on series and parallel component connections. This method is simple and straightforward to implement, and is a good way to gain some familiarity and insight into system reliability. The mathematics involved in network modeling are primarily algebraic and there is usually a strong resemblance between the network model and the distribution system that it represents.

Network modeling is a component-based technique rather than a state-based technique. Each component is described by a probability of being available, P, and a probability of not being available, Q. Since components are assumed to be either available or not available, Q and P are arithmetic complements: $Q = 1 - P$. If a component is described by an annual failure rate (λ) and a Mean Time To Repair in hours (MTTR), The probability of being available can be computed as follows:

$$P = \frac{8760 - \lambda \cdot \text{MTTR}}{8760} \quad \text{; component availability} \qquad (5.2)$$

Network modeling is based on two fundamental types of component connections: series and parallel. Two components are in series if both components must be available for the connection to be available. Two components are in parallel if only one of the components needs to be available for the connection to be available. The schematic representations of series and parallel connections are shown in Figure 5.4. The left sketch shows two overhead feeder sections connected in series. If either of the sections fails, power to downstream customers is interrupted. The right sketch shows two transformers connected in parallel. If either transformer is de-energized, the downstream bus remains energized.

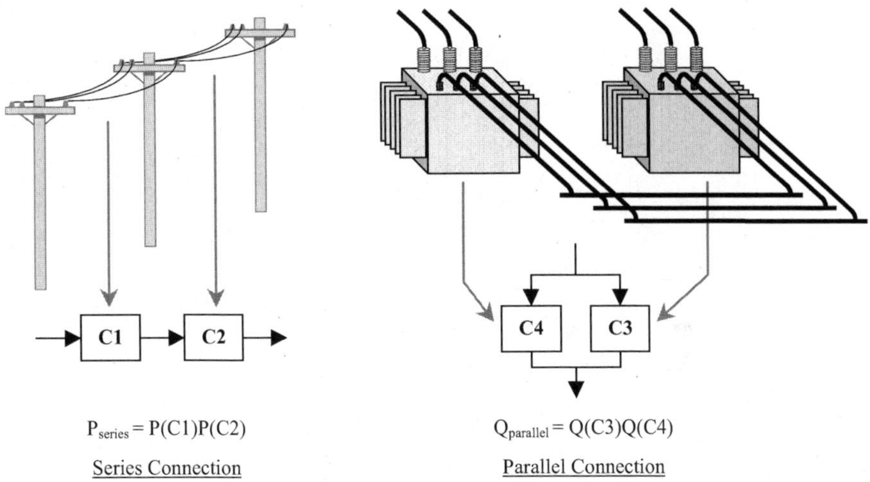

Figure 5.4. Examples of a series connection and a parallel connection. The series connection represents two overhead line sections; if either line section in unavailable, the system is unavailable. The parallel connection represents two transformers supplying power to a bus; if either transformer is available, the system is available.

Figure 5.4 shows two simple network models, each consisting of two components. The probability of a network model being available is equal to the probability of an available path existing from the entrance of the network to the exit of the network. For a series network, the probability of an available path is equal to the product of the individual component availabilities. For a parallel network, the probability of an unavailable path is equal to the product of the individual component unavailabilities.

$$P_{series} = \prod P_{component} \tag{5.3}$$

$$Q_{parallel} = \prod Q_{component} \tag{5.4}$$

Components in series reduce availability and components in parallel improve availability. Graphical examples are shown in Figure 5.5. The graph on the left represents a series network of components with individual availabilities from 0.999 to 0.97. As can be seen, high component availability does not necessarily translate into high system availability. If 20 components with availabilities of 0.99 are in series, system availability is about 0.8. If 100 components are in series, system availability drops below 0.4. The effect is opposite for parallel connections. Components with individual availabilities of 0.5 produce a system with an availability of 0.999 if ten are connected in parallel.

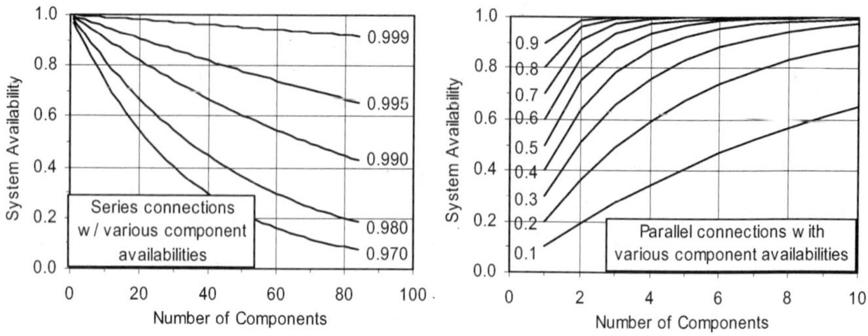

Figure 5.5. Availability impact of series connections and parallel connections. Connecting components in series make a system less reliable and connecting components in parallel makes a system more reliable.

Many systems consist of both series and parallel connections. In these instances, it is necessary to reduce the network in order to compute its overall availability. Network reduction is accomplished by repeatedly combining sets of parallel and series components into equivalent network components until a single component remains. The availability of the last component is equal to the availability of the original system. An example of network reduction is shown in Figure 5.6. In this system, a line section (component 1) provides power to a distribution transformer (component 2). The distribution transformer can also receive power from an alternate path consisting of three line sections in series (components 3, 4, and 5). Three steps of network reduction are required to compute the availability of the system.

An alternative to network reduction is the minimal cut set method. A cut set is a group of components that, when unavailable, causes the system to be unavailable. A minimal cut set will only cause the system to be unavailable if all of the cut set components are unavailable.

> **Minimal Cut Set** — a set of *n* components that cause the system to be unavailable when all *n* components are unavailable but will not cause the system to be unavailable if less than *n* components are unavailable.

There are three advantages to using cut sets. First, the technique is easily implementable on a digital computer. Second, the technique can handle "bridges" in a network that cannot be characterized by either a series connection or a parallel connection. Third, minimal cut sets can give insight into critical component dependencies.

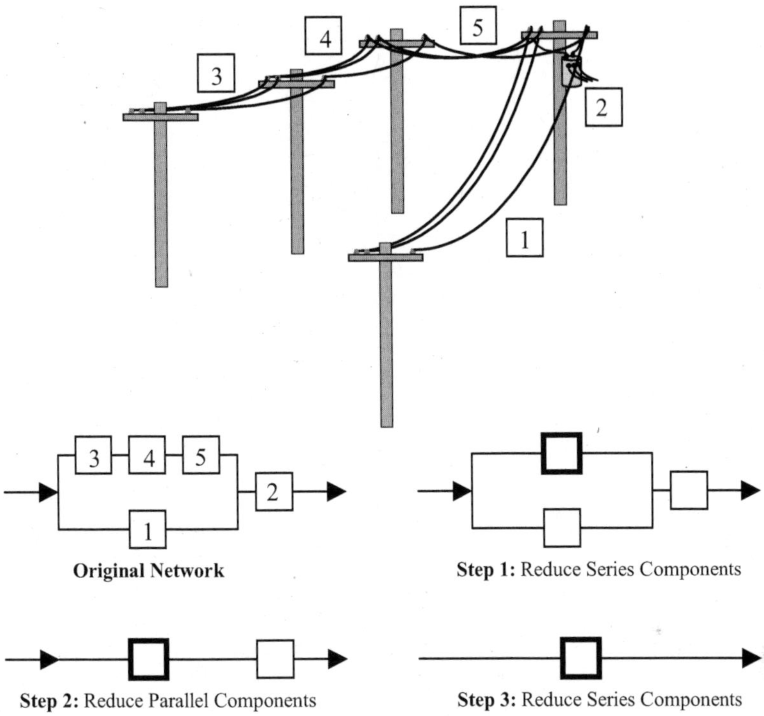

Figure 5.6. An example of network reduction. Sets of series and parallel components are combined into equivalent components until a single component remains. The availability of the last component is the availability of the network.

An example of minimal cut sets for a five component system is shown in Figure 5.7. This system represents a substation with two transformers (components 1 and 2) feeding dedicated buses that are connected by a tie breaker (component 3). Two dedicated feeders serve a critical load. These feeders are protected by breakers (components 4 and 5). There are four minimal cut sets for this system: (1) both transformers, (2) both feeder breakers, (3) components 1, 3, and 5, and (4) components 2, 3, and 4. The unavailability of the system is equal to the sum of the unavailabilities of these minimal cut sets.

Most of the time computer programs only approximate the unavailability of a network model. The primary reason is because algorithms do not typically identify all minimal cut sets, but only those consisting of n or less components. IEEE Std 493-1990 recommends that n be one greater than the lowest order minimal cut set.[1] Since most distribution systems will have minimal cut sets consisting of one component, the standard implies that $n = 2$ is appropriate (notice that $n = 2$ would not identify the third-order cut sets shown in Figure 5.7).

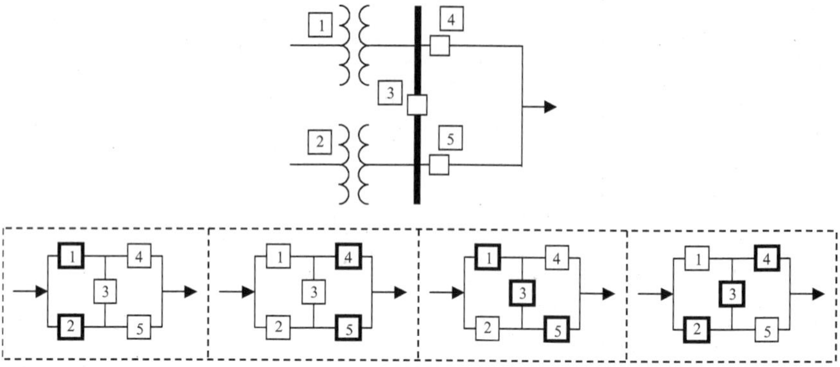

Figure 5.7. Minimal cut sets of a simple system. Minimal cut sets are groups of components that, when unavailable, result in the system being unavailable.

The network reliability models that have been considered up to this point have not been able to respond to system contingencies. In reality, distribution systems will respond to events in a complicated manner including recloser operations, fuse tripping, partial service restoration, and so forth. When the system is operating in such a manner, reliability measures other than availability (such as interruption frequency) become important. Network modeling can handle very simple switching models, but is generally not suitable for modeling complex distribution system behavior. For more detailed analyses, state-based methods such as Markov modeling, analytical simulation, and Monte Carlo simulation are needed.

5.4 MARKOV MODELING

Markov modeling is a powerful method based on system states and transition between these states. Though computationally intensive, Markov techniques are well suited to a variety of modeling problems and have been successfully applied to many areas related to reliability analysis.[2-5]

Markov models make two basic assumptions regarding system behavior. The first assumption is that the system is memoryless. This means that the future probability of events is solely a function of the existing state of the system and not what has occurred prior to the system entering the present state. The second assumption is that the system is stationary, which means that transition probabilities between states are constant and do not vary with time.

Markov models can be either discrete or continuous. Discrete models have state transitions that occur at specific time steps, while continuous models have constant state transitions. Although most reliability modeling applications utilize continuous Markov models, discrete models are easier to understand and will therefore be introduced first.

5.4.1 Discrete Markov Chain

A discrete Markov chain characterizes a system as a set of states and transitions between states that occur at discrete time intervals. Consider a simple system that consists of a single component and two possible states, an available state, A, and a failed state, F. If the component is available, it has a 10% chance of failing in the next time interval. If the component is failed, it has a 60% chance of being repaired in the next time interval. Such a system is graphically represented in Figure 5.8. Circles represent states, arrows represent state transitions, numbers next to arrows represent transition probabilities and numbers below circles represent the probability of being in each state.

The system begins in Stage 0 with a 100% probability of being State A. When transitioning to Stage 1, there is a 90% probability of remaining in State A and a 10% chance of transitioning to State F. The reader should note that the sum of all transitions out of any state must equal unity. Stage 2 can be reached from a transition from either State A or State F. As such, there are two possible paths to reach each state and the probability of being in each state is the sum of each path. For example, State A can be reached by the path A→A→A or by the path A→F→A. The first path will occur 81% of the time, the second path will occur 6% of the time, and the probability of being in State A in Stage 2 is 81% + 6% = 87%. Similarly, State F will be reached via path A→A→F with a 9% probability and via path A→F→F 4% of the time for a total probability of 13%. Notice that the probability of being in State A plus the probability of being in State B is equal to 100%.

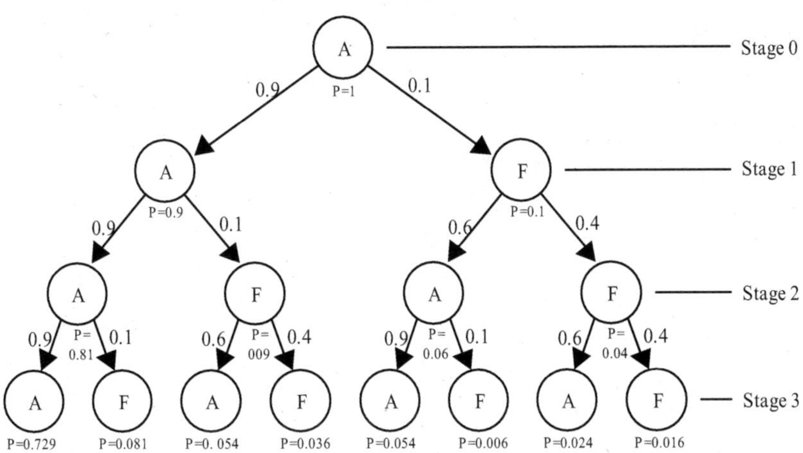

Figure 5.8. An example of a discrete Markov chain consisting of a single component with two possible states (available, A, and failed, F). The chain begins in Stage 0 with the component in an available state. When available, the component has a 90% probability of remaining available in the next stage. When unavailable, there is a 40% chance of remaining unavailable in the next stage. The result of the Markov chain is the probability of being in each state at each stage.

State probabilities in Stage n+1 can be computed based on a state transition matrix and the state probabilities in Stage n. Mathematically this is expressed as follows:

$$\begin{bmatrix} P_{n+1}(A) \\ P_{n+1}(F) \end{bmatrix} = \begin{bmatrix} 0.9 & 0.6 \\ 0.1 & 0.4 \end{bmatrix} \begin{bmatrix} P_n(A) \\ P_n(F) \end{bmatrix} \tag{5.5}$$

Typically, reliability models are not concerned with the stage-by-stage probabilities associated with each state. Rather, they are concerned with the steady state probabilities that occur when the stage number is large. If each state converges to a constant value, the system is referred to as ergodic and satisfies the following system of equations:

$$\begin{bmatrix} P(A) \\ P(F) \end{bmatrix} = \begin{bmatrix} 0.9 & 0.6 \\ 0.1 & 0.4 \end{bmatrix} \begin{bmatrix} P(A) \\ P(F) \end{bmatrix} \tag{5.6}$$

Equation 5.6 represents an underdetermined set of equations, since each equation reduces to the form $P(A) = 6 \cdot P(F)$. In fact, all Markov chains are underdetermined when expressed in this manner and require an additional equation to ensure a unique solution. This equation recognizes that the sum of the probabilities of being in each state must be equal to unity. Replacing one of the rows in Equation 5.6 with this identity creates a fully determined set of equations that can be solved by standard methods.

$$\begin{bmatrix} 1 \\ P(F) \end{bmatrix} = \begin{bmatrix} 1 & 1 \\ 0.1 & 0.4 \end{bmatrix} \begin{bmatrix} P(A) \\ P(F) \end{bmatrix} \tag{5.7}$$

For this example, State A converges to a probability of 85.7% and State F converges to a probability of 14.3%. The example is simple, but can easily be extended to distribution systems consisting of a large number of states and complex transition behaviors between these states.

5.4.2 Continuous Markov Process

In the field of reliability assessment, most references to Markov modeling refer to the continuous Markov process. Like a Markov chain, the Markov process is described by a set of states and transition characteristics between these states. Unlike a Markov chain, state transitions in a Markov process occur continuously rather than at discrete time intervals. Rather than state transition probabilities, Markov processes use state transition rates.

Markov processes are easily applied to distribution system reliability models since failure rates are equivalent to state transition rates. As long as equipment failures are assumed to be exponentially distributed, failure rates are constant and Markov models are applicable. Other values that are important include switching rate (σ) and repair rate (μ). Assuming exponential distributions, switching rate is equal to the reciprocal of mean time to switch and repair rate is equal to the reciprocal of mean time to repair.

$$\lambda \quad \text{; failure rate} \tag{5.8}$$

$$\sigma = \frac{1}{\text{MTTS}} \quad \text{; switching rate} \tag{5.9}$$

$$\mu = \frac{1}{\text{MTTR}} \quad \text{; repair rate} \tag{5.10}$$

Although constant failure rates can generally be considered a good assumption,[6] switching times and repair times are generally not exponentially distributed and, therefore, cannot accurately be described by constant state transition rates.[7-8] Regardless, research has shown that switching rates and repair rates can be modeled as constant values with only a small sacrifice in result accuracy.[9]

States in a Markov process are characterized by transitions into the state (represented as positive values) and transitions out of the state (represented as negative values). Consider the simple distribution system shown in Figure 5.9. This system consists of two substations (SS1, SS2), a breaker (B1), two line sections (L1, L2), a normally closed switch (SW1) and a normally open switch (SW2). The system is normally in State 0. If LS1 fails, B1 opens and the system transitions into State 1a. To restore customers, crews will then open SW1 and close SW2, transitioning the system into State 1b. When the line is repaired, the system transitions back to State 0. A similar sequence of events occurs for failures on LS2. The Markov model representing this system is also shown in Figure 5.9, with λ_1, μ_1, and σ_1 corresponding to LS1 and with λ_2, μ_2, and σ_2 corresponding to LS2.

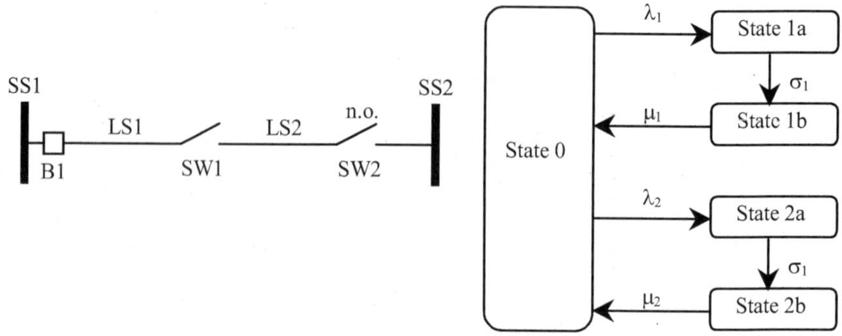

Figure 5.9. A simple distribution system and corresponding Markov model. The model accounts for failures on LS1 and LS2 including customer restoration and equipment repair. λ refers to failure rate, σ refers to switching rate, and μ refers to repair rate.

Markov processes are solved in a manner similar to Markov chains except that differential equations rather than difference equations are utilized. The transition rate into each state is equal to the sum of probability of transferring in from external states minus the probability of transitioning out from the considered state. The set of equations for the system shown in Figure 5.9 is:

$$\begin{bmatrix} dP_0/dt \\ dP_{1a}/dt \\ dP_{1b}/dt \\ dP_{2a}/dt \\ dP_{2b}/dt \end{bmatrix} = \begin{bmatrix} -(\lambda_1+\lambda_2) & 0 & \mu_1 & 0 & \mu_2 \\ \lambda_1 & -\sigma_1 & 0 & 0 & 0 \\ 0 & \sigma_1 & -\mu_1 & 0 & 0 \\ \lambda_2 & 0 & 0 & -\sigma_2 & 0 \\ 0 & 0 & 0 & \sigma_2 & -\mu_2 \end{bmatrix} \begin{bmatrix} P_0(t) \\ P_{1a}(t) \\ P_{1b}(t) \\ P_{2a}(t) \\ P_{2b}(t) \end{bmatrix} \tag{5.11}$$

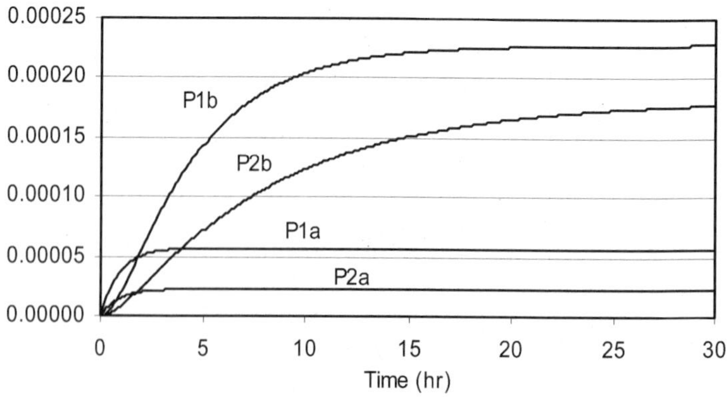

Figure 5.10. A time sequential simulation of the Markov model represented in Figure 5.9 and Equation 5.11. The probability of being in State 0 is initially assumed to be 1 (not shown), and the probabilities of the remaining states gradually converge to steady state values as time approaches infinity.

Equation 5.11 can be used to run a simulation. Similar to the Markov chain, the probability of initially being in State 0 is assumed to be 100% and the probability of future states is computed by considering linearized changes in probabilities associated with small time steps. A simulation of state probabilities for the following parameter values are shown in Figure 5.10. State probabilities asymptotically converge as time approaches infinity, indicating an ergodic system.

Parameter Values for Sample Markov Model

λ_1=0.5 /yr (0.0000571/hr)	λ_2=0.2 /yr (0.0000228/hr)
σ_1=1 /hr (MTTS = 1 hr)	σ_2=1 /hr (MTTS = 1 hr)
μ_1=0.25/hr (MTTR = 4 hr)	μ_2=0.125/hr (MTTR = 8 hr)

Steady state solutions for Markov processes are computed by setting all state transition derivatives (the left part of Equation 5.11) equal to zero. Since the set of equations is underdetermined, one of the rows must also be replaced with an equation indicating that the sum of all state probabilities must equal unity. For this example, the equation becomes:

$$\begin{bmatrix} 1 \\ 0 \\ 0 \\ 0 \\ 0 \end{bmatrix} = \begin{bmatrix} 1 & 1 & 1 & 1 & 1 \\ \lambda_1 & -\sigma_1 & 0 & 0 & 0 \\ 0 & \sigma_1 & -\mu_1 & 0 & 0 \\ \lambda_2 & 0 & 0 & -\sigma_2 & 0 \\ 0 & 0 & 0 & \sigma_2 & -\mu_2 \end{bmatrix} \begin{bmatrix} P_0(t) \\ P_{1a}(t) \\ P_{1b}(t) \\ P_{2a}(t) \\ P_{2b}(t) \end{bmatrix} \tag{5.12}$$

Equation 5.12 can be solved using standard techniques. For the parameters listed above, state probabilities converge to P_0 = 0.999511, P_{1a} = 0.000057, P_{1b} = 0.000228, P_{2a} = 0.000023 and P_{2b} = 0.000181. As expected, the system spends a majority of its time in its normal operating state (State 0).

Before a Markov model reliability assessment is complete, state probabilities must be converted into standard reliability measures. To do this, the reliability implications associated with each customer must be computed for each state (e.g., interrupted or not interrupted). Once complete, the unavailability of a customer's service can be computed by adding up the probabilities of all states associated with the customer being interrupted. Computing customer interruption frequencies is a bit more complicated, but can be achieved by examining the probability of transitioning between states where the customer is not interrupted to states where the customer is interrupted. For example, customers associated with LS1 are interrupted in State 1a, State 1b and State 2a, and have an unavailability of $P_{1a} + P_{1b} + P_{2a}$. Interruption frequencies are associated with a transition between State 0 and either State 1a or State 2a, and have a corresponding frequency of $P_0 \cdot (\lambda_1 + \lambda_2)$.

5.5 ANALYTICAL SIMULATION FOR RADIAL SYSTEMS

Analytical simulation techniques model system responses to contingencies, allowing the impact that the contingency has on each component to be calculated. Once calculated, the impact of the contingency is weighted by its probability of occurrence, resulting in the expected annual reliability impact of the contingency on each component. The expected annual reliability characteristics for each component are obtained by summing the individual contributions of each contingency. A summary of the analytical simulation process is:

Analytical Simulation

1. Select a contingency with probability of occurrence λ.
2. Simulate the system's response to the contingency and compute the impact on all components.
3. Weight the impact of the contingency by λ.
4. Have all contingencies been simulated? If not, select a new contingency and go to step 2.
5. End.

The result of an analytical simulation is the expected number of annual momentary interruptions, sustained interruptions and interruption hours for each component. It can also compute expected annual operational frequencies such as protection device operations and switching operations. Analytical simulation is generally the preferred method for assessing expected distribution system reliability if the system is in its normal state a large percentage of the time and if a large majority of contingencies are independent and mutually exclusive.[10] This technique is intuitive, is capable of modeling detailed physical and operational characteristics, and has the ability to reflect small changes in output for small changes in input.

The remainder of this section will discuss the details of implementing an annual simulation suitable for radial distribution networks. Since a majority of US distribution systems are operated radially, the techniques described in this section are appropriate for most US applications. The radial assumption simplifies implementation and reduces algorithm execution time, but is not, by definition, suitable for nonradial systems. Since many non-US distribution systems (and a small but growing number of US systems) are not radial, Section 5.6 will discuss the implementation of an analytical reliability simulation for nonradial distribution system topologies.

5.5.1 Radial Structure and Navigation

A radial distribution system is defined as a system where each component has a unique path to a source of energy. The direction of power flow in a radial system is, therefore, unambiguous and always flows away from the source. Some distribution systems are built with a strictly radial topology. Most distribution systems, however, are highly interconnected but are operated radially by selectively choosing normally open points.

A generic radial tree structure is shown in Figure 5.11. Each component, C, is characterized by its unique path to a source of power, P. All components on P are closer to the source of power and are said to be upstream of C. Components that receive power by a path of which P is a subset are said to be downstream of C. The component immediately upstream of C is referred to as the parent of C and the components immediately downstream of C are referred to as children of C. A summary of key definitions related to radial distribution systems is:

Radial — each component has a unique path to a source of power.

Radially Operated — the system is highly interconnected, but radial conditions are satisfied by the use of normally open switches.

Upstream — towards the source of power.

Downstream — away from the source of power.

Parent — the first upstream component.

Children — the set of first downstream components.

Siblings — components with the same parent.

Radial structures are often referred to as trees, with the source of power being the root of the tree, components being nodes of the tree, and component connections being branches of the tree. Since many generic algorithms utilize this terminology, familiarity with the following nomenclature is desirable:

Tree — a system consisting of arcs and nodes.

Node — a point on a tree.

Arc — a connection between two nodes.

Root — the beginning node on a tree.

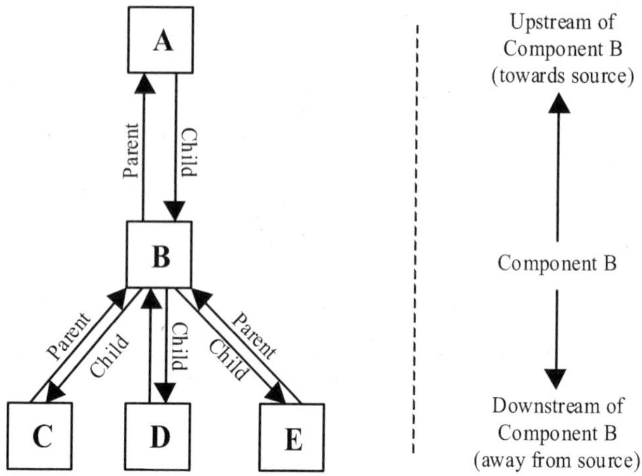

Figure 5.11. Generic radial tree structure. Each component has a single parent in the direction of the source of power (upstream). Each component also has a set of children in the direction away from the source of power (downstream).

The reader is cautioned to note that typical power system applications represent series components as arcs between two nodes. In this section, all components are represented by nodes. Arcs represent component interconnectivity and do not correspond to components on the system.

Tree navigation algorithms are fundamental to distribution system reliability modeling. The two primary navigation requirements for radial systems are upstream searches and downstream searches. Upstream searches start from a component and trace towards the source of power. They are useful for identifying sources of power, identifying protection devices, identifying fault isolation points, and performing a variety of other functions. Downstream searches start from a component and trace away from the source of power. They are useful for identifying affected customers, identifying switches for customer restoration, and a host of other tasks.

An upstream search starts from a component and traces to subsequent parents until stopping criteria are met or a source is encountered. This simple algorithm is summarized in the following steps:

Upstream Search

1. Clear all component flags
2. Start at component C
3. Set the flag of component C
4. If stopping criteria have been met, go to step 6.
5. If C has a parent, set C equal to its parent and go to step 3
6. End

Basically, the upstream search traces from parent to parent until a source is reached. At the termination of an upstream search, all components on the path to the source will be flagged, allowing further analysis and/or operations to be based on this component set. In addition, the last encountered component will either be the source of power or the component that satisfied the stopping criteria, allowing specific components to be identified. For example, the primary protection device of a component can be found by initialing an upstream search that terminates when the first protection device is encountered. Similarly, the nearest upstream switching point of a component (the switch used to isolate a fault occurring on this component) can be found by initiating an upstream search that terminates when the first switching device is encountered.

A downstream search starts at a component and traces subsequent sets of children until normally open points are encountered, no more children are found, or other stopping criteria are met. There are two basic algorithms to perform this type of search: breadth-first and depth-first. Generally, breadth-first searches utilize less memory but are computationally expensive, while depth-first searches require large amounts of memory but execute in a shorter amount of time.

A breadth-first search traces downstream components one layer at a time, first identifying components that are 1 arc downstream on the starting component, next identifying all components that are 2 arcs downstream on the starting component, and continuing until no further downstream components are encountered. The methodology is appropriate for finding components known to be near the starting component, but becomes computationally intensive for large systems (especially systems that have a small number of children per component). The breadth-first algorithm is summarized as follows:

Breadth-First Downstream Search

1. Set the depth of the starting component equal to one and the depth of all other components equal to zero.
2. Set depth of search: D = 1.
3. Examine component C. If the depth of C = D, set the depth of the children of C equal to D+1.
4. Repeat step 3 for all components.
5. If one or more component have a depth equal to D + 1, set D = D + 1 and go to step 3.
6. End.

A depth-first search follows subsequent children to the end of a radial path and recursively initiates new depth-first searches for all children that were ignored on previous iterations. A depth-first search is computationally efficient, but is memory intensive since each recursive call requires current function data to be stored in stack partitions. The depth-first algorithm, which produces the same results as a breadth-first algorithm, is summarized as follows:

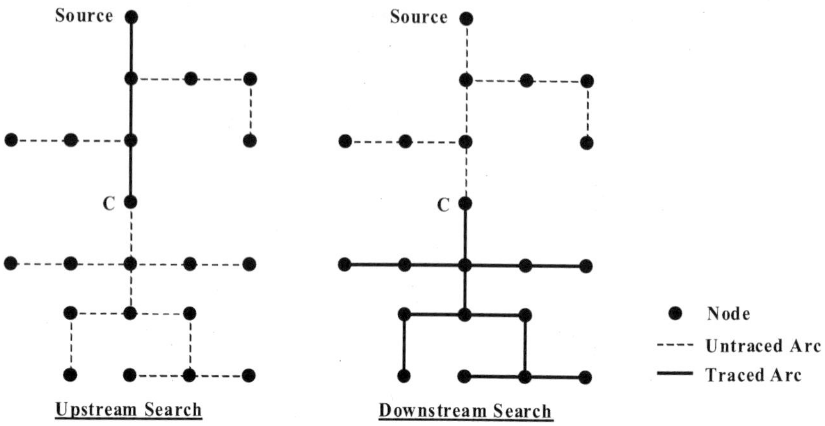

Figure 5.12. Results of an upstream and downstream search starting at component C. The upstream search traces arcs upstream towards the source and the downstream search traces arcs downstream away from the source. These two algorithms form the basis for many radial reliability assessment functions.

Depth-First Downstream Search

1. Clear all component flags.
2. Select a starting component, S.
3. Call function Depth_First_Search(S).
4. End.

F1. Function Depth_First_Search(P).
F2. Set the flag of parent P.
F3. For each child C of P, call Depth_First_Search(C).

The results of an upstream and downstream search applied to a sample tree structure are shown in Figure 5.12. Notice that the union of an upstream and downstream search does not span the entire tree because branches off of the upstream path do not get traced by either algorithm. Notice also that a downstream search will generally trace a greater number of arcs than will an upstream search.

Upstream and downstream searches are basic building block for many algorithms required for contingency simulations. The remainder of Section 5.5 will summarize the contingency simulation process and give detailed descriptions of how to implement each step of this process.

5.5.2 Contingency Simulation

A contingency occurring on a distribution system is followed by a complicated sequence of events. Because of this, each contingency may impact many different customers in many different ways. In general, the same fault will result in momentary interruptions for some customers and varying lengths of sustained interruptions for other customers depending on how the system is switched and how long the fault takes to repair. The key to an analytical simulation is to accurately model the sequence of events after a contingency to capture the different consequences for different customers. A generalized sequence of events for a fault contingency is:

Analytical Simulation: Sequence of Events After a Fault

1. **Contingency** — A fault occurs on the system.
2. **Reclosing** — A reclosing device opens in an attempt to allow the fault to clear. If the fault clears, the reclosing device closes and the system is restored to normal.
3. **Automatic Sectionalizing** — Automatic sectionalizers that see fault current attempt to isolate the fault by opening when the system is de-energized by a reclosing device.
4. **Lockout** — If the fault persists, time overcurrent protection clears the fault. Lockout could be the same device that performed the reclosing function, or could be a different device that is closer to the fault.
5. **Automated Switching** — Automated switches are used to quickly isolate the fault and restore power to as many customers as possible. This includes both upstream restoration and downstream restoration. In upstream restoration, a sectionalizing point upstream from the fault is opened. This allows the protection device to reset and restoration of all customers upstream of the sectionalizing point. In downstream restoration, other sections that remain de-energized are isolated from the fault by opening switches. Customers downstream from these points are restored through alternate paths by closing normally open tie switches.
6. **Manual Switching** — Manual switching restores power to customers that were not able to be restored by automated switching (certain customers will not be able to be restored by either automated or manual switching). As in automated switching, manual switching has both an upstream restoration component and a downstream restoration component.
7. **Repair** — The fault is repaired and the system is returned to its pre-fault state.

The seven steps outlined above generate a set of system states for each contingency. These states are characterized by switches and protection devices being open or closed, with corresponding customers being energized or interrupted. The following sections describe the process of contingency simulation in more detail. The first item discussed is how to model the protection system response (including reclosing) to faults. The second item discussed is how to model fault isolation and system reconfiguration. Section 5.5 concludes by discussing several contingency simulation enhancements including protection failures, switching failures, and various applications of distributed generation.

5.5.3 Protection System Response

The protection system response of a radial distribution system is straightforward since fault energy can be assumed to flow downstream from the source of power to the fault location. If the protection system is properly coordinated and operates correctly, the protection device nearest to the fault will operate before other upstream devices. Slight modifications of this rule occur when reclosing devices are utilized in an attempt to allow temporary faults to clear. The two basic reclosing schemes, discussed previously in Chapter 1, are referred to as fuse saving and fuse clearing. Fuse saving allows all temporary faults a chance to automatically clear and results in fewer sustained interruptions but more momentary interruptions. Fuse clearing allows lateral fuses to clear all downstream faults and results in fewer momentary interruptions but more sustained interruptions. Fuse saving and fuse clearing can be summarized as follows:

> **Reclosing with Fuse Saving** — After a fault occurs, the nearest upstream reclosing device opens without any intentional delay (initiated by an instantaneous relay). After a brief delay, the device recloses. If the fault persists, the device may operate several more times with increasing delays. If the fault still persists, the reclosing device will lock out. If a temporary fault occurs on a fused lateral with a reclosing device upstream of the fuse, the reclosing device will operate first and "save" the fuse. Fuse-saving schemes are also referred to as feeder selective relaying.

> **Reclosing (Fuse Clearing)** — In this scheme, reclosing devices have their instantaneous relays disabled. If a temporary fault occurs on a fused lateral with a reclosing device upstream of the fuse, the fuse will blow before the recloser's time overcurrent relay operates. Fuse-clearing schemes are also referred to as instantaneous relay blocking.

Reclosing schemes have a substantial impact on customer reliability.[11-12] As such, it is critical to model them accurately when simulating a contingency. Rarely does such a simple and inexpensive act (i.e., enabling or blocking an instantaneous relay) have such a large impact on system performance, with some customers experiencing improved reliability and other customers experiencing degraded reliability. A protection system response algorithm that accounts for both fuse-saving and fuse-clearing schemes is summarized as follows:

Protection System Response

1. Fault occurs on component F with frequency λ and repair time MTTR.
2. Is there a reclosing device, R, upstream of F? If no, go to step 7.
3. Is there a fuse upstream of F and downstream of R that is coordinated with R (fuse-clearing scheme)? If yes, go to step 7.
4. Increment reclosing operations of R by λ.
5. Increment momentary interruptions of all components downstream of R by λ.
6. Is the fault temporary in nature? If yes, end.
7. Find the nearest upstream protection device, P.
8. Has a reclosing device, R, operated? If no, go to step 10.
9. Decrement momentary interruptions of all components downstream of P by λ.
10. Increment sustained interruptions of all components downstream of P by λ and increment the protection operations of P by λ.
11. Increment sustained interruption duration of all components downstream of P by $\lambda \cdot$MTTR.

The protection system response initiates after a fault occurs. Assuming that a fuse-saving scheme is in place, the nearest reclosing device will trip open and cause a momentary interruption for all downstream customers. If the fault is permanent, reclosing will be ineffective and eventually allow the nearest upstream protection device to clear the fault. If this device is the same as the reclosing device, all of the previously recorded momentary interruptions are converted to sustained interruptions. If this protection device is downstream of the reclosing device, only those momentary interruptions downstream of the protection device are converted to sustained interruptions.

The reader should note that this protection system response algorithm will only record one interruption per customer per contingency. If a customer experiences a momentary interruption followed by a sustained interruption, only the sustained interruption is recorded. Also, since the algorithm cannot tell how many reclosing operations will be required, all temporary faults are assumed to clear on the first reclosing attempt (another way to interpret this is to view the algorithm as counting momentary events rather than momentary interruptions as defined in IEEE 1366[13]).

This algorithm assumes that all reclosing devices operate simultaneously on all phases (e.g., a single phase to ground fault on a three-phase system will cause all three phases of a recloser to trip open). If three single-phase reclosing units are used or if a three-phase unit has independently controlled phases, only the faulted phases will typically trip. This single-phase behavior can be incorporated into the algorithm if several additional pieces of information are available. First, the faulted phases associated with the contingency must be known since the reclosing units associated with these phases will trip while the other phases will not trip. Second, the phases associated with each customer must be known since only reclosing associated with these phases will cause interruptions. Last, information must be known about whether a single-phase lockout can occur or whether all three phases are forced to lock out (single-phase lockouts can result in safety problems for systems with delta connected primary transformer windings and can result in damage to three-phase motors).

5.5.4 System Reconfiguration

After a fault has been cleared, the system can be reconfigured to isolate the fault and restore power to certain customers. This reconfiguration is performed by sectionalizing devices. An obvious example is a switch, but other devices such as elbow disconnects, line reclosers, and fused cutouts can also be used as sectionalizing devices. System reconfiguration efforts, accomplished by opening and closing switches, can be categorized into upstream restoration and downstream restoration.

Upstream restoration restores service to as many customers as possible by using their normal electrical path to the tripped protection device. The normal path is restored by opening the first sectionalizing device upstream from the fault and resetting the tripped protection device. This simple algorithm, visualized in Figure 5.13, can be summarized in the following steps:

Upstream Restoration

1. Fault occurs on component F with frequency λ and repair time MTTR.
2. Identify the first upstream sectionalizing device, S, with switching time MTTS (this includes the time required to reset the tripped protection device).
3. Is MTTS > MTTR? If yes, end.
4. Is MTTS greater than the maximum duration momentary interruptions? If yes, go to step 7.
5. Increment momentary interruptions of all components upstream of S (including branches) by λ.
6. Decrement sustained interruptions of all components upstream of S (including branches) by λ.
7. Decrement sustained interruption duration of all components upstream of S (including branches) by $\lambda \cdot$(MTTR-MTTS).
8. End.

If extended repair times are expected, additional interrupted customers can be restored by downstream restoration. The purpose of downstream restoration is to restore as many customers as possible using alternate electrical paths. Downstream restoration is visualized in Figure 5.14 and can be described by the following algorithm:

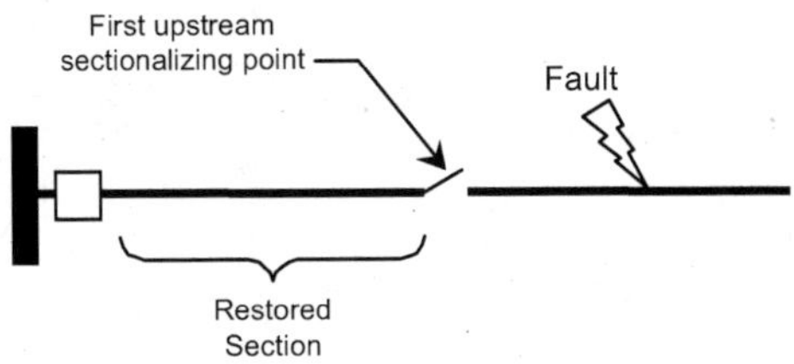

Figure 5.13. Single-stage upstream restoration. After the feeder breaker clears the fault, interrupted customers call the utility and crews are dispatched to locate the fault. Once the fault is found, the crew opens the nearest upstream switch. This allows the feeder breaker to be closed and all customers upstream of the switch to be restored.

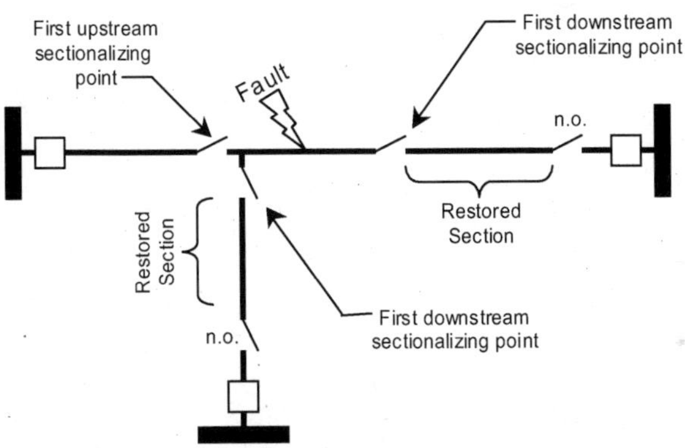

Figure 5.14. Single-stage downstream restoration. After the upstream sectionalizing point is opened, a downstream search for sectionalizing switches is performed. In this case, two sectionalizing locations are identified. When these switches are opened, downstream components are isolated from the fault location. This allows a normally open (n.o.) point to be closed, restoring service to customers.

Downstream Restoration

1. Fault occurs on component F with frequency λ and repair time MTTR.
2. Identify the first upstream sectionalizing device, S_{US}.
3. Identify a normally closed sectionalizing device, S_{NC}, which is outaged and downstream of S_{US}, with switching time $MTTS_{NC}$.
4. Identify a normally open sectionalizing device, S_{NO}, which is downstream of S_{NC} with switching time $MTTS_{NO}$.
5. Will opening S_{NC} and closing S_{NO} restore customers downstream of S_{NC} without exceeding the emergency ratings of equipment or violating voltage constraints? If no, go to step 13.
6. Set $MTTS = MTTS_{NC} + MTTS_{NO}$.
7. Is MTTS > MTTR? If yes, go to step 13.
8. Is MTTS greater than the maximum duration momentary interruptions? If yes, go to step 11.

9. Increment momentary interruptions of all components downstream of S_{NC} by λ.
10. Decrement sustained interruptions of all components downstream of S_{NC} by λ.
11. Decrement sustained interruption duration of all components downstream of S_{NC} by $\lambda \cdot$(MTTR-MTTS).
12. Go to step 14.
13. Are there any other normally open sectionalizing devices downstream of S_{NC}? If yes, go to step 4.
14. Are there any other normally closed sectionalizing devices downstream of S_{US} and still outaged? If yes, go to step 3.
15. End.

The algorithm just described assumes that downstream restoration time is equal to the sum of the switching times of the normally open switch and the normally closed switch (see Step 6). From this perspective, the switching time of the normally open switch can be viewed as the incremental time to perform downstream restoration when compared to upstream restoration. This is a simplification, but proves to be adequate for most analysis requirements. More complicated switching time calculations (e.g., making the incremental switching time a function of geographic distance) are easily implemented by modifying Step 6 to account for any desired modeling characteristic.

Downstream restoration is a more complicated algorithm than upstream restoration since multiple sets of switching combinations can be used to restore service to multiple sets of customers. In addition, switching actions must not be allowed to violate system operating constraints such as exceeding equipment emergency ratings or violating low voltage constraints. In fact, much more complicated algorithms have been developed in an attempt to restore as many customers as possible with as few switching actions as possible.[14-17] Though these algorithms may result in more restoration than the algorithm just detailed, they are operationally focused and are much more computationally intensive.

The upstream and downstream algorithms just described are referred to as single-stage restoration since customers are restored in a single switching sequence. Single-stage restoration can significantly reduce average interruption duration (as measured by SAIDI), and is the best restoration strategy if all sectionalizing devices are identical (true for both upstream and downstream restoration).

In general, distribution systems that contain sectionalizing devices with different switching times will result in situations where one restoration strategy will restore more customers while a competing strategy will restore fewer customers more quickly and/or more dependably. This is the case when a distribution system contains both manual and automated switches.

Two-stage restoration is a viable switching and restoration strategy when a feeder contains automated devices—devices that operate in a very short time, minutes or less, after a fault occurs. Device operation may be automatic, or may be performed by a dispatcher through a communications channel. Most distribution systems either have no automated devices or are partially automated with a combination of manual and automated devices.

When a fault occurs on a partially automated system, the general concepts of upstream restoration and downstream restoration still apply. There is not, however, a single-stage strategy that will result in restoration of the greatest number of customers in the least amount of time. This premise is illustrated in Figure 5.15 (automated switches are indicated by an attached circle). When a fault occurs at the location indicated, the breaker will clear the fault and interrupt all customers on the feeder. If the first upstream manual switch is opened, both Section A and Section B can be restored. However, since the switch is manual, restoration of these feeder sections may not occur for an hour or more. Alternatively, the first automated switch can be opened allowing Section A to be quickly restored (in a few minutes or less), but leaving Section B without power. Choosing the manual device restores more customers, and choosing the automated device restores fewer customers more quickly.

An obvious way to reduce SAIDI is to use both the automated and manual devices in a two-stage restoration strategy. In the first stage, the upstream automated device is opened and Section A is quickly restored. In the second stage, the upstream manual device is opened, the automated device is closed, and Section B is restored. In this scenario, Section A experiences a short interruption and Section B experiences a longer interruption, but both sections are restored before the fault is repaired.

The two-stage switching strategy also applies to downstream restoration. After the fault is cleared, automated switching can quickly restore some customers via alternate paths, and manual switching can restore additional customers at a later time. Figure 5.16 illustrates two-stage downstream restoration.

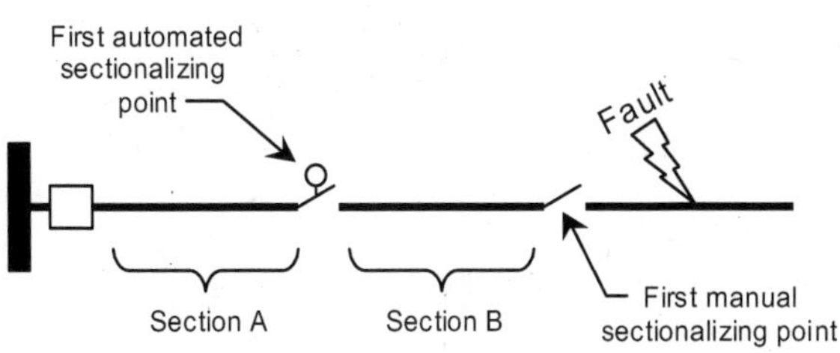

Figure 5.15. Two-stage upstream restoration. When the fault occurs, the breaker will clear the fault and interrupt all customers on the feeder. In Stage 1, the first automated switch is opened, allowing Section A to be quickly restored (in a few minutes or less), but leaving Section B without power. In Stage 2, the upstream manual device is opened, the automated device is closed, and Section B is restored. In this scenario, Section A experiences a short interruption and Section B experiences a longer interruption.

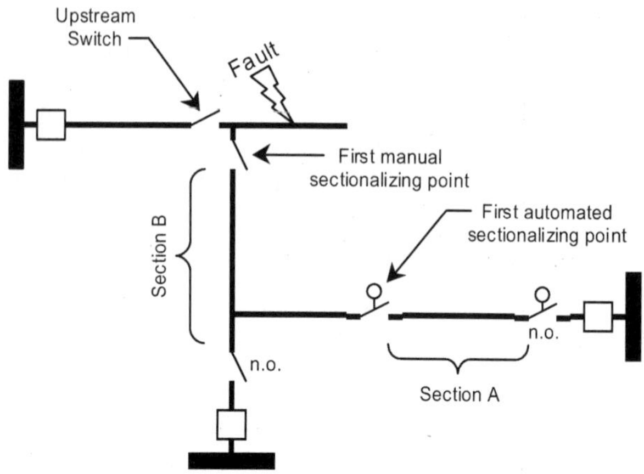

Figure 5.16. Two-stage downstream restoration. When the fault occurs, each downstream path is first searched for automated restoration strategies and then searched for manual restoration strategies. In this case, two-stage restoration allows Section A to be quickly restored in Stage 1, and Section B to be restored in Stage 2 after an additional delay.

Once the fault is cleared, each downstream path is searched for automated sectionalizing points. In Figure 5.16, an automated switch is encountered in the downstream path immediately prior to Section A. In addition, an automated normally open switch is located that can supply Section A with power through an alternate path. With these two criteria satisfied, Section A can be rapidly restored.

After the first stage of switching has restored power to certain customers, the second stage of downstream switching can begin. This is done by returning to the faulted section and searching for all downstream sectionalizing points. If any of these points are not automated, and a downstream open point capable of supplying power to the outaged section is found, additional customers can be restored by opening the closed switch and closing the open switch. In Figure 5.16, two-stage restoration allows Section A to be quickly restored, and Section B to be restored after an additional delay.

Implementing a two-stage switching algorithm is straightforward. First, the single-stage algorithm is applied in a modified form such that only automated switches are considered in the restoration process. Next, the single-stage algorithm is re-applied in another modified form that considers all switches, but only adjusts outage information for components that were not previously restored by the first stage.

The reliability benefits of two-stage restoration can be substantial. Although there is no difference between the two strategies for systems with no automation or full automation, two-stage restoration can reduce the SAIDI of typical systems with partial levels of automation by 8% to 10%.[18] Consequently, the full potential of partial feeder automation cannot be quantified by predictive reliability models that only model a single-stage restoration process.

5.5.5 Transfer Switches

Critical loads are often connected to more than one source through the use of a transfer switch. The load is normally connected to a primary feeder and is transferred to a secondary feeder if the primary feeder experiences an outage. Typically, transfer switches consist of two interlocked switches that prevent both switches from being closed at the same time. Transfer switches can generally be classified as manual, automatic, high-speed mechanical, and static.

Manual transfer switches require a human operator. As such, switching will generally require a minimum of several minutes to accomplish (possibly much more if there is nobody located at the site of the manual transfer switch). These devices are relatively inexpensive, but may not provide adequate reliability to loads and processes vulnerable to extended interruptions. A schematic of a mechanical transfer switch is shown in Figure 5.17.

Automatic transfer switches have the ability to immediately sense loss of voltage on the primary feeder and switch over to the secondary feeder in a matter of seconds. They are more expensive than manual transfer switches since they require transducers and actuators, but can substantially improve customer reliability since many types of electric equipment can ride through an interruption of this duration (e.g., motors with sufficient rotational inertia, electronics with sufficient capacitance). Further reliability improvements can be achieved through the use of high-speed mechanical transfer switches that are typically able to switch within six cycles.

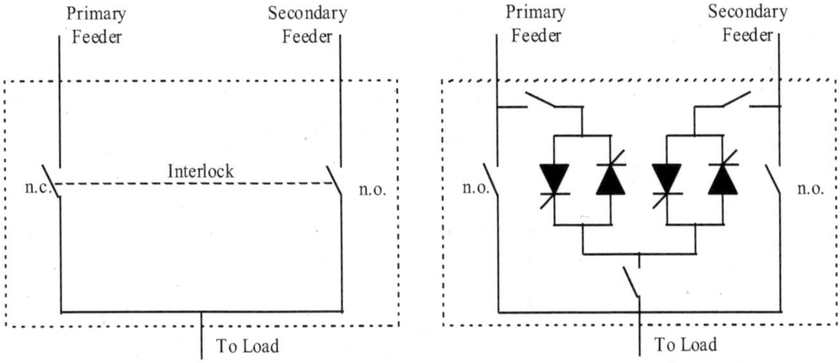

Figure 5.17. One-line diagram of a mechanical transfer switch (left) and a static transfer switch (right). The mechanical transfer switch has an interlock mechanism to ensure that only one switch is closed at any instant, preventing the sources from being operated in parallel. The static transfer switch, capable of switching in less than half of a cycle, is outfitted with mechanical switches to bypass power electronic problems should they occur.

Many types of sensitive electronic equipment are not able to withstand interruptions lasting more than a half cycle. Attaining this speed of operation with a transfer switch can only be attained through the use of power electronic switching equipment. A static transfer switch (STS) consists of two power electronic switches, each connected to a source. Under normal conditions, the power electronic switch connected to the primary feeder will be turned on (closed) and the power electronic switch connected to the secondary feeder will be turned off (open). If a disturbance occurs on the primary feeder (e.g., a loss of voltage or a sag in voltage), the two power electronic switches toggle their positions so that the load is fed from the secondary feeder, with switching typically occurring in less than a quarter of a cycle. This load transfer will occur if the secondary feeder is within specified voltage and phase angle tolerances. Bypass switches are typically included as part of the circuit so that the load can be fed during maintenance periods. STSs have been available for many years at 600 volts and below, and are becoming more common (though expensive) at voltage levels as high as 35 kV.[19] A one-line diagram of a static transfer switch can be seen in Figure 5.17.

The transfer configuration shown in Figure 5.17 is referred to as dual service topology. An alternative implementation is referred to as a bus tie topology. In this configuration, two busses are normally fed from separate sources, but are connected by a normally open power electronic switch. If one source experiences low voltage or a voltage dip, it can be disconnected by its primary power electronic switch and fed from the other source by closing the tie switch.

Modeling the reliability of transfer switches is a bit different from modeling the reliability of other switches since transfer switches are three terminal devices (two source terminals and a load terminal) whereas typical switches are two terminal devices (a source terminal and a load terminal for normally closed switches and two source terminals for normally open switches). Because of this difference, it is convenient to model the reliability impact of transfer switches implicitly in downstream interruption allocation algorithms rather than explicitly in system reconfiguration algorithms.[20]

When a protection device or a sectionalizing switch is operated, the reliability of all customers downstream of the device will be impacted. This is handled by a recursive search routine referred to as Adjust_Interruptions. A description of Adjust_Interruptions and its parameters is:

Adjust_Interruptions(C, ΔMAIFI, ΔSAIFI, ΔSAIDI)

In this subroutine, C is the component from which the depth-first recursive scan begins. ΔMAIFI, ΔSAIFI, and ΔSAIDI correspond to the amounts to increment or decrement the number of momentary interruptions, sustained interruptions and interruption duration, respectively. For example, when a recloser, R, operates with frequency λ_R, all customers downstream will experience a momentary interruption as reflected by Adjust_Interruptions(R, λ_R,0,0). If a protection device, P, downstream of the recloser then clears the fault with a frequency of λ_P, and a duration of d_P, the customers downstream of P can have the momentary interruption reclassified as a sustained interruption by the following function call: Adjust_Interruptions(P, $-\lambda_P$, λ_P, $d_P \cdot \lambda_P$).

When modeling a transfer switch in this framework, two options exist. The first is to model the transfer switch event explicitly as a new step and rebate customers using the Adjust_Interruptions function. This method, however, becomes complicated and computationally intensive due to the large number of possible system responses to any given contingency (due to the possibility of any number of protection device actions or switching actions failing for each contingency). The alternative is to model the device implicitly within the Adjust_Interruptions function.

To model a transfer switch implicitly, the Adjust_Interruptions() function checks for a transfer switch connection when performing its downstream recursive scan. If the primary feeder connection of a transfer switch is encountered, the following steps are taken:

Implicit Transfer Switch Model Embedded in Adjust_Interruptions

1. Start when S is a transfer switch within the recursive routine Adjust_Interruptions(S, ΔMAIFI, ΔSAIFI, ΔSAIDI), with S having a switching time MTTS.
2. Can the secondary feeder of S serve the load downstream of S without violating system constraints. If no, end.
3. Is MTTS > ΔSAIDI? If yes, end.
4. Is MTTS = 0? If yes, terminate recursive search.
5. Is ΔMAIFI = 0 and MTTS less than the momentary interruption threshold? If yes, call Adjust_Interruptions(S, ΔSAIFI,0,MTTS). End.
6. Call Adjust_Interruptions(S, ΔMAIFI,ΔSAIFI, MTTS). End.

Step 4 models a static transfer switch by preventing all interruptions from occurring downstream (in this case, a static transfer switch is modeled as a transfer switch with a switching time of zero. Step 5 models an automatic transfer switch by reclassifying sustained interruptions as momentary interruptions and by reducing outage time to the switching time. Step 6 models a manual transfer switch by simply reducing outage time to the switching time.

5.5.6 Distributed Generation

In the past, most distribution feeders were energized from a single source and were operated in a radial configuration. At any point on the feeder, power was guaranteed to flow away from the substation, during both normal operation and fault conditions. Under the 1935 legislation of PUHCA (the Public Utilities

Holding Company Act), utilities were granted the exclusive right to generate commercial electricity in their franchise area and could prevent customers from injecting power into the distribution system.

In 1978, the Public Utilities Regulatory Policy Act (PURPA) allowed qualified facilities to generate and sell electricity, which the utility was obligated to purchase at its avoided cost. These small and scattered generators, referred to as distributed generation (DG), meant that associated distribution feeders had multiple sources and it was now possible for power flow to flow upstream from the DG location towards the substation.

Qualified facilities typically utilize large generating units capable of producing 1 MW or more. These devices are complex, expensive, and closely coordinated with utility protection systems and operational protocols to ensure system reliability and safety. The next wave of DG will consist of many small units rather than a small number of large units.[21] Microturbines are already able to generate energy at competitive rates in sizes as small as 20 kW (see Figure 5.18). Other DG technologies such as fuel cells and photovoltaics are coming down in price and will become increasingly popular where noise levels and strict emission requirements are a concern.

The widespread proliferation of DG has many implications for distribution systems.[22-25] Since voltage regulation control is usually based on radial power flow assumptions, DG can lead to both overvoltage and undervoltage conditions. Turning a DG unit on or off can cause noticeable voltage flicker, and changes in DG output can cause voltage regulators to hunt. DG units can have a substantial impact on short circuit characteristics, causing equipment ratings to be exceeded and protection schemes to become miscoordinated. In addition, both single-phase and three-phase DG units can exacerbate problems associated with unbalanced loads and impedances. Positive possibilities include reactive power compensation to achieve voltage control and reduce losses, regulation and load power consumption tracking to support frequency regulation, spinning reserve to support generation outages, and dynamic braking to improve network stability.

Figure 5.18.. A 100-kW microturbine and its internal components. This particular microturbine is a combined heating and power unit (CHP) that utilizes high-temperature exhaust gases to heat buildings. The microturbine has only one moving part, its rotor (bottom right).

If DG units are coordinated properly, they can have a positive impact on distribution system reliability. A simple example is backup generation, which starts up and serves customers when a utility supply interruption occurs. Online DG units can also reduce equipment loading and enable load transfers from adjacent feeders experiencing outages.[26-27] The remainder of this section focuses on radial reliability models capable of quantifying these reliability impacts. Although there are many DG technologies that utilize renewable energy (e.g., photovoltaic, wind, small hydro), this paper will focus on technologies that utilize a continuous fuel source (e.g., reciprocating engines, microturbines, fuel cells). In addition to being dispatchable, these technologies are much less expensive and will dominate the DG market for many years.

Today, the most common DG application by far is backup generation, with an estimated 14 GW of reciprocating generators being sold each year. The DG unit remains offline during normal operation and is started up to serve critical loads after a utility service interruption occurs. Just as an actual backup generator is connected to the distribution system through a transfer switch, DG backup generators can be easily modeled with the use of a transfer switch (Figure 5.19, left). In this model, the DG unit can be treated as a voltage source since it will never be operated in parallel with the utility source. After an interruption occurs, the transfer switch shifts the downstream load from the utility to the DG unit. Since the DG unit does not start instantaneously, loads will still experience a short interruption. This can be modeled by setting the switching time of the transfer switch equal to the starting time of the DG unit.

For critical and sensitive loads, backup generators can be combined with batteries and inverters to ensure uninterruptible power. After an interruption occurs, loads are instantly transferred to the batteries and inverter. The batteries are sized to serve the critical loads until the generator can accelerate to full speed. To model this application, reliability assessment tools must be able to model power electronic transfer switches capable of switching critical loads from the utility source to the DG source without registering an interruption (Figure 5.19, right).

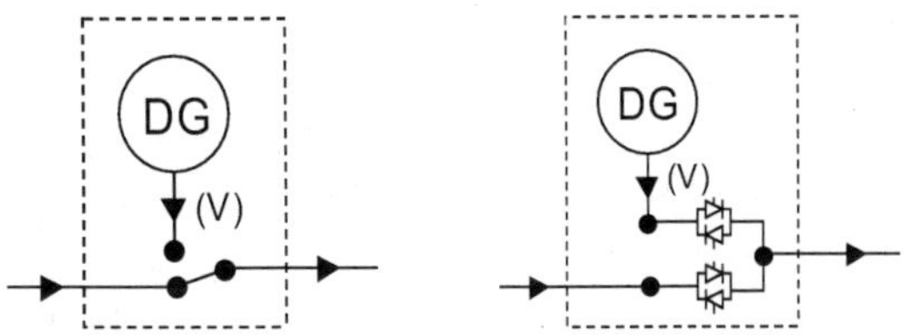

Figure 5.19. Distributed generation models for backup applications. The left model connects the secondary terminal of a mechanical transfer switch to a distributed generator modeled as a voltage source. After a primary source failure, power will be restored after a switching delay. The right model represents an uninterruptible power supply by connecting the distributed generator to the secondary terminal of a static transfer switch.

The alternative to backup applications is to have a distributed generator operating in parallel with the utility source during normal conditions. This can take the form of peak shaving, where distributed generation output is always less than its local load, or net metering, where distributed generation output can exceed its local load.

A peak shaving strategy starts up distributed generators during time periods of high energy demand and/or high energy prices to serve part of the on-site load. In addition to reducing customer energy bills, peak shaving can improve system reliability by reducing overall feeder loading.

A net metering strategy runs distributed generators on a more continual basis and allows on-site generation to exceed site demand, resulting in power being fed back into the distribution system. Energy that is fed back into the system is metered and a customer's energy bill will be based on the difference between energy taken from the distribution system and energy supplied to the distribution system.

Net metering impacts distribution reliability because it changes the power flow characteristics of distribution feeders.[28] Consider a 10-mile feeder serving 10 MW of uniformly distributed load. The full 10 MW will flow from the distribution substation and will gradually decrease until the end of the feeder is reached (Figure 5.20, top left). If a 4-MW unit is placed at the midpoint of this feeder, power metered at the beginning of the feeder is 6 MW rather than the total load of the feeder (Figure 5.20, top right). This can be deceptive since load transfers are often based on metered data at the substation, but may be constrained by more heavily loaded sections downstream of the DG unit. Further, distributed generation can mask load growth and cause load forecasting and planning difficulties.[29-31] If feeder load growth is not recognized due to the presence of DG, and is allowed to grow too large, loss of a DG unit during peak loading can result in equipment overloads and customer interruptions.

If the output of a DG unit is more than the downstream feeder load, power will flow from the DG location towards the substation (Figure 5.20, bottom). Somewhere along this path there will be a point where zero current flows. The opportunity to improve reliability is higher as the zero point becomes closer to the substation, but the probability of operational and protection coordination difficulties increases as well. Having a zero point upstream of the substation is generally unacceptable since it will result in reverse power flow through the substation transformers. If power flow from a DG unit towards the substation is large enough, equipment near the DG unit may experience higher loading than if no DG were present. Consider the example in Figure 5.20, where there is normally 2 MW of load 8 miles away from the substation. If an 8-MW DG unit is placed at this point, 6 MW of load will flow towards the substation—three times the normal loading. If equipment is not sized properly, equipment can become overloaded and cause reliability problems.

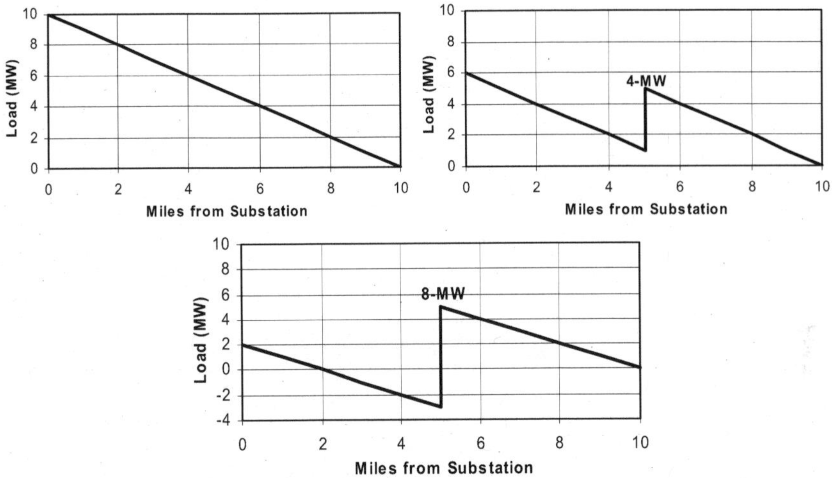

Figure 5.20. Loading profile of a 10-MW feeder without distributed generation (top left), with 4-MW of distributed generation at the midpoint (top right), and with 8-MW of distributed generation at the midpoint (bottom).

For a radial analysis, by definition, distributed generation cannot be modeled explicitly as a source in parallel with the utility source. To do so would create a nonradial system incapable of being analyzed by radial algorithms. The simplest work-around is to model DG as a negative load that injects real and reactive power into the system independent of system voltage (Figure 5.21, left). As long as the DG unit is connected to a utility source, it will deliver power to the system in either peak shaving or net metering mode (depending upon whether the DG output is greater than the load it is associated with). From a utility perspective, modeling DG as a negative load is reasonable since utilities will probably require DG units to disconnect from the system in the absence of a utility source to ensure safety, allowing faults to clear and avoiding the problems associated with islanded operation.[32]

Modeling DG as a negative load can have a positive impact on reliability if the reliability assessment model considers capacity constraints during post-fault system reconfiguration. Consider the left one-line diagram of Figure 5.21. When a fault occurs downstream of Source A, the breaker will open and interrupt all customers on the feeder. Opening the normally closed downstream switch will potentially allow customers in the switchable section to be restored by closing the normally open tie point. For this transfer to be allowed, power flowing into the normally closed switch must be supplied from a transfer path starting at the normally open switch and ending at Source B. If any component in this path cannot handle the additional load without exceeding its ratings, the load transfer is blocked and reliability is reduced. If DG is located in the switchable section, load transfer requirements may be reduced, allowing previously blocked load transfers to occur and improving overall system reliability.

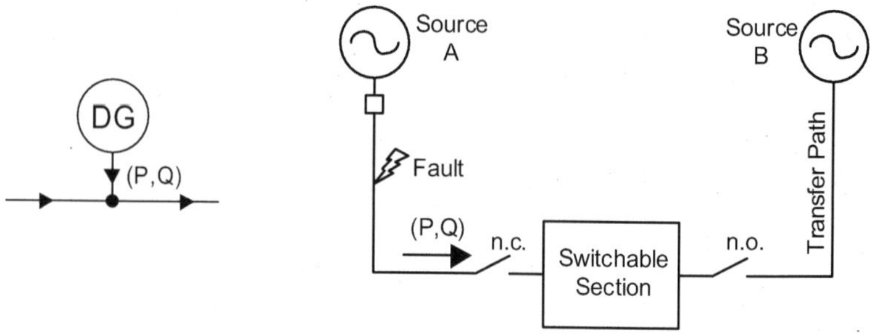

Figure 5.21.. The left diagram shows how distributed generation can be modeled as a negative load. This type of model can improve reliability by unloading feeders and enabling load transfers. This will occur when the load in a switchable section cannot be served via a transfer path without overloading equipment, but can be served if equipment loading is reduced sufficiently by distributed generation.

In many cases, customers will want to operate DG in a net metering mode during normal conditions, but will want to disconnect from the utility and operate as an island during utility service interruptions. To analyze this behavior with radial algorithms, a composite model consisting of a negative load and a voltage source is used. During normal operation, the negative load reduces overall feeder loading and improves system reliability. During an interruption, a transfer switch disconnects the customer load from the negative load and connects it to an alternate source. Similar to backup models, this transfer can be modeled with a delay (Figure 5.22, left) or without a delay (Figure 5.22, right).

When using composite models consisting of both a negative load and a voltage source, care must be taken to ensure that DG units are not "double counted." Double counting will occur if the reliability assessment algorithm reruns a complete power flow after the transfer switch has toggled to the voltage source. If this computational sequence occurs, the negative load will artificially reduce demand and may allow other load transfers that would otherwise be capacity constrained.

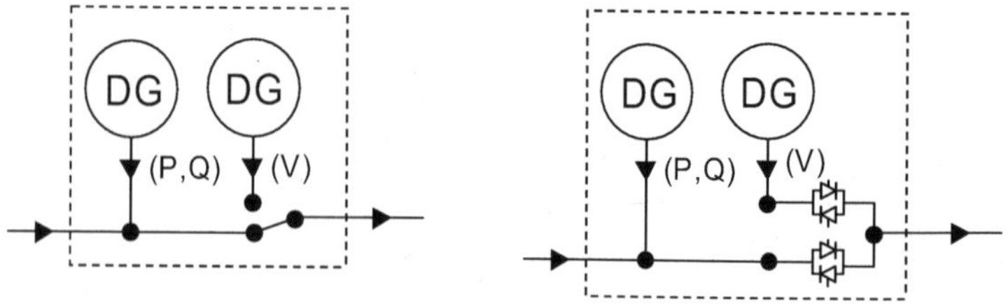

Figure 5.22.. Composite distributed generation models that allow online and backup operation without violating radial power flow constraints. During normal operation, DG is modeled as a negative load. After a utility source outage, a transfer switch connects the load to a DG unit modeled as a voltage source.

The author has examined the impact of DG on reliability by applying it to a heavily loaded feeder on a model based on an actual distribution system.[33] In this case, DG resulted in improved the reliability of both the feeder with connected DG and the other interconnected feeders without DG. Since the feeder with DG is able to supply part of its own load, it can more effectively transfer customers to adjacent feeders after a fault occurs. Similarly, reduced feeder loading allows surrounding feeders to more effectively transfer load to the feeder with DG. For this particular system, DG resulted in a SAIDI improvement of 22% on the feeder with DG and SAIDI improvements ranging from 5% to 14% on interconnected feeders without DG (peak loading assumed). This analysis assumes constant loading at peak levels.

Of more interest is the impact of DG on system with varying loading levels. The test system used to examine this issue is shown in Figure 5.23. It consists of two looped feeders, each with three switchable sections that are two miles in length. Each switchable section also has a fused lateral also two miles in length. Failures are assumed to be 0.1 per mile per year and repair times are assumed to be four hours. Switches are assumed to be fully automated and able to consider pre-fault loading levels so that load transfers are prevented if it will result in emergency loading levels being exceeded.

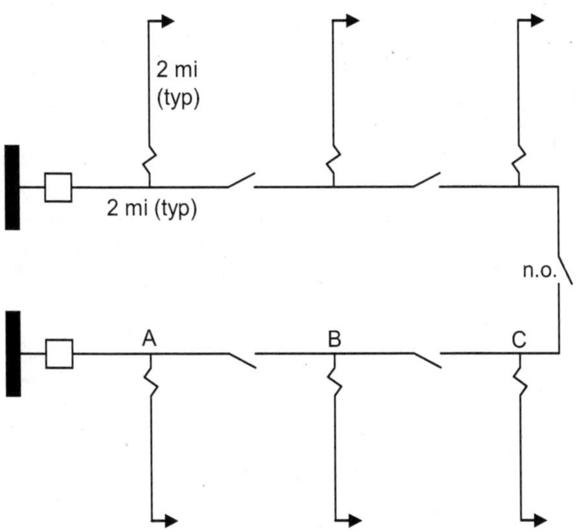

Figure 5.23. Test system to examine the impact of DG on feeder reliability. Each feeder is identical and DG penetration is assumed to be evenly divided among all load locations. Switches are assumed to be fully automated and able to consider pre-fault loading levels so that load transfers are prevented if it will result in emergency loading levels being exceeded.

When lightly loaded, each customer (the arrows) will experience a sustained interruption for faults on the associated lateral and for faults on the associated switchable section. For all other faults, cither no interruption will occur or automated switching will quickly restore power. In this lightly-loaded case, each customer can expect 1.6 hours of interruption duration per year.

As the system becomes more heavily loaded, there comes a point when Sections B and C cannot be restored after a fault on Section A. For even more heavily loaded systems, Section C cannot be restored for a fault on Section B. These constraints occur when the required load transfer will result in the connected feeder exceeding its emergency rating (in all cases, all loads are treated as equal and both feeders are assumed to have the same emergency rating).

To explore the reliability impacts of DG on heavily loaded distribution feeders, the peak loading of a feeder is measured as the feeder peak load divided by the feeder emergency rating. For example, if the emergency rating is 600 amps and the peak load is 450 amps, the feeder peak load is, by definition, 450 ÷ 600 = 75%.

Of course, feeders are rarely loaded to peak values. The percentage of time that a feeder is loaded at or below a particular value is called a load duration curve. A set of actual feeder load duration curves for a large US utility has previously been provided in Figure 1.24. This includes feeder load duration curves for residential, central business district (CBD), and commercial and industrial (C&I).

The load duration curve of a feeder can potentially impact reliability, especially if the peak load of the feeder is high. If all feeders are always lightly loaded, post-fault load transfers are unconstrained and reliability is highest. If all feeders are loaded near their emergency rating all the time, post-fault load transfers are not possible and reliability is lowest.

This phenomenon is investigated by computing reliability at each loading level, and weighting the result by the probability of the feeder being at this loading level. This calculation is then repeated for each feeder ranging from light peak loading to heavy peak loading. The result is a relationship of reliability to the ratio of peak load divided by emergency feeder rating. Reliability is measured by feeder SAIDI (System Average Interruption Duration Index), which is the total customer interruption hours on a feeder divided by the number of customers on the feeder. These reliability versus feeder loading curves are shown in Figure 5.24.

As can be seen, the reliability of residential feeders is not highly sensitive to peak loading (at least in terms of load transfer capability). This is because the feeder is only heavily loaded for a few hours per year; DG does not significantly impact the reliability of residential feeders due to load transfer constraint reduction, even when heavily loaded.

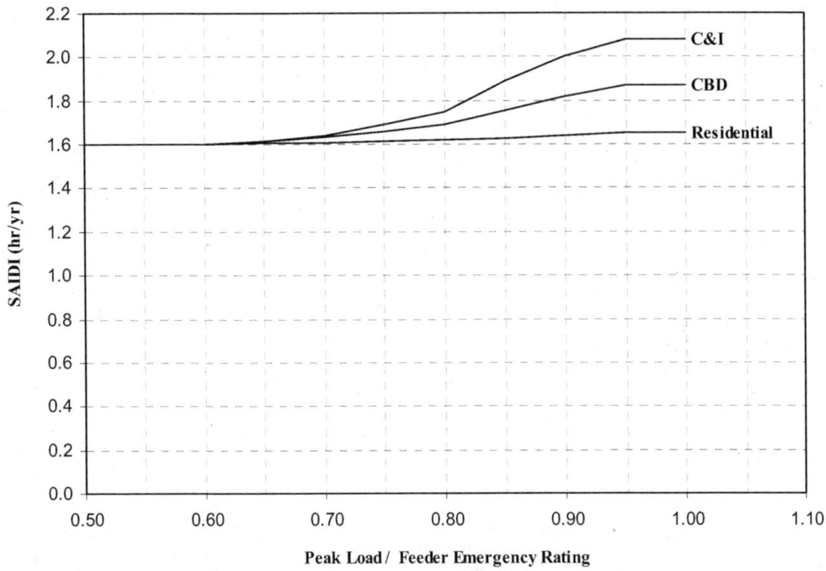

Figure 5.24.. Reliability versus feeder loading for residential, central business district (CBD), and commercial and industrial (C&I) feeders. The reliability of heavily loaded C&I feeders is aided by the presence of DG since they are loaded near peak levels a large percentage of the time. In contrast, the reliability of residential feeders is not highly sensitive to peak loading since they are only heavily loaded for a few hours per year.

The situation is different for C&I feeders. In the case of the test system, feeders loaded to 60% of emergency ratings at peak have a SAIDI of 1.6 hours, while feeders loaded near their emergency capacity at peak have a SAIDI of 2.1 hours. This represents a more than 30% increase in SAIDI due to heavy loading. The percentage of increase will be higher for the case of feeders with a low amount of lateral exposure.

The impact of DG on C&I feeders can be estimated by reducing the peak load of the feeder by the amount of DG. Consider the C&I test system and assume that both feeders have a peak load equal to 90% of emergency ratings without any DG. Figure 6 indicates that this system has an expected SAIDI of 2.0 hours. Now consider the same system with DG amounting to 15% of the emergency rating. Figure 5.24 indicates that the reliability impact of the DG is to reduce SAIDI to 1.7 hours, which correspond to a 15% SAIDI reduction. This reliability improvement occurs because each feeder is able to more effectively transfer loads to the other feeder after a fault occurs during heavy loading conditions.

There are no typical feeders, and generalizations about feeder reliability characteristics must always be heavily qualified. However, based on the test system and reliability assumptions, the following rules of thumb can be observed.

For feeders systems that are loaded near their emergency ratings at peak, DG will improve SAIDI by about 1% per 1% penetration in commercial and industrial areas, by about 1% for every 2% of penetration in central business districts, and will not materially improve reliability in residential areas. These reliability improvements are due to the reduction of post-fault load restoration, hold true for about 15% DG penetration, and assume that all DG is available during peak loading conditions. In specific cases, the expected reliability improvements due to DG should be computed by modeling actual feeder systems in predictive reliability assessment software.

Most existing US distribution systems were designed to be operated radially. A small amount of distributed generation will not substantially alter system behavior in areas such as protection, control, voltage regulation, and reliability. Increasing amounts of distributed generation, however, will eventually cause the distribution system to behave in ways not consistent with radial assumptions. This amount of distributed generation, referred to as the penetration factor, is defined as the aggregate DG kVA rating on a feeder divided by the peak feeder kVA load. Radial algorithms can easily accommodate systems with penetration factors up to 15%, and the IEEE interconnection standard defines low penetration as less than 30% and high penetration as greater than 30%.[34]

5.5.7 Operational Failures

The analytical simulation sequence of events becomes more complicated if operational failures are considered. Operational failures occur when a device is supposed to operate, but fails to do so. The probability of such an event, as previously discussed in Chapter 4, is termed probability of operational failure, POF. Operational failures cause the simulation sequence to split. One path assumes that the device fails to operate and has a weight of POF, the other path assumes that the device operates and has a weight of 1 – POF. This path splitting is illustrated in Figure 5.25 by considering a fuse that is supposed to clear a fault.

The result of simulation path splitting is an enumerative consideration of all possible system responses to each contingency (in the context of operational failures). Enumerative consideration is important since some states may be rare, but have a major impact on the system when they do occur. During restoration, path splitting associated with the enumerative consideration of possible outcomes is important when intended switching fails and customers that would otherwise have been restored are not restored.

Operational failures can be modeled by extending previously presented algorithms. For example, primary protection system failure can be modeled by simulating both a primary protection sequence of events and a secondary protection sequence of events. The algorithm steps are as follows:

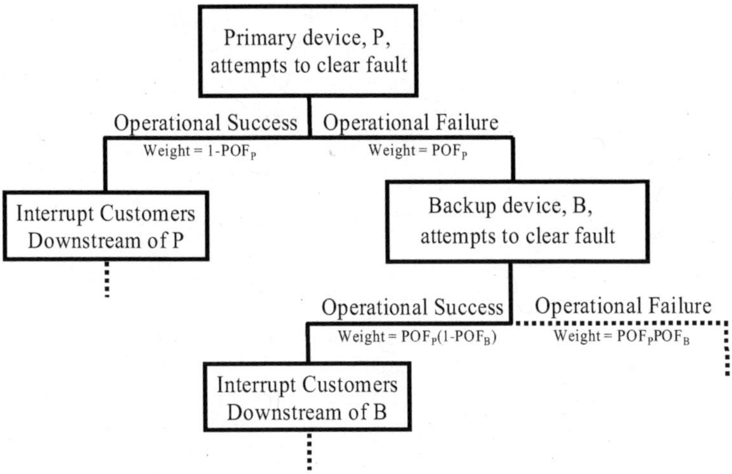

Figure 5.25. An example of simulation path splitting due to operational failures. In this case, a primary protection device can either clear a fault or fail to clear a fault. If successful, all customers downstream of the device are interrupted. If unsuccessful, a backup device will attempt to clear the fault. The result of each path is weighted by its probability of occurrence.

Protection System Response with Backup

1. Fault occurs on component F with frequency λ and repair time MTTR.
2. Find the nearest upstream protection device, P, with a probability of operational failure POF.
3. Find the backup protection device, B, which is the nearest upstream protection device from P.
4. Increment sustained interruptions of all components downstream of P by $\lambda\cdot(1\text{-POF})$.
5. Increment sustained interruption duration of all components downstream of P by $\lambda\cdot\text{MTTR}\cdot(1\text{-POF})$.
6. Increment sustained interruptions of all components downstream of B by $\lambda\cdot\text{POF}$.
7. Increment sustained interruption duration of all components downstream of B by $\lambda\cdot\text{MTTR}\cdot\text{POF}$.
8. End.

The above algorithm assumes that backup protection is 100% successful. Backup and higher order protection failures can be modeled by recursively applying the algorithm to each operational failure path. For example, if primary protection has a POF of 10% and backup protection has a POF of 20%, primary protection will operate (100% - 10%) = 90% of the time, backup protection will operate 10% x (100% - 20%) = 8% of the time and tertiary protection will operate 10% x 20% = 2% of the time. Notice that the sum of the probabilities of all possible outcomes is equal to 100%.

Operational failure modeling becomes a bit more complicated when modeling reclosing devices. If a reclosing device is not backed-up by another reclosing device, a temporary failure will not be given a chance to clear itself and will result in a sustained outage. Similarly, a sustained failure will be immediately cleared without reclosing operations, potentially reducing the number of momentary interruptions seen by customers.

Operational failure of switches during restoration is modeled in a manner similar to protection devices, except that backup switching may or may not be considered. If backup switching is not modeled, the simulation simply has to weight the impact of restoration by its probability of success. If backup switching is modeled, its impact must be weighted by the probability of primary switching failure. Transfer switch failures are included in this category, but are typically modeled within the interruption allocation function (as previously discussed) rather than as a separate restoration algorithm.

5.6 ANALYTICAL SIMULATION FOR NETWORK SYSTEMS

The previous section described how to implement an analytical simulation for systems operated with a radial topology. Although applicable to most primary distribution systems in the US, many subtransmission, substation, and secondary systems may not operate according to a radial philosophy. Further, many primary distribution systems in Europe are operated in closed loop or more complicated network topologies. As such, analyses of these systems require algorithms that are not based on the radial assumption.

In this section, a network refers to a system that has one or more components with multiple paths to a source or sources of power. Hence, the direction of power flow is not known *a priori* and must be computed with network power flow algorithms such as Gauss iteration and Newton-Raphson. Implications of a network are perhaps even greater for reliability analysis since a component outage will not necessarily interrupt downstream customers (the concept of downstream is no longer relevant), multiple protection devices may be required to clear faults and reconfiguration strategies become combinatorially numerous and difficult to compare.

Although implementing an analytical simulation for network reliability assessment is more complicated than for radial reliability assessment, many of the radial concepts still apply. In general, the sequence of events following a fault is the same, except that the system must respond to flows of power that can come from more than one direction, which can be both an advantage and a disadvantage. The following sections do not attempt to develop a network reliability assessment from scratch. Rather, they build upon the material presented in the radial section and spend a majority of time discussing how to compensate for the nonradial assumptions.

5.6.1 Network Structure and Navigation

Networks consist of a set of components and connections between components. A network structure is fundamentally different from a radial structure since there is no built-in hierarchy among components (i.e., upstream and downstream relationships). Components in networks simply have a set of other components to which they are connected. Connectivity can be represented locally by assigning each component a list of connected components or can be represented globally through a binary connectivity matrix.

The connectivity of a network consisting of N components can be represented by connectivity matrix **A,** consisting of N rows and N columns. If component x is connected to component y, **A**(x,y) and **A**(y,x) are set to one. If they are not connected, these values are set to zero. All connectivity matrices have diagonal symmetry and are typically sparse, allowing the use of alternative data structures that greatly reduce memory requirements. A sample network structure and associated connectivity matrix are shown in Figure 5.26.

Network navigation algorithms are similar to downstream radial search algorithms in that they can be categorized into breadth-first and depth-first. The differences are (1) all search directions are allowed (as opposed to downstream), and (2) traces are only allowed to visit a component once (ensuring algorithm termination). The same advantages and disadvantages for breadth-first and depth-first approaches apply. Breadth-first searches generally require longer execution times but do not have large memory requirements. Depth-first searches generally have faster execution times but are memory intensive due to their recursive nature. Network algorithms for these two types of searches are:

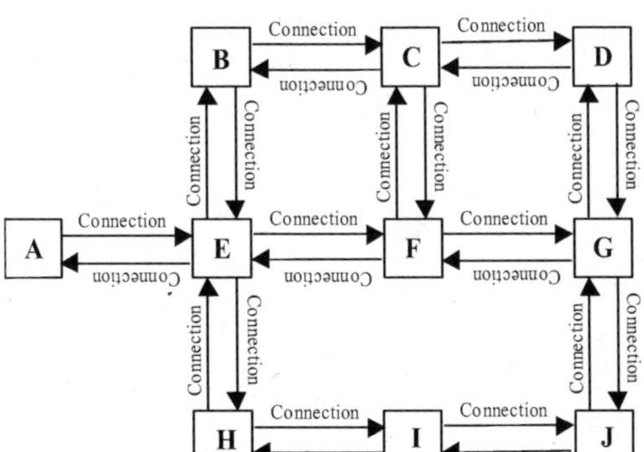

Network

	A	B	C	D	E	F	G	H	I	J
A	0	0	0	0	1	0	0	0	0	0
B	0	0	1	0	1	0	0	0	0	0
C	0	1	0	1	0	1	0	0	0	0
D	0	0	1	0	0	0	1	0	0	0
E	1	1	0	0	0	1	0	1	0	0
F	0	0	1	0	1	0	1	0	0	0
G	0	0	0	1	0	1	0	0	0	1
H	0	0	0	0	1	0	0	0	1	0
I	0	0	0	0	0	0	0	1	0	1
J	0	0	0	0	0	0	1	0	1	0

Connectivity Matrix

Figure 5.26. A sample network structure and its corresponding connectivity matrix. Because a network can have multiple routes between certain pairs of components, there is no unique path to a source of power and the direction of power flow is ambiguous. Notice that the connectivity matrix is diagonally symmetric with value (r,c) being equal to value (c,r).

Breadth-First Network Search

1. Set the depth of the starting component equal to one and the depth of all other components equal to zero.
2. Set depth of search: D = 1.
3. Examine component C. If the depth of C = D, set the depth of all connected components with a current depth of zero to D + 1.
4. Repeat step 3 for all components.
5. If one or more components have a depth equal to D + 1, set D = D + 1 and go to step 3.
6. End.

Depth-First Network Search

1. Clear all component flags.
2. Select a starting component, S.
3. Call function Depth_First_Search(S).
4. End.

F1. Function Depth_First_Search(P) .
F2. Set the flag of component P.
F3. For each connected component C of P that has not previously been flagged, call Depth_First_Search(C).

The above algorithms begin at a component and start tracing all connected components, with the breadth-first algorithm tracing all closer connections before exploring further connections and with the depth-first algorithm following a single path to its terminus before going back and tracing missed branches. Both algorithms are similar in that they leave a "bread crumb trail" to prevent any connection from being followed more than once. Without this memory of previously traced paths, each algorithm would cycle indefinitely about closed-loop paths.

5.6.2 Network Simulation Algorithms

The contingency simulation of a network reliability assessment is similar to the contingency simulation of a radial reliability assessment in that it consists of a protection system response, system reconfiguration, and repair. The differences occur due to the possibility of multiple paths to a source of power, resulting in (1) contingencies not necessarily resulting in customer interruptions, (2) contingencies possibly requiring multiple protection devices to operate, and (3) the possible use of multiple sources and/or paths to sources when reconfiguring the system to restore interrupted customers.

Because of the possibility of multiple paths to a source of power, it is not immediately clear whether a system topology change will result in customer interruptions. In a radial system, unidirectional power flow makes it clear that downstream customers are interrupted if a protection device or switch is opened. In a network system, an algorithm must be executed to identify the state of the system. A simple algorithm that identifies components not connected to a source of power is:

Identifying De-energized Components

1. Clear all component flags.
2. Identify a source of power, S.
3. Perform a network search, starting at S, flagging all traced components and stopping at normally open components.
4. Have traces been performed for all sources? If no, go to step 2.
5. End.

Because the above algorithm must be called many times, a network reliability assessment tends to require substantially more computation time than a radial reliability assessment. To further complicate matters, the above algorithm cannot identify violations of operating constraints such as equipment overloading and low voltage conditions. Full consideration of these constraints requires a network power flow to be integrated into the network reliability assessment—not a desirable option due to extreme computational load and the tendency of certain types of contingencies to cause algorithm convergence problems.

There is a reasonable compromise between no consideration of operational constraints and full consideration with a network power flow. This compromise if referred to as a "capacity flow" and allocates each load to one specific path to a source (typically using a breadth-first search). If a path cannot be found without violating equipment loading constraints, either the load must be shed or the system configuration disallowed. When implementing a capacity flow, care must be taken to include the directionality of flows through equipment so that power flowing in the same direction can be treated as additive and power flowing in opposite directions can be treated as subtractive.

When a fault occurs on a network system, protection devices must clear all possible paths to a source of power. A simple method of accomplishing this task is to perform a network search starting at the fault location that terminates at all encountered protection devices (these protection devices are opened). Although this strategy will correctly calculate equipment outages and customer interruptions, it will not correctly calculate protection device operations since protection devices will not operate unless fault current flows through them.

To determine whether an encountered protection device will operate, a network search must be performed (not allowing traces back towards the fault). The device operates if it is connected to a source and will not operate if it is not connected to a source (e.g., on a radial branch). Obviously, performing such searches further increases the computational burden of network assessments when compared to radial assessments—a characteristic that is exacerbated when operational failures of the protection system are considered. A basic algorithm to simulate network protection system response following a fault is:

Network Protection System Response

1. Clear Flag A and Flag B for all components.
2. Fault occurs on component F with frequency λ and repair time MTTR.
3. Perform a network search starting at F, setting Flag A of all traced components and stopping at protection devices and normally open points.
4. Perform network searches starting at each source, setting Flag B of all traced components.
5. Increment by λ the operations of all protection devices with both Flag A and Flag B set.
6. Increment by λ the sustained interruptions of all components without Flag B set.
7. Increment by $\lambda \cdot$MTTR the sustained interruption duration of all components without Flag B set.
8. End.

Reclosing is modeled by executing a similar algorithm prior to the protection system response described above. In the reclosing algorithm, only consider reclosing devices and increment momentary interruptions instead of sustained interruptions. Components that experience momentary interruptions must then be rebated if they are subsequently followed by a sustained interruption.

Post-fault system reconfiguration and customer restoration is much more complicated for network systems than for radial systems, primarily because there are many feasible solutions that will restore the same set of customers. These solutions can be ranked based on other criteria such as the number of required switching operations, losses, equipment loading and voltage regulation,[35-36] but such rankings do not generally add value to reliability assessment simulation algorithms.

A simple method of implementing post-fault reconfiguration is to follow a two-step approach consisting of fault isolation and load restoration. Fault isolation is achieved through the same basic algorithm as protection device operation. After a fault has been cleared, perform a breadth-first search and open all normally closed switches that are first encountered. For each such switch, identify all normally open switches and operated protection devices in directions other than towards the fault location. If there are operated protection devices, they can be closed to restore load. If only normally open switches are found, start to close them in order of shortest switching time until the load can be restored without overloading any piece of equipment. If load cannot be restored, the associated normally closed switch should not be operated. The amount of outage duration rebated to a customer is equal to the repair time of the fault minus the switching time required to restore that customers (weighted by the frequency of occurrence of the fault).

This general process of fault isolation and load restoration can be expanded to consider two-stage switching. Similar to the radial case, two-stage switching is implemented by simulating a first stage that only considers switches with a short switching time and subsequently simulating a second stage that considers all switches. Careful bookkeeping is needed to properly reclassify sustained outages as momentary outages when the switching time required to restore a customer is less than the momentary interruption threshold.

5.6.3 Higher Order Contingencies

Higher order contingencies refer to events that occur at the same time, typically faults that are not mutually exclusive. Radial reliability assessments do not typically consider higher order contingencies since the majority of customer interruptions will be due to single contingencies. This is not the case for many networks. In fact, some networks are specifically designed to withstand any single contingency without interrupting a single customer. If higher order contingencies are not considered, these systems will appear to have perfect reliability.

Higher order contingencies are best understood by first examining second order contingencies. Consider contingencies A and B, characterized by annual failure rates and repair times. The probability of being in contingency A, P(A), the probability of being in contingency B, P(B), and the probability of being in both contingencies simultaneously, $P(A \cap B)$, are computed as follows:

$$P(A) = (\lambda_A \cdot MTTR_A) / 8760 \quad (5.13)$$

$$P(B) = (\lambda_B \cdot MTTR_B) / 8760 \quad (5.14)$$

$$P(A \cap B) = P(A) \cdot P(B) \quad (5.15)$$

To perform a second-order reliability assessment each individual contingency must be simulated and each combination of two contingencies must be simulated. The total number of simulations for a system of N components is proportional to N^2. This relationship results in a quickly growing computational burden as system size increases. If a system increases in size by a factor of 1000, a second-order assessment will require nearly one million times the number of simulations. In comparison, a first-order assessment will only require 1000 times the number of simulations.

When performing higher order reliability assessments, the impact of each simulation must be weighted according to its probability of occurrence. Consider the second-order case. The weight associated with B occurring while an A contingency is in progress is equal to $\lambda_B \cdot P(A)$. Similarly, the weight associated with A occurring while a B contingency is in progress is equal to $\lambda_A \cdot P(B)$. To preclude simulating both cases, the contingencies can be assumed to happen simultaneously and the weight taken to be the average of these values. To be precise, this overlapping event must be subtracted from the probability of single contingencies occurring exclusively. If A and B are the only contingencies possible, the associated weights are summarized as:

Weights Associated with a Two Contingency Case	
Weight of A exclusively:	$\lambda_A \cdot [1 - P(A \cap B)]$
Weight of B exclusively:	$\lambda_B \cdot [1 - P(A \cap B)]$
Weight of A and B:	$[\lambda_A \cdot P(B) + \lambda_B \cdot P(A)]\ /2$

When more than two contingencies are considered, the weights of each single contingency must exclude the probability of all possible overlapping contingencies. Practically, such bookkeeping is prone to error and an approximation is typically adequate. This approximation involves simulating each single contingency and weighting its impact by its full failure rate. Next, higher order contingencies are simulated, but only considering the impact of customers not impacted by a lower order subset of contingencies. For example:

Weights Associated with a Two-Contingency Case (Approximation)	
Weight of A exclusively:	λ_A
Weight of B exclusively:	λ_B
Weight of A and B:	
- Customers impacted by A exclusively:	0
- Customers impacted by B exclusively:	0
- All other customers:	$[\lambda_A \cdot P(B) + \lambda_B \cdot P(A)]\ /2$

Another advantage of utilizing the approximate contingency weights relates to contingency ranking. In this process, higher order contingencies are ranked based on their probability of interrupting customers that would not be impacted by lower order contingencies. Higher order contingencies are simulated in order of their contingency ranking until a certain number are found to not interrupt additional customers. In many cases, only a small percentage of contingencies will need to be simulated, greatly reducing computation time without significantly impacting solution quality.

5.6.4 Common Mode Failures

Common mode failures refer to failures that are caused by the same physical root cause. They are similar to higher order contingencies in that they involve the simultaneous failure of multiple components. They differ from higher order contingencies (as described in Section 5.6.3) in that the simultaneous failure of multiple components is not purely coincidence.

Common mode failures come in two distinct varieties: deterministic and stochastic. Deterministic common mode failures are associated with events that necessarily result in the failure of multiple components. Examples include vault fires that cause faults on all circuits routed through the vault and pole failures that cause faults on all circuits associated with the pole. Stochastic common mode failures are associated with events that do not necessarily result in equipment failures, but increase the probability of failure such that overlapping events become likely. Examples include wind storms and lightning storms.

> **Deterministic Common Mode Failures** — Multiple failures that occur necessarily as a result of an external event, such as a vault catching fire or a multiple circuit tower collapsing.

> **Stochastic Common Mode Failures** — Failures associated with external events, such as severe weather, that increase component failure rates to a degree such that overlapping equipment outages become likely.

Deterministic common mode failures are easily incorporated into network reliability assessment algorithms by simply simulating the multiple contingency and weighting the impact of the contingency by the probability of the external event occurring. To accomplish this, a user interface must allow common mode failures to be entered and associated with a failure rate, a repair time, and the failure of a specific set of components.

Stochastic common mode failures are more problematic to incorporate into analytical models since, by definition, their behavior is stochastic rather than deterministic. For this reason, these types of failures are typically treated separately using Monte Carlo techniques. Stochastic common mode failures are treated in more detail in Section 5.7.5.

Another type of common mode event is referred to as a cascading failure. Cascading failures are characterized by the failure of one component leading directly to the failure of another component. This component, in turn, can lead to the failure of a third component, and so forth. An example is equipment overloading. If an overloaded piece of equipment in a networked system trips off-line, it can cause other pieces of equipment to become overloaded. When these components trip off-line, the situation becomes even worse and the cycle repeats.

> **Cascading Failures** — When the failure of one component directly leads to the failure of a second component, perhaps leading to the failure of a third component, and so forth.

Cascading failures can be modeled in several ways. The easiest is to assume that all failures occur in a short amount of time and with 100% certainty. With these assumptions, cascading failures can be modeled as deterministic common mode failures. If significant time delays occur between failures, the sequence of failures must be modeled explicitly in the analytical simulation. If each subsequent failure has less than a 100% chance of occurring, simulation tree splitting techniques such as those described in Section 5.5.7 must be used.

5.6.5 Distributed Generation

Distributed generation can be easily modeled in network reliability algorithms since there are no restrictions on locating sources of power. Typically, the only difference between a distributed generator and a generic source will be a restriction on the maximum power output. Care must also be taken to protect the distributed generator, a source of fault current, with a fuse or circuit breaker so that it will be isolated from faults after they occur.

The simplest distributed generation model is simply a source of power and voltage connected to the system by a protection device (Figure 5.27, left). When a fault occurs on the connected system, the unit is disconnected until the fault is isolated and the system can be reconfigured. A more sophisticated model places an additional protection device at the service entrance of a customer (Figure 5.27, middle). When a fault occurs on the system, the service entrance protection device opens and allows the distributed generator to deliver uninterruptible power to its associated load. A similar scheme places a breaker between the customer load and the distributed generator connection Figure 5.27, right), allowing the distributed generator to supply uninterrupted power to the feeder in the event of a customer-side fault.

The only difficulty in using distributed generation in network reliability models is the tendency of many algorithms to treat them as dispatchable utility sources and aggressively use them during system reconfiguration. This may not be the case if utility customers own the units. Using radial distributed generator configurations with transfer switches can help address this problem.

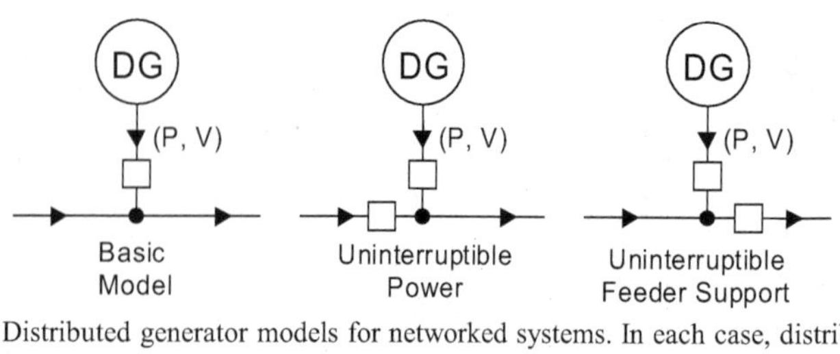

Figure 5.27. Distributed generator models for networked systems. In each case, distributed generators are modeled as a source of real power and voltage. Various protection schemes can isolate the unit for all feeder faults, provide uninterruptible power to customers in the event of a feeder-side fault, or provide uninterruptible power to feeders in the event of a customer-side fault.

5.7 MONTE CARLO SIMULATION

A Monte Carlo simulation utilizes random number generators to model stochastic event occurrences. As such, the results of two Monte Carlo simulations with identical inputs will generally not result in the same output. This variation is typically welcome because repeated simulations will eventually produce a distribution of results from which the mean, median, variance and other statistical measures can be computed.

When applied to distribution system reliability assessment, a Monte Carlo simulation typically analyzes system behavior for a specific period of time (such as a year). Because each simulation will produce different results, many simulations are typically needed. Theoretically, the expected value of the simulation is equal to the average of the results as the number of simulations, N, approaches infinity:

$$\overline{x} = \lim_{N \to \infty} \left(\frac{1}{N} \sum_{i=1}^{N} x_i \right) \tag{5.16}$$

$\overline{x}$ = expected value

x_i = result of simulation i

N = number of simulations

When utilizing a Monte Carlo approach, it is necessary to decide upon the number of simulations. If the desired result is the expected value, simulations can be performed until the mean of all results converges to a stable value. If the inclusion of rare events is required, the number of years should be large enough to give the rare event a high probability of occurring (e.g., if a component is expected to fail once per 100 years, 1000 simulations will have a high probability of simulating several failures while 50 simulations will not).

A Monte Carlo simulation has several advantages when compared to an analytical simulation. One, as previously mentioned, is the ability of a Monte Carlo simulation to produce a distribution of possible results rather than the expected value alone. Another is the ability to easily model component parameters as random variable characterized by probability distribution functions rather than as constant values. A Monte Carlo simulation can also more easily model complex system behavior such as nonexclusive events, cascading failures, conditional probabilities, and so forth.

A Monte Carlo simulation also has several disadvantages when compared to an analytical simulation. Perhaps the most significant is computational intensity. While an analytical simulation only needs to simulate a single expected year, a Monte Carlo simulation typically needs to simulate hundreds of random years. Another disadvantage is imprecision. Even with a large number of simulation years, multiple Monte Carlo simulations will still produce slightly different answers. This lack of precision hampers the ability to perform sensitivity analyses, compute gradients, and quantify the impact of small changes to large systems.

Monte Carlo simulations can be categorized into two basic varieties: sequential and nonsequential. Both techniques are useful for distribution reliability assessment and will be discussed in detail below. Before presenting these topics, a brief introduction to random number generation is presented.

5.7.1 Random Number Generation

Random number generation forms the basis for all Monte Carlo simulations. If the random number generator is not truly random, the results of the Monte Carlo simulation are not valid. To be valid, a random number generator must produce the promised distribution of random numbers. It must also avoid trends while generating this distribution. Consider a random number generator simulating a coin toss. After a large number of simulations, a valid random number generator will produce an equal number of "heads" and "tails." While doing this, it will avoid trends such as producing 1000 heads in sequence, 1000 tails in sequence, and so forth (See Figure 5.28).

Perhaps the most commonly used random number generator is the congruential generator, which produces a uniformly distributed random number with a value between zero and one. The congruential generator is the default algorithm used in most programming languages and spreadsheet algorithms. It is characterized by three parameters and computes each new number based on the previously generated number. An initial number, referred to as a seed, must be specified or automatically assigned (a common technique is to assign the seed based on the current time). Formulae associated with the congruential generator are:

$$x_{i+1} = [A \cdot x_i + C] \bmod B \tag{5.17}$$

$$r_i = \frac{x_i}{B} \quad ; \text{Uniform Random Number, } 0 \le r < 1 \tag{5.18}$$

A = nonnegative integer (called the multiplier)
B = nonnegative integer (called the modulus)
C = nonnegative integer (called the increment)
x_0 = no-negative integer (called the seed)

The modulo notation, A mod B, refers to the division remainder of the quotient of A and B, which can also be represented as A-B·int(A/B).

The random numbers generated by the above equations, in fact, can be shown to repeat with a pattern no greater in length than the modulus. If the modulus is smaller than the number of required random numbers, repetition of the pattern is guaranteed. Implementers and users of Monte Carlo simulations should take care that this repetition does not compromise the integrity of results.

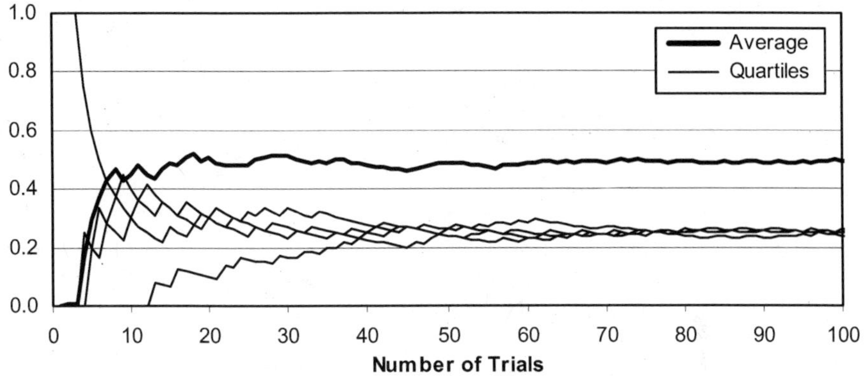

Figure 5.28. The results of a congruential random number generator with a multiplier of 3333, a modulus of 30000 and an increment of 333. Since the frequencies of average and quartile values converge to their proper values with no visible trends, the first 100 generated numbers can be considered random.

Using the same seed will produce the same sequence of random numbers. Using the same sequence of random number can be beneficial for testing and reproducing results, but should generally be avoided since rare results will automatically be repeated. Special care must be taken with seeds when using multiple microprocessors (such as distributed or parallel computing) to execute Monte Carlo simulations. If the sequences of random numbers used by different microprocessors are correlated in some way, the information produced by the different processors will be redundant, degrading algorithm performance.[37-38]

Oftentimes, random number of distributions other than uniform are required. To do this, the uniform random number is mapped to another value by taking the inverse transform of the desired cumulative distribution function (see Figure 5.29). Sometimes there is a closed-form solution to this function, but other times there is not (e.g., the normal distribution) and an integration calculation or a table lookup must be performed.

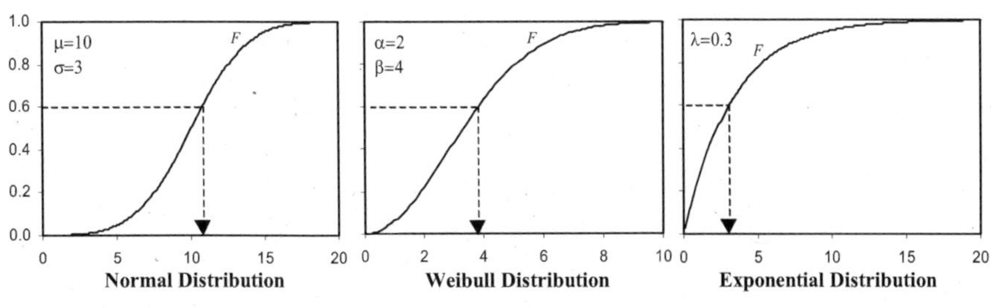

Figure 5.29. Commonly available random number generators produce numbers uniformly distributed between zero and one. These numbers can be mapped to other distributions by taking the inverse transform of the desired cumulative distribution function. The three examples shown in this figure map the randomly generated value of 0.6 to corresponding values for normal, Weibull, and exponential distributions.

To illustrate random number mapping, consider a situation where a random number with an exponential distribution is desired. In this case, a random variable with uniform distribution, r_u, must be mapped to a random variable with exponential distribution, r_e. This is easily done by setting r_u equal to the exponential cumulative distribution function:

$$r_u = 1 - \exp(-\lambda r_e) \tag{5.19}$$

This equation is easily solved for, resulting in a closed-form expression for generating the new random number:

$$r_e = \frac{\ln(1 - r_u)}{-\lambda} \tag{5.20}$$

Eq. 5.20 can be used to verify the graphical solution shown in Figure 5.29. With r_u equal to 0.6 and λ equal to 0.3, r_e computes to be a value of 3.0543, which corresponds to the graph. This technique generalizes to other distribution functions, but, as previously stated, may not result in a closed-form solution.

5.7.2 Sequential Monte Carlo Simulation

A sequential Monte Carlo simulation attempts to model system behavior precisely as it occurs in reality—as a sequence of random events that build upon each other as the system progresses through time. Some system contingencies are modeled by probability distributions and can randomly occur at any point in the simulation. Other contingencies are conditional in that their probability of occurrence depends upon prior events and the present state of the system (such as cascading failures). In a similar fashion, system responses to contingencies are probabilistically modeled, allowing for a system to respond in many different ways. If implemented with a high level of detail, a sequential Monte Carlo simulation produces a highly realistic simulation, almost comparable to a physical experiment.

A sequential Monte Carlo simulation is implemented by dividing the simulation time period into small slices. Starting at the first, each time slice is simulated in sequence to identify new contingencies (such as faults) and respond to prior unresolved contingencies (see Figure 5.30). Simulation accuracy increases as time slices become smaller, but at the expense of increasing computation time. To model a year with a one-hour time resolution requires 8760 time slices. To model a year with a one-minute time resolution requires more than half a million time slices. Computational efficiency can be improved by utilizing variable time slice resolution, which utilizes long time slices during normal conditions and switches to shorter time slices while the system is responding to a contingency.

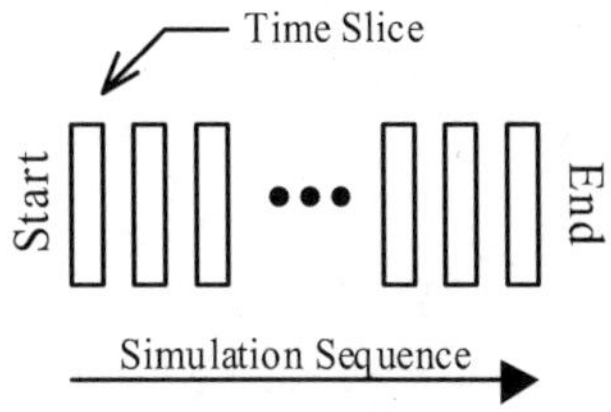

Figure 5.30. A sequential Monte Carlo simulation divides the simulation into small time slices. In order (starting with the first), each time slice is tested to see if random events occur and to determine how the system responds to prior events. This type of algorithm is computationally expensive, but is intuitive to implement and allows complicated component and system behavior to be modeled.

The probability of a contingency occurring during a time slice is equal to the probability of its having occurred at the end of the time slice minus the probability of its having occurred at the beginning of the time slice. This is approximately equal to the value of the probability density function at the beginning of the time slice multiplied by the time slice duration. Formulae are:

$$P(t) = F(t + \Delta t) - F(t) \tag{5.21}$$

$$P(t) \approx \Delta t \cdot f(t) \tag{5.22}$$

P(t) = probability of occurrence in time slice bounded by t and Δt
F(t) = probability distribution function (cumulative)
f(t) = probability density function
Δt = duration of time slice

Some probability distribution functions, like the exponential distribution, are stationary and do not vary in time. These events will have the same likelihood of occurring in each time slice, equal to the rate of occurrence multiplied by the time slice duration. Consider a component with 0.1 failures per year and a time slice duration of 5 minutes. Since 0.1 failures per year corresponds to 0.0000002 failures per minute, the probability of failing in a five-minute time slice is equal to $5 \cdot 0.0000002 = 0.000001$. Other types of events are nonstationary and have probability functions that vary with time and must be individually computed for each time slice.

Using a sequential Monte Carlo simulation is appropriate when the system response at a given time is highly dependent upon prior events. Storms are a good example, since there is a good chance that multiple faults will occur at or near the same time and location and influence each other (a "fault" occurring downstream of an already tripped protection device will not cause fault current to flow). Applications that do not have this characteristic can still utilize a sequential Monte Carlo simulation, but can greatly reduce computational requirements without sacrificing accuracy by utilizing other techniques such as nonsequential Monte Carlo simulations.

5.7.3 Nonsequential Monte Carlo Simulation

A sequential Monte Carlo simulation is not needed if contingencies are mutually exclusive and system behavior does not depend upon past events. Instead, contingencies can be probabilistically selected and simulated in any arbitrary order. Since this process does not necessarily simulate contingencies in order of occurrence, it is referred to as a nonsequential Monte Carlo simulation. Nonsequential Monte Carlo simulations tend to be much less computationally intensive than sequential simulations since (1) only a small percentage of each simulation period is actually simulated, and (2) simulation rules are simpler since contingencies are assumed not to interact with each other.

A nonsequential Monte Carlo simulation starts with a pool of possible contingencies. For each contingency, a random number is used to generate the number of times that the contingency will occur in the simulation time period. For a typical year-long distribution system reliability assessment, most contingencies will not occur, some contingencies will occur once, and a few contingencies may occur multiple times. The precise number of occurrences will depend upon the probability distribution. For example, an exponentially distributed failure function (characterized by a constant failure rate) will follow a Poisson process. If a component has a constant failure rate of λ per year, the probability of its failing x times in a year is:

$$\text{Prob. of failing x times} = \frac{\lambda^x e^{-\lambda}}{x!} \tag{5.23}$$

To determine the number of times that a component fails in a simulated year, a random number between zero and one is generated. If this number is less that $e^{-\lambda}$, no failure will occur in the year being simulated. If the random number is between $e^{-\lambda}$ and $\lambda \cdot e^{-\lambda}$, one failure will occur. If the random number is greater than $\lambda \cdot e^{-\lambda}$, multiple failures will occur. To further illustrate, consider a component with a failure rate of 0.1 per year. The probability of this component failing multiple times per year is:

None: 0.9048 Once: 0.0905 Twice: 0.0045 Thrice: 0.0002

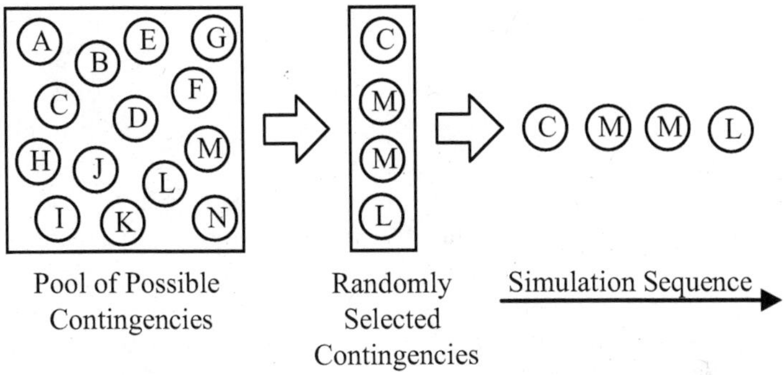

Figure 5.31. A nonsequential Monte Carlo simulation determines all contingencies that will occur before the simulation begins. Contingencies are randomly selected from a pool of possible contingencies based on contingency probabilities (a contingency can be selected more than once). The selected contingencies are then simulated in any order, assuming that all contingencies are mutually exclusive.

The randomly determined number of occurrences of each contingency can be thought of as the score of the contingency. To continue the nonsequential Monte Carlo simulation, contingencies with nonzero scores are simulated using the same techniques described in the analytical simulation sections of this chapter. Instead of weighting the impact of the contingencies by the probability of the contingency occurring, the impact is weighted by the score of the contingency. To illustrate, consider distribution transformer with a failure rate of 0.1 per year. When this transformer fails, downstream customers are interrupted for 4 hours. In an analytical simulation, downstream customers are assigned the expected impact of transformer failures: 0.1 interruptions per year and 0.4 interrupted hours per year. In a nonsequential Monte Carlo simulation, the transformer is assigned a score, s, and customers are assigned the simulated impact of a transformer failure: s interruptions per year and 4·s interrupted hours per year.

Like a sequential Monte Carlo simulation, a nonsequential Monte Carlo simulation is typically repeated many times to produce a distribution of results. In fact, a large number of years can be simulated without substantially more computing power than an analytical simulation. Like in an analytical simulation, each contingency is simulated to determine its impact on system reliability. Unlike an analytical simulation, the impact of each contingency is stored so that whenever a contingency occurs, its impact is already known (this technique, though computationally efficient, is memory intensive).

Because a nonsequential Monte Carlo simulation, as described in this section, combines elements of both Monte Carlo and analytical techniques, it is sometimes referred to as a hybrid technique. The occurrence of contingencies is randomly determined, but the impact of a contingency is analytically determined. The combination allows traditional analytical techniques that produce expected results to be easily extended to produce other statistically useful results such as median, mode, variance, and confidence intervals.

5.7.4 Risk Assessment

In many cases, decisions are determined not by considering expected outcomes, but by considering the risk associated with rare events. To do this, a risk assessment must identify all possible outcomes and the probability of each outcome occurring. When possible, this is done through analytical methods such as function convolution. Usually this is not feasible and Monte Carlo techniques are required.[39]

Risk assessment is appropriate when considering performance-based rates. Consider a system with an expected SAIDI of 3 hours per year. This system is subject to a simple performance-based rate that assesses a $10 million penalty if SAIDI exceeds 6 hours in a given year. In this situation, the utility should not care about the system's expected reliability. Rather, the utility cares about the probability of incurring the $10 million penalty, that is, the probability of SAIDI exceeding 6 hours per year (Figure 5.32, left). If efforts are made to reduce SAIDI, the probability of incurring the penalty may actually increase (Figure 5.32, center). Similarly, SAIDI may be allowed to degrade as long as the probability of exceeding 6 hours is reduced (Figure 5.32, right). Instead of focusing on the expected outcome, risk assessment focuses on the shape of the distribution function.

To focus on the shape of the distribution function, it must be calculated. It is usually sufficient to use a nonsequential Monte Carlo simulation of the type described in Section 5.7.3. Once a sufficient number of sample years are simulated, results can be fitted to a distribution curve and statistical measures related to risk can be treated in a rigorous manner.

The left graph in Figure 5.33 shows a histogram of SAIDI outcomes produced from 1000 nonsequential Monte Carlo simulations of an actual US distribution system (approximately 200 MVA of peak load). The same graph shows a lognormal curve fit produced by direct parameter extraction.

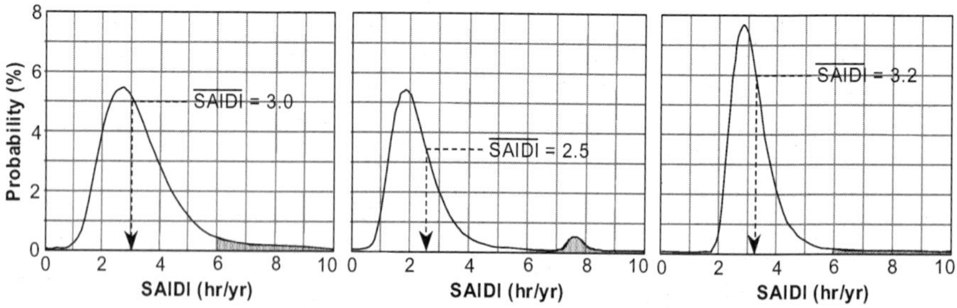

Figure 5.32. Three SAIDI probability distribution functions and their associated risk when subjected to a performance-based penalty of $10 million for all outcomes greater than 6 hours (the risk of incurring the penalty is proportional to the shaded region). The left figure shows the original system with an expected SAIDI of 3 hr/yr. The middle figure shows an improved expected value of SAIDI, but an increase in risk—an unintuitive result. The right figure shows a degraded expected value of SAIDI, bus a decrease in risk—also an unintuitive result.

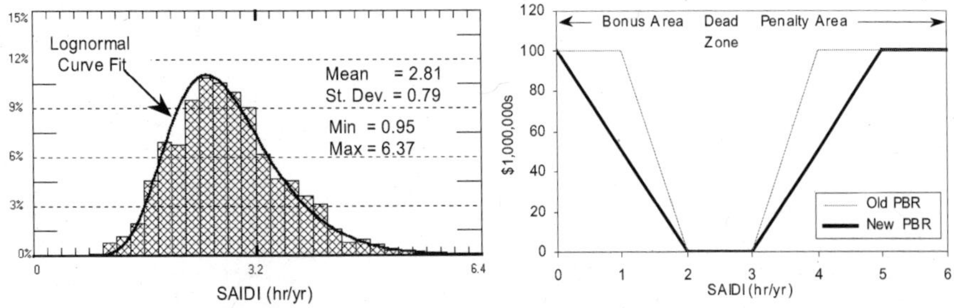

Figure 5.33. The left graph shows a histogram (and associated curve fit) of 1000 nonsequential Monte Carlo simulations. The distribution of results becomes important when the system is subject to performance-based rates such as those shown in the right graph.

The right graph of Figure 5.33 shows a hypothetical performance-based rate structure (light line). From this graph, the bonus/penalty corresponding to any SAIDI is easily determined. The probability of each SAIDI occurring is also known, and can be computed from the lognormal curve fit. Using these two functions, statistics about financial risk can be easily determined. For example, the expected penalty will be equal to:

$$\text{Penalty} = \int_0^{\infty} \text{PBR(SAIDI)} \cdot \text{f(SAIDI)}\, \text{dSAIDI} \tag{5.24}$$

The probability of certain financial outcomes can also be computed. For example, the probability of landing in the dead zone will be equal to the probability of having a SAIDI between 2 hours and 3 hours. This is mathematically represented by:

$$\%\text{ in dead zone} = 100\% \cdot \int_2^3 \text{f(SAIDI)}\, \text{dSAIDI} \tag{5.25}$$

The expected outcome of the test system is 14 million dollars in penalties per year. Penalties will occur 32% of the time, bonuses will occur 12% of the time, and the utility will be in the dead zone 57% of the time. The distribution of all possible outcomes and their associated probabilities is shown in the left graph of Figure 5.34. For a cash-strapped utility, this represents a risky situation. Even though the average penalty of $14 million may be acceptable, this outcome will rarely happen. A penalty of $50 million or more will occur once every 7 years and the maximum penalty of $100 million will occur once every 12 years. Faced with this situation, this utility would be wise to negotiate a less risky PBR.

Possibilities for adjusting the PBR to mitigate risk include: (1) make the reward and penalty slopes less steep, (2) widen the dead zone boundaries, (3) move the dead zone to the right, and (4) reduce bonus and penalty caps. Each of these options can be used alone, or a combination can be negotiated.

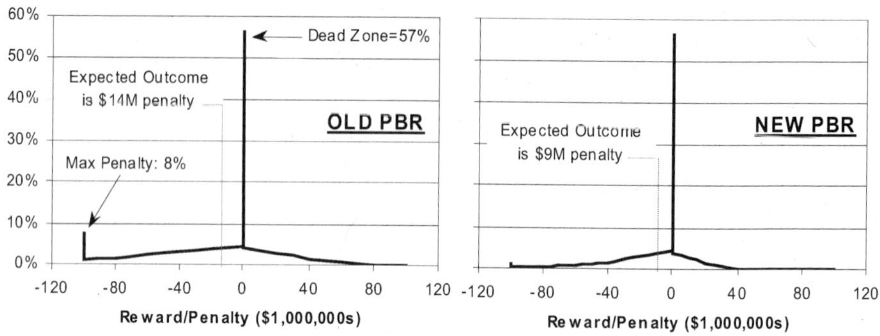

Figure 5.34. Outcome profiles associated with two possible performance-based rate structures. The left graph represents a relatively risky situation where the maximum penalty will occur 8% of the time. The right graph reflects both a lowered risk of maximum penalty assessment (from 8% to less than 1%) and a lowered expected outcome (from $14 million to $9 million).

A suggested change to mitigate the risk of the test system is shown in the right graph of Figure 5.33 (heavy line). This change stretches out the bonus and penalty transition zones from one hour to two hours. The impact of the new PBR is shown in the right graph of Figure 5.34. On the surface, it may seem like not much has changed. Penalties will still occur 32% of the time, bonuses will still occur 12% of the time, and the dead zone will still occur 57% of the time. The good news is that the expected annual penalty is reduced from $14M to $9M. More importantly, risk is reduced. Penalties will occur just as often as they did before, but in much smaller amounts. Penalties will only exceed $50M once every 16 years, and a maximum penalty will occur only once per century.

Although this section has focused on performance-based rates, risk assessment is applicable when analyzing many other scenarios such as reliability guarantees and inventory levels—whenever statistical information beyond expected values is required.

5.7.5 Storm Assessment

Analytical methods that can predict the reliability of distribution systems typically assume that all faults are mutually exclusive. This is a reasonable assumption in nonstorm conditions, but is definitely not true during adverse weather. The main difficulty during storms is that a common failure mode results in many overlapping fault events. Consequently, sufficient crews may not exist to attend to each fault as it occurs.

This section describes a framework in which distribution system reliability during storms can be modeled with a sequential Monte Carlo simulation. The framework includes weather modeling, associating component failure rates to weather conditions, defining a storm criterion, and developing the sequential storm simulation. The framework will first be stated in general, and will then be applied to the assessment of linear wind storms.

Most people view a storm as a period of adverse weather such as a few hours of high winds and rain. A storm simulation must begin by probabilistically modeling such adverse conditions. Each type of adverse weather to be assessed must be characterized by a set of probability distribution functions related to storm severity. Common storm features include the number of storms per year, the duration of the storm, and the magnitude of key storm parameters such as wind speed for wind storms and flash density for lightning storms.

Since utilities have precise definitions for storm events, a period of adverse weather may or may not be classified as a storm. For this reason, a randomly generated period of adverse weather is referred to as a *potential storm event.* If the potential storm event satisfies a particular utility's definition of a storm, it is reclassified as a *storm event.* In some cases, reclassification of a potential storm event can be checked without a full storm simulation, especially if the definition is based strictly on weather severity. In other cases, a sequential simulation must be performed before this decision can be made.

Since a typical storm consists of many overlapping and interrelated events, an accurate storm assessment model is typically based on a sequential Monte Carlo simulation. The simulation can be broken down into two stages, a first stage to simulate the period of adverse weather and a second stage to simulate residual restoration and repairs after the adverse weather has abated.

During the first stage of the storm simulation, adverse weather increases the probability of component failures. Therefore, one of the most critical (and difficult) aspects of a storm simulation is modeling the relationship of storm severity to component failure rates. Ideally, a utility will have a large amount of data from previous storms so that relationships between storm severity and equipment failures can be determined. Alternatively, failure rates models can be inferred from physical models, such as tree pressure being proportional to the square of wind speed.

After generating storm models and component failure models, storm behavior can be modeled with a sequential Monte Carlo simulation. During the simulation, each component is tested for failure in each time slice. If an energized component fails, the protection system and automated switching system must also be simulated. Crew movement must also be tracked. In each time slice, crews may be waiting for dispatch instructions, on-route to a fault location, on-route to a switching location, or repairing a fault. If all crews are occupied, additional faults must go unattended until a crew becomes available. During major storms, additional crews may become available as they arrive from neighboring utilities.

After the period of adverse weather is over, many pieces of equipment may yet need to be repaired and many customers may yet need to be restored. A second-stage simulation can model this event in a manner similar to the first-stage simulation, but without the possibility of additional component failures. Basically, crews continue to repair and reconfigure the system until all customers are restored and the system is returned to its normal operating state.

To achieve meaningful results, a storm assessment must be repeated many times. First, simulations must be repeated to reflect multiple storms in a given year. Second, multiple years must be simulated so that a distribution of the total reliability impact of storms can be generated. The entire storm assessment process can be summarized as follows:

Generic Storm Assessment Algorithm

1. Model a storm event by a set of probability density functions $f_1, f_2, \ldots, f_N$.
2. Determine the parameters of the probability density functions based on historical storm event data.
3. Model component failure rates, λ, as a function of $f_1, f_2, \ldots, f_N$: $\lambda=\lambda(f_1, f_2, \ldots, f_N)$.
4. Determine a storm event criterion as a function of $f_1, f_2, \ldots, f_N$ and other system parameters such as the number of dispatchable crews.
5. Probabilistically determine the number of potential storm events that will occur in the simulation period.
6. Probabilistically generate the severity parameters of a potential storm event.
7. Is the potential storm event a storm event (based on the criterion developed in step 4)? If no, go to step 10.
8. Perform a sequential Monte Carlo simulation of the period of adverse weather based on the failure rate relationships, $\lambda=\lambda(f_1, f_2, \ldots, f_N)$.
9. Perform a sequential Monte Carlo simulation from the time the adverse weather ends until the system is restored to its normal operating state (assuming that no additional equipment failures occur during this time).
10. Have all potential storm events for the simulation year been examined? If no, go to step 6.
11. Repeat steps 5-10 for many years to generate a distribution of reliability outcomes due to storms.
12. End.

To further illustrate storm assessment, the generic storm assessment algorithm will now be applied to wind storms. Specifically, it will be developed to determine the contribution of linear wind storms to SAIDI for a utility located in the Pacific Northwest.

Calculating the contribution of wind storms to SAIDI requires information about wind speed and storm durations. In this case, a potential storm event is characterized by two values: a duration and an rms wind speed. RMS wind speed is chosen since the pressure exerted on poles and trees is proportional to the square of wind speed.[40] To characterize these values, historical wind data should be analyzed and modeled. To achieve this goal, potential storm event starting and stopping criteria are defined. The chosen starting criterion is wind speed staying above a specified speed for a minimum duration. Similarly, the chosen stopping criterion is wind speed staying below a specified speed for a minimum duration. This algorithm can be summarized in the following steps:

Algorithm for Identifying Potential Storm Events

1. Set the following parameters so that the number of potential storm events found in a given year is approximately an order of magnitude greater than the expected number of storm events in a given year:
   k_{min}: the minimum storm wind speed considered (in knots).
   d_{min}: the duration of low wind speeds for a potential storm event to be considered over (hours).
   E^2_{min}: the minimum wind energy considered ($knots^2$-hours).
2. Consider hour h with a wind speed of s.
3. Is the wind blowing at a minimum speed of k_{min} for a minimum duration starting at hour h? If not, $h = h + 1$, go to step 2.
4. Set the starting time of the potential storm event, h_{start}, equal to h.
5. Set the accrued squared wind speed, E^2 equal to s^2.
6. Does the wind speed fall and stay below k_{min}, for a minimum duration of d_{min},? If so, set the ending hour, h_{end}, equal to h and go to step 8.
7. $h = h + 1$, find new value of s, $E^2 = E^2 + s^2$, go to step 6.
8. Is E^2 less than the minimum value of E^2_{min}? If so, $h = h + 1$, go to step 2.
9. Record a potential storm event with a duration of $(h_{end} - h_{start})$ and an rms wind speed equal to $(E^2/(h_{end} - h_{start}))^{0.5}$.

The potential storm event analysis for this example uses historical wind data obtained from a local airport. This historical data consists of hourly wind speed measurements from the years of 1974 to 1993. The average wind speed over this period was 7.46 knots and the maximum sustained wind speed over this period was 45.10 knots. The following parameters are used in the calculations:

$$k_{min} = 15 \text{ knots}$$
$$d_{min} = 12 \text{ hours}$$
$$E^2_{min} = 1000 \text{ knot}^2\text{-hours}$$

Using these parameters, the potential storm event algorithm identifies an average of 24.17 potential storm events per year. Using these events, it is now necessary to find probability density functions that can accurately describe the potential storm event durations and rms wind speeds. In this example, lognormally distributed random variables are chosen since they accurately fit the historical wind data characteristics of the distributions system. The probability density functions for potential storm event duration (f_d) and potential storm event rms wind speed (f_s) are:

$$f_d(t - t_{min}) = \frac{1}{\sigma_d (t - t_{min})\sqrt{2\pi}} \exp\left[-\frac{(\ln(t - t_{min}) - \mu_d)^2}{2\sigma_d^2}\right] \tag{5.26}$$

t = duration of potential storm event (hours)
t_{min} = minimum duration of potential storm event (hours)
μ_d = lognormal parameter
σ_d = lognormal parameter

$$f_s(s - s_{min}) = \frac{1}{\sigma_s (s - s_{min})\sqrt{2\pi}} \exp\left[-\frac{(\ln(s - s_{min}) - \mu_s)^2}{2\sigma_s^2}\right] \tag{5.27}$$

s = rms wind speed of potential storm event (knots)
s_{min} = minimum rms wind speed value of potential storm event (knots)
μ_s = lognormal parameter
σ_s = lognormal parameter

A hill-climbing search has been used to identify the following lognormal parameters that minimize the chi squared error of the distribution function (plots of these functions are shown in Figure 5.35):

RMS Wind Speed Parameters	Duration Parameter
s_{min} = 0.70 knots	t_{min} = 2.49 hours
μ_s = 2.70 knots	μ_d = 2.33 hours
σ_s = 0.20 knots	σ_d = 1.22 hours

As can be seen from Figure 5.35, the lognormal density function is a very good representation of historical rms wind speed data, and a reasonable representation of potential storm event duration. Both models satisfy the minimum chi squared error criterion and are sufficient for use in the Monte Carlo simulation.

Since faster wind speeds will increase overhead line failure rates, storm overhead line failure rates must be modeled as a function of wind speed, not as a fixed value. Since these failures are largely due to trees and poles blowing over, and the pressure exerted on trees and poles is proportional to the square of wind speed, overhead line failure rates during storms are assumed to be proportional to wind speed squared. Each overhead line, then, will be characterized by a base storm failure rate, λ_{base}, at a base storm wind speed, s_{base}. The storm failure rate of this overhead line at wind speed *s* can then be described as:

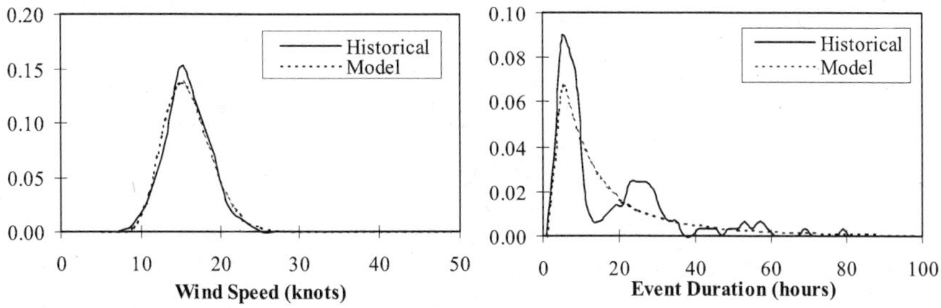

Figure 5.35. Probability density functions for potential storm event rms wind speed and duration. Historical data represents 20 years of weather in the Pacific Northwest, which can be reasonably modeled with a lognormal distribution function.

$$\lambda(s) = \lambda_{base}\left(\frac{s}{s_{base}}\right)^2 \qquad (5.28)$$

The wind dependent failure rates of overhead lines are perhaps the most challenging data to obtain (this is largely due to the absence of storm-related failure data). The overhead line failure rate, λ_{base}, is therefore computed from historical storm event data. The result obtained is:

λ_{base} = 0.0065 failures per mile per hour

This value is determined by first setting the base storm wind speed, s_{base}, to twice the average area wind speed (in this case, 15 knots). Next, the overhead line storm failure rates are adjusted until the number of potential storm events classified as storm events per year corresponds the historical number of storms per year experienced by the utility. In this case, a storm event is simply defined as an event where the number of faults on a system exceeds the number of crews, resulting in a fault that is unattended.

Given a potential storm event with a duration of d and an rms wind speed of s, it is necessary to determine if it should be classified as a storm event. The expected number of faults that a particular line will experience during a storm is approximately equal to its failure rate multiplied by the duration of the storm. The total expected number of faults on a system, E(faults), can therefore be determined by summing this product over all of the line sections on a system. This relationship is:

$$E(\text{faults}) = d\left(\frac{s}{s_{base}}\right)^2 \sum_{\text{lines}} \lambda_{base} \qquad (5.29)$$

Each line section will have a mean-time-to-repair, MTTR, associated with it. The total expected repair time associated with this line will then be equal to its expected number of faults multiplied by its MTTR. The total expected repair time of all faults occurring during a storm, E(repair), can then be computed by summing up the expected repair times of all lines on the system. This is mathematically expressed as:

$$E(\text{repair}) = d\left(\frac{s}{s_{\text{base}}}\right)^2 \sum_{\text{lines}} \left[\text{MTTR} \cdot \lambda_{\text{base}}\right] \tag{5.30}$$

If a potential storm event is to be classified as a storm event, the expected number of faults must exceed the number of crews, c. In addition, the total expected repair time must exceed the total available crew time, c·d. These two conditions are mathematically represented by the following pair of equations:

$$\begin{aligned} &E(\text{faults}) > c \quad \text{and} \\ &E(\text{repair}) > c \cdot d \end{aligned} \tag{5.31}$$

To simulate the storm events of a given year, the following process is followed. First, generate a number of potential storm events equal to the average number per year found in historical data. Each potential storm event will have a characteristic duration and rms wind speed probabilistically determined according to probability density functions, which have been fitted to match historical data. Second, test each potential storm event against the storm event criteria. If these criteria are satisfied, classify the potential storm event as a storm event. The distribution system can now be simulated to determine the impact of the storm events on each customer during the particular year.

After a potential storm event has been classified as a storm event, it needs to be analyzed in detail. The goal of this analysis is to determine the expected reliability impact that each customer will experience due to the storm event. As previously discussed, this can be accomplished by a two-stage sequential Monte Carlo simulation.

The first stage of the simulation begins when a storm event begins, and ends when the storm event ends (when the weather calms down). For the purposes of this example, it is assumed that no downstream isolation or backfeeding takes place. This is generally a good assumption for two reasons. First, crew constraints force dispatchers to focus on system repair. Second, unobserved equipment damage occurring on de-energized parts of the system can pose a safety problem during system reconfiguration. The first-stage simulation algorithm can be summarized in the following steps:

First Stage of Wind Storm Simulation (during Adverse Weather)

1. Set the time of the storm event, t, equal to zero.
2. Select an initial line section.
3. Does this line section have an existing fault? If not, go to step 8.
4. If the fault is unattended and a crew is available, dispatch a crew to the fault.
5. If the dispatched crew has reached the nearest upstream switch, then attempt to open the switch and close the associated protection device.
6. If the dispatched crew has completed repairs, then close the associated upstream switch and free the crew.
7. Go to step 11.
8. Probabilistically check to see if the line section fails. If a failure does not occur, go to step 11.
9. Determine the repair time of the failure.
10. If the fault was on an energized line, find and trip an upstream protection device (this will usually be the nearest upstream device, but device failures may force a backup protection device to operate).
11. Have all line sections been considered? If no, select next line section, go to step 3.
12. Increase the outage duration of all interrupted customers by Δt.
13. $t = t + \Delta t$. If t is less than the storm duration, go to step 2.
14. End.

The first stage of the simulation begins with the system operating normally. It is then checked to see if any failures occur during the first time interval. If a fault occurs and a crew is available, a crew is dispatched. If all crews are already dispatched, the next available crew will attend the fault. When a crew is dispatched to a fault, it is assumed that the crew will first open the nearest upstream switch. The crew can then replace the blown fuse (or reset the tripped breaker) and restore service to those customers located between the protection device and the switch. The crew will then begin repairs on the fault. When the fault is repaired, the crew will close the upstream isolation switch (if possible) and be dispatched to another fault. This process continues until the adverse weather subsides and the second stage begins.

After the high winds have calmed down, it is likely that one or more faults still remain on the system. The second stage of simulation considers this time period. In the second stage, downstream isolation and back feeding are considered, but it is assumed that no additional faults will occur on the system.

This part of the simulation begins with the system in the exact state that first stage left off; unrepaired faults exist on the system, crews are dispatched, fuses are blown, isolation switches are open, and so forth. The first task that is done is to check and see if the time required to back feed any customers has been exceeded (remember that no back feeding was allowed in the first stage). If this time has been exceeded for certain customers, the system is immediately reconfigured to restore service to these customers through alternate paths. The simulation continues in a manner similar to the first stage except that no additional faults will occur on the system and dispatched crews are allowed to back feed as many customers as possible. The second stage continues until all faults have been repaired and all customers have been restored service. The second-stage simulation algorithm can be summarized in the following steps:

<u>**Second Stage of Wind Storm Simulation (after Adverse Weather)**</u>

1. Configure system to backfeed all sections whose required time to backfeed has been exceeded.
2. Select an initial line section.
3. Does this line section have an existing fault? If not, go to step 8.
4. If the fault is unattended and a crew is available, dispatch a crew to the fault.
5. If the dispatched crew has reached the nearest upstream switch, then attempt to open the switch and close the associated protection device.
6. If the dispatched crew has had time to isolate all downstream paths, then close the appropriate tie switches and back feed as many customers as possible.
7. If the dispatched crew has completed repairs, then close the associated isolation switches, open the associated tie switches, and free the crew.
8. Increase the outage duration of all interrupted customers by Δt.
9. Have all line sections been considered? If no, select next line section and go to step 3.
10. $t = t + \Delta t$.
11. Are all faults repaired? If no, go to step 2.
12. End.

When applied to the utility distribution system at the location from which the weather data was taken, wind storms are shown to contribute an expected 24.4 hours per year to customer interruptions. This expected value, however, will vary widely from year to year because of variations in annual weather patterns (the computed standard deviation is 160% of the mean). This 24.4 hours of storm hours compares to 3.8 hours of nonstorm interruptions, showing that storm events can sometimes contribute much more to poor customer reliability than nonstorm events. This is an important point to keep in mind since many utilities exclude storms when computing historical reliability indices.

5.8 OTHER METHODOLOGIES

The first seven sections of this chapter have presented the techniques necessary to create distribution system reliability models. This section presents several additional techniques that are commonly applied to generic reliability assessment problems and may find occasional use for distribution system applications.

5.8.1 Fault Tree Analysis

Fault tree analysis is an enumerative method that examines each initiating event and the system's response to this event. To do this, each initiating event must be associated with a sequence of system responses, with each response having a probability of succeeding or failing. An initiating event with n associated system responses will then be associated with a fault tree with 2^n possible outcomes. The probability of each outcome is equal to the product of the probability of each stage of the fault tree.

To illustrate, consider a fault that occurs on a distribution system with a probability of P_A. After the fault occurs, the system responds by attempting to clear the fault with a probability of P_B, restoring upstream customers with a probability of P_C, and restoring downstream customers with a probability of P_D. Each combination of successful and unsuccessful system responses corresponds to a path on the fault tree, and the probability of any outcome is computed by multiplying the associated probabilities of each individual system response. The fault tree for this example is shown in Figure 5.36.

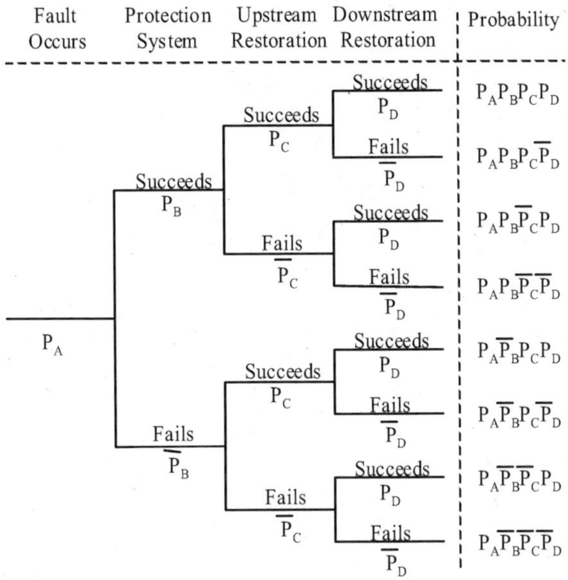

Figure 5.36. Sample fault tree. After a fault occurs, the system responds through a sequence of actions that can either succeed or fail. The probability of each outcome is equal to the product of the probability of each stage of the path that led to the outcome.

The reader may notice that the concept of fault trees is similar to the concept of event path splitting discussed in Section 5.5.7. In fact, simulation path splitting is simply a method of automatically generating the fault tree associated with the operational failures of devices. The traditional method of fault tree generation is much more time consuming and requires manually identifying all initiating events and all system responses associated with these events.

The results of a fault tree analysis are the probabilities of all system outcomes associated with initiating events. These results can be translated into a reliability analysis by determining the system impact of each outcome and weighting this impact by the probability of the associated outcome occurring.

5.8.2 Failure Modes and Effects Analysis

Failure Mode and Effects Analysis, commonly referred to as FMEA, is an inductive technique that seeks to identify all possible equipment failure modes and their associated impact on system reliability. For each component, the following information is required:

- List of failure modes
- Possible causes of each failure mode
- Possible system effect of each failure mode
- Probability of each failure mode occurring
- Possible actions to reduce the failure rate or effect of each failure mode

To illustrate, consider some possible failure modes for a relay:[41] contacts stuck closed, contacts slow in opening, contacts stuck open, contacts slow in closing, contact short circuit, contacts chattering, contacts arcing, coil open circuit, coil short circuit, improper coil resistance, coil overheating, and coil overmagnetization. FMEA requires that each of these failure modes be examined for each relay in the system—a thorough analysis, but only practicable for small systems.

A natural extension of FMEA is to consider criticality information associated with each failure mode. The result, Failure Mode Effects and Criticality Analysis (FMECA), is similar to a manual form of the analytical simulation techniques described in Sections 5.5 and 5.6 except that criticality is generally quantified in vague terms based on a combination of effect severity and effect probability. FMECA has been commonly applied to military applications, and a department of defense document recommends severity classifications of catastrophic, critical, marginal, and minor.[42]

5.8.3 RAM Analysis

RAM, an acronym for Reliability, Availability and Maintainability, describes the scope of an analysis rather than a specific methodology. Typically, a RAM analysis will utilize network modeling and/or FMECA techniques to compute system reliability, availability, and component criticality.[43-44] In addition, it will identify the vulnerability of the system while maintaining critical pieces of equipment. A RAM analysis may also include laboratory and/or field investigations such as accelerated life testing, fault tolerance testing, and environmental testing.

When a RAM analysis includes a risk and safety analysis, it is referred to as a RAMS analysis (Reliability, Availability, Maintainability, and Safety). The risk and safety analysis will typically investigate the likelihood of death or injury associated with all possible failure modes.

5.9 STUDY QUESTIONS

1. What is meant by event independence?
2. Explain what is meant by a minimal cut set.
3. What are some advantages and disadvantages of using Markov modeling for distribution system reliability calculations?
4. What post-fault sequence of events should be considered in an analytical simulation?
5. What are some of the algorithmic differences required for radial versus network reliability analysis?
6. Why is it important to consider operational failures in a reliability assessment?
7. Explain some of the potential reliability advantages and disadvantages of using distributed generation on distribution systems.
8. Name the two basic types of Monte Carlo simulation. How are they different?
9. When is it appropriate to use a Monte Carlo simulation for distribution system reliability calculations? What are some of the disadvantages of a Monte Carlo simulation?
10. What are the two types of common mode failures? Give several examples of each.

REFERENCES

1. *IEEE Recommended Practice for the Design of Reliable Industrial and Commercial Power Systems,* IEEE Std 493-1990.
2. M. L. Shooman, *Probabilistic Reliability: An Engineering Approach*, McGraw-Hill, 1968.
3. G. H. Sandler, *System Reliability Engineering*, Prentice-Hall, 1963.
4. R. Ramakumar, *Engineering Reliability: Fundamentals and Applications*, Prentice-Hall, Inc., 1993.
5. R. E. Brown, S. Gupta, R. D. Christie, S. S. Venkata, and R. D. Fletcher, "Distribution System Reliability Analysis Using Hierarchical Markov Modeling," *IEEE Transactions on Power Delivery*, Vol. 11, No. 4, Oct. 1996, pp. 1929-1934.
6. R. E. Barlow and F. Proschan, *Statistical Theory of Reliability and Life Testing: Probability Models*, Holt, Rinehart and Winston, Inc., 1975.
7. S. Asgarpoor and M. J. Mathine, "Reliability Evaluation of Distribution Systems with Non-Exponential Down Times," *IEEE/PES 1996 Winter Meeting,* Baltimore, MD, IEEE, 1996.
8. C. C. Fong, "Reliability of Transmission and Distribution Configurations with Duration-Dependent Effects," *IEE Proceedings*, Part C, Vol. 136, No. 2, March 1989, pp. 64-67.
9. R. Billinton and G. Lian, "Monte Carlo Approach to Substation Reliability Evaluation," *IEE Proceedings-C,* Vol. 140, No. 2, March 1993, pp. 147-152.
10. R. E. Brown, "Probabilistic Reliability and Risk Assessment of Electric Power Distribution Systems," *DistribuTECH Conference*, San Diego, CA, Feb. 2001.
11. R. Brown, S. Gupta, S.S. Venkata, R.D. Christie, and R. Fletcher, "Distribution System Reliability Assessment: Momentary Interruptions and Storms," *IEEE Transactions on Power Delivery,* Vol. 12, No. 4, Oct 1997, pp. 1569-1575.
12. C. M. Warren, "The Effect of Reducing Momentary Outages on Distribution Reliability Indices," *IEEE Transactions on Power Delivery,* Vol. 7, No. 3, July, 1992, pp. 1610-1615.
13. *IEEE Guide for Electric Power Distribution Reliability Indices*, IEEE 1366-2003.
14. K. Aoki, T. Satoh, M. Itoh, H. Kuwabara, and H. Kanasashi, "Voltage Drop Constrained Restoration of Supply by Switch Operation in Distribution Systems," *IEEE Transactions on Power Delivery,* Vol. 3, No. 3, July 1988, pp. 1267-1274.
15. D. Shirmohammadi, "Service Restoration in Distribution Networks Via Network Reconfiguration," *IEEE Transactions on Power Delivery,* Vol. 7, No. 2, April 1992, pp. 952-958.
16. C. Ucak and A. Pahwa, "An Analytical Approach for Step-by-Step Restoration of Distribution Systems Following Extended Outages," *IEEE Transactions on Power Delivery,* Vol. 9, No. 3, July 1994, pp. 1717-1723.
17. D.S. Popovic and R. M. Ciric, "A Multi-Objective Algorithm for Distribution Network Restoration," *IEEE Transactions on Power Delivery,* Vol. 14, No. 3, July 1999, pp. 1134-1141.
18. R. E. Brown and A. P. Hanson, "Impact of Two Stage Service Restoration on Distribution Reliability," *IEEE Transactions on Power Systems*, Vol. 16, No. 4, Nov. 2001, pp. 624-629.
19. R. W. DeDonker and J. A. Maranto, "Application of a Mobile Medium Voltage Sub-Cycle Transfer Switch," 1996 *IEEE Transmission and Distribution Conference Proceedings,* Los Angeles, CA, September, 1996.
20. R. E. Brown, J. R. Ochoa, "Impact of Sub-Cycle Transfer Switches on Distribution System Reliability," *IEEE Transactions on Power Systems*, Vol. 15, No. 1, Feb. 2000, pp. 442-447.
21. G. Cler, N. Lenssen, and C. Manz, *Residential Distributed Generation: Options and Prospects,* Report DE-6, E Source, 1999.
22. P. P. Barker and R. W. de Mello, "Determining the Impact of Distributed Generation on Power systems: Part 1—Radial Distribution Systems," *IEEE Summer Power Meeting*, July 2000, Vol. 3., pp. 1645-1656.
23. M. K. Donelly, J. E. Dagle, D. J. Trudnowski, and G. J. Rogers, "Impacts of the Distributed Utility on Transmission System Stability," *IEEE Transactions on Power Systems,* Vol. 11, No. 2, May 1996, pp. 741-746.

24. F. Li, R. Broadwater, J. Thompson, and F. Goodman, "Analysis of Distributed Resources Operating in Unbalanced Distribution Circuits," *IEEE Summer Power Meeting*, July 2000, Vol. 4., pp. 2315-2319.
25. G. Jóos, B. T. Ooi, D. McGillis, F. D. Galiana, and R. Marceau, "The Potential of Distributed Generation to Provide Ancillary Services," *IEEE Summer Power Meeting*, July 2000, Vol. 3., pp. 1762-1767.
26. I. Krepchin, *Distributed Generation: A Tool for Power Reliability and Quality,* Report DE-5, E Source, 1998.
27. G. W. Ault and J. R. McDonald, "Planning for Distributed Generation Within Distribution Networks in Restructured Electricity Markets," *IEEE Power Engineering Review*, Feb. 2000, pp. 52-54.
28. H. L. Willis and G. B. Rackliffe, *Introduction to Integrated Resource T&D Planning*, ABB Power T&D Company, Inc., 1994.
29. H. L. Willis and W. G. Scott, *Distributed Power Generation: Planning and Evaluation*, Marcel Dekker, Inc., 2000.
30. T. M. Taylor, H. L. Willis, and M. V. Engel, "New Considerations for Distributed Network Planning and Design," *CIRED 97*, Conference Publication No. 438, June 1997, pp. 6.1.1-6.1.5.
31. R. C. Dugan, T. E. McDermott, and G. J. Ball, "Distribution Planning for Distributed Generation," *Rural Electric Power Conference*, May 2000, pp. C4.1-C4.7.
32. J. E. Kim and J. S. Hwang, "Islanding Detection of Distributed Generation Units Connected to Power Distribution Systems," *Proceedings of the International Conference on Power System Technology*, Dec. 2000, pp. 643-647.
33. R. E. Brown and L. A. A. Freeman, "Analyzing the Reliability Impact of Distributed Generation," *IEEE PES Summer Power Meeting*, Vancouver, BC, Canada, July 2001.
34. IEEE Std. P1547/D07, *Standard for Distributed Resources Interconnected with Electric Power Systems (Draft 7)*, Feb. 2001.
35. V. Borozan and N. Rajakovic, "Application Assessments of Distribution Network Minimum Loss Reconfiguration," *IEEE Transactions on Power Delivery,* Vol. 12, No. 4, Oct. 1997, pp. 1786-1792.
36. R. Takeski and D. Rajicic, "Distribution Network Reconfiguration for Energy Loss Reduction," *IEEE Transactions on Power Systems,* Vol. 12, No. 1, Feb. 1997, pp. 398-406.
37. C. L. T. Borges, D. M. Falcão, J. C. O. Mello, and A. C. G. Melo, "Composite Reliability Evaluation by Sequential Monte Carlo Simulation on Parallel and Distributed Processing Environments," *IEEE Transactions on Power Systems,* Vol. 16, No. 2, May 2001, pp. 203-209.
38. G. Fox, M. Johnson, G. Lyzenga, S. Otto, J. Salmon, and D. Walker, *Solving Problems on Concurrent Processors,* Vol. 1, Prentice Hall, 1988.
39. R. E. Brown and J. J. Burke, "Managing the Risk of Performance Based Rates," *IEEE Transactions on Power Systems,* Vol. 15, No. 2, May 2000, pp. 893-898.
40. R. H. Stillman, "Probabilistic Derivation of Overstress for Overhead Distribution In-Line Structures," *IEEE Transactions on Reliability,* Vol. 43, No. 3, Sept. 1994, pp. 366-374.
41. H. Kumamoto and E. J. Henley, *Probabilistic Risk Assessment and Management for Engineers and Scientists*, 2nd Edition, IEEE Press, 1996.
42. *Procedures for Performing a Failure Mode, Effects and Criticality Analysis*, Department of Defense, MIL-STD-1629A.
43. A. Villemeur, *Reliability, Availability, Maintainability and Safety Assessment*, Vol. 1, John Wiley & Sons, 1992.
44. A. Villemeur, *Reliability, Availability, Maintainability and Safety Assessment*, Vol. 2, John Wiley & Sons, 1992.

6
System Analysis

The first five chapters in this book have provided a framework for understanding distribution system reliability. The first chapter presented a high-level overview of distribution systems, distribution equipment, and distribution operations. The second chapter discussed ways to quantify reliability, and the third chapter summarized the causes of poor reliability. The fourth and fifth chapters, in turn, discussed how to model the reliability of distribution components and systems so that reliability measures can be predicted.

By themselves, reliability models do not have much value. To be useful, they must accurately reflect the historical performance of actual systems, be small enough to enable the generation and analysis of many scenarios, and be incorporated into a process that can consistently identify reliability problems and consistently identify good solutions to mitigate these problems. This chapter discusses each of these issues in detail and ends by presenting a comprehensive system analysis example based on an actual utility distribution system.

6.1 MODEL REDUCTION

Detailed distribution system models of large utilities can easily consist of millions of components. Performing a reliability assessment on a model of this magnitude is problematic at best and impossible at worst. It is true that reliability planning is an off-line function and computation speed is less critical than for real time applications, but it is also true that reliability planning is an iterative process that becomes increasingly effective as the number of feasible analyses per unit of time increases. Consider a distribution planner exploring reliability improvement options for an area experiencing poor reliability. If each assessment requires one hour to execute, the planner can only explore a handful of alternatives per day. If each assessment requires one minute to execute, the planner can potentially explore hundreds of alternatives per day. This issue becomes more pronounced when the reliability assessment model is used as a kernel for computer optimization techniques like those described in the next chapter. Since optimization algorithms can require millions of reliability assessments, fast execution time is of paramount importance.

There are three basic ways to improve the speed of a reliability analysis: use a faster computer, use more efficient algorithms, and reduce the size of the model. The first two alternatives are independent of any particular system analysis process and will not be presented. The correct application of model reduction, however, depends upon the topology of the system under consideration and the desired results. When used appropriately, model reduction can result in profound increases in analysis speed—up to a 4-fold increase in speed for each 50% reduction in model size—without significantly affecting the accuracy of results.

With nearly all power system analysis tools, it is important to keep the models as small as possible. Motivations include user interface speed, computation speed, data maintenance, interpretation of results, and so forth. A reliability analysis, however, also benefits from making models as large as possible so that geographic result patterns can be identified and examined. In the end, it is somewhat of an art form to identify the correct balance between model size, model detail, and the disadvantages that are associated with each. The rest of this section describes techniques to reduce the size and complexity of reliability models. The correct degree to which these techniques should be employed, however, will vary depending upon the purpose of each modeling effort and is not discussed further. The reader, however, should be aware of these issues and carefully consider modeling requirements before blindly applying model reduction techniques.

6.1.1 Analysis Areas

The most effective method of limiting the model size of large distribution systems is to divide them into study areas. A study area can be as small as a single feeder, but more typically consists of one or more substation service territories. In general, it is desirable to have analysis areas as large as possible without causing reliability assessment execution time to become prohibitively long. A simple rule of thumb is to limit computation time to one minute for manual analyses and one second for computer optimization.

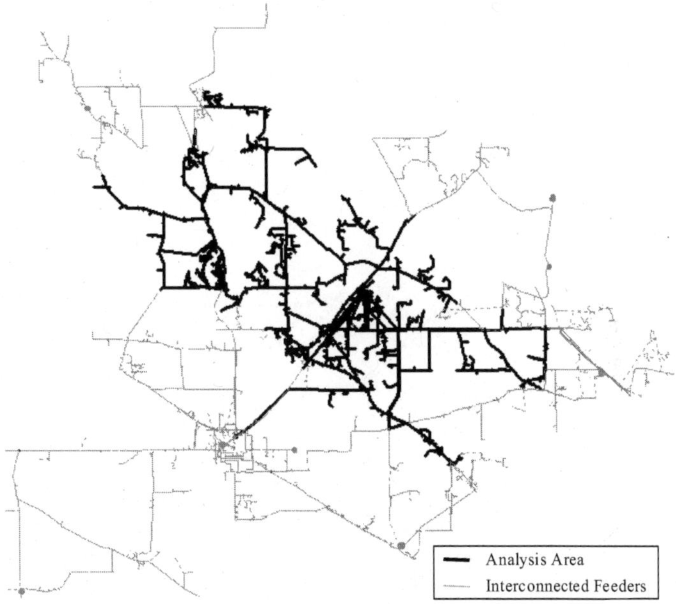

Figure 6.1. A system model consisting of an analysis area and feeders that interconnect the analysis area with the rest of the system. This model will produce accurate results for the analysis area, but not for the interconnected feeders.

In general, a model that can accurately assess the reliability of an analysis area must include some parts of the system not included in the analysis area. Since reliability calculations are impacted by the ability of interrupted feeders to transfer loads to interconnected feeders, an accurate model must include the analysis area and all feeders that interconnect the analysis area with the rest of the system (see Figure 6.1). Doing so ensures that the system is able to properly restore customers after contingencies occur in the analysis area.

Though the reliability results of the analysis area will be correct if all interconnected feeders are modeled, the reliability of the interconnected feeders themselves will not be correct. This is because, in general, not all of the feeders connected to the interconnected feeders are modeled. Because of this inaccuracy, it is desirable to mask results of interconnected feeders so that they are not unintentionally interpreted as correct results. In fact, since faults occurring on the interconnected feeders will not impact the reliability of the analysis area, they do not need to be simulated, further reducing computational requirements.

In many cases, the size of the interconnected feeders associated with an analysis area is bigger than the analysis area itself. If details about the interconnected feeders are not needed, their impact on total system size can be virtually eliminated by replacing each with a reduced equivalent, which consists of a source and a capacity-limited line section connected to a normally open point connected to the analysis area (see Figure 6.2).

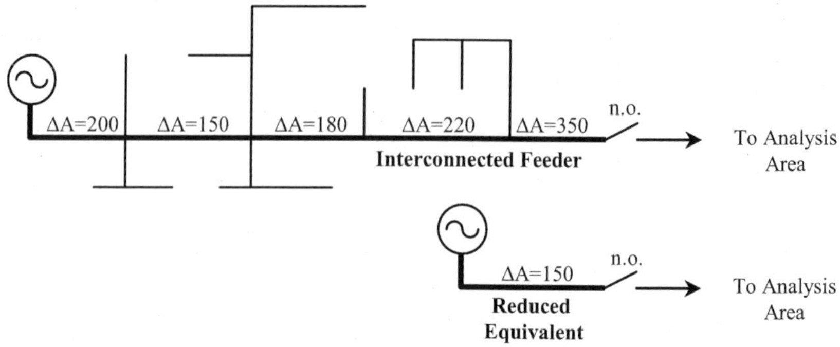

Figure 6.2. An interconnected feeder and its reduced equivalent. Since the amount of current that can be transferred to an interconnected feeder is equal to the difference between its emergency ampacity and its normal load current, ΔA, a reduced equivalent can be modeled with a single element with ΔA equal to the smallest difference found on the path from the normally open point to the interconnected feeder source.

The purpose of an interconnected feeder is to provide a path to restore load to the analysis area that has been interrupted. This path starts at the normally-open tie point and traces upstream to the source of power, but a load transfer is not allowed to overload any piece of equipment on this path. The amount of capacity that each piece of equipment has available to accommodate load transfers is equal to its emergency rating minus its normal loading. The amount of available transfer capacity on the transfer path, therefore, is equal to the minimum spare capacity along this path. If a single component has 50 amps of capacity available for load transfers and all other components have 300 amps, the transfer capacity of the path is limited to 50 amps. In a similar manner, the rating of the reduced equivalent feeder is equal to the smallest difference between emergency rating and loading found on the path from the normally open point to the interconnected feeder source.

6.1.2 Substation and Bulk Power Equivalents

A typical distribution substation will receive one or more transmission lines, step the voltage down to distribution levels, and serve as a source for several distribution feeders. For typical systems in developed nations, bulk power and substation events account for approximately 20% of customer interruptions.[1] This is an important factor to consider, but the contribution of bulk power and substation events can often be decoupled from distribution system reliability assessment.[2] Not only does this decoupled approach reduce system size, but it allows the use of specialized tools for bulk power and substation reliability assessment.

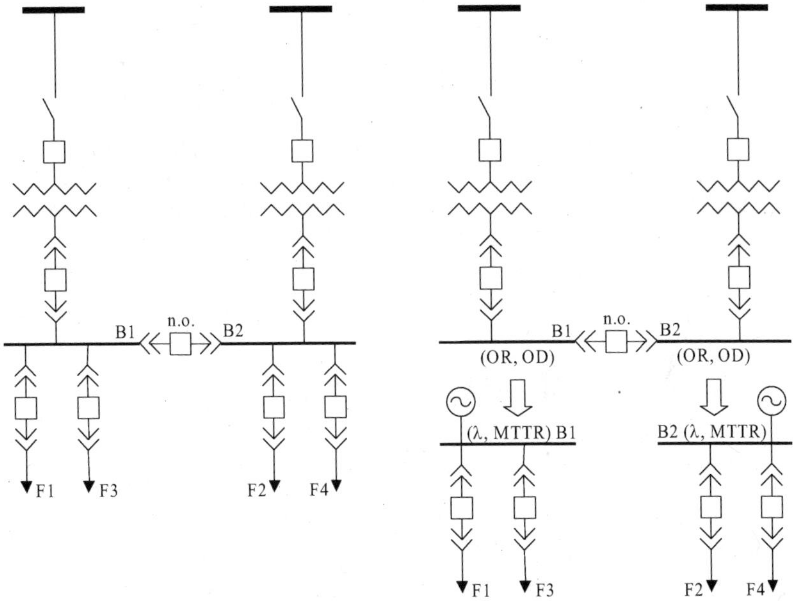

Figure 6.3. Bulk power and substation reliability can be incorporated into a distribution reliability model through secondary bus equivalents based on bus outage rates and outage duration. This technique allows bulk power and substation reliability to be computed using different tools designed especially for the task, and can substantially reduce the size of distribution models.

An equivalent substation model can be generated by making a detailed model that excludes all feeder breakers. A reliability analysis will produce an annual outage rate and an annual outage duration for each low voltage bus of the substation. The following formulae can then be used to compute equivalent failure rates and repair times for equivalent buses in the distribution reliability model:

$$\lambda_{equiv} = OR \tag{6.1}$$
$$MTTR_{equiv} = OD / OR \tag{6.2}$$

λ_{equiv} :	Equivalent bus failure rate (failures per year)
$MTTR_{equiv}$:	Equivalent bus mean time to repair (hours)
OR:	Computed bus outage rate (outages per year)
OD:	Computed bus outage duration (hours per year)

Consider the example in Figure 6.3, which consists of two transmission lines, two transformers, two buses (B1, B2), and four feeders. An equivalent substation/bulk power model is created by computing the reliability of the system excluding the feeders, resulting in the annual outage rate (OR) and outage duration (OD) for each bus. For each bus in the bulk power model, Equations 6.1

and 6.2 can be used to create an equivalent bus in the distribution system reliability model.

In lieu of an equivalent model, the impact of substations and bulk power can be eliminated entirely from a distribution reliability model. In this case, the reliability model will not compute expected customer interruptions. Rather, it will compute the contribution of customer interruptions due to the distribution system. This approach will result in the same absolute changes when comparing the reliability impact of system alterations, but will result in different percentages in change. Therefore, care must be taken to ensure that reliability measurements and targets are adjusted to only reflect the distribution system.

6.1.3 Switchable Section Equivalents

Switchable sections are subsets of a distribution system bounded by switchable devices. Because all components in a switchable section are always electrically connected, each will have identical reliability characteristics. In addition, any failure within the switchable section will have the same system impact regardless of where it occurs (assuming varying fault current levels do not result in different protection device response). For these reasons, switchable sections can be reduced to single component equivalents (see Figure 6.4).

The first step in creating a switchable section equivalent is to lump all loads into an equivalent load. This lumped load represents the aggregated load and aggregated customers so that capacity constraints can be accurately reflected and reliability indices can be accurately computed.

The second step in creating a switchable section equivalent is to lump all component failures into an equivalent component. An equivalent failure rate is computed by taking the sum of all component failure rates. An equivalent mean time to repair is computed by taking the weighted average of all component repair times:

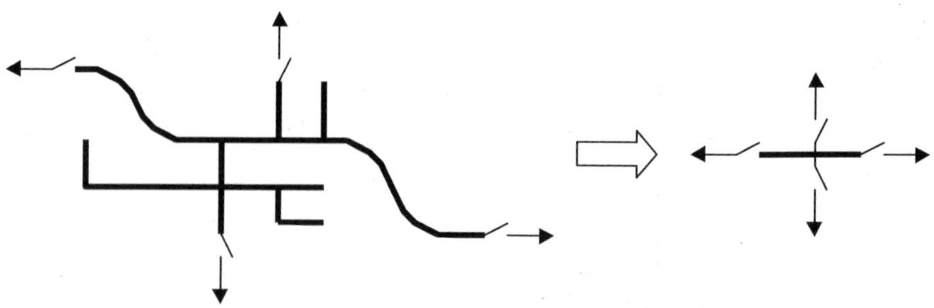

Figure 6.4. A switchable section is defined as all components bounded by switchable devices (left). Since all components within a switchable section are electrically connected, they will experience identical levels of reliability. In addition, any fault within the switchable section will cause the system to respond in the same manner. For these reasons, switchable sections can be reduced to a single equivalent component (right).

Reducing a Switchable Section

$$\lambda_{equiv} = \sum_{i=1}^{N} \lambda_i \qquad (6.3)$$

$$MTTR_{equiv} = \frac{1}{\lambda_{equiv}} \sum_{i=1}^{N} \lambda_i \cdot MTTR_i \qquad (6.4)$$

λ_{equiv}: Equivalent failure rate (failures per year)
$MTTR_{equiv}$: Equivalent mean time to repair (hours)
λ_i: Failure rate of component i (failures per year)
$MTTR_i$: Mean time to repair of component i (hours)
N: Number of components in switchable section

Although switchable section equivalents will greatly reduce system size without degrading solution quality, they can reduce the usefulness of reliability models if used in excess since (1) the system is no longer a geographical representation of the actual system, and (2) equivalent components no longer correspond to actual components. These limitations can be overcome by temporarily reducing models just before an engine is run and cleverly integrating results back into the original model. These limitations can also be minimized by selective and judicious application of model reduction.

When reducing switchable sections, it is often desirable to preserve the general geographic representation of lines and cables. This can be done by combining pairs of connected lines only if they are the same line type and are nearly co-linear (have an included angle near 180°). Consider a straight-line stretch of feeder consisting of 50 segments. These segments can easily be combined into a single equivalent component with near-equivalent routing, as long as there are no branches between the first and last segment. Combining line segments is done using Equations 6.3 and 6.4.

In many cases, switchable section reduction applies to radial taps and radial secondary systems. If a radial tap is protected by a fuse (or other device) and does not have downstream switching capability, it can be reduced to an equivalent component representing both the failures associated with the tap and the customers associated with that tap. Similarly, a protected distribution transformer model that only serves radial service drops can include all downstream secondary exposure and all downstream load.

If components within the switchable section are subject to open circuits, electrical separation can occur and the use of equivalents can produce errors. Typically, the only components that experience a large number of open circuits are switchable protection devices, and this caution can be ignored. Switchable sections that may experience high-impedance line faults or self-extinguishing cable faults should not be reduced to an equivalent.

6.2 SYSTEM CALIBRATION

After a reliability model has been created, it is desirable that the results of the model reflect the historical reliability of the system it represents. Consider an actual distribution system with a SAIFI of 1 interruption per year and a SAIDI of 2 hours per year. If a reliability model of this system predicts a SAIFI of 5 interruptions per year and a SAIDI of 10 hours per year, planners and engineers will not trust other results such as the impact of potential reliability improvement options.

Since the accuracy of system geography and connectivity are relatively easy to verify, most modeling errors result from unrepresentative component data. Because of this, careful and systematic treatment of data is a critical part of building a credible reliability model. Usually, utilities will not have enough historical data to directly compute the reliability characteristics of all types of distribution components. To compensate, reliability models should be based on a careful selection of default data, selective use of customized data and, finally, model calibration to ensure that predicted results match all available historical information.

6.2.1 Default Data

One of the first steps in creating a reliability model is to assign default data to all component parameters. This data should be based on components or systems that are as similar to the system being modeled as possible, with sources ranging from books to journals to conference proceedings to manufacturer's test data (e.g., the component reliability data provided in Chapter 4 of this book).

Most reliability assessment applications allow default data to be automatically assigned to components when they are placed in a model. Doing so ensures that all of the information required to perform a reliability assessment is available and allows a quick preliminary analysis to be performed after the system topology has been entered.

A convenient way to handle default data is to create a library of default component templates, with each template representing a hypothetical component with associated reliability parameters. When a component is placed in the system model, it is automatically assigned a default template and inherits all corresponding reliability parameters. This method allows many types of equipment to be conveniently managed. For example, a separate default template can be created for different types of cable sizes, with each template having a different capacity. Similarly, a separate template can be created for different switch types, with manual switches having a long switching time and automated switches having a short switching time.

Using default component templates provides several advantages besides reducing modeling time. Reliability parameters for all components sharing the same default template can be modified simultaneously and are assured of being the same value (e.g., increasing the failure rate of all "15-kV Hydraulic Reclosers" by 20%). Queries can be made for components with specific default templates (e.g., determining how many miles of "4/0 Copper with 8' Crossarms" are on the system). Even entire default template libraries can be replaced to reflect different global assumptions (e.g., replacing library "Well Maintained Equipment" with "Poorly Maintained Equipment").

Although much insight can be achieved through systems based entirely on default data, increased confidence is obtained by replacing some default data with customized data, and modifying some default data through system calibration techniques.

6.2.2 Customized Data

Customized data is typically based on historical failure data and is used to replace default data for one or more pieces of equipment. The most common type of data customization is based on equipment failure history, and allows the default failure rates for certain types of equipment to be replaced. Consider an operating region with 500 fuse cutouts, with 30 fused cutout failures having occurred over the last five years. These numbers result in an expected failure rate of 30 / 500 / 5 = 0.012 failures per year. Since this value is more likely to reflect fuse cutout performance than data taken from other sources, all fuse cutout failure rates in this operating region should be customized to reflect this value.

Customization based on historical data is only valid if the data set is statistically significant. For this reason, there are typically limits as to how many different failure rates can be assigned to equipment for various characteristics. Consider a utility with 1000 power transformers and 10 years of data describing 200 failures. On average, these transformers have a failure rate of 0.02 per year. These transformers, however, vary widely based on a number of characteristics related to reliability including age, voltage class, and loading history. Now consider classifying the transformers based on five age categories (0-10, 10-20, 20-30, 30-40, and 40+ years), three voltage classes (5 kV, 15 kV, and 35 kV), and three levels of loading history (light, medium, and heavy). This results in 5 x 3 x 3 = 45 subcategories with an average of 4.4 failures per subcategory—not enough to be statistically significant. In general, each customized reliability value should be based on at least 20 historical failures, so that a random variation of ± 1 failure will only impact failure rates by approximately 10%.

This said, it is often useful to customize component data based on intuition rather than historical data. This allows heuristic human knowledge to be reflected in the reliability model. For example, consider a system with a calculated overhead line failure of 0.2 per mile per year. The number of trees surrounding these lines varies greatly, but historical data is not detailed enough to allow failures rates to be customized based on tree density. Regardless, it is common knowledge that failures occur much more frequently in heavily treed areas when compared to lightly treed areas. In this case, it is perfectly acceptable to disproportionally allocate failures (as long as all assumptions and techniques are clearly stated). For example, 25% of lines with lightest vegetation can be assigned a failure rate of 0.1 per miles per year, and 25% of lines with heaviest vegetation can be assigned a failure rate of 0.3 per mile per year. Average line failure rates have not changed, but heavily treed areas now have three times as many failures as lightly treed areas.

Historical component data has been scarce in the past, but is becoming increasingly more available and accurate as more utilities utilize outage management systems. This data can profoundly increase the precision and usefulness of reliability models and can ultimately become a basis of competitive advantage for achieving cost-effective system reliability.[3]

6.2.3 Calibrated Data

Most utilities do not have a substantial amount of historical component reliability data. Nearly all utilities, however, have historical system reliability data in the form of reliability indices. When a study area is modeled, the reliability indices predicted by the assessment tool should agree with these historical values. If so, a certain level of confidence in the model is achieved and more specific reliability results (e.g., the reliability of a specific load point or the impact of a design change) can be trusted to a higher degree. The process of adjusting component reliability data so that predicted reliability matches historical reliability is referred to as *model calibration*.

When calibrating a study area, component reliability parameters should be adjusted based on the following two factors: (1) confidence that parameters accurately reflect component behavior, and (2) sensitivity of reliability indices to changes in the parameters. Based on these two factors, it is usually appropriate to calibrate study areas by adjusting the failure rates and repair times of overhead lines and underground cables.[4]

This section describes how to adjust line section default parameters so that predicted values of MAIFI, SAIFI, and SAIDI agree with historically computed values. The specific default reliability parameters that will be adjusted are temporary failure rate (λ_T, in failures per mile per year), permanent failure rate (λ_P, in failures per mile per year), and Mean Time To Repair (MTTR, in hours per year).

Determining the appropriate values of default line λ_T, λ_P, and MTTR can be broken down into two steps. This is possible because MTTR has no effect on

MAIFI and SAIFI. Consequently, λ_T and λ_P can be chosen so that predicted MAIFI and SAIFI values match historical values. After this has been done, MTTR can be adjusted to control SAIDI without impacting MAIFI or SAIFI. The relationship of SAIFI and MAIFI to adjustments of λ_T and λ_P ($\Delta\lambda_T$ and $\Delta\lambda_P$) is:

$$\begin{bmatrix} \text{SAIFI}_t \\ \text{MAIFI}_t \end{bmatrix} = \begin{bmatrix} \frac{\partial \text{SAIFI}}{\partial \lambda_P} & \frac{\partial \text{SAIFI}}{\partial \lambda_T} \\ \frac{\partial \text{MAIFI}}{\partial \lambda_P} & \frac{\partial \text{MAIFI}}{\partial \lambda_T} \end{bmatrix} \begin{bmatrix} \Delta\lambda_P \\ \Delta\lambda_T \end{bmatrix} + \begin{bmatrix} \text{SAIFI}_i \\ \text{MAIFI}_i \end{bmatrix} \tag{6.5}$$

SAIFI_t : Target SAIFI value
MAIFI_t : Target MAIFI value
SAIFI_i : Initial SAIFI value
MAIFI_i : Initial MAIFI value
λ_P : Permanent Line Failure Rate (/km/yr)
λ_T : Temporary Line Failure Rate (/km/yr)

In the equation, the partial derivatives can be approximated by perturbation methods. This allows the values of $\Delta\lambda_T$ and $\Delta\lambda_P$ to be computed. New values of λ_P and λ_T are then found by adjusting the old value by the computed change: $\lambda_{new} = \lambda_{old} + \Delta\lambda$. Once these new values are computed, the relationship between SAIDI and MTTR can be used to find the new value of MTTR. This relationship is:

$$\text{SAIDI}_t = \frac{\partial \text{MTTR}}{\partial \text{MTTR}} \Delta\text{MTTR} + \text{SAIDI}_i \tag{6.6}$$

SAIDI_t : Target SAIDI value
SAIDI_i : Initial SAIDI value
MTTR : Mean Time To Repair (hours)

If all parameter/index relationships are linear within the ranges considered, Equations 6.5 and 6.6 will results in predicted MAIFI, SAIFI, and SAIDI values that exactly match historical values. If an exact match is not found, the process can be repeated iteratively until a certain error tolerance is reached.

The process of calibrating three default parameters based on three reliability indices can be extended through the concept of default data multipliers. A default component parameter multiplier is simply a number that corresponds to a component parameter. The parameter value that will be assigned to the component parameter is equal to the default component parameter multiplied by the corresponding default component parameter multiplier. For example, if a specific line section has a MTTR multiplier of 1.5 and the default line MTTR is 4 hours, then the line will be assigned a MTTR of (1.5 x 4) = 6 hours. Default component multipliers allow a system to have a single set of default parameters, but different calibrated component parameters.

Sometimes it is necessary to calibrate more than one type of component. A good example is a system that consists of both overhead lines and underground cables. In this case, there are six parameters to calibrate based on three reliability indices, three for overhead lines (λ_{P1}, λ_{T1}, $MTTR_1$) and three for underground cables (λ_{P2}, λ_{S2}, $MTTR_2$). This leads to an underdetermined set of equations with multiple solutions. A solution is typically found by selecting an initial guess and identifying the solution nearest to this guess by performing a gradient descent in the error space. For the initial part of the problem, error is defined as:

$$\text{Error} = \frac{1}{2}\left[(\text{SAIFI} - \text{SAIFI}_t)^2 + (\text{MAIFI} - \text{MAIFI}_t)^2\right] \tag{6.7}$$

To solve the equation, each parameter is adjusted in small amounts in proportion to its error sensitivity. For example, the sensitivity of error to overhead line failure rates and the corresponding adjustment are shown in Equations 6.8 and 6.9, respectively. η determines convergence speed and resolution, is typically between 0.01 and 0.1, and is the same for all parameters.

$$\frac{\partial \text{Error}}{\partial \lambda_{P1}} = (\text{SAIFI} - \text{SAIFI}_t)\frac{\partial \text{SAIFI}}{\partial \lambda_{P1}} + (\text{MAIFI} - \text{MAIFI}_t)\frac{\partial \text{MAIFI}}{\partial \lambda_{P1}} \tag{6.8}$$

$$\lambda_{P1} = \lambda_{P1} + \eta \frac{\partial \text{Error}}{\partial \lambda_{P1}} \tag{6.9}$$

After each small adjustment to every parameter, gradients are recomputed and the process repeats until error falls below a specified threshold. Repair times can then be adjusted to calibrate SAIDI based on the following error and gradient formulae:

$$\text{Error} = \frac{1}{2}(\text{SAIDI} - \text{SAIDI}_t)^2 \tag{6.10}$$

$$\frac{\partial \text{Error}}{\partial \text{MTTR}_1} = (\text{SAIDI} - \text{SAIDI}_t)\frac{\partial \text{SAIDI}}{\partial \text{MTTR}_1} \tag{6.11}$$

$$\frac{\partial \text{Error}}{\partial \text{MTTR}_2} = (\text{SAIDI} - \text{SAIDI}_t)\frac{\partial \text{SAIDI}}{\partial \text{MTTR}_2} \tag{6.12}$$

The importance of model calibration cannot be understated. Unless the predictions of a reliability model correspond to actual historical performance, the model cannot be trusted and results are of questionable value. Proper calibration is also an enabling process for utilities without historical component failure data. Once convinced that component failure rates and repair times can be accurately inferred from reliability indices, utilities without a large amount of historical component reliability data can justify the incorporation of reliability analysis into their planning process and begin to realize the associated benefits.

6.3 SYSTEM ANALYSIS

Once a distribution system is modeled and calibrated, meaningful analyses can be performed. Typically the first step is to assess the reliability of the existing system without any modifications. Doing so can identify the reliability that each customer can expect, as well as the geographic distribution of reliability. Further, it can identify the major contributing factors to reliability as well as the sensitivity of reliability to various system aspects. The remainder of this section divides system analysis into four aspects: reliability analysis, risk analysis, sensitivity analysis, and root-cause analysis.

6.3.1 Reliability Analysis

A reliability analysis computes the expected reliability of the system. Typically, results consist of the momentary interruption frequency, sustained interruption frequency, and total annual interruption duration of each customer. Many analyses will also compute the expected number of operations for each switch and protection device. Regardless, a reliability analysis returns an enormous amount of data that must be turned into information for it to be useful.

The most common method of turning a large amount of reliability data into reliability information is through the use of reliability indices such as MAIFI, SAIFI, and SAIDI. In essence, these represent the average reliability for a group of customers. Since a typical reliability model is calibrated to historical reliability indices, no insight is derived from computed indices representing areas equal to or greater than the areas represented by this historical data. For example, if a substation is calibrated to a historical SAIDI of 120 min/yr, it is uninteresting to see that the calibrated model also predicts a SAIDI of 120 min/yr. It may be interesting, however, to look at reliability indices for areas smaller than the calibration areas. For instance, it is often beneficial to look at predicted feeder indices after calibrating a model by substation service territory.

Visualization is a powerful method of representing a large amount of data in a meaningful manner. Techniques can range from graphs and charts to geographic-based techniques, and their effectiveness for conveying various types of data are well documented.[5-6] The most common types of visualization used in reliability assessment are data maps and histograms, with the former suitable for representing geographic distributions and the latter suitable for representing quantitative distributions of results.

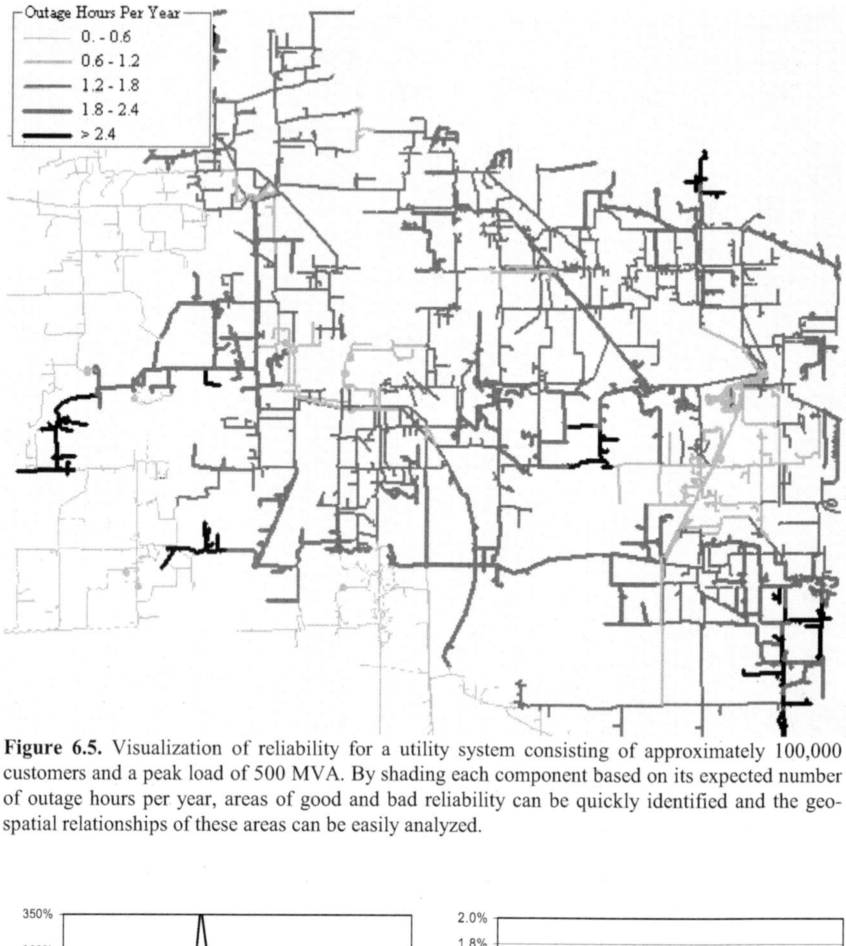

Figure 6.5. Visualization of reliability for a utility system consisting of approximately 100,000 customers and a peak load of 500 MVA. By shading each component based on its expected number of outage hours per year, areas of good and bad reliability can be quickly identified and the geospatial relationships of these areas can be easily analyzed.

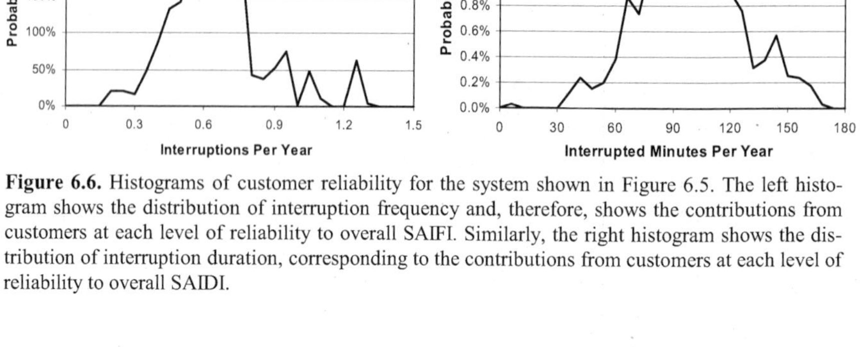

Figure 6.6. Histograms of customer reliability for the system shown in Figure 6.5. The left histogram shows the distribution of interruption frequency and, therefore, shows the contributions from customers at each level of reliability to overall SAIFI. Similarly, the right histogram shows the distribution of interruption duration, corresponding to the contributions from customers at each level of reliability to overall SAIDI.

Data maps visualize information by shading geographically located information. For reliability assessment visualization, each component is typically shaded based on a particular result such as outage frequency or outage duration. Because reliability results do not vary continuously with geography, visualization is best accomplished by grouping results in two to ten bins, with each bin being assigned a shade, weight, and color that easily distinguishes it from other bins. Figure 6.5 shows an example data map that visualizes interruption duration information by using a combination of line weighting and gray scale.

Data maps are a powerful tool, but are not suitable for data visualization of information that is not strongly related to system geography. A typical example is customer interruption information. Since customer density can vary greatly on different parts of the system, it is difficult to infer the distribution of customers with different levels of reliability from a data map. A more appropriate method is to display this type of data in a histogram. These histograms make it easy to identify the percentage of customers that are experiencing each level of reliability, and provide much more information than simple statistical data such as averages and variances. Example histograms for customer interruption frequency and customer interruption duration are shown in Figure 6.6.

6.3.2 Risk Analysis

Reliability varies naturally from year to year. Some years will prove lucky and reliability will be higher than expected. Other years will be unlucky and reliability may be much worse than average. This natural variation in reliability is an important aspect of distribution systems and can be addressed in an analytical manner by using the techniques presented in Section 5.7.4.

At a minimum, a risk assessment should compute the mean and standard deviation of reliability indices as they naturally vary on an annual basis. A more comprehensive approach will compute the probability distribution of reliability outcomes and identify the scenarios associated with years of unusually poor reliability. From this perspective, reliability improvement projects can be justified on the basis of reducing risk rather than reducing expected value. For example, it may be acceptable to make the expected value of SAIDI slightly worse if the risk of having a very bad year is reduced.

The variability of reliability tends to become less as the system size increases. For example, the variability of SAIDI for the entire service territory of a large utility will be relatively small, whereas the variability of reliability for an individual customer on this system will tend to be much higher. This phenomenon is illustrated in Figure 6.7, which is based on a reliability model of a 250-MVA urban service territory in Florida. This figure shows the SAIDI distribution for the entire system as well as the SAIDI distribution for the three substation service territories within the system.

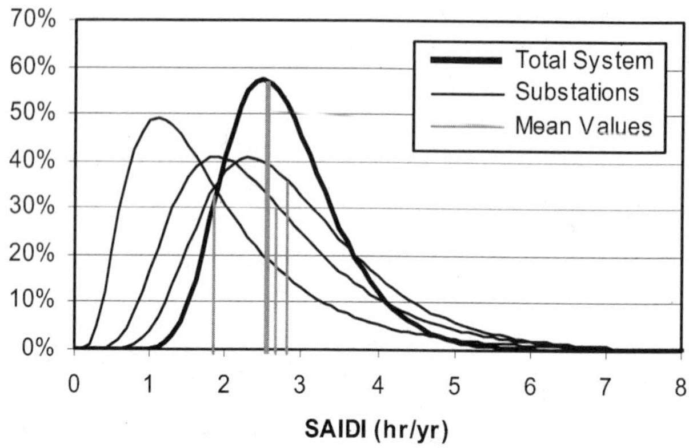

Figure 6.7. SAIDI risk profiles for a 250-MW, three substation service territory in southern Florida. Each of three substation curves is seen to have a greater variance that the total system curve. This is a reflection of the principle that, in general, reliability becomes more predictable as system size increases.

In Figure 6.7, the total system curve does not resemble the curves of the three substations that make up the system. Two substations have a worse mean reliability, and one has a better mean reliability. Further, the substation curves have shorter modes, indicating greater variance. Comparisons have also been made with feeder curves and load curves. Load curves tend to be the most variable, followed by feeder curves, substation curves, and system curves.[7] There are exceptions, but as the number of loads in an area increases, reliability tends to become more predictable from year to year.

6.3.3 Sensitivity Analysis

The sensitivity of a function to a parameter is defined as the partial derivative of the function with respect to that parameter. This is a measure of how much the value of the function will change if the parameter is perturbed, and can be approximated by actually perturbing the parameter (keeping all other parameters fixed) and measuring how much the function changes.

$$\frac{\partial F(x_1, x_2, ..., x_n)}{\partial x_1} \approx \frac{F(x_1 + \Delta x_1, x_2, ..., x_n) - F(x_1, x_2, ..., x_n)}{\Delta x_1} \tag{6.13}$$

For example, consider a system with a default overhead line mean time to repair (MTTR) of 400 minutes, and the predicted SAIDI of the system with this assumption is 100 min/yr. The sensitivity of SAIDI to MTTR is computed by slightly perturbing MTTR and re-computing SAIDI. In this case, MTTR is increased by 4 minutes and the new SAIDI is computed to be 100.5 min/yr. The sensitivity of the system is equal to (100.5 min/yr. 100 min/yr) / 4 min = 0.125 min/yr/min. That is, the linearized projection of SAIDI will increase by 0.125 min/yr for each 1-min increase in MTTR.

Sensitivities can also be computed as percentages. For the example in the previous paragraph, SAIDI increases 0.5% (100 min/yr to 100.5 min/yr) for a 1% increase in MTTR. Therefore, the sensitivity of SAIDI to default overhead line MTTR is (0.5% ÷ 1%) x 100% = 50%.

Sensitivity analyses are useful for many aspects of system analysis. First, they can be used to test the sensitivity of results to default reliability data. Concern over an uncertain data assumption may be mitigated if results are found to be insensitive to these assumptions. Second, sensitivity results can be used to efficiently calibrate systems to historical reliability data (see Section 6.2.3). Last, sensitivity analyses can be used to help anticipate whether certain actions can be expected to have a significant reliability impact on the system. For example, if system SAIDI is highly sensitive to overhead line failure rates, reducing overhead line failures will probably be an effective strategy for reducing SAIDI (but not necessarily cost effective).

Figure 6.8 shows a sensitivity analysis where system sensitivities vary widely depending on the reliability index being examined. This shows that in certain cases, sensitivity scores for certain components can vary widely depending upon the reliability metric that is chosen. In this example, SAIDI is very sensitive to substation transformer failure rates while SAIFI is very insensitive to the same parameter.

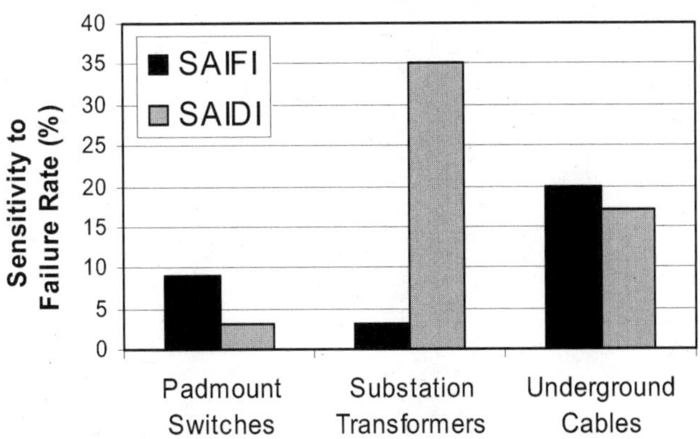

Figure 6.8. Sensitivities of reliability indices to equipment failure rates for the distribution system of a military base in the southern US. For this system, SAIFI is least sensitive and SAIDI is most sensitive to the failure rate of substation transformers. Depending on reliability improvement goals, addressing substation transformer failures could be very effective (reduce SAIDI) or very ineffective (reduce SAIFI).

6.3.4 Root-Cause Analysis

Knowing the expected reliability of a system is valuable, but knowing the major factors that contribute to poor reliability can be just as valuable. A root-cause analysis accomplishes this by determining the contribution of each component to reliability indices. For example, if reliability is measured using SAIDI, a root-cause analysis will identify the components that are having a large impact on SAIDI and the components that are having a minimal impact on SAIDI. Results can be displayed graphically to quickly identify problem areas of the system (see Figure 6.9).

A predictive root-cause analysis is different from a physical root-cause analysis. While a physical root-cause analysis identifies actual cause of a component outage, a predictive root-cause analysis identifies the expected total contribution of a component to reliability indices or other measures.[8] If a component has a high failure rate, it does not necessarily have a high root-cause score. The calculation must also consider the number of customers impacted by the failure, the duration of the failure, and the ability of the system to restore interrupted customers before the failure is repaired.

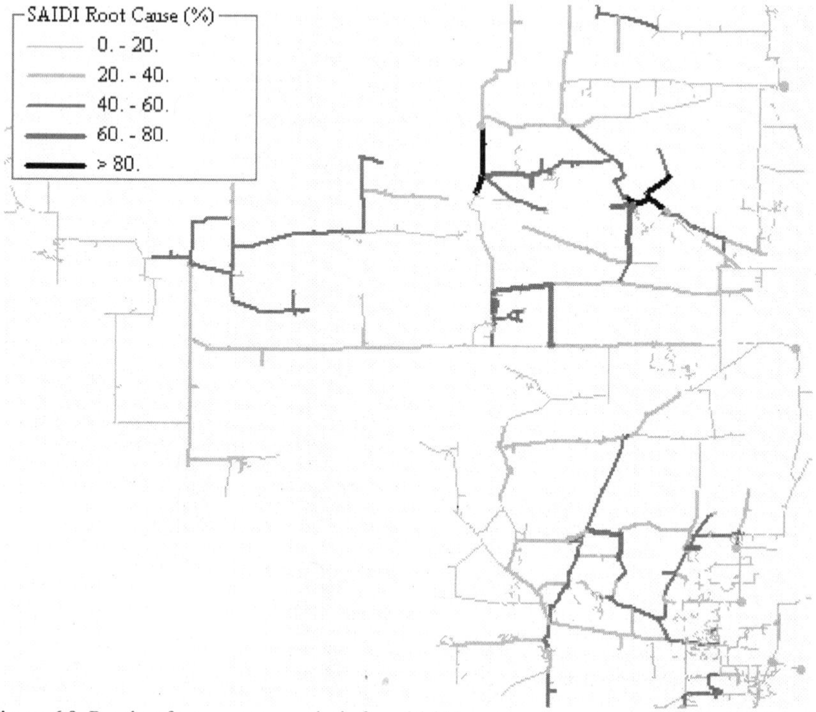

Figure 6.9. Results of a root-cause analysis for a 34,000-customer area in the US Midwest. Heavily shaded components contribute the most to SAIDI and are good places to start when looking for effective ways of improving system reliability.

To illustrate how component root-cause scores are computed, consider a component with a failure rate of 0.02 failures per year on a system with 10,000 customers. When this component fails, 8000 customers do not experience an interruption, 1500 customers experience a 1-hr interruption, and 500 customers experience a 4-hr interruption. The SAIDI root-cause score for this component is equal to the weighted sum of customer interruptions multiplied by the frequency of these interruptions. In this case, the SAIDI root-cause score is (0 hr x 8000 + 1 hr x 1500 + 4 hr x 500) x 0.02 /yr = 70 customer hours per year. Once the root-cause score is computed for each component, all scores can be normalized by expressing them as a percentage of the highest scoring component.

Each root-cause analysis addresses a single specific reliability metric. A component with a high root-cause score for SAIDI will not necessarily have a high root-cause score for SAIFI or $MAIFI_E$. Because of this, a separate root-cause analysis is commonly performed for all reliability indices of interest (often $MAIFI_E$, SAIFI, and SAIDI). In addition, a root-cause analysis can be performed for a separate "meta-index" that is a function of the more common indices. For example, if a utility wants to improve $MAIFI_E$, SAIFI, and SAIDI with the most emphasis on SAIDI and the least emphasis on $MAIFI_E$, a root-cause analysis can be performed for a weighted sum such as $0.1 \cdot MAIFI_E + 0.3 \cdot SAIFI + 0.6 \cdot SAIDI$. Results of this particular root-cause analysis will weight contribution to SAIFI three times as much as contributions to $MAIFI_E$, and will weight contribution to SAIDI twice as much as contributions to SAIFI.

6.3.5 Loading Analysis

System reliability is a function of system loading. This is especially true for interruption duration measures such as SAIDI. If a system is lightly loaded, system operators can freely reconfigure the system to restore customers after a fault occurs. Since equipment is loaded well below emergency ratings, there is no danger of system reconfiguration resulting in equipment overloads. As a system becomes more heavily loaded, some system reconfiguration options become constrained and SAIDI begins to deteriorate. If all equipment is normally loaded above emergency ratings, no load transfers can occur and SAIDI is at its worst.

Since reliability varies as a function of load, it is often beneficial to compute system reliability for a range of loading levels. Doing so can give insight on how expected reliability will change according to weather severity (e.g., a mild summer with a low peak demand or a hot summer with a high peak demand), and will also give insight on how reliability will change with load growth. Reliability versus loading curves, such as those seen in Figure 6.10, are easily generated by multiplying all loads by a global scaling factor. It is usually sufficient to compute reliability for load levels ranging from 0% to 200% of peak, with steps ranging from 5% to 25% depending upon the resolution required.

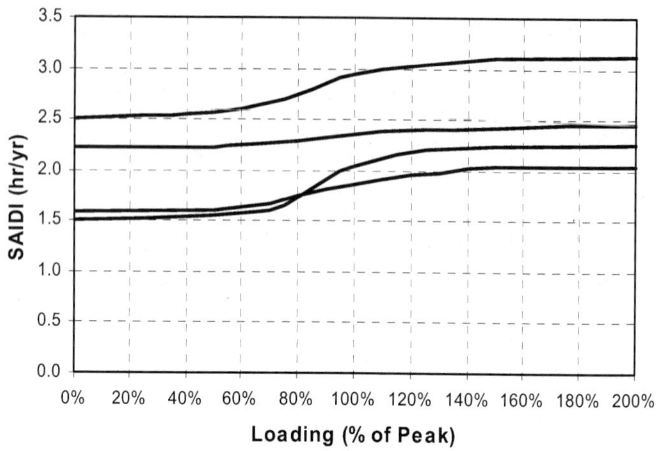

Figure 6.10. Reliability versus loading curves for four actual US utility substations. SAIDI is seen to increase with loading in each case, but the rate and degree of increase depend on a host of factors including the severity of peak loading, the number of normally closed switches, and the number of normally open switches.

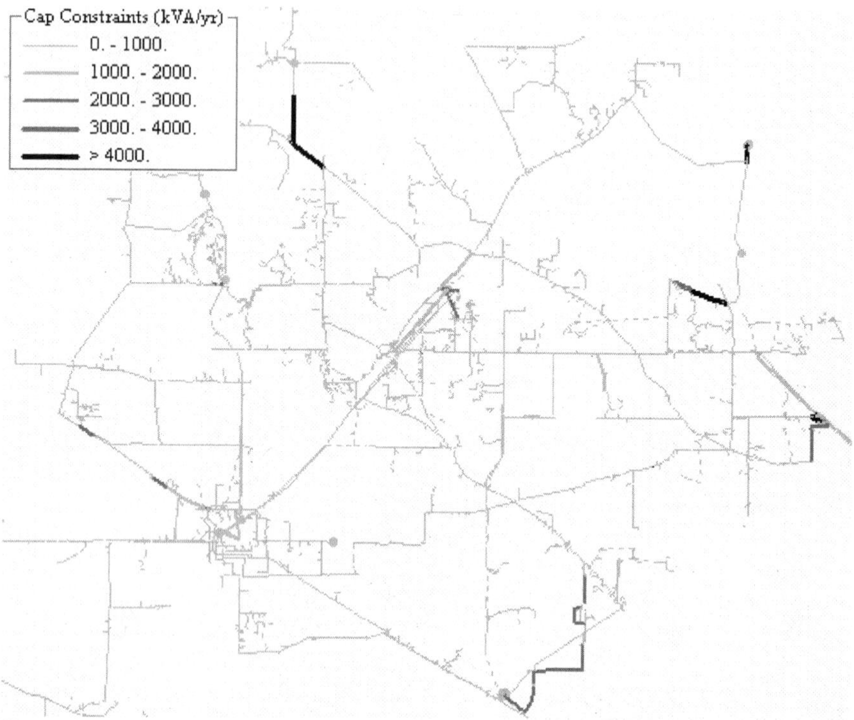

Figure 6.11. Reliability assessment algorithms can easily keep track of equipment capacity constraints that restrict post-contingency load restoration. Dark shading reveals components that are expected to constrain the most kVA of load restoration per year.

SAIDI versus loading curves are typically characterized by a sigmoidal shape. At 0% loading, SAIDI is at a minimum and remains at this minimum until equipment capacity constraints begin to occur. At this point, SAIDI will begin to increase until a maximum is reached and all beneficial load transfers are constrained. The point at which SAIDI begins to increase primarily depends upon how heavily loaded the system is at 100% of peak. A very lightly loaded system may not see SAIDI begin to deteriorate until 200% of peak or greater, while a heavily loaded system may experience SAIDI deterioration at 30% of peak or lower. The range of best to worst SAIDI is independent of peak loading and primarily depends upon the number of normally open and normally closed switches on the system.

It is often helpful to be able to identify the components that are contributing most to load transfer constraints. To do this, the reliability assessment algorithm keeps track of all components that constrain the potential transfer of a block of load. Consider a contingency that results in an interrupted switchable section. Unconstrained, this switchable section can be transferred to another source through a transfer path. Each component on the transfer path has a loading margin equal to its emergency rating minus its normal loading. If the loading on the switchable section is greater than the loading margin of the component, a capacity constraint occurs. To keep track of these occurrences, each component causing a capacity constraint keeps a running total of the frequency of each constraining event and the severity of each event (equal to the frequency multiplied by the load on the switchable section). Components that result in a large number of large capacity constraints can be identified in a sorted list or through the use of visualization techniques (see Figure 6.11).

6.4 IMPROVING RELIABILITY

Although reliability assessment models can provide much insight into the state of a distribution system, their primary value is in their ability to quantify the impact of design improvement options.[9] In fact, the temptation of quickly identifying effective design improvement options often results in an abbreviated system analysis effort, which is not recommended. Taking the time to perform a thorough system analysis of the existing system will generally allow for higher quality design improvement options to be identified in a shorter amount of time.

Defining criteria and constraints is another important step to complete before exploring reliability improvement options. Both criteria and constraints may consist of one or more of the following: reliability indices, individual customer reliability, risk profiles, and cost. Other factors, such as redundancy, equipment loading, and voltage can usually be ignored since they are not usually binding.

The remainder of this section focuses on reliability improvement strategies and tactically applying these strategies in reliability assessment models. When considering each strategy, it is helpful to understand how it will impact the reliability of each customer in terms of momentary interruption frequency, sustained interruption frequency, and sustained interruption duration. Oftentimes, the reliability of certain customers can be improved at the expense of other customers. At other times, one aspect of a particular customer's reliability can be improved at the expense of another aspect. Distribution planners have always known about these tradeoffs, but reliability models allow them to be assessed and compared with analytical rigor.

6.4.1 Protection Devices

Adding protection devices is one of the most straightforward and effective methods for improving distribution system reliability. Assuming proper coordination, increasing the number of protection devices reduces the number of customers that experience interruptions after a fault occurs. Stated differently, increasing the number of protection devices increases the selectivity of the protection system[10-11].

The first step towards improving reliability is to place a protection device, typically a fuse, on all radial branches. Both field experience and reliability studies show conclusively that laterals should be fused.[12] The only compelling reasons not to fuse a lateral are nuisance fuse blowing (which can generally be avoided by specifying larger fuses) and the inability to coordinate. Three-phase laterals may require devices with 3ϕ lockout capability if they serve large motors, which may be damaged by unbalanced voltages, or transformers with primary delta-connected windings, which may create safety problems due to the possibility of backfeeding.

The effectiveness of lateral fusing increases as total lateral exposure increases and as average lateral length decreases. Assuming perfect fuse operation, a fault on an unfused lateral will interrupt the entire feeder while a fault on a fused lateral will only interrupt customers on that lateral. Figure 6.12 shows the sensitivity of SAIFI to lateral exposure for various lateral lengths (expressed as a ratio of average lateral length to main trunk length, L/M). When lateral lengths are short when compared to the main trunk, SAIFI decreases nearly in proportion to total lateral exposure—if short laterals make up X% of total feeder exposure, lateral fusing will generally reduce SAIFI by X% when compared to an equivalent unfused system. This rule of thumb is based on a homogeneous feeder and actual improvement for actual feeders can, of course, be computed by adding lateral fuses to a reliability model.

Main trunk protection, typically in the form of a recloser, can also be an effective method of improving feeder reliability. Although exceedingly effective, many utilities have not systematically considered the application of reclosers for reliability improvement (they are most commonly used to protect areas of the feeder with low levels of fault current). Rather, line reclosers for reliability are limited to the protection of radial branches and a few critical customers.

Strategically placing a line recloser on the main trunk of a feeder can be one of the simplest methods to achieve substantial reliability improvements. Consider a line recloser placed in the center of a homogeneous feeder. Since this device will prevent interruptions to 50% of customers for 50% of all faults, $MAIFI_E$ and SAIFI for the feeder will be reduced by 25%. Similarly, two evenly spaced reclosers will reduce these indices by 33% and three will reduce them by 37.5%. Reclosers are also effective to use where the main trunk splits into two or more branches, effectively isolating each branch from faults occurring on other branches.

Some utilities do not consider main trunk protection due to the difficulty of maintaining protection coordination as the distribution system is reconfigured. Although a valid concern in the past, modern electronic reclosers can easily incorporate adaptive protection schemes to insure coordination in all possible system states.

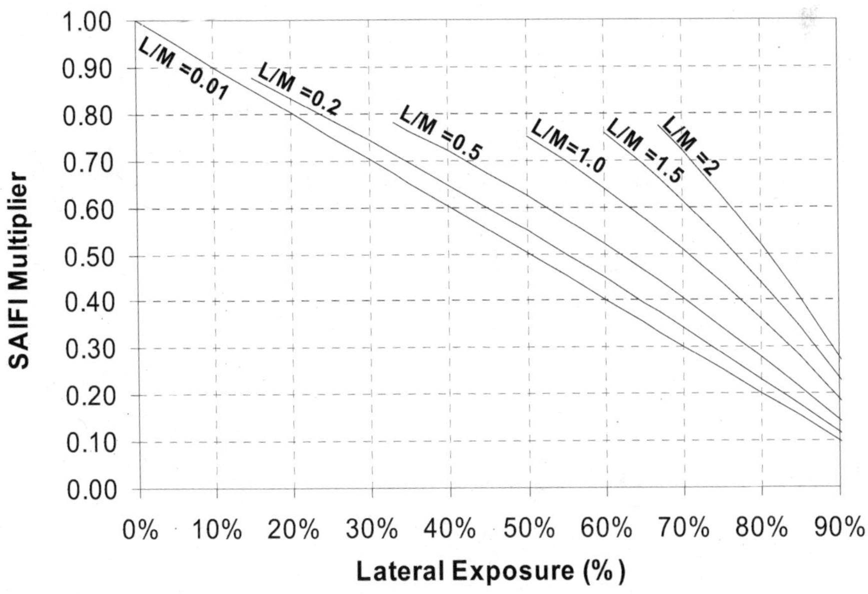

Figure 6.12. Lateral fusing can greatly reduce feeder SAIFI. Effectiveness depends upon the amount of lateral exposure (as a percentage of total exposure) and average lateral length (expressed as a ratio of average lateral length to main trunk length, L/M). For feeders with short laterals, fusing X% of total exposure will decrease SAIFI by nearly X%.

6.4.2 Reclosing Devices

Reclosing devices are most commonly used to allow temporary faults on overhead systems to self-clear. Since 70% to 80% of overhead faults are temporary in nature,[13] any feeders with primarily overhead exposure should be protected by a reclosing relay on its main circuit breaker.

Placing a line recloser on a feeder will improve the reliability of all upstream customers by protecting them from downstream faults. As such, an effective method of improving reliability for a specific customer is to place a recloser just downstream of its service connection. Though effective for targeted improvements, this type of placement strategy is myopic and reactive and may not result in the best placement from an overall system reliability perspective.

Many utility engineers are hesitant to place reclosing devices on feeders with a large amount of underground cable since cable faults are generally permanent and reclosing into a cable fault puts mechanical stress on cable insulation. In these situations, overhead sections can still be protected with line reclosers (see Figure 6.13). Doing so will not only allow temporary faults to self-clear (preventing a potential sustained interruption for the entire feeder), but will clear all faults more quickly, reducing cable insulation stress and lowering the probability of future cable faults.

There has been much confusion in recent years concerning the application of fuse saving and fuse clearing. Since fuse-saving schemes tend to result in a higher number of customer momentary problems, some utilities switch to fuse-clearing schemes when customers begin to complain about short interruptions in power. Doing this will result in less momentary interruptions for all customers but more sustained interruptions for customers on fused laterals (see Section 1.6.3). Other utilities have switched to fuse clearing since fuses in areas of high fault current tend to blow anyway, resulting in both a recloser operation and a fuse operation.

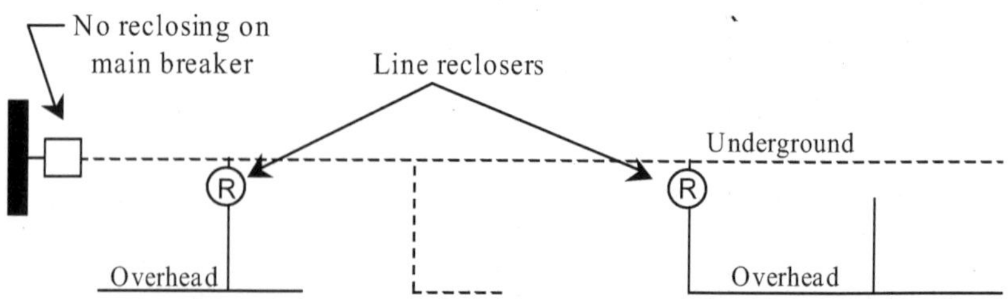

Figure 6.13. The reliability of mostly underground systems can be improved by protecting overhead sections with reclosers. In this figure, sections of underground cable are not protected by reclosing, but overhead line sections are protected with line reclosers so that temporary faults can be cleared without resulting in sustained customer interruptions.

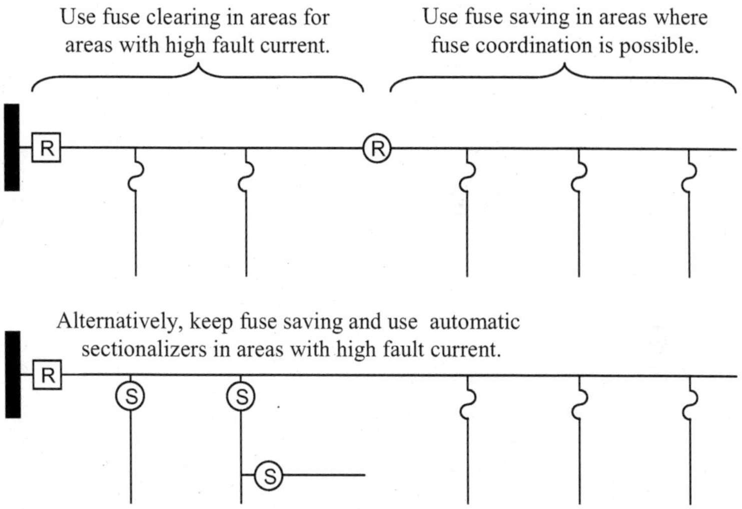

Figure 6.14. The high fault current levels near a substation may make fuse-saving schemes impossible since the main circuit breaker is not fast enough to open before the fuse melts. In these situations, fuse clearing can be used for the areas of high fault current, and fuse saving, through the application of a line recloser, can be used when fault current levels allow for proper coordination. Alternatively, fuse saving can be retained at the main breaker and fuses in areas of high fault current can be replaced with automatic sectionalizers.

The difficulties associated with fuse saving can be addressed, and the systematic use of fuse clearing is not typically recommended (although exceptions do occur). If fuses near the substation cannot be coordinated with instantaneous relays due to high fault current, fuse clearing can be used by the substation breaker and a fuse-saving line recloser can be placed out on the feeder where fault currents are lower and proper coordination can be achieved (see Figure 6.14).

Automatic sectionalizers provide another opportunity to overcome recloser/fuse coordination problems near substations. These devices detect fault current flowing through them. After a pre-set fault current count is reached, a sectionalizer will open at the next zero voltage condition. This allows fuses that cannot be coordinated to be replaced with sectionalizers while retaining fuse saving at the substation breaker (see Figure 6.14). Multiple sectionalizers can be coordinated in series by assigning them different count thresholds, and inexpensive sectionalizers are available that fit in a standard fuse cutout. Last, inexpensive single-phase reclosers that fit in a standard fuse cutout are now available.

Quantifying the impact of reclosing, fuse saving, fuse clearing, and sectionalizing requires proper algorithms in distribution reliability assessment tools. The reader is encouraged to thoroughly test algorithm behavior on simple systems to ensure that algorithm results correspond to actual system behavior. A short checklist of test scenarios includes:

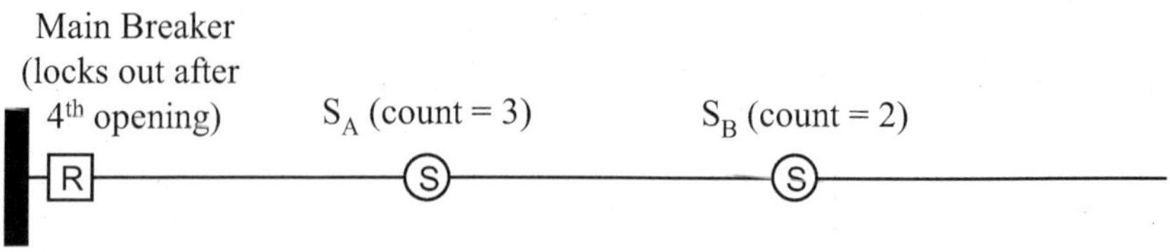

Figure 6.15. Reliability can be improved by placing 3-phase automatic sectionalizers in the main feeder trunk. In this figure, the sectionalizing switches will sequentially open from right to left, with one switch opening per reclosing cycle, until the fault is isolated or the main breaker locks out.

Test Scenarios for Reclosing Algorithms

- A temporary fault occurs at a location that is not protected by an upstream recloser. Is this fault treated like a sustained fault?
- A temporary fault occurs downstream of a fuse, which is downstream of a fuse-saving recloser. Do all customers downstream of the recloser experience a momentary interruption event?
- A permanent fault occurs downstream of a fuse, which is downstream of a fuse-saving recloser. Do all customers downstream of the fuse experience a permanent interruption, and all other customers downstream of the recloser experience a momentary interruption event?
- A temporary fault occurs downstream of a fuse, which is downstream of a fuse-clearing recloser. Do all customers downstream of the fuse experience a permanent interruption, and all other customers downstream of the recloser experience no interruption?

The algorithm being tested should result in answers of "yes" to all of the above questions if it is designed properly.

Reclosers can also be used in conjunction with 3ϕ automatic sectionalizers on the main trunk. These are similar in function to the 1ϕ devices used in place of fuses, but are typically used with more than one in series so that permanent faults can be automatically isolated near their location. Consider an overhead feeder with a 9-mi main trunk and a reclosing relay on its main breaker (see Figure 6.15). To improve reliability, 3ϕ sectionalizers can be placed at miles 3 (S_A) and 6 (S_B), dividing the main trunk into thirds. In this situation, S_B is assigned a count of 2, S_A is assigned a count of 3, and the recloser is programmed to lockout after four attempts. If a permanent fault occurs downstream of S_B, the main circuit breaker will open and close in an attempt to clear temporary faults, and open a second time when the fault persists. At this point, S_B has counted two fault currents and will open. When the breaker closes, the fault is isolated and customers upstream of S_B are automatically restored. If the fault occurs upstream of S_B but downstream of S_A, the breaker will open a third time and S_A will open on the count of three. If the fault occurs upstream of S_A, the main breaker will lock out on its fourth opening.

6.4.3 Sectionalizing Switches

Sectionalizing switches have the potential to improve reliability by allowing faults to be isolated and customer service to be restored before the fault is repaired. The effectiveness of this process depends upon how much of the feeder must be switched out to isolate the fault, and the capability of the system to reroute power to interrupted customers via normally open tie points. Generally more manual normally closed and normally-open switches will result in reduced duration-oriented indices like SAIDI and will not impact frequency-oriented indices like SAIFI. However, since each switch has a probability of failure, placing more and more switches on a feeder will eventually result in a degradation of system reliability.

The basic motivation of switching for reliability improvement is to restore interrupted customers prior to repair of the cause of the interruptions. From this perspective, the inherent value of switching is the difference between mean time to repair (MTTR) and mean time to switch (MTTS). For example, if a single switch is placed in the middle of a feeder, the 50% of customers upstream of this switch can be restored for the 50% of feeder faults occurring downstream of the switch, resulting in a SAIDI reduction of 0.25 x (MTTR - MTTS) for the feeder. If a normally open tie switch is placed at the end of the feeder, all customers can be restored for 50% of faults and SAIDI reduction is doubled to 0.5 x (MTTR - MTTS). The marginal SAIDI reduction decreases as the number of switches increases.

Figure 6.16 shows this relationship for a homogeneous feeder (full backfeeding implies that all customers outside of an isolated fault area can be restored in all cases).

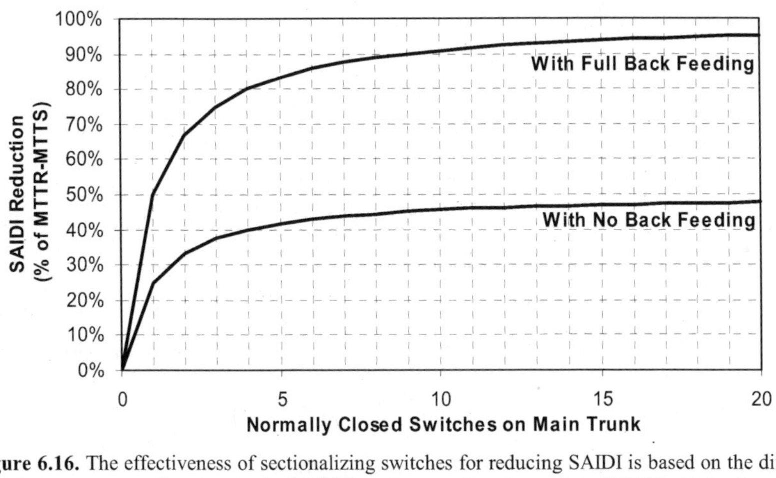

Figure 6.16. The effectiveness of sectionalizing switches for reducing SAIDI is based on the difference between mean time to repair (MTTR) and mean time to switch (MTTS). Reductions are relatively large for the first few switches and become smaller as the number of switches grows. The ability to backfeed can double SAIDI reductions when compared to no backfeeding.

There are several strategies that are helpful when placing new switches in an effort to improve reliability. The first is to keep the amount of load bounded by a set of switches limited to a value that can be reasonably fed from another source. This criterion can be tested by simulating a fault just downstream of each switch and determining whether all customers outside of the fault isolation area are able to be restored during peak loading conditions. The second strategy is to place switches just downstream of large branches or radial taps. In general, customers upstream of a switch benefit when downstream faults occur and customers downstream of a switch benefit when upstream faults occur. Since fewer reliability gains are realized for upstream faults (due to the longer time required for restoration and the possibility of not being able to backfeed), placing switches just downstream of large customer blocks is generally recommended.

6.4.4 Automation

Automation refers to remote monitoring and control of equipment. In terms of reliability improvement, automation typically refers to remotely operated substation and feeder switches. Since the main advantage of automated switches is their ability to be opened and/or closed much more quickly than manual switches, their reliability impact can be easily modeled by modifying their mean time to switch (MTTS). Users should be careful to make sure that the impact of fast switching time is appropriately reflected in momentary interruption measures such as $MAIFI_E$ and sustained interruption measures such as SAIFI. Users should also be aware that the dependability of automated switches may be lower than that of manual switches, accounted for by an increase in the probability of operational failure (POF).

Previous sections have described single-stage and two-stage restoration strategies and have discussed their qualitative differences. This section quantifies the differences between the two methods by applying predictive reliability assessment techniques to a test feeder. This test feeder, shown in Figure 6.17, is representative of a US overhead feeder with a single three-phase trunk and fused single-phase laterals.[14] The feeder is characterized by 10 MVA of peak load, 47 mi of single-phase lateral, and a 3.6-mi three-phase trunk. The trunk is divided into twelve switchable sections and is connected to an alternate source at its furthest downstream point by a normally open switch. Component reliability data is derived from "typical value" entries in the component reliability tables provided in Chapter 4. Manual switching times are 1 hr, automated switching times are 2 min, and the switching times of normally open switches are incremental so that downstream restoration may take longer than upstream restoration.

The benefits of automation are examined for both single and two-stage restoration. Test cases include no automation, full automation, and five scenarios of partial automation. The first partial scenario automates only the normally open tie switch. The second scenario automated the tie switch and the middle switch dividing the main trunk into two automated switchable sections. The remaining scenarios divide the main trunk into 3, 4, and 6 switchable sections using 3, 4, and 6 automated switches (in addition to the tie switch), respectively. The SAIDI for each scenario was computed for both single-stage switching and two-stage switching. Results are shown in Table 6.1 and Figure 6.17.

The difference in height of the two curves in Figure 6.17 is the added benefit that automation can achieve through the use of two-stage restoration. There is no difference in reliability for the case with no automation, automation of the tie switch only and full automation. Tie switch automation only results in the elimination of the incremental time required to close the normally open point, reducing the average interruption duration experienced for all faults. For partial levels of automation, two-stage restoration reduces SAIDI by 8% to 10% on the test system when compared with single-stage restoration.

Table 6.1. The impact of automation on SAIDI and the sensitivity of SAIDI to switching failures.

Automated Switches	SAIDI (hr/yr) Single-stage	SAIDI (hr/yr) Two-stage	Sensitivity of SAIDI to Switch POF
None	2.10	2.10	2.15%
Tie Switch	1.87	1.87	3.61%
Tie Switch + 1	1.84	1.69	5.27%
Tie Switch + 2	1.81	1.63	5.98%
Tie Switch + 3	1.78	1.60	6.38%
Tie Switch + 5	1.72	1.57	6.81%
All	1.54	1.54	7.27%

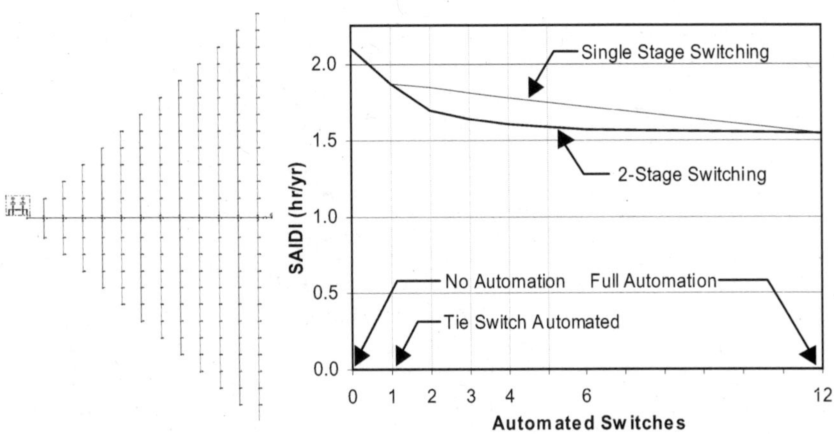

Figure 6.17. The impact of automation on SAIDI. Results are derived from the test system shown to the left, which consists of a 3.6-mi main trunk, 24 fused laterals, and switches located between laterals. The biggest improvement in SAIDI is derived from automating the tie switch. Further gains are achieved from automating a single normally closed switch, with the benefits of 2-stage switching substantially exceeding single-stage switching. Benefits noticeably diminish as the number of automated switches increases.

The sensitivity of SAIDI to switch POF detailed in Table 6.1 provides insight into how reliability will change if automated switches are more or less dependable than manual switches. For example, if switch POF is increased by 1%, and the resulting SAIDI value increases by 0.5%, the sensitivity of SAIDI to switch POF is (0.5 / 1) × 100% = 50%. As the level of feeder automation increases, SAIDI becomes more sensitive to switch POF because the impact of a switch failing is, on average, more severe. For example, if a fault takes 4 hours to repair, and manual switching would have restored some customers in 1 hour, the customer impact is 3 additional interruption hours. However, if automated switching would have restored the customer in a few minutes, the impact to these customers is 4 additional hours. The maximum sensitivity of 7.27% for the fully automated trunk is low compared to other reliability parameters such as line failure rates and should not be a concern except in extreme situations, where the POF of automated switches is much lower than the POF of manual switches. In such cases, it is possible that SAIDI can actually worsen if the impact of failed automation outweighs resulting benefits of faster switching.

In general, the systematic use of widespread automation is not recommended as an initial strategy to improve reliability. Automating tie switches and one or two additional switches can certainly improve reliability, but is generally expensive when compared to other reliability improvement alternatives. That said, targeted feeder automation can sometimes be cost-effective, and widespread automation is generally required for dramatic reliability improvements. A simple but effective approach is to test the effectiveness of switch automation in the order of expected switching frequency. Presumably, a switch expected to operate a larger number of times per year has a greater chance of benefiting from automation than a switch expected to operate a small number of times per year. In any event, reliability models can easily quantify the reliability gains for various automation scenarios to help ensure that the best number and location of automated switches are identified.

6.4.5 Faster Crew Response

An obvious way to improve duration related reliability indices such as SAIDI is to speed up service restoration and speed up repair times. Though easily accounted for in reliability models by reducing the mean time to switch (MTTS) of switches and the mean time to repair (MTTR) of equipment, these reductions should be justified by actual projects that improve operational performance. Several reliability improvement projects that have the potential to improve crew response time are:

- Outage Management Systems
- Faulted Circuit Indicators

- Automatic Fault Location Devices
- Increased Number of Crews and Dispatch Centers

In their most basic form, outage management systems contain system connectivity information, switch position information and customer location information. When an interrupted customer call the utility, the outage management system automatically identifies the geographic location of the customer as well as the location of the upstream protection device that has likely tripped. If many interrupted customer calls are received, the outage management system infers which upstream device has likely tripped based on predefined rules. In any case, trouble calls are immediately brought to the operator's attention and can immediately be acted upon by sending crews to the expected fault location. Subsequent switching can also be performed faster due to the possibilities of visual representation of switch positions, superimposition of crew locations on electronic maps, and automatically generated switching sequences. When modeling the impact of outage management systems, it is appropriate to reflect decreases in both MTTR and MTTS.

Faulted circuit indicators set a visible flag when they detect fault current. When placed regularly along a distribution circuit, they allow crews to follow the activated flags directly to the fault. These devices are effective for both overhead and underground systems, and can be modeled by reducing the MTTR and MTTS of feeder devices by the expected reduction in time from the crew being dispatched to the fault being located.

Automatic fault location devices measure fault magnitudes and infer fault location based on system topology and line impedances.[15-17] Though uncertainties in pre-fault current, fault impedance and circuit impedance make precise fault location impossible,[18] advanced techniques using protection device, and reclosing characteristics (location and time-overcurrent characteristics) can often reduce likely fault locations to a few possibilities (see

Figure 6.18). Further accuracy can be achieved by combining fault location information with trouble call locations obtained from outage management systems. Knowing the likely location of a fault allows dispatchers to send crews directly to this position, which results in reduced repair and switching times.

Another method of improving crew response time, and thereby reducing MTTS and MTTR, is to improve crew management and dispatch. Improvement can be obtained by hiring more crews, utilizing more service centers, optimally siting service centers, tracking crews with global positioning systems, incentivizing crews based on response time, and a variety of other methods. The difficulty is in quantifying the expected improvements associated with various strategies so that their reliability impact can be modeled and quantified. Typically, improvements in response time are associated with reductions in travel time from the initial crew location to the fault location.

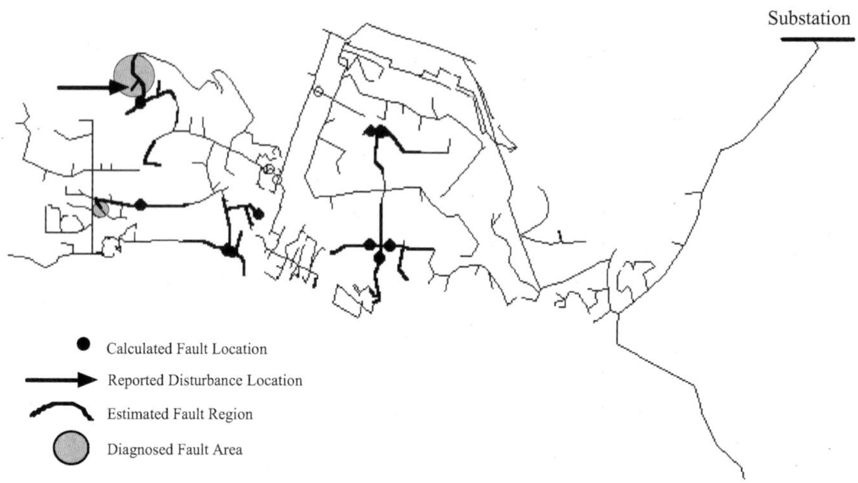

Figure 6.18. Automatic fault location devices can help improve reliability by allowing crews to respond more efficiently to trouble calls. The above example, taken from Reference 19, shows the effectiveness of different methods for an actual fault. A simple method, computing possible locations based on fault current magnitude and line impedances, results in possible spot locations (black dots). Probabilistic methods incorporate uncertainty to produce feasible regions (heavy lines). Protection system diagnosis prunes these possibilities by eliminating locations that would not cause the protection system to respond like the recorded event (shaded circles).

Consider the impact of service center location on reliability. Assuming that crews are dispatched from the service center, the expected travel time that it will take for a crew to arrive at a fault location is as follows:

$$\text{Travel Time} = \left(\left|S_x - F_x\right| + \left|S_y - F_y\right|\right) \cdot s \tag{6.14}$$

(S_x, S_y) = Coordinate of service station
(F_x, F_y) = Coordinate of fault location
s = Average travel speed

The above equation assumes that roadways are orthogonal and uses the "taxicab distance" from the service center to the fault location when computing travel time. Using this equation, equipment repair and switching times can be customized based on their distance to the nearest service center, allowing the reliability impact of various quantities and locations of service centers to be quantified.

6.4.6 Fewer Equipment Failures

Perhaps the most effective way of improving reliability is to prevent contingencies from occurring in the first place. In terms of reliability modeling, such improvements are reflected by reducing component failure rate values. The difficult part is in knowing how much improvement can be expected for various reliability improvement efforts.

A good approach when modeling failure rate reduction programs is to decompose equipment failure rates into physical root-cause categories. Once this is done, the impact of each failure rate reduction program can be applied to appropriate categories and the aggregate failure rate can be re-computed. For example, consider an overhead feeder with root cause categories as shown in Figure 6.19. Based on historical data, it is known that 20% of the 0.1 failures per mile per year are due to animals. The utility wishes to assess the reliability impact of installing animal guards on equipment bushings. Since historically, animal guards have typically cut animal-related failures in half, it is assumed that animal-related failures will be reduced from 0.02 /mi/yr to 0.01 /mi/yr. The overall effect is to reduce the overhead line failure rate from 0.1 /mi/yr to 0.09/mi/yr.

There are a virtually unlimited number of ways to reduce equipment failure rates, and it is impossible to provide a comprehensive treatment in this section. Regardless, each is similar in that it will mitigate one or more physical root-causes of failure rates and can be modeled by a reduction in aggregate component failure rate. Some of the more common failure reduction programs include:

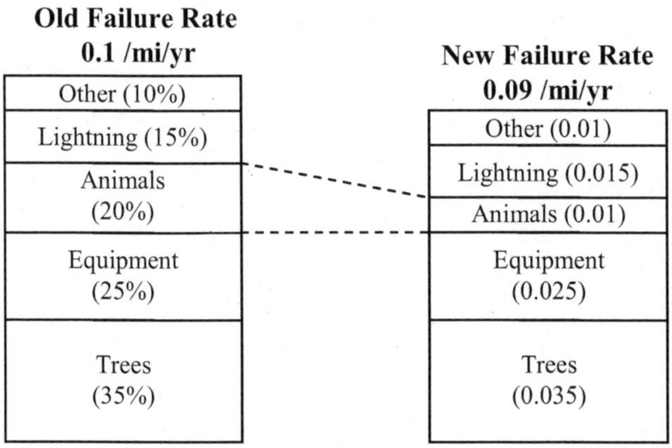

Figure 6.19. When examining the impact of failure rate reduction programs, it is advisable to decompose equipment failure rates into physical root causes based on historical data. This example shows a possible failure rate decomposition of an overhead line. If installing animal guards will cut animal failures in half, the aggregate failure rate will be reduced from 0.1 /mi/yr to 0.09 /mi/yr.

Common Failure Reduction Programs

- Increased Inspection and Monitoring
- Substation Transformer Programs
- Tree Trimming Programs
- Use of Covered Wire
- Infrared Feeder Inspection Programs
- Wire Replacement Programs
- Cable Replacement Programs
- Increased Lightning Protection
- Transformer Load Management Programs

Substation power transformers are one of the most expensive components on a distribution system and can have a large impact on system reliability when they fail. As such, most utilities perform regular inspection and maintenance on their transformer fleet in an effort to minimize failures. In theory, substation transformer failures are a function of many factors including age, manufacturer, loading history, number of through faults, maintenance history, the presence of on-load tap changers, etc. Although models and data are immature at present, utilities, service companies, and manufacturers are beginning to compile databases of historical failure data so that these relationships can be quantified. This will allow the failure rate reduction of various transformer programs to be compared so that questions such as the following can be answered: "Should this old transformer be kept, rejuvenated, or replaced?" "How frequently should this transformer be inspected and maintained?" "Should this transformer be outfitted with condition monitoring equipment?" and "How should I treat each transformer in the fleet differently?"

Tree trimming programs are vital to distribution reliability and can have a profound effect on the failure rates of overhead lines. In the past, tree trimming programs have attempted to identify appropriate cycle times, with every right-of-way being trimmed every 2 to 6 years. Figure 6.20 shows the impact of cycle time on vegetation failures for a utility in the midwestern US. More sophisticated programs may have different cycle times for main trunks and laterals, and the most sophisticated methods utilize reliability-centered maintenance techniques to identify the best cycle times for each right-of-way.

Another way to reduce the number of vegetation-related failures is to replace bare overhead conductor with covered conductor. Doing so will reduce both momentary failure rates and sustained failure rates, but may increase repair times.

Often times line hardware causes more feeder failures than actual wire problems. Many potential problems can be identified and fixed through infrared inspection programs (and, more recently, ultraviolet inspection). The expected failure rate reduction associated with such programs will be utility specific and should be based on historical improvements due to past inspection programs.

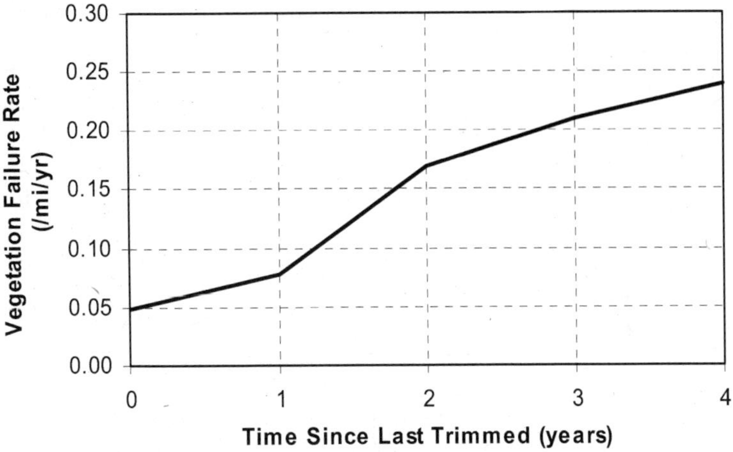

Figure 6.20. Tree trimming cycle time can have a profound impact on vegetation-related failure rates on overhead lines. The above data, obtained from a utility in the midwestern US,[20] shows that vegetation-related failures increase by nearly a factor of five from the time a right-of-way is trimmed through four years without additional trimming.

Many old feeder systems have been extended and reconfigured many times and, consequently, have sections containing old and small wire that has a lower than desired ampacity. The failure rate of old wire tends to be higher than that of new wire due to oxidation, cumulative annealing, and fatigue. In addition, the failure rate of small wire tends to be higher due to burn-down during short circuits. Replacing these sections of old and small wire will reduce associated failure rates, and will have the additional reliability advantage associated with increased capacity (see Section 6.3.5).

One of the biggest concerns for many utilities is the increasing failure rate associated with aging solid dielectric cable, typically underground residential distribution cable with cross-linked polyethylene insulation. As more historical failure data becomes available, it is becoming easier to customize the failure rates of individual cable sections based on age and vintage. This information can be used to identify cables that are good candidates for testing, rejuvenation, or replacement. The system impact of these efforts, of course, are reflected by the appropriate reduction in cable failure rates.

Augmenting lightning protection is an important failure rate reduction method for areas with high keraunic activity. Although it is typically not feasible to protect distribution feeders against direct lightning strikes due to their relatively low BILs, the increased application of arresters and static wire can mitigate the impact of nearby strikes. Surge arresters mounted on equipment will reduce the failure rate of that equipment by clamping induced voltages. Surge arresters at normally open points will reduce flashovers caused by the voltage doubling of traveling waves. Static wires, when used in conjunction with low-impedance grounding points, allows lightning energy to be dissipated into the ground, potentially preventing a flashover and reducing the energy required to be absorbed by surge arresters.

Transformer load management programs infer the peak loading of distribution transformers from billing data, transformer connection data, and typical customer load curves. If a utility is experiencing a large number of distribution transformer failures due to overloading, this sort of program can be effective in reducing failure rates by automatically identifying overloaded transformers so that they can be proactively replaced.

6.4.7 System Configuration

One of the most powerful aspects of distribution system reliability modeling is the ability to easily calculate the reliability associated with different system configurations. At the most basic level, this may consist of reconfiguration of the system by changing the location of normally open tie points. More involved configuration strategies might include adding tie points within or between feeders, adding new feeders, or adding substations of simple or complex design.

A distribution system can be reconfigured by changing the location of normally open switches, effectively changing the allocation of customers and the flow of power for the affected feeders. It is relatively common to use feeder reconfiguration methods to minimize losses[21-22] and operational costs.[23] This technique can also be used to improve system reliability. The basic strategy is to transfer customers presently receiving poor reliability to nearby feeders with better reliability (of course, this transfer will tend to make the original feeder more reliable and the new feeder less reliable). The effectiveness of this technique primarily depends upon the number of available tie switches and the number of normally closed switches on the system, with dense urban areas typically realizing the most benefit and sparse rural areas seeing far less improvement.

Table 6.2. Impact of system reconfiguration on SAIDI.

System	Feeders	Load (MVA)	Circuit Miles	Switches	Old SAIDI (hr/yr)	New SAIDI (hr/yr)	Change
Urban 1	29	223	20.4	295	1.81	1.55	14.4%
Urban 2	31	193	86.3	242	0.86	0.74	13.8%
Suburban 1	15	212	545.7	102	4.31	4.08	5.3%
Suburban 2	21	267	545.6	151	4.24	4.05	4.5%
Suburban 3	13	85	204.9	61	10.84	10.31	4.9%
Rural	8	144	496.0	46	5.72	5.68	0.7%
Institutional	7	108	52.1	345	1.23	1.20	2.4%

Actual modeled reliability improvements for seven distribution systems are shown in Table 6.2. For these test cases, urban areas saw SAIDI improvements well above 10%, suburban areas saw improvements of around 5%, and the rural system saw an improvement of less than 1%. Results for specific systems will vary, but the potential for substantial reliability improvements through system reconfiguration is evident. For more information on how these results were obtained, see Section 7.4.3 of the next chapter.

Additional reliability gains can be made by adding new tie switches between feeders (often in combination with the addition of normally closed switches). Good feeders to target have a small percentage of load that they are able to transfer to other substations during peak conditions. Other factors to help identify attractive potential locations include the number of capacity-constrained load transfers and the proximity of lightly loaded feeders. Of course, the reliability impact of each new tie switch can be computed so that the optimal number and locations can be identified.

Sometimes the best way to improve the reliability of heavily loaded or heavily exposed systems is to construct new feeders. Reliability models can help to identify the best number of new feeders, the best routing, the best number and location of protection devices, the customers that should be transferred to new feeders, the number and location of tie switches, and so forth. They can also help to site and size potential new substations for these circuits. Siting, routing and sizing new substations and feeders is a pure planning function and, as such, should account for future load growth patterns, future uncertainty, and multiple scenarios.

6.5 STORM HARDENING

Most utilities have goals to maintain or improve reliability and use the techniques discussed in the previous section in attempts to achieve these goals. However, most measures of reliability exclude outages and interruptions that occur during major storms such as hurricanes, tornadoes, and ice storms (see Section 2.2.7).

Distribution systems are rarely designed explicitly for extreme weather. As a result, reliability efforts typically focus on customer interruptions during normal weather and efficient restoration during major storms. Storm-related interruptions may still be tracked, but are typically not used for examining system reliability characteristics.

Efficient restoration may no longer be enough. Since the turn of the new century there have been a number of highly visible extreme weather events that have caused extensive damage to electric power systems and have received high public awareness. Some of these recent weather events are listed in Table 6.3.

Table 6.3. Some major weather events since the turn of the century.

2002 Events	January Central Plains Ice Storm
2003 Events	Hurricane Isabel
2004 Events	Hurricane Charlie Hurricane Frances Hurricane Ivan Hurricane Jeanne Hurricane Dennis
2005 Events	Hurricane Emily Hurricane Katrina Hurricane Rita Hurricane Wilma December Southern States Ice Storm
2006 Events	December Pacific Northwest Wind Storm
2007 Events	January North American Ice Storm

Consider the 2006 wind storm in the Pacific Northwest, which was the worst in the region in more than a decade. This storm knocked out power to more than 1.5 million homes and businesses and killed at least six people. Wind gusted to a record 69 mph at Seattle-Tacoma International Airport (SEATAC), breaking the old mark of 65 mph set in 1993. Wind was clocked at 113 mph near Mount Rainier. Power was knocked out at one of the SEATAC concourses, canceling dozens of flights. Flights were also canceled at Portland International Airport in Oregon. Widespread electric distribution system damage occurred in Oregon, Washington State, and British Columbia.

Consider also 2004 and 2005, in which nine hurricanes made landfall in the southeastern states. These storms caused repeated widespread damage to utility distribution systems, millions of customer interruptions, and billions of dollars in restoration costs.

At a certain point, customers and regulators begin asking themselves questions such as, "Wouldn't it be better to make the system stronger rather than rebuild it after every severe storm?" To answer this question, it is necessary to look at typical design criteria, and the ability of these criteria to withstand major storms.

6.5.1 Storm Design Criteria

There are three primary justifications for design standards: safety, economy, and reliability. At a minimum, electrical equipment must be designed and maintained so that they ensure worker and public safety. The safety standard used by most US states for utility electrical facilities is the National Electrical Safety Code, or

NESC. Details of the NESC as it relates to distribution design criteria is provided in Section 3.3.1.

Distribution structures built to minimum NESC requirements (Grade C) often experience extensive damage during major storms. This is expected since major storms often exceed Grade C strength. Transmission structures generally perform well during major storms. This is also expected since they are built to withstand extreme summer and winter weather.

A utility can justify exceeding minimum safety standards if the additional cost is offset by future financial benefits. For example, a utility may decide to standardize on a single pole size to simplify inventory and to receive quantity discounts. In many cases, this pole size will exceed minimum NESC requirements. The structure is stronger and safer, but these benefits are incidental. The motivation for using the stronger pole is to reduce cost.

It is rare for utilities to exceed minimum safety standards for distribution structures to achieve economic efficiencies. Most costs associated with installed distribution facilities do not depend upon design standards, or increase with stronger design standards. For example, pole inspections are a fixed cost per pole. If a distribution system uses stronger poles, pole inspection costs remain the same. If a distribution system uses more poles of the same size, pole inspection costs increase.

Some utilities are beginning to use nonwood poles for distribution system. These poles are more expensive than wood, but have economic advantages. For example, a California utility has started using composite distribution poles in certain locations to control maintenance expenses and to prevent woodpecker damage. An Arizona utility has adopted steel poles as the standard for all overhead distribution construction to reduce transport costs (steel is lighter than wood) and to increase expected life. In both cases, the driver for change is economic, but does result in a strengthened system. The same loading criteria are used so that the nonwood poles are essentially the equivalent strength of wood poles.

Utilities also have the option to spend more money to achieve higher reliability during major storms. However, there are not standard reliability measures for a system during extreme weather. Perhaps the closest is construction grade. For example, several US utilities build their distribution system to Grade B, which is 50% to 100% stronger than Grade C. Utilities also have the option of using extreme wind ratings as described in Section 3.3.1.

Extreme wind ratings allow a utility to examine each structure and manage its ability to withstand extreme winds. For example, a utility may choose to build all of its coastal structures to an extreme wind rating of 130 mph, even though this exceeds NESC requirements. A Florida utility has divided its service territory into three extreme wind zones corresponding to 105 mph, 130 mph, and 145 mph. Similar ratings can be developed for extreme ice with concurrent wind for utilities with severe winter conditions.

6.5.2 Tree Damage during Storms

For many major storms, a large percentage of damage is related to trees. All utilities affected by the 2006 wind storm in the Pacific Northwest cite trees as the primary cause of damage. Unfortunately, normal tree trimming practices are of limited effectiveness for preventing damage during major storms because most damage results from entire trees falling over into lines. One utility has focused its vegetation management on tree removal, and has noticed reduced damage during wind storms and ice storms.

Unfortunately, many trees that fall into lines during storms are outside of the utility right-of-way. A utility generally does not have authority to remove trees outside of its right-of-way, and most tree owners will not agree to tree removal or replacement when asked.

6.5.3 Storm Hardening

State utility commissions are increasingly investigating the response of utilities after a major storm and investigating infrastructure damage that occurs during the major storm. Utilities are taking notice and are beginning to consider the possibility of exceeding minimum safety standards so that structures will be less likely to fail during extreme winds. This process is typically referred to as "storm hardening."

Hardening infrastructure for severe storms is an emerging but important topic. Ideally, a utility can compute the expected damage that will occur in future storms, compute the cost of various hardening options, and compute the expected damage reduction that will result from each of these options. This process allows for decisions to be made based on quantifiable costs and benefits, and goes far beyond the design of a structure to a specific wind speed.

There are many possible approaches to harden distribution systems against hurricane damage, each with advantages, disadvantages, and interrelationships with other approaches. A brief description of the major distribution hardening approaches is now provided.

Stronger Poles – Pole strength is one of the most important factors for extreme wind rating. This is true for new construction, where stronger poles allow for longer spacing between poles, and upgrading of existing construction, where extreme wind ratings can be increased by upgrading existing poles with stronger poles. When selecting a pole, there are several important factors that must be considered. These factors include weight, visual impact, wind performance, insulating qualities, corrosion, and climbability. The most promising alternatives to wood for strong distribution poles are steel and composite (concrete tends to be too heavy for typical digger-derrick trucks).

Upgraded Poles – There are several ways to increase the strength of an existing pole. This includes using an extended-length steel brace that is driven below the groundline and extends above any third-party attachments. This can typically increase the strength of the pole by two to three pole classes. Another approach is to increase the strength of the pole with a fiberglass wrap, although this is much more expensive.

Shorter Spans – Shorter spans directly result in a higher extreme wind rating. Using shorter spans also allows hardened systems to use standard construction practices and materials. For this reason, shorter spans should always be considered as an approach to hardening. However, sometimes it is not practical to shorten spans in certain areas, and in many places the span length required to meet extreme wind criteria would result in many close-spaced poles and a corresponding high visual impact.

Conductor Size – A large percentage of wind force on a pole is due to wind blowing on conductors. Therefore, conductors with smaller diameters will reduce the wind loading on poles. It is worth considering the possibility of using conductors with smaller diameters in an attempt to increase extreme wind rating without requiring exceedingly short spans or very strong poles. In contrast, very small conductors tend to break during storms, and the replacement or removal of this small wire (often secondary mains) can result in less wire damage.

Storm Guying and Push Braces – Adding transverse guys to existing poles (one on each side) serves to transfer some or all of the stress from wind forces from the pole to the guy wires, thus enhancing the overall ability of the installation to survive the storm event. Adding push braces to existing poles can provide similar benefits to adding storm guys.

Pole-Mounted Equipment – Wind forces on pole-mounted equipment transmit force to the pole in addition to forces generated by conductor, attachments, and the pole itself. Therefore, wind forces on pole-mounted equipment must be considered in the hardening analysis, especially for higher gust speeds. Equipment mounted on poles can significantly impact the maximum allowed span, especially for the higher extreme wind ratings. Therefore, it is important to understand this effect and potentially leverage it when considering hardening alternatives (e.g., converting a three-phase pole-mounted transformer bank to a pad-mounted unit).

Third-Party Attachments – For hardening purposes, the benefits of fewer attachments are reflected in the extreme wind rating of the overall distribution design including pole height, pole strength, span length, conductors, attachments, and other pole loading considerations. All else equal, fewer and/or smaller attachments will result in a higher extreme wind rating, which will result in a reduced probability of failure during a hurricane. Removing third-party attachments can be an effective way to increase extreme wind ratings from an engineering perspective. The practicality of removing third-party attachments will vary for each specific situation.

Pole Hardware – Wind forces can have adverse effects on framing materials such as insulators, crossarms, conductor ties/clamps, brackets, and other asso-ciated hardware. Use of stronger design standards can reduce damage in these areas.

Undergrounding – The conversion of overhead distribution to underground removes extreme wind as a design factor. This is a complicated subject and is discussed in detail in Section 6.6.

6.5.4 Hardening Roadmap

Increased performance expectations for major storms will result in many utilities choosing to exceed safety standards in an effort to reduce storm damage. This decision to "harden" the distribution system is potentially expensive and politically sensitive. It is therefore desirable to define a clear strategy for hardening and to translate this strategy into a hardening roadmap that identifies anticipated actions, costs, and benefits. This roadmap must address two key questions. First, how should circuits be prioritized for hardening? Second, what should be done on each circuit to be hardened?

When a severe storm hits an area and results in widespread damage, there are certain critical facilities that typically have priority for restoration. Circuits serving these facilities should be targeted first for hardening. Examples include circuits serving hospitals, nuclear plants, and dispatch centers. In addition, it is desirable to harden circuits serving large numbers of essential services such as gas stations, grocery stores, restaurants, and home improvement stores. These types of critical facilities are "no brainers" and present a good opportunity to become familiar with the rest of the hardening process.

After circuits serving critical facilities and essential services are hardened, the remaining circuits must be prioritized. A pragmatic method for this is to allow individual counties to prioritize circuits within each county, and to equalize hardening effort across all counties. A more analytical approach for circuit prioritization is to direct hardening efforts where it will result in the most benefits including damage reduction and customer impact. The best approach to circuit ranking will vary by utility, but keeping the above factors in consideration will help to add some rigor to the process and some added justification to the final prioritized list.

After a utility identifies a circuit that it wishes to harden, a specific set of projects must be identified. It is beneficial for this set of projects to be based on specific criteria and to be the result of a documented process. Three promising ways to approach circuit hardening are full hardening, critical pole hardening, and value hardening.

Full hardening ensures that the extreme summer and winter storm ratings of each structure meets NESC requirements (even though this is not required for safety reasons). For example, the extreme wind level for much of Southern Florida is 145 mph. Full hardening of a circuit in this area requires each structure to have an extreme wind rating of at least 145 mph. Full hardening of existing systems is very expensive, and will typically only be practical on three-phase main trunks.

A minimalist approach to hardening is to identify all critical poles on a circuit and ensure that all critical poles meet NESC extreme storm criteria. Critical poles will generally be poles that impede restoration when down, are difficult to repair, or are expensive to repair. For example, at a southern US utility it is critical to energize the substation feeder breaker as the first step in restoration. This means that the first switch downstream of the feeder breaker is considered a critical pole and is given a high priority for hardening.

Value hardening attempts to prioritize potential hardening projects on a circuit in terms of cost-to-benefit analysis. For example, an existing Class 5 pole may have an extreme wind rating of 100 mph and has a 5% chance of failing in a Category 3 Hurricane. Replacing it with a Class 2 pole may increase the extreme wind rating to 130 mph, and reduce the chance of failing to 1%. The benefit of this potential project is the number of man-hours required to repair a broken pole multiplied by the failure rate reduction of 4%. The cost-to-benefit ratio of other potential projects for this pole and all other poles can then be calculated and used to create a ranked list. Projects for this circuit can then be approved until a threshold cost-to-benefit ratio is reached.

Achieving the proper level of infrastructure performance during major storms at the lowest possible cost is a challenging task, but will increasingly be demanded of utilities by their regulators and customers.

6.6 CONVERSION OF OVERHEAD TO UNDERGROUND

The topic of converting overhead distribution system to underground has been extensively examined from many perspectives including reliability improvement, storm hardening, aesthetic improvement, and others. These examinations can generally be summarized as follows:

The conversion of overhead electric distribution systems to underground is expensive and, except in occasional targeted situations, cannot be fully justified based upon quantifiable benefits.

The literature on undergrounding conversion generally falls into the following categories: consultant reports, state regulatory reports, municipal reports, and property value studies. A brief discussion of each of these categories is now provided.

Consultant Reports – Consultant reports typically to provide a comprehensive overview of undergrounding issues with regards to costs, benefits, regulatory issues, previous work, case studies, and implementation issues. Major consultant reports on undergrounding have been sponsored by the Edison Electric Institute.[24] Long Island Power Authority.[25] and the electric utilities in Florida.[26]

State Regulatory Reports – Several state regulatory commissions have performed investigations on the costs and benefits of undergrounding all electric utilities in their corresponding state (sometimes limited to investor-owned utilities). Examples include Virginia,[27] North Carolina,[28] Maine,[29] Maryland,[30] and Florida.[31]

Municipal Reports – A number of towns, cities, and local authorities have either hired consultants or assigned municipal staff to produce studies of the costs and benefits of undergrounding utilities in their franchise area. Examples include: Fort Pierce, Florida;[32] Palm Beach, Florida;[33-35] Tallahassee, Florida;[36] Davis Island, Florida;[37] Tahoe Donner Subdivision in Truckee, California;[38] Honolulu, Hawaii;[39] and Washington, D.C.[40] Most of these reports focus on the cost implications of undergrounding, rather than making a specific recommendation on whether undergrounding should be undertaken.

Property Value – There is very little data available on the impact of underground conversion on property values. A literature review performed for Reference 39 found no references in either *The Appraisal Journal* or the *Journal of Real Estate Research* that specifically consider the impact of overhead electric distribution lines on property values. This report concludes that data on changes in property value due to undergrounded distribution facilities are inconclusive as to whether there is actually a measurable impact.

A review of the literature indicates that there are three primary issues related to the undergrounding of distribution systems, each with common misconceptions. These issues are related to cost, positive effects, and negative effects. These issues are discussed in detail in subsequent sections.

6.6.1 Costs of Undergrounding

The initial cost of converting existing overhead distribution systems to underground is very high. A summary of initial costs for based on a variety of studies is shown in Table 6.4 (adapted from Ref. 24). These costs do not include the costs of converting or modifying each individual customer's private service equipment or the cost for undergrounding third-party utilities such as telephone, cable television, and broadband fiber.

Table 6.4. Estimate of initial utility costs for underground conversion.

Scope of Estimate	$ / Mile
State of Florida	814,929
Virginia Investor Owned Utilities	1,195,050
Long Island Power Authority	1,578,976
Tahoe-Donner, California	1,191,176
Allegheny Power	764,655
Baltimore Gas & Electric	952,066
PEPCO	1,826,415
Conectiv	728,190
Virginia Power	950,000
California	500,000
Georgia Power	950,400
Puget Sound Energy	1,100,000

The costs cited in Table 6.4 provide a rough estimate of undergrounding costs of about one million dollars per mile, with a low of $500,000 and a high of $1.8 million. The differences in estimates stem from three primary factors: (1) differences in construction standards, (2) differences in geography, and (3) differences in accounting methods for recording and allocating costs.

Construction Standards – There are many standards-related issues affecting cost such as voltage level, number of phases, and circuit ampere capacity. There are also certain engineering standards (such as the requirement to operate underground systems in loops) that can have a large impact on cost. However, the largest impact of standards with regards to cost is whether the underground system will be directly buried (less expensive) or be placed in a system of manholes and conduit (much more expensive).

Geography – It is much less expensive per mile to underground a feeder following a rural country road than to underground a feeder in a central business district. However, the cost per customer could be substantially higher because of the lower customer density. In addition, installing underground facilities in underdeveloped areas can be problematic when the area is later developed. Other potentially expensive areas to underground are through rocky terrain, inaccessible mountains, swampland, and other difficult terrains. These geographic issues are often correlated with customer classes such as urban, suburban, and rural. A study by Dominion Virginia Power breaks down cost by both customer class and construction type is shown in Table 6.5 (adapted from Reference 24).

Table 6.5. Dominion Virginia Power undergrounding estimates for initial cost.

Construction Type	Heavy Commercial/ Urban Residential	Suburban	Rural	Units
3-phase bulk feeder	3.1	2.5	2.7	$M per mile
3-phase tap	3.1	2	2.1	$M per mile
1-phase tap	1.4	1.4	1	$M per mile
Service drop	4,269	4,269	7,092	$ per service

Accounting – Different companies account for and allocate nonmaterial costs in different ways. For example, some utilities may include equipment depreciation costs in overall construction costs, while others may have a separate vehicle account that does not impact construction cost accounts. Differences in accounting treatment can easily vary per-mile undergrounding cost estimates by 100% or more, for precisely the same construction activities. Therefore, care must be taken when comparing undergrounding costs across utilities.

One way to measure the cost of undergrounding is to determine the impact on rates if all undergrounding costs were funded through rate increases. Rate impact results for several large areas are summarized in Table 6.6.

Table 6.6 shows a range of calculated rate impacts. Possible reasons for this range include differences in the following areas: initial percentages of underground distribution, percentages of rates allocated to distribution costs, customer density, amount of overhead distribution in urban areas, terrain, book value of existing overhead distribution assets, and many others.

Costs for Undergrounding Nonelectric Pole Attachments – Electric lines are typically not the only equipment installed on utility poles. Also commonly found are cables for telephone, cable television, and broadband fiber. Since it is almost never acceptable to just underground the electric facilities, costs for undergrounding these "third-party attachments" must also be considered. Cost estimates for these activities from Reference 38 is shown in Table 6.7.

Table 6.6. Estimated rate impact of undergrounding.

Area to Underground	Estimated Rate Increase
Florida	81%
North Carolina	125%
Long Island	126%
Virginia	$3,577 per customer per year

Table 6.7. Tahoe Donner cost estimates for undergrounding all utilities.

	Initial Cost	% of Total Undergrounding Cost
General Contractor	76,881,639	65.75%
Electric	11,207,828	9.59%
Cost for Electric Only	88,089,467	75.34%
Telephone	11,858,615	10.14%
Cable Television	9,697,150	8.29%
Broadband Fiber	7,276,709	6.22%
Adder for Third-Party Attachments	28,832,474	24.66%
Total	$ 116,921,941	100.00%

Costs of Undergrounding Service Drops – There is yet another cost of undergrounding that customers will typically have to pay. This includes all costs required to prepare customer-owned facilities to accept underground service. For example, meter sockets designed to accept overhead service are typically not suitable for underground service. Since these devices are the property of the customer and not the utility, the customer will typically have to directly pay for the new socket and installation cost. For example, the North Carolina study[28] estimates an average service drop cost of $1,481 per suburban customer and $2,346 per rural customer. The Dominion Virginia Power study[27] estimates an average service drop cost of $4,269 per suburban customer and $7,092 per rural customer.

Undergrounding cost issues can be succinctly summarized as follows: (1) undergrounding is expensive; (2) broad undergrounding initiatives would have a significant impact on rates; and (3) many undergrounding cost estimates do not include the cost of undergrounding other utilities or the cost of customer-related work.

6.6.2 Positive Effects of Undergrounding

There are many potential benefits that may result from undergrounding existing overhead electrical facilities. These can generally be grouped into economic benefits for utilities, aesthetic benefits, health and safety benefits, and reliability benefits. In typical underground conversion studies, economic benefits for utilities are typically quantified, and aesthetic benefits are typically treated qualitatively. The treatment of safety and reliability ranges from qualitative to quantitative.

Table 6.8. O&M costs per mile in North Carolina.

	Overhead	Direct Buried Under-ground	Duct Bank Urban Underground
High	$1,064	$1,160	$6,404
Low	$757	$614	$1,700
Average	$917	$920	$4,052

Undergrounding can potentially result in a savings to the electric utility operating the system, due to reduced operations and maintenance costs, reduced vegetation management costs, reduced storm restoration costs, and reduced lost revenue due to customer interruptions. A brief discussion of each of these potential benefits is provided below.

Operations & Maintenance (O&M) – Contrary to common perception, depending on the type of specifications and design, underground distribution is somewhere between slightly more expensive to much more expensive to operate and maintain than equivalent overhead facilities. Table 6.8 presents the results of a cost comparison study done in North Carolina.[28] It shows that overhead and direct buried underground have about the same O&M cost. However, underground duct bank systems, the type required in urban areas or where subsurface conditions may damage direct-buried lines, are from two to five times more expensive to operate and maintain as compared to overhead. The higher cost for duct-bank systems is due to the requirement of manhole and vault inspections and the difficulties associated with manhole and vault access (e.g., traffic diversion, manhole flooding) in the areas where they are typically installed.

Vegetation Management – Tree trimming is one of the most expensive activities related to overhead distribution systems. Actual tree trimming costs can range from $7,000 to $70,000 per mile depending on the size and height of trees, the climate and annual rate of growth, the number of trees removed per mile, accessibility by necessary equipment, and whether the work is being done in rural or urban locations. In an extreme situation a utility would need to spend $70,000 per mile every two years on vegetation management for a portion of its system. This results in $35,000 per mile per year for tree trimming as compared to about $1 million per mile in capital cost to underground, which has a corresponding carrying cost of about $110,000 per year (assuming an 11% annual carrying charge). In this extreme situation, reduced tree trimming costs will offset about 30% of the cost of undergrounding. In less extreme cases, the offset will be less.

Storm Restoration – One of the primary motivations for undergrounding electric facilities is the potential for far less damage and interruption of electric service during major storms. Less damage translates directly into lower restoration cost and faster restoration time. The Virginia study[27] concludes that the economic benefits for Virginia utilities would be about $40 million per year. This assumes the elimination of all storm damage, one "100-year storm" every 50 years (one hurricane and one ice storm over a 100-year period), and an expected underground system life of 30 years. This $40 million per year in savings compares to an estimated initial capital outlay of $75 billion. These types of calculations require important assumptions about future weather events that have a strong impact on the estimated benefits of reduced storm damage.

Lost Revenue – An electric utility can make no sale of electricity when its electric system is out of service. Thus, if undergrounding results in fewer customer hours of interruption, utilities will lose less revenue. This will occur during hurricanes, and lost revenue during storms is a factor in most underground cost benefit analysis. It is debatable whether lost revenue will lessen during normal weather. In nonstorm conditions, underground systems tend to fail less often but take longer to restore and are more difficult to reconfigure. Despite this, the Virginia study[27] calculates that if 80% of all nonstorm outage hours could be eliminated via undergrounding, annual saving would be about $12 million per year (compared to initial capital outlay of $75 billion). Even with the extremely high assumption of an 80% reduction, savings due to avoided lost revenue are close-to-negligible.

Improved Aesthetics – One of the most often-cited benefits of undergrounding utilities is an improvement in aesthetics. This includes the elimination of unsightly distribution poles and overhead wires, and the possibility of more aesthetically-pleasing tree locations, types, and pruning methods. Such aesthetic benefits are extremely difficult to quantify with any degree of accuracy, but they are almost always an important part of any justification for an actual undergrounding projects. Improved aesthetics are commonly expected to result in improved property values, and anecdotal evidence suggests this is the case because almost all developers commonly pay premiums to put distribution systems in new neighborhoods and in new business parks underground. However, there are no published studies that provide evidence that undergrounding distribution facilities results in increased property values.

Improved Tree Canopies – The preservation of existing trees can be considered an extension of improved aesthetics; the interaction of undergrounding with the tree canopy is often discussed separately, too. When overhead power lines have been removed, existing trees no longer have to be trimmed frequently and can thus grow into more pleasing, full shapes. This also creates an opportunity to replace every pole with a new tree, to have taller trees, and to plant faster growing types of trees.

Improved Customer Relations Due To Reduced Tree Trimming – Many utility customers do not appreciate the trimming of trees on or in sight of their property, regardless of the need to remove trees that pose a potential hazard to electric lines and that may affect their service reliability. These issues of customer dissatisfaction are largely eliminated when overhead systems are placed underground.

Reduced Motor Vehicle Accidents – Undergrounding completely eliminates the risk of vehicular pole collisions if all equipment is relocated to the sub-surface (although streetlight poles are still often required). Pad-mounted equipment is still subject to vehicle collisions, but there are typically fewer pieces of pad-mounted equipment on underground systems, though this is somewhat offset by their larger footprint. In general, pad-mounted equipment tends to be located farther from traffic areas where collisions can be less likely.

Reduced Electrical Contact Injuries – Overhead lines will occasionally "burn down" and fall to the ground. If the line remains energized, human contact with the line can result in electrical contact injury. Undergrounding minimizes this type of incident, but replaces it with the risk of electrical contact injury due to dig-in contact with the underground facilities. Detailed analyses of public health cost differences between overhead and underground distribution are not available. Overhead lines are also subject to contact from tall objects such as mobile cranes and boat masts. Undergrounding eliminates these events, but detailed economic analyses are similarly not available.

Increased Reliability during Severe Weather – Underground systems are not immune from hurricane damage; flooding and storm surges can cause equipment failures and outages. However, underground equipment will typically not fail due to high winds alone. This means that wind-related hurricane damage will be greatly reduced for an underground system, and areas not subjected to flooding and storm surges will experience minimal damage and interruption of electric service. This topic is discussed in detail in Section 6.5.

Fewer Outages During Normal Weather – The failure rates of overhead lines and underground cables vary widely, but typically underground cable outage rates are about half that of their equivalent overhead line types. This will generally result in fewer faults per mile for underground systems.

Far Fewer Momentary Interruptions – Momentary interruptions are those lasting only a very short time. The most common cause of momentary interruptions on power systems are lightning, animals, and tree branches falling on wires. These events cause an interrupting device, such as a circuit breaker or recloser, to de-energize the circuit and then automatically re-energize the circuit a moment later. These temporary faults occur far less frequently on underground equipment when compared to overhead equipment, thus the practice of reclosing is rarely used on pure underground distribution systems, and therefore momentary interruptions will typically be far fewer.

6.6.3 Negative Effects of Undergrounding

Converting overhead systems to underground is not without some negative consequences. These negative effects can be broadly grouped into economic, environmental, health and safety, reliability, and miscellaneous. Summaries of the major potential negative effects of underground distribution conversion are now provided.

Regulatory Affairs – The initial cost of undergrounding is discussed in detail in Section 6.6.1, including the possibility of a significant rate increase. When a utility proposes large spending increases, interested parties typically use any means available to challenge the overall economic efficiency of the requesting utility and any of its initiatives, forthcoming projects, or policies, even if only tangentially associated with the current request. Dealing with these issues can be expensive and distracting for the utility, government agencies, and municipal franchise authorities involved in the hearings.

Funding – If and when the increased spending is approved, cost increases must be funded. Financing through large amounts of new debt can potentially increase the utility's cost of borrowing money. In addition, there is often a contentious issue of how the additional cost is to be divided among customers by area or type through rate increases, based on benefits and other associated issues. All of this means that in addition to the very real cost of conversion of overhead to underground, there are many other costs which must be borne by the utility, regulatory commission, and other governmental authorities.

Environmental Impact – Although underground systems have improved aesthetics when compared to overhead, there are often negative environmental impacts. The trenching or boring required for undergrounding can damage tree roots which can kill trees directly, structurally weaken trees, and make trees more susceptible to disease. Open trenching techniques are commonly used in underground construction. This process destroys surface vegetation and can result in an increased susceptibility to soil erosion. Distribution systems will sometimes have to traverse ecologically-sensitive land such as wetlands, streams, and rivers. Overhead systems in these areas will typically place poles so as to minimize impact and may use wide spans to traverse particularly sensitive terrain such as the wetlands along a stream. In contrast, trenching and placing an underground cable in such areas has the potential to disrupt the local ecosystem, especially during construction.

Safety Concerns – Although underground systems are typically safer than overhead systems, there are several safety concerns that are associated with underground systems. The first relates to vault inspections. When underground systems are installed in conduit, manholes, and vaults, regular equipment inspection and maintenance must be done in the manholes and vaults. This exposes workers to potential electric contacts, arc flash burns, and vault explosions, to a higher degree than when similar equipment is examined on overhead facilities. The second relates to dig-ins. Underground systems are susceptible to damage from digging activities from backhoes and excavators and even hand-operated equipment like powered post-hole diggers. Underground service drops are also subject to damage from shovels and pickaxes. Not only do dig-in events constitute a reliability problem, but they also pose the risk of electric contact to the workers involved.

Susceptibility to Dig-ins – It is common perception that underground systems are more reliable than overhead systems. This perception is oversimplified. A good example is dig-ins. Beyond any safety impacts, a dig-in on underground electrical facilities causes a fault that normally results in interruption of electric service to customers. Additional reliability concerns are now discussed.

Longer Duration Interruptions – It is relatively difficult to locate and fix an underground fault. Repair times are system specific, but as a general rule an underground fault will take at least twice as long to locate and repair when compared to an overhead fault.

More Customers Impacted Per Outage – Due to the nature of cable and underground equipment, it is much more expensive and difficult to install fuses, circuit interrupters, and sectionalizing switches in underground systems as compared to overhead systems. As a result, underground systems tend to have less protection selectivity, which means that a fault or failure in an underground system will interrupt service to more customers than an equivalent problem in an overhead system.

Constrained Post-fault Reconfiguration – After a fault or failure occurs on a distribution system, it is desirable to reconfigure the system so that service is quickly restored to as many customers as possible while repairs are taking place. As mentioned in the paragraph above, underground systems typically have fewer switching locations. In addition, in order to avoid "three-way splices" and other weakness, they tend to utilize mostly loop configurations instead of branching as in overhead circuits. The result is less flexibility in field-reconfiguration for restoration while awaiting repairs.

Limited Emergency Overload Capability – For a variety of reasons related to how different types of conductor, cable, and equipment respond to high levels of loading, underground equipment and systems are less tolerant of short periods of heavy overloads than equivalent overhead equipment. Thus, where overhead equipment can be loaded from 20% to 50% above normal peak ratings for up to several hours to maintain service after a storm or major failure, underground equipment cannot be loaded to these levels without risking damage.

Flooding – Underground systems, especially those in manholes and ductbanks, are susceptible to flooding. Flooding can cause interruption of and damage to nonwaterproof equipment, and leave contamination residue on equipment that increases the risk of future failures. Water exposure can also increase the rate of electrochemical treeing (a major failure mode) in underground cable insulation. Flooding can slow restoration activities since flooded manholes and vaults must be pumped out before being entered.

Storm Surges – When a hurricane approaches a coastline, winds push a wall of water ashore called a *storm surge*. A storm surge will pick up debris push it inland in a wall of wreckage that can batter pad-mounted equipment and bury equipment in sand. A receding storm surge can also severely erode topsoil and sand and leave previously undergrounded equipment exposed.

Bulldozer Damage during Cleanup – After a hurricane, there are often large piles of debris that must be cleaned up with heavy machinery such as bulldozers and front-end loaders. Padmounted utility equipment near or under these piles of debris have proven susceptible to damage from this cleanup activity.

Reduced Flexibility for Upgrading and Reconfiguring Circuits – It is much easier to modify, extend, and add equipment to an overhead circuit when compared to an underground circuit. In this sense, operational and planning flexibility is more limited for underground systems. This is especially relevant in rural and underdeveloped areas that are subject to future development.

Increased Business Costs – Large undergrounding initiatives will likely result in higher electricity rates, and higher electricity rates will result in higher local business costs due to higher electricity bills. These higher business costs will result in lower competitiveness for electricity-intensive businesses. They will also result in increased prices (or lower profits) for all local businesses.

6.6.4 Summary of Undergrounding

There are a number of potential negative effects that must be weighed against the positive effects when investigating the possibility of undergrounding. None is necessarily reason enough to not consider the possibility of undergrounding, but all demonstrate the very complicated nature of any decision to convert overhead distribution facilities to underground.

In summary, the conversion of overhead electric distribution systems to underground is expensive and, except in targeted situations (such as undergrounding as part of a road widening), cannot be fully justified based quantifiable benefits. Therefore, justification almost always must rely on qualitative and often intangible aesthetic benefits. This conclusion is reached consistently throughout a large body of published literature.

6.7 ECONOMIC ANALYSIS

Economic analysis is quickly becoming an inextricable part of reliability assessment. In its most general form, economics deals with the allocation of scarce resources and answers questions such as "How much should we spend to improve reliability?" and "What is the most cost-effective way to make these improvements?" In its most capitalistic form, economics deals with maximizing shareholder value through the generation of free cash flow. In either case, reliability engineers must become familiar with the basic language and techniques of economic analysis, presented in the remainder of this section. The reader is referred elsewhere for comprehensive treatment of the subject.[41-45]

6.7.1 Cost Considerations

Cost typically refers to cash payment obligations related to business expenses such as products, services, labor, real estate, insurance, interest, and taxes. In the reliability analysis process, anticipated costs are commonly grouped according to reliability improvement projects. It is important to accurately capture all costs when examining economic attractiveness, and to understand the concepts of time value of money, sunk costs, avoided costs, and opportunity costs.

Sunk Cost – a cost that has already been incurred. From an economic perspective, sunk costs should not impact future decisions. In reality, sunk costs often influence later decisions due to noneconomic issues such as attempts to justify these prior expenditures.

Avoided Cost – a cost that would have normally been incurred, but is not as a result of a decision. Avoided costs are equivalent in value to cash costs and should be included in all economic decisions.

Opportunity Cost – the cost of the next best economic decision. In business, opportunity cost is typically considered to be the cost of buying back bank loans, bonds, or stock. The weighted expected return of these cash sources is referred to as the weighted average cost of capital (WACC), and is the minimum opportunity cost hurdle rate of a company.

Costs can also be divided into initial costs and recurring costs (sometimes referred to as one-time costs and annual costs). Initial costs are typically incurred during the procurement and construction of a capital project, and do not repeat once they are paid. Recurring costs are the carrying charge of an asset and must be periodically paid as long as the asset is still owned. Table 6.9 provides a list of commonly considered initial and recurring costs.

Depreciation of an asset on a company's general ledger is treated as a negative recurring cost since it reduces a company's tax burden. Since depreciation lowers earnings by an equal amount, the negative recurring cost is computed by multiplying the amount of depreciation by the company's marginal tax rate.

Table 6.9. Common initial and recurring costs.

Initial Costs	Recurring Costs
Procurement Cost	Operating Cost
Material Cost	Maintenance Cost
Shipping Cost	Income Tax
Labor Cost	Property Tax
Equipment Rental Cost	Insurance Premiums
Commissioning Cost	(Depreciation)

The main difficulty in comparing the cost of different projects occurs when initial and recurring costs are different and/or occur at different points in time. The remainder of this sections deals with ways to compare different payment schedules including payback period, net present value, internal rate of return, and annualized cost. It ends by discussing how to compare projects with different economics and different benefits by introducing a process called *benefit-to-cost analysis*.

6.7.2 Payback Period

The simplest method to examine the economics of projects with both one-time and recurring costs is to examine actual cash payments. This is easily done by using cash flow diagrams such as those shown in Figure 6.21. These diagrams represent cash paid or cash received as arrows with length proportional to the amount of money for each cash transaction. The location of the arrow corresponds to when the payment takes place, with the start of the project corresponding to the leftmost point on the diagram and the project completion corresponding to the rightmost point. Multiple transactions occurring at the same time are represented by stacking arrows on top of each other, and a reduced cash flow diagram can be generated by summing all of the arrows corresponding to the same time period.

When making an investment, many investors simply want to know how long it will take for them to recover their initial outlay. This length of time is referred to as the payback time or payback period, and can be computed by summing values on a reduced cash flow diagram until the total becomes positive. If the project is characterized by an initial fixed cost (C_i), an annual recurring cost (C_a), and annual revenues (R_a), payback time can be computed as follows:

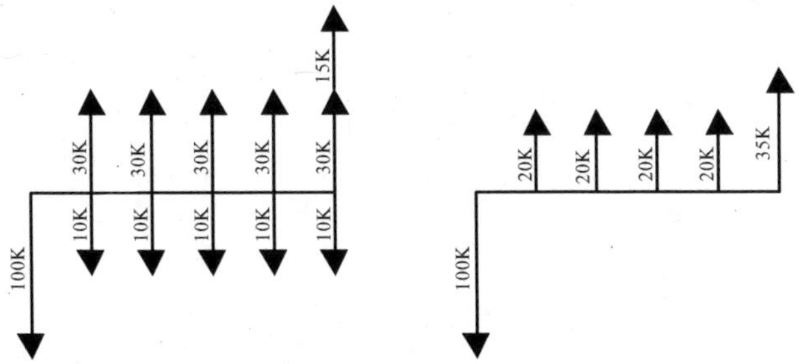

Figure 6.21. Cash flow diagrams represent cash transactions as arrows on a time line. Downward pointing arrows correspond to cash payments and upward pointing arrows correspond to cash received. In the above example, each step on the time line corresponds to a year. In the first year, \$100K is spent constructing the project. In the next 5 years, the project generates \$30K in revenue and consumes \$10K in recurring costs. In the last year, the equipment is sold for \$15K. The left figure shows all cash transactions and the right figure, a reduced cash flow diagram, shows the net cash generated or spent each year.

$$\text{Payback Time} = \frac{C_i}{R_a - C_a} \tag{6.15}$$

Payback time is useful in deciding whether or not to do a project (e.g., a company might choose not to do a project with a payback time of 10 years). Care must be taken, however, if payback time is used to compare projects. This is because payback time does not consider cash flow and the time value of money. For example, one project might cost \$50 and provide \$10 in revenues for 5 years. A second project might also cost \$50 and provide no revenue for 4 years but \$50 in revenues in year 5. Both of these projects have a payback time of 5 years, but the first project is more valuable than the second project. This is because the utility will receive revenues earlier. These revenues can then be invested in other projects that have value. To quantify the differences in these projects, metrics such as net present value are needed.

6.7.3 Net Present Value

Net present value is a method of measuring and comparing costs that occur at different points in time. For example, a utility might value being paid \$100 today just as much as it would value being paid \$110 in one year. In this case, \$110 in one year is said to have a present value of \$100, and \$100 is said to have a future value of \$110 in one year.

Present value is based on the fundamental economic principle that money received today is more valuable than money received tomorrow, and is calculated based on a discount rate, d. This is the perceived rate of depreciation in one year. For example, if \$115 paid in one year is worth only \$100 if it is paid today, the discount rate is said to be 15% per year. Present value for a cost, C, at year t can then be computed as follows:

$$\text{Present Value} = C \cdot (1 + d)^{-t} \tag{6.16}$$

The net present value of an entire project is computed by summing the present values of all individual cash transactions. This can include both costs, which have a negative present value, and revenues, which have a positive present value.

$$NPV = \sum_{i=1}^{m} R_i (1 + d)^{-t_{r,i}} - \sum_{j=1}^{n} C_j (1 + d)^{-t_{c,j}} \tag{6.17}$$

NPV = Net Present Value
R = Revenue
C = Cost
$t_{r,i}$ = Year that revenue i is realized
$t_{c,j}$ = Year that cost i is incurred

Net present value is a powerful economic metric that is commonly used in reliability and other engineering analysis. It is, however, somewhat subjective since the value chosen for the discount rate can have a major impact on the relative attractiveness of various reliability improvement projects. In addition, net present value is not a good measure of financial return since revenues are not expressed as a percentage of costs. Consider a project in which costs have a present value of \$10 million and in which revenues have a present value of \$11 million. Compare this to a project in which costs have a present value of \$1 million and in which revenues have a present value of \$1.2 million. The first project has a net present value of \$1 million. This is five times greater than the second project, which has a net present value of \$200,000. However, the second project costs 5 times less. To eliminate the need for an assumed discount rate, and to account for differences in costs and revenues, metrics such as internal rate of return are often used.

6.7.4 Internal Rate of Return

Present value calculations require the use of a discount rate. The discount rate chosen can have a profound impact when comparing solutions. This is because a higher discount rate will heavily favor options that defer investments until later years. This would not be a problem, except that discount rates are fairly "soft" values. Many intangible factors go into assigning a discount rate, and opinions can vary widely depending upon who is asked. This problem can be avoided by using the internal rate of return.

Internal rate of return (IRR) is defined as the discount rate that results in a zero net present value for a set of cash transactions in time. Stated another way, the IRR is the discount rate that will make the present value of all costs equal to the present value of all revenue.

IRR is computed iteratively. To begin, an arbitrary discount rate is chosen and the resulting net present value computed. Next, the discount rate is modified and a second net present value is computed. The next discount rate is chosen by either interpolating or extrapolating from these two values and the process repeats until a net present value of zero is obtained.

As an example, consider a project that has an initial cost \$100 of million and annual costs of \$5 million. This project is expected to produce revenues of \$20 million per year and the life of the project is 10 years. The iterative IRR calculation for this example is shown in Table 6.10, with each row corresponding to a net present value calculation based on the stated discount rate.

Table 6.10. Sample iterative calculation for internal rate of return.

Discount Rate (%)	PV of Costs ($ millions)	PV of Revenues ($ millions)	Net Present Value ($ millions)
0.0	150.0	200.0	50.0
10.0	130.7	122.9	-7.8
8.0	133.6	134.2	0.7
8.2	133.3	133.0	-0.3
8.1	133.4	133.6	0.2

The IRR calculation in Table 6.10 starts with a discount rate of zero, resulting in a net present value of $50 million. A discount rate of 10% is then chosen, and results in a net present value of –7.8 MUSD. Since the value moves from positive to negative, the IRR must lie somewhere between 0% and 10%. Further computations show that the IRR for this example is 8.1%.

IRR can be used to decide whether to perform a project or not. If the IRR is above a utility's weighted average cost of capital, the project should be considered for implementation. If multiple solutions are being considered, the solution with the highest IRR is a more attractive option from an economic perspective.

6.7.5 Annualized Cost

Often times competing projects have different expected lifetimes, and comparing their economic attractiveness is problematic from a present value or IRR perspective. Consider the purchase of a wood distribution pole with a 30 year life costing $500 versus a steel distribution pole with a 60 year life costing $1000, with no recurring costs for either. Even if the steel pole is expected to last twice as long as the wooden pole, the wooden pole has a present value that is half the price of the steel pole. The situation becomes more complicated when recurring costs are considered, since projects with longer lifetimes will appear worse due to the increased number of recurring charges.

Projects with different lifetimes can be compared by converting their net present value into equal annual payments spread over the lifetime of the project. These payments, referred to as annualized costs or levelized costs, are essentially an annuity with an interest rate equal to the discount rate of the project. For example, the $500 wood pole can be financed over 30 years at an interest rate of 15% for $76.15 per year (assuming no financing charges), corresponding to an annualized cost of $76.15 per year at a discount rate of 15%. Similarly, the $1000 wood pole can be financed over 60 years at an interest rate of 15% for $150.03 per year. From this analysis, it is clear that the wooden pole costs much less per year than the steel pole considering the time value of money.

The annualized cost of a project is computed by calculating its net present value and then converting this result into an annuity over the life of the project. If the life of a project is t years and the annual discount rate is d, the annualized cost is computed as follows:

$$\text{Annualized Cost} = \text{Present Value} \cdot \frac{d(1+d)^t}{(1+d)^t - 1} \tag{6.18}$$

Annualized cost is simpler to compute when a project consists of an initial cost plus equal annual recurring costs. For projects with these cost structures, the annualized cost is equal to the annual recurring cost plus the annualized initial cost.

$$\text{Annualized Cost} = \text{Recurring Cost} + \text{Initial Cost} \cdot \frac{d(1+d)^t}{(1+d)^t - 1} \tag{6.19}$$

To summarize, project economics are based on three factors: cash transactions at specific points in time, the time value of money, and the expected life of the project. The payback period only requires cash transaction information, but does not consider the fact that a dollar today is worth more than a dollar tomorrow, and does not consider the expected life of the project. Net present value considers the time value of money, but requires the use of a discount rate and still does not consider the expected life of the project. Internal rate of return reflects the time value of money without requiring a discount rate, but is non-intuitive and requires an iterative process to compute. Annualized cost is similar in function to net present value, but can reflect the different economics associated with different expected project lifetimes. Figure 6.22 summarizes the different ways to value a set of cash transactions in time.

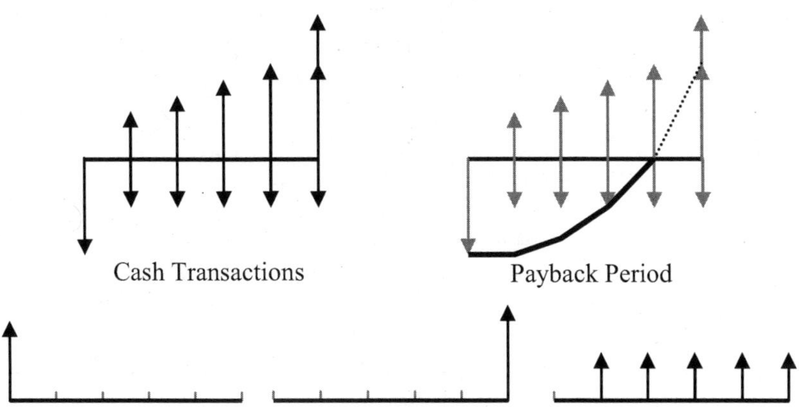

Figure 6.22. Projects consist of a set of cash transactions in time. The payback period is the time required for cash received to equal or exceed cash paid. The present and future value convert all payments to a specific point in time based on a discount factor, and an annualized cost is the required annual payment required to meet all cash payments over the life of the project.

6.7.6 Benefit-to-Cost Analysis

Benefit-to-cost analysis is a measure of how much value a project will produce relative to how much the project costs. Similar to internal rate of return, a benefit-to-cost ratio takes into account the fact that large benefits are not as attractive when they are associated with large costs. The benefit-to-cost ratio (B/C) is generally not a pure economic measure since project benefits may be measured in noneconomic terms such as reliability improvements. For example, a utility may wish to choose between a \$5 million animal guard program expected to improve SAIDI by 10 minutes per year and an \$8 million lightning protection program expected to improve SAIDI by 15 minutes per year. To compare these projects from a B/C perspective, the following formula is used:

$$B/C = \frac{\text{Expected Benefit}}{\text{Net Present Value}} \tag{6.20}$$

In this case, the animal guard program has a B/C of 2 minutes per year per million dollars and the lightning protection program has a B/C of 1.875 minutes per year per million dollars. Between these two options, the animal guard program is more cost effective and should be implemented before the lightning protection program. It should be noted that annualized cost can be substituted for net present value in the B/C calculation without changing the relative ranking of projects.

Benefit-to-cost analysis is a good way to prioritize projects and many utilities are making use of B/C measures in their budgeting process. It is especially valuable when budgets are constrained and only a limited number of proposed projects can be approved. In such situations, all projects can be ranked based on their B/C scores and can be approved in order until budget limits are reached.

Table 6.11. Benefit-to-cost ratios for three North American utilities.

	B/C (customers · minutes ÷ dollars)		
Reliability Improvement Project	**Utility A**	**Utility B**	**Utility C**
Tree Trimming Modifications	—	—	142.9
Faulted Circuit Indicators	100.0	76.9	100.0
SCADA with Breaker Control	20.0	—	20.0
Infrared Feeder Inspection	1.5	1.5	1.5
URD Cable Replacement	1.5	—	0.7
Reclosers and Sectionalizers	1.1	1.0	—
Lightning Protection	0.8	0.8	0.8
Sectionalizing Switches	—	0.5	—
Feeder Automation	—	0.2	—

Table 6.11 shows the B/C scores for a variety of reliability improvement projects for three different North American utilities. B/C scores for particular project types do not seem to vary substantially from utility to utility, but B/C scores vary by more than two orders of magnitude for different project types. Viewed from this perspective, some projects are very cost effective at improving reliability and others do not have much bang for their buck.

Benefit-to-cost analysis is not able to address a major problem associated with project ranking—the gold plating projects. Consider a tree trimming program that is competing with the animal guard program and lightning protection program described above. If trees are cut back 8 feet from lines, the program will cost \$3 million and improve SAIDI by 12 minutes per year. This corresponds to a B/C of 4 minutes per year per million dollars, which far exceeds the other proposed programs. Knowing this, the vegetation manager can propose a clearance of 12 feet instead of 8 feet, which increases cost to \$5 million, but only increases SAIDI improvement to 15 minutes per year. The B/C of the gold-plated proposal, 3 minutes per year per million dollars, is still better than the other projects, but the marginal value of going from 8 feet to 12 feet of clearance is actually lower than the animal guard and lightning protection program (1.5 minutes per year per million dollars). To prevent gold plating and help ensure that the most value is being purchased for the lowest cost, benefit-to-cost analysis is insufficient and marginal benefit-to-cost analysis is required.

6.8 MARGINAL BENEFIT-TO-COST ANALYSIS

In the past, utilities have operated with a guaranteed return on their rate base. This motivated utilities to aggressively tackle capacity and reliability problems with capital intensive projects, and gave little incentive to reject projects with small value. Further, capital projects had a tendency to be gold plated since there was no motivation to keep costs low. This cost-plus system resulted in very reliable, but very expensive distribution systems.

In response to deregulation, nearly all investor-owned utilities went through massive cost cutting efforts starting in the early 1980s. These programs successfully reduced budgets, but did not allow engineers and planners to maintain, upgrade, and expand the system as they did in the past. The impact of this is becoming apparent as the reliability of neglected areas begins to degrade.

How can a utility improve reliability while cutting costs? Maybe it cannot, but it can assure that each dollar spent on the distribution system is buying the most reliability possible. It may seem obvious, but a competitive utility with a fixed budget must spend this budget in the best possible manner. A technique to help accomplish this goal is referred to as marginal benefit-to-cost analysis[46].

Marginal benefit-to-cost analysis (MBCA) is not magic. It simply states that dollars will be spent one at a time, with each dollar funding the project that will result in the most reliability benefit, resulting in an optimal budget allocation that identifies the projects that should be funded and the level of funding for each. This process allows service quality to remain as high as possible for a given level of funding—allowing electric utilities to be competitive, profitable, and successful in the new environment of deregulation.

6.8.1 Projects and Project Options

This section applies MBCA to the process of approving reliability improvement projects. Projects, however, are not treated as a single set of activities that will either all be approved or all be rejected. Rather, projects are defined by functional goals that may be achieved to different degrees and through different means. For the purposes of MBCA, a project is defined as follows:

> **Project** – A functional goal aimed at improving the reliability of a specific set of customers. A project may succeed to varying degrees (as measured by reliability improvement), and can be achieved through different means.

A project addresses the reliability of customers, but does not address how to fix the problem. For example, if the area "Squirrel Valley" is experiencing poor reliability, a project might be "Improve the Reliability of Squirrel Valley." System expansions are also considered projects since new customer connections will greatly improve the reliability of these customers (from a baseline of zero availability).

Fundamental to MBCA is the concept of project options, which identify potential ways to achieve the functional goals of the associated project. These options should range from low-cost, low-benefit solutions to high-cost, high-benefit solutions.

> **Project Option** – A method of potentially achieving the functional goal of an associated project. Each project option has an associated cost and an associated benefit.

The concept of project options is the most critical component of MBCA, but tends to be resisted at many utilities. This is due to the dominance of standard design practices at most utilities. Engineers and planners are often reluctant to submit project options that are less expensive and less aggressive than what would have typically been proposed in the past. This attitude is not desirable given the present needs of the industry—it may be better to address a project in a low-cost manner than to not address the project at all. For example, a utility may not have the budget to connect a new customer with an underground primary selective system in concrete ductbanks. It may also be undesirable to connect this customer with a radial overhead system. Consequently, many project options should be identified so that the best option can be selected and implemented.

6.8.2 Problem Formulation

Before any problem can be optimized, an objective function must be defined and constraints must be identified. For now, the objective function will be referred to as Benefit, B. Benefit is a function of the project options that are approved, and can generally be thought of as the sum of the benefits (although the benefits associated with projects may not be independent of each other). Candidate projects will be referred to as $\{p_1, p_2, \ldots, p_n\}$, where n is the number of projects under consideration. If project k is rejected, then $p_k = 0$. If option q of project k is approved, then $p_k = q$. These project values can then be combined into a project vector, **P**. Consider three projects, each with 5 options. If the 2nd option of the first project is approved, the second project is rejected, and the 5th option of the third project is approved, then:

$p_1 = 2$	Project 1 has option 2 approved
$p_2 = 0$	Project 2 is rejected
$p_3 = 5$	Project 3 has option 5 approved
$\mathbf{P} = [2, 0, 5]$	Project vector = $[p_1, p_2, p_3]$

The first constraint considered in MBCA will, of course, be the budget. This constraint simply states that the cost of all approved project options, C, must be less than or equal to the budget. It is important to note that cost, like benefit, is a function of **P**.

The next constraint that will be considered will put a lower limit on a project option's value. This value will be measured by a benefit-to-cost ratio, B/C. If B/C is high for a project option, then the money spent on the option has good value. The minimum B/C that will be tolerated is referred to as $(B/C)_{min}$. Any project option that has a B/C ratio below $(B/C)_{min}$ will not be considered for approval.

Similar to B/C, MBCA must ensure that a project option's marginal value is good. This means that the additional money spent above the next cheapest option (ΔC) must buy its fair share of benefit above the cheaper option (ΔB). This additional value is referred to as marginal benefit-to-cost, $\Delta B/\Delta C$. The minimum $\Delta B/\Delta C$ that will be tolerated is referred to as $(\Delta B/\Delta C)_{min}$.

Knowing the objective function and constraints, MBCA can be formulated as the following optimization problem:

MBCA Problem Formulation (Budget Constraint) (6.21)

Maximize:

Benefit(**P**) $\mathbf{P} = [p_1, p_2, \ldots, p_n]$

Subject to:

Cost(**P**) $\leq$ Budget

$(B/C)_i \geq (B/C)_{min}$ $i = \{1,2,\ldots, n\}$

$(\Delta B/\Delta C)_i \geq (\Delta B/\Delta C)_{min}$ $i = \{1,2,\ldots, n\}$

n = number of projects considered

MBCA Problem Formulation (Reliability Constraint) (6.22)

Minimize:

Cost(**P**) $\mathbf{P} = [p_1, p_2, \ldots, p_n]$

Subject to:

$\text{Benefit}_i(\mathbf{P}) \leq \text{Target}_i$ $i = \{1,2,\ldots,m\}$

$(B/C)_i \geq (B/C)_{min}$ $i = \{1,2,\ldots,n\}$

$(\Delta B/\Delta C)_i \geq (\Delta B/\Delta C)_{min}$ $i = \{1,2,\ldots,n\}$

m = number of reliability benefit targets

n = number of projects considered

This problem formulation states that a utility wants to choose projects so that it can maximize its benefit without going over its budget. Further, it is not going to waste money by approving projects without good value or by gold plating projects, even if there is sufficient budget to do so.

6.8.3 Selecting Project Options

MBCA operates on the philosophy that each additional dollar spent must be justified based on the value it adds to the project. To do this, each project must have options ranging from cheap to expensive. If a project is approved, its cheapest option is approved first. A more expensive option will only be approved if its increased benefit compared to its increased cost is high compared to other projects and project options. More expensive options are allocated until constraints become binding.

Table 6.12. Example of projects, project options, and marginal benefit-to-cost.

	C	B	ΔB/ ΔC		C	B	ΔB/ ΔC
Project 1				**Project 3**			
→ 1.1 Do Nothing	0	0	---	→ 3.1 Nothing	0	0	---
1.2 Cheap	20	40	2.00	3.2 Cheap	10	25	2.50
1.3 Moderate	25	48	1.92	3.3 Moderate	12	40	3.33
1.4 Expensive	30	55	1.83	3.4 Expensive	20	45	2.25
Project 2				**Project 4**			
→ 2.1 Do Nothing	0	0	---	4.1 Nothing	0	0	---
2.2 Cheap	50	150	3.00	→ 4.2 Cheap	50	120	---
2.3 Moderate	60	165	2.75	4.3 Moderate	60	130	1.00
2.4 Expensive	70	175	2.50	4.4 Expensive	65	150	2.00

→ is the project set point, C is cost, B is benefit, and ΔB/ΔC is the marginal benefit-to-cost.

The MBCA problem is roughly analogous to the economic dispatch problem of power generation facilities.[47] In economic dispatch, generators must be started up and increased in output until electric demand is satisfied. In MBCA, projects must be approved and upgraded until reliability targets are reached or budget constraints become binding. Because of these similarities, the following MBCA methodology is referred to as MBCA project dispatch process.

MBCA project dispatch optimally allocates funding to projects so that the net benefit of these projects is maximized. Certain projects will be rejected, and certain projects will be approved. For those projects that are approved, a specific option associated with that project will be identified. Each option will consist of a cost and a benefit, and all options will be sorted from the cheapest option to the most expensive option. Each project starts with a "do nothing" option that will typically have zero cost and zero benefit. An example of four projects, each with four options, is shown in Table 6.12.

The first thing to do when performing a MBCA is to assign a set point to each project. This is analogous to the set point of a generator in the economic dispatch problem. The set point of each project should initially be set to "do nothing." Exceptions occur when a project needs to be done for safety or legal reasons. In Table 6-12, Project 4 must be done for safety reasons. Instead of "Nothing," its initial set point is assigned to the next least expensive option. All other projects are initialized to a set point of "Nothing."

The next step in MBCA is to determine the marginal benefit-to-cost ratio (ΔB/ΔC) of each possible project upgrade. This is simply the additional benefit of moving to this option (as compared to the currently dispatched option) divided by the additional cost of moving to this option. Table 6-12 shows the marginal benefit-to-cost ratios corresponding to the initial project set points.

The MBCA process allocates additional money by selecting the project upgrade with the highest ΔB/ΔC. In this case, the best option is to upgrade Project 3 from "Nothing" to "Moderate." Notice that the "Cheap" option was skipped over. This illustrates that all possible upgrades must be considered—not just the next most expensive option. If a particular option has a low ΔB/ΔC and is not allowed to be skipped over, it may block other possible project upgrades and prevent the optimal combination of project options from being identified.

After the option with the highest $\Delta B/\Delta C$ is identified, the MBCA checks to see if all reliability requirements are satisfied, if budget constraints have been reached or if any other constraints will prevent further project options from being approved. If not, the project containing the approved option must have its set point updated and its $\Delta B/\Delta C$ values for the more expensive options recomputed based on the new set point. After this, the budget is updated based on the marginal cost of the approved project and the process is repeated. A summary of the MBCA process is:

Marginal Benefit-to-Cost Analysis Process

1. Identify all projects and project options.
2. Identify the cost and benefit for all project options.
3. Initialize the set point of all projects to the lowest cost option. This will be "Do Nothing" in most cases.
4. Determine the remaining budget.
5. Compute $\Delta B/\Delta C$ for all potential project upgrades.
6. Identify the project upgrade that has the highest $\Delta B/\Delta C$ without violating any constraints.
7. Upgrade this project and re-compute $\Delta B/\Delta C$ for potential future upgrades based on the new set point.
8. Update the budget.
9. Have reliability goals been achieved or has the budget been exhausted? If yes, end.
10. Are there any project upgrades that do not violate constraints? If yes, go to step 6. If no, end.

This algorithm is not complicated or difficult to understand. It is a pragmatic way for utilities to systematically address reliability issues with constrained budgets. Large problems with many projects and project options cannot easily be solved by hand, but this algorithm can easily be implemented by using computer software or by using a standard spreadsheet application.

6.8.4 Utility MBCA Application

The marginal benefit-to-cost analysis just described was used to allocate the 1999 discretionary capital projects fund at a large investor-owned utility in the Midwestern US. Due to reductions in staffing levels and high work loads, it was important that the MBCA process be such that it did not burden the planning engineers with a lot of additional work. For this reason, only options that offer viable solutions were included with a project. Option zero for a project was always "do nothing." Option 1 for a project was always the least cost solution that met the load and voltage requirements for at least 6 years. A six-year planning horizon was selected to avoid options being submitted that only provided for temporary (1 or 2 year) solutions. Project option benefits were quantified in terms of kVA-hrs of reduced customer interruptions.

Districts submitted a total of 75 project proposals, with nearly 300 associated project options, for discretionary budget funding. This is about 50% fewer projects than had been submitted in recent years, which was indicative of district engineers realizing that certain projects are not cost effective and have no chance of being funded. Some examples of project proposals are:

- Underground cable replacement
- Conversion of 4-kV system to 12-kV system
- New distribution substation
- Conductor upgrades
- Transformer upgrades
- Convert 1-phase and/or 2-phase line to 3-phase line
- Load transfer to adjacent substation
- Convert overhead feeder to underground
- Install SCADA controlled switches on feeder
- Substation expansion
- Pole replacement
- Replace bad/inoperable switches
- Construct new feeders
- Shunt capacitor banks at substation
- Shunt capacitor banks on feeder
- New substation
- Second supply to critical load
- Install shield wire
- Galloping prevention
- Aerial infrared inspection of subtransmission
- Install surge arresters
- Install SCADA on rural substations

As can be seen, a wide variety of projects were submitted for funding approval. In the past, it was quite difficult to make a value judgment as to which projects would be approved and which projects would be rejected. The MBCA dispatch process eliminated this problem by valuing each project by the same metric: marginal benefit-to-cost. In addition, the engineers were forced to look at various methods to address problems since project funding requests with a single project option were highly discouraged.

In years prior, the discretionary capital projects had been funded at levels greater than $30 million per year. For this reason, a MCBA was performed for budget levels of up to $30 million. The left graph of Figure 6.23 shows how the marginal benefit-to-cost decreases as the budget is spent. Initially, reliability improvements are inexpensive. After about $10 million is spent, the benefit of improving reliability begins to drastically decrease.

In addition to the marginal price of reliability, it is essential to know how much total reliability is being purchased at each budget level. This relationship is easily computed by integrating the marginal benefit-to-cost curve. For this example, the cumulative benefit as a function of total expended budget is shown in the right graph of Figure 6.23. The curve reveals that nearly all of the reliability improvements are realized after the first $10 million is spent. Additional spending does not substantially improve reliability.

When deciding upon the amount of money to spend on capital projects, this particular utility decided to set a budget limit of $20 million and a marginal B/C limit of 0.14 kVA-hr. The marginal B/C constraint was reached after approving project options worth $12.4 million, well under the approved budget limit. In addition to not spending the entire budget, the optimal budget allocation approved different types of projects than in the past. Historically, reliability-based projects were generally the first to be cut from the budget when reductions were made. With the MBCA approach, which ranks projects based on added service availability, 36 out of the 75 reliability projects had options that were approved and 39 projects were rejected. Some of the approved projects included:

- Infrared testing of distribution system using helicopters
- Worst performing feeders program
- Pole replacement on 34.5-kV system
- Replace bad 34.5-kV switches
- Build a new substation
- Expand an existing substations
- Construct new feeders

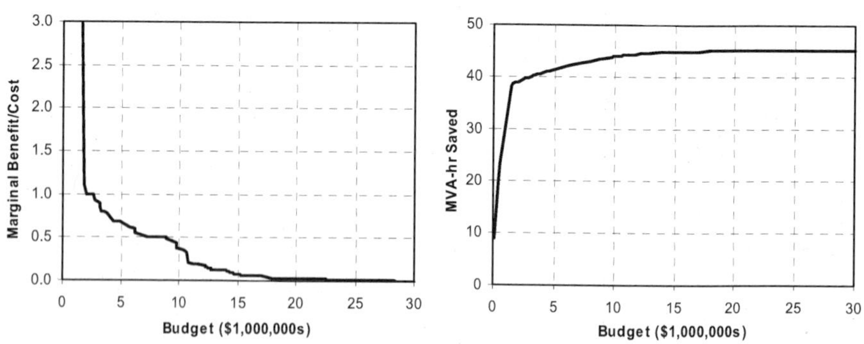

Figure 6.23. Marginal benefit-to-cost curves for an investor-owned US utility based on 75 project proposals containing approximately 300 project options (benefit is defined as kVA-hours of interruption saved and cost is defined as net present value). The left graph shows that marginal value drops off sharply as total expenditure increases. The right graph shows that almost all reliability improvements are obtained by the time $10 million dollars are spent.

Marginal benefit-to-cost analysis is a useful tool for assisting utilities in prioritizing expenditures as well as optimizing reliability within a given budget. MBCA does not in itself result in budget reductions, but does shows how to best spend within a defined budget, and how to achieve reliability targets for the lowest possible cost given a pool of projects and project options.

6.9 COMPREHENSIVE EXAMPLE

The analysis techniques presented in the previous sections of this chapter are of no use unless they are applied to actual distribution systems. Although all of the required tools have now been presented, the reader may still not be confident in how to approach a distribution system reliability analysis, what process to follow, and what results should look like. In an attempt to address these concerns, a comprehensive system analysis example based on an actual utility system is now presented. The process begins with data collection and identification of objectives, continues with a systematic analysis of the existing system and potential reliability improvement projects, and ends with a set of recommendations that can cost-effectively achieve all reliability objectives.

6.9.1 Scope and Objectives

The goal of this analysis is to identify ways to cost-effectively improve the non-storm reliability of a 40 square mile area of a US electric utility. Specifically, this analysis should identify the least-cost way to reduce $MAIFI_E$, SAIFI, and SAIDI by 20%, with most improvements targeted at worst performing feeders.

The territory for this analysis is served by 6 substations and 25 feeders. The substations are referred to as A, B, C, D, E, and F, and the feeders are designated by their associated substation and a number (e.g., B1, B2, B3). In total, this analysis contains about 280 circuit miles of primary feeder, serves nearly 25,000 customers, and has a peak demand of more than 270 MVA.

A map of feeder routings and substation service territories is shown in Figure 6.24. This model does not show the feeders that are interconnected to the analysis areas, which are reduced to equivalent sources in the model reduction process according to the guidelines set forth in Section 6.1.1.

A summary of feeder information is shown in Table 6.13. In addition to exposure, load and customer data, this table shows average customer size and three-year reliability statistics. Average customer size is a simple way to infer the dominant type of customers on a feeder, whether it be residential, commercial, or industrial. Reliability statistics can be used to compute reliability indices, which will later be used for calibration purposes (due to natural variations, three years of reliability data should be considered the minimum amount for feeder-level reliability calibration).

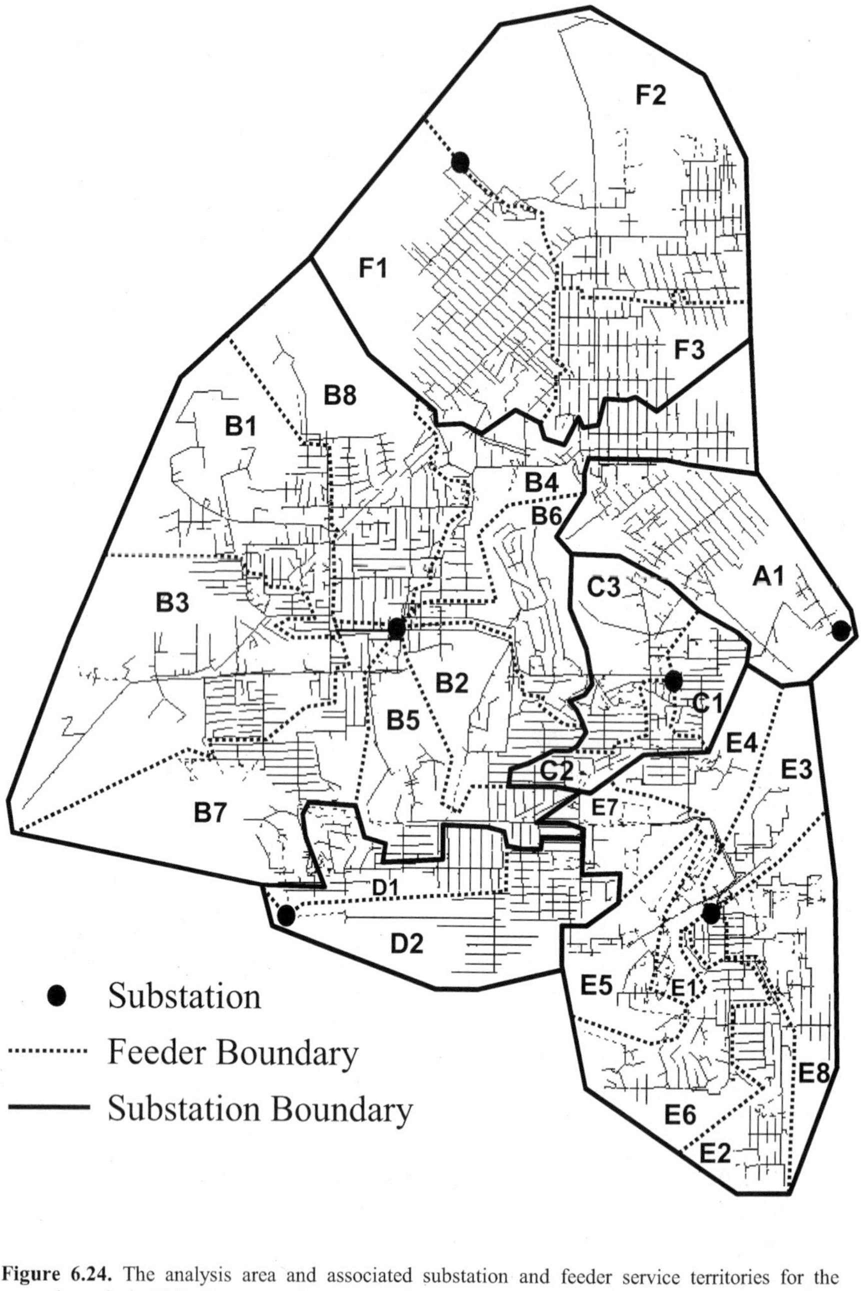

Figure 6.24. The analysis area and associated substation and feeder service territories for the example analysis. This 40 square mile area consists of 25 feeders serving 25,000 customers and 270 MVA of peak demand. Feeders that are interconnected to this analysis area are not shown since they are modeled as reduced equivalents.

Table 6.13. Feeder data summary including three years of reliability data.

Feeder	Exposure (miles)	Peak Load (kVA)	Customers	kVA per Customer	1997-1999 Outage Data Outages	Cust Int	Cust Hr	Outage Min
A1	17.1	10438	1933	5.4	226	10856	6709.16	18829
B1	16.6	13781	1271	10.8	157	2977	5821.02	18829
B2	9.3	9263	992	9.3	109	3955	4773.56	12875
B3	21.6	13163	1369	9.6	162	8905	12945.59	24177
B4	17.1	13786	1702	8.1	209	13401	12618.44	22720
B5	12.4	13880	1235	11.2	121	9717	14885.1	16116
B6	10.4	8425	978	8.6	150	5791	8758.12	18838
B7	15.9	11870	1212	9.8	118	3392	3665.51	13375
B8	18.4	14084	1898	7.4	178	8534	13106.1	22628
C1	5.3	9097	317	28.7	60	2215	1326.14	5640
C2	5.9	8532	628	13.6	58	4019	4075.56	4660
C3	9.2	13099	1373	9.5	75	1844	1813.01	9034
D1	8.5	10934	1047	10.4	99	4562	4170.89	9787
D2	12.6	11669	1668	7.0	102	2374	1387.09	10116
E1	4.0	6020	59	102.0	12	22	35.14	1215
E2	10.0	11486	1383	8.3	96	9661	6683.62	8705
E3	4.8	16366	87	188.1	14	141	122.44	1633
E4	6.7	10071	326	30.9	29	870	733.86	2797
E5	6.7	9142	165	55.4	24	1463	1508.37	2009
E6	12.3	10524	881	11.9	120	5052	8632.62	16280
E7	6.5	4497	143	31.4	35	549	635.69	3123
E8	6.1	12756	675	18.9	64	3805	2768.1	6590
F1	17.6	13350	1874	7.1	227	5212	8310.72	25166
F2	21.7	12177	1371	8.9	210	3645	6834.39	27717
F3	19.1	15902	2185	7.3	266	3565	7003.74	30569
TOTAL	**278.9**	**273874**	**24839**	**11.0**	**2921**	**116527**	**129324**	**337138**

The first step in the modeling process is to obtain geographic circuit information from the utility's geographic information system so that it can be loaded into the reliability assessment program. This includes information about overhead feeder sections, underground cable sections, distribution transformers, recloser, fuses, switches, capacitors, and so forth. A list of data fields supplied by the utility is shown in Table 6.14.

Substation information is also required for the reliability model. Since the substation data are not part of the geographic information system data, they are obtained from one-line diagrams and manually entered into the reliability model. For this system analysis, detailed substation models are not required since only feeder reliability improvement projects are being considered.

Table 6.14. Distribution system data required to build feeder models.

Feeder Sections	Distribution Transformers	Other Feeder Devices
Start and end location	Location	Recloser locations
Circuit length	Number of customers	Fuse locations
Overhead or underground	Nameplate rating	Closed switch locations
Conductor size and type	Peak load from TLM data	Open switch locations
Conductor impedance	Phase connections	Capacitor sizes and locations
Normal and emergency rating	Associated feeder and s/s	Associated feeder and s/s (all)
1ϕ, 2ϕ, or 3ϕ		
Associated feeder and s/s		

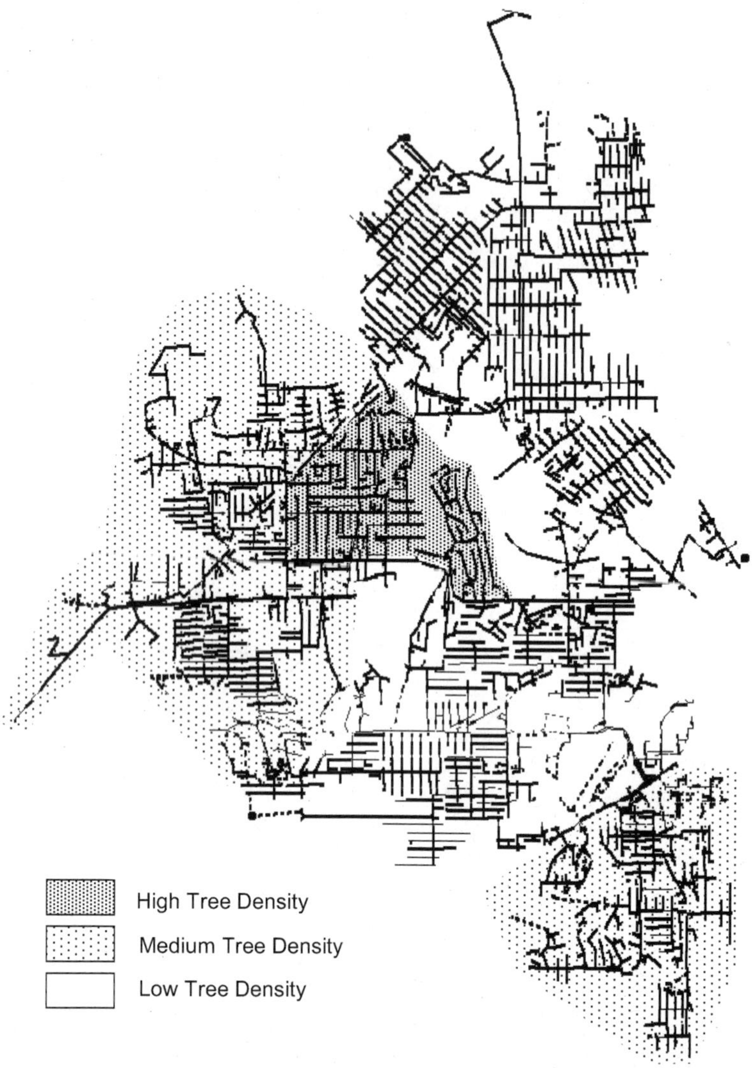

Figure 6.25. Tree density for the 25-feeder analysis area. This information, obtained from satellite imagery, is used to increase the failure rates of overhead lines in areas of high tree density and to decrease the failure rates of overhead lines in areas of low tree density.

Load data is entered as peak kVA for each transformer (obtained from transformer load management data). Since all transformers do not typically peak at the same time, feeder peak demand is less than the sum of individual transformer peak demand. To reflect this diversity effect, distribution transformer loads are proportionally scaled down on a feeder by feeder basis so that the feeder peak loading of the model matches metered feeder peak loading.

All components are assigned default reliability data when they are first entered into the reliability model. These default values are based on the typical values provided in Section 4.5. These default values, however, do not sufficiently capture the reliability variations for overhead lines due to varying tree density (see Figure 6.25 for the geographic distribution of tree density). To reflect this information, the failure rate of overhead lines in heavily treed areas is set to twice the typical default value and the failure rate of overhead lines in lightly treed areas is set to half the typical default value.

At this point, a geographically representative model of the analysis area has been created. This model is unwieldy and rough, but could still be used to gain much insight regarding the reliability characteristics of the analysis area. Before going further in the system analysis process, however, the model should be reduced in size and calibrated to historical reliability data.

6.9.2 Reduction and Calibration

The 25 feeders in the analysis area are connected to 24 additional feeders outside of the analysis area via 32 normally open tie switches. Each of these additional feeders also needs to be modeled so that the reliability analysis can accurately account for post-contingency load transfers. As a result, the initial reliability model consists of nearly 30,000 components, or about 600 components per feeder. For the particular reliability assessment program used for this analysis, this number of components is undesirably high and should be generally kept below 15,000 to keep computation time at a desirable level. The next step, therefore, is to use model reduction techniques to lower the number of components without sacrificing the potential quality of results.

The first step in lowering the number of components is to reduce the 24 interconnected feeders initially included in the analysis area to equivalent sources, with the capacity of each source equal to the smallest transfer capacity margin found on its path to tie points. Doing so cuts the model size almost in half without having any impact on reliability results in the analysis area. After this step, the model is at a tolerable size of 14,160 components, but further reductions will speed up the analysis process and will therefore be pursued.

The next step in model reduction is to combine distribution transformers within the same switchable section into a single component. Doing so reduces the model by more than 1,500 components. Last, adjacent line segments of the same type are combined if they have an included angle greater that 150° (lower thresholds can be used, but the geographic representation of the system will tend to become distorted). In the end, the initial system of 28,904 components is reduced to only 9,050 component (see Table 6.15), with the reduced model producing identical results for all components within the analysis area.

Table 6.15. Impact of model reduction on system size.

	Model Reduction Step	# of Components	
		Reduced	Total
0	Unreduced Analysis Area + Interconnected Feeders	—	28,904
1	Reduce Interconnected Feeders to Equivalent Sources	14,160	14,744
2	Combine distribution transformers in switchable sections	1,576	13,168
3	Combine adjacent segments with a 150° to 180° included angle	4,118	9,050

The next step in the analysis process is to calibrate the model to historical reliability data. In this case, SAIFI and SAIDI values can be computed for each feeder based on the three years of outage data described in Table 6.13. This is done by adjusting the default failure rates and repair times for overhead lines as described in Section 6.2.3. Since there is no data available to compute $MAIFI_E$, the temporary failure rates of overhead lines are calibrated by assuming they are always three times the value of the sustained failure rate of the same line.

Table 6.16 shows calibration results for the 25 feeders in this analysis area. The uncalibrated indices reflect initial results based on the original default component parameters. Sensitivities of reliability indices about this initial condition are also shown. Historical indices are based on 3 years of historical data, and calibrated indices are the indices produced from the calibrated model. To limit computation time, calibration iterations were restricted to two per feeder, resulting in a RMS error of about 7% for both SAIFI and SAIDI. These error levels are sufficient for the objectives of this analysis and would be obtainable through manual calibration if the software did not have automatic calibration features.

Table 6.16. Calibration results for each feeder in the analysis area.

	Uncalibrated			Sensitivities to Line FR (%)			Historical		Calibrated		
Feeder	MAIFIe	SAIFI	SAIDI	dMAIFIe	dSAIFI	dSAIDI	SAIFI	SAIDI	MAIFIe	SAIFI	SAIDI
A1	6.11	1.10	4.28	16.48	61.38	65.74	1.87	1.16	6.63	1.88	1.16
E1	0.99	0.49	1.79	15.49	30.51	45.45	0.12	0.20	0.16	0.12	0.20
E2	3.46	0.82	3.15	16.67	46.91	53.88	2.33	1.61	3.75	2.29	1.60
E3	1.55	0.57	2.04	11.47	50.76	54.77	0.54	0.47	1.58	0.60	0.59
E4	1.57	0.75	2.71	12.08	43.13	53.09	0.89	0.75	1.69	0.77	0.70
E5	2.30	1.10	2.97	10.33	43.54	53.83	2.96	3.05	5.93	2.71	3.01
E6	3.20	0.77	2.36	16.34	49.33	54.65	1.91	3.27	3.41	1.69	3.23
E7	1.99	0.81	3.41	11.41	49.07	59.40	1.28	1.48	3.16	1.40	1.52
E8	2.03	0.70	2.47	13.21	51.14	55.92	1.88	1.37	2.69	1.75	1.42
F1	6.39	0.92	3.07	18.15	61.99	67.16	0.93	1.48	6.39	0.91	1.48
F2	7.12	1.08	3.59	17.31	62.20	66.60	0.89	1.66	6.17	0.92	1.73
F3	6.79	1.15	3.43	17.34	58.73	64.37	0.54	1.07	5.72	0.40	1.03
B1	5.45	1.19	5.35	15.25	63.09	67.41	0.78	1.53	3.07	0.85	1.38
B2	3.28	0.72	2.80	16.70	56.81	61.38	1.33	1.60	6.25	1.30	1.68
B3	8.50	1.33	4.56	17.96	57.96	63.70	2.17	3.15	14.15	2.08	3.11
B4	5.89	1.46	4.61	15.29	55.61	60.38	2.62	2.47	11.09	2.68	2.54
B5	4.11	1.05	4.00	15.89	51.66	58.78	2.62	4.02	10.34	2.49	3.99
B6	3.77	0.66	2.57	17.96	56.01	61.49	1.97	2.99	12.57	1.95	2.86
B7	5.55	0.98	3.60	18.26	51.51	58.20	0.93	1.01	4.49	0.93	1.07
B8	6.65	1.20	4.13	17.76	55.48	61.11	1.50	2.30	8.20	1.52	2.27
D1	3.10	0.61	1.97	17.14	49.13	54.10	1.45	1.33	7.70	1.49	1.31
D2	4.19	0.95	3.50	14.04	64.61	71.90	0.47	0.28	2.04	0.53	0.33
C1	1.73	0.58	1.65	11.51	56.28	56.54	2.33	1.39	7.34	2.36	1.42
C2	1.71	0.72	2.08	10.64	46.82	51.58	2.13	2.16	4.59	2.10	2.09
C3	2.94	0.63	2.46	17.37	45.36	52.15	0.45	0.44	2.12	0.46	0.49

Now that the distribution system model has been reduced to a manageable size and calibrated to historical data, the system analysis process can begin. There will be a temptation at this time to explore the impact of reliability improvement options in an attempt to achieve the stated reliability goals (a 20% reduction in $MAIFI_E$, SAIFI, and SAIDI). This temptation should be resisted until a systematic investigation of the existing system is completed and the general reliability characteristics of the system are fully understood.

6.9.3 System Analysis

One of the best ways to start a system analysis is to generate data maps reflecting the geographic distribution of reliability on the system. This allows areas of high and low reliability to be quickly identified and the geospatial relationships of these areas to be examined. Figure 6.26 shows data maps for momentary interruption frequency, sustained interruption frequency, and sustained interruption duration, which correspond to the three reliability indices targeted for improvement, $MAIFI_E$, SAIFI, and SAIDI.

The data maps reveal several interesting reliability characteristics. First, momentary events are highest on feeders running to the east and west out of substation B (feeders B3 and B6). High sustained outage frequencies, in contrast, appear throughout the system except in the region served by substation F. Last, there are several small pockets of high outage duration associated with substations B, D, and E.

Feeders that have high SAIFI and SAIDI indices are given the highest priority for reliability improvement. These feeders include B3, B4, B5, B6, C2, and E5. Some feeders have high SAIFI but acceptable SAIDI. These feeders can be targeted for inspection, maintenance, and increased protection selectivity, and include C1 and E2. The last group of priority feeders is characterized by high SAIDI but acceptable SAIFI. These feeders can be targeted for increased sectionalizing and automated switches, and include B8 and E6. A summary of priority feeders is:

High SAIFI and SAIDI:	B3, B4, B5, B6, C2, and E5
High SAIFI Only:	C1, E2
High SAIDI Only:	B8, E6

Figure 6.26 also shows a data map of equipment loading at peak, which helps to identify vulnerable equipment during peak loading conditions and possible bottlenecks during post-contingency load restoration. The data map reveals that many feeder getaways are heavily loaded at peak, and offer possible opportunities for reliability improvement projects.

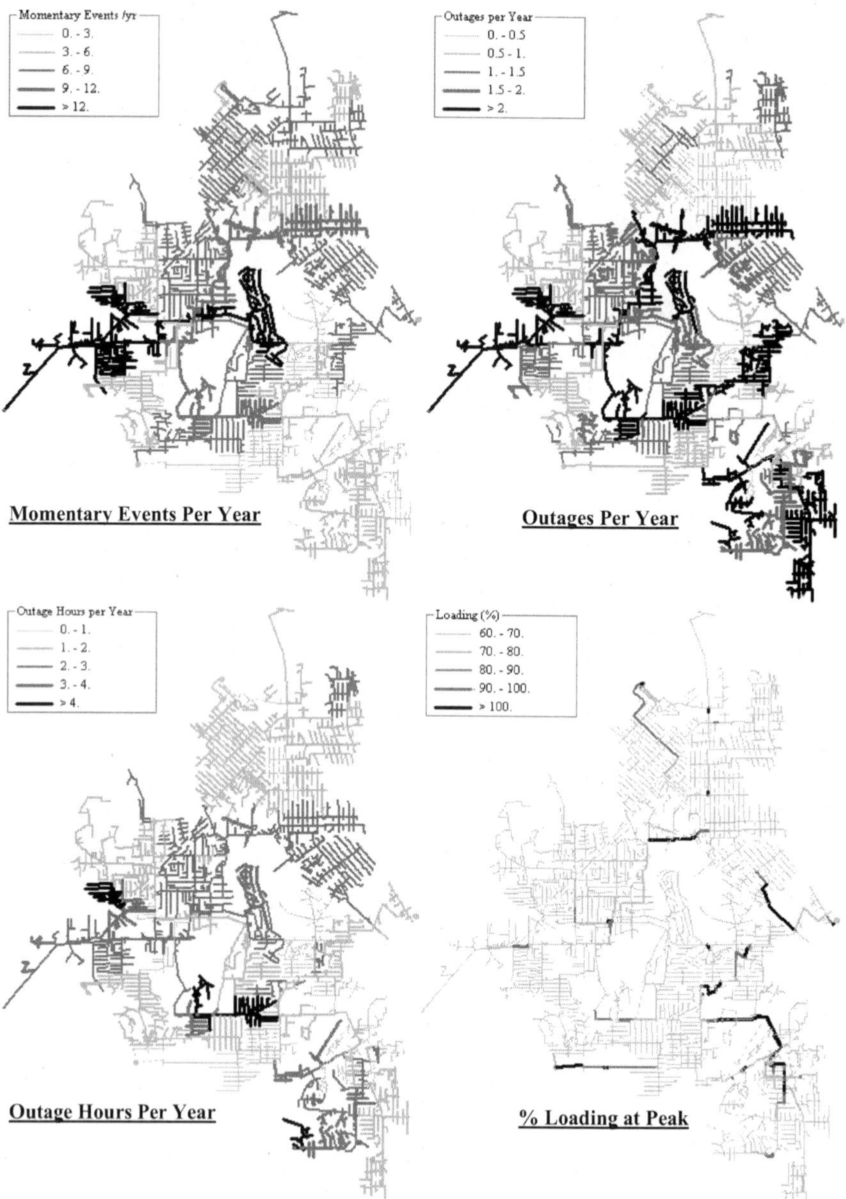

Figure 6.26. Data maps showing various reliability results shaded by component. Momentary events are related to $MAIFI_E$, outage frequency is related to SAIFI, and outage hours are related to SAIDI. Loading is not directly related to reliability indices, but identifies possible locations for capacity increases and possible locations that constrain post-contingency load restoration.

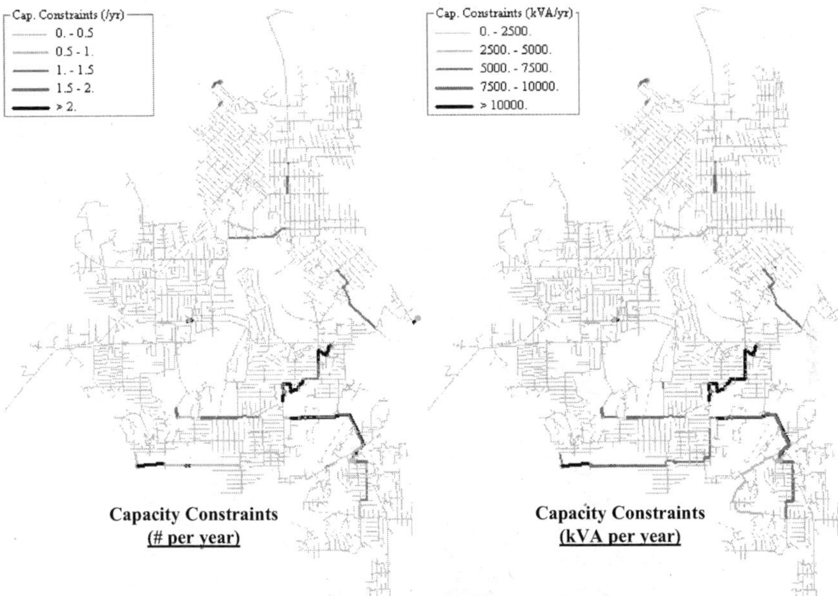

Figure 6.27. Frequency of capacity-constrained load transfer. These are the components that are preventing load from being restored after a fault because they do not have enough spare capacity. The severity of each component is measured by both the number of times per year that constraints occur (left) and the total amount of power that is constrained per year (right).

Visualizations of these capacity constraints are shown in Figure 6.27, which help to identify components that are preventing post-contingency load restoration due to a lack of spare capacity. Severity is measured in two ways: (1) by the expected frequency that load transfers will be constrained, and (2) by the total expected amount of kVA that will be constrained. For this system, each measure identifies roughly the same set of components as highly constraining.

The most critical capacity constraints occur near the substations on feeders E7 and C2. To a lesser extent, this problem occurs on feeders D2 and E6. Perhaps more interestingly, capacity constraints also significantly occur on a stretch of 2/0 copper conductor on F3 and a stretch of 4/0 ACSR on A1, small sections of legacy conductor that have substantially less capacity that adjacent conductors.

Reliability indices are not necessarily a good indication of customer satisfaction. System averages may be good while many customers experience reliability far worse than average. For this reason, histograms of individual customer $MAIFI_E$, SAIFI, and SAIDI are shown in Figure 6.28. These distributions are acceptable, with no significant customer set experiencing far worse than average reliability.

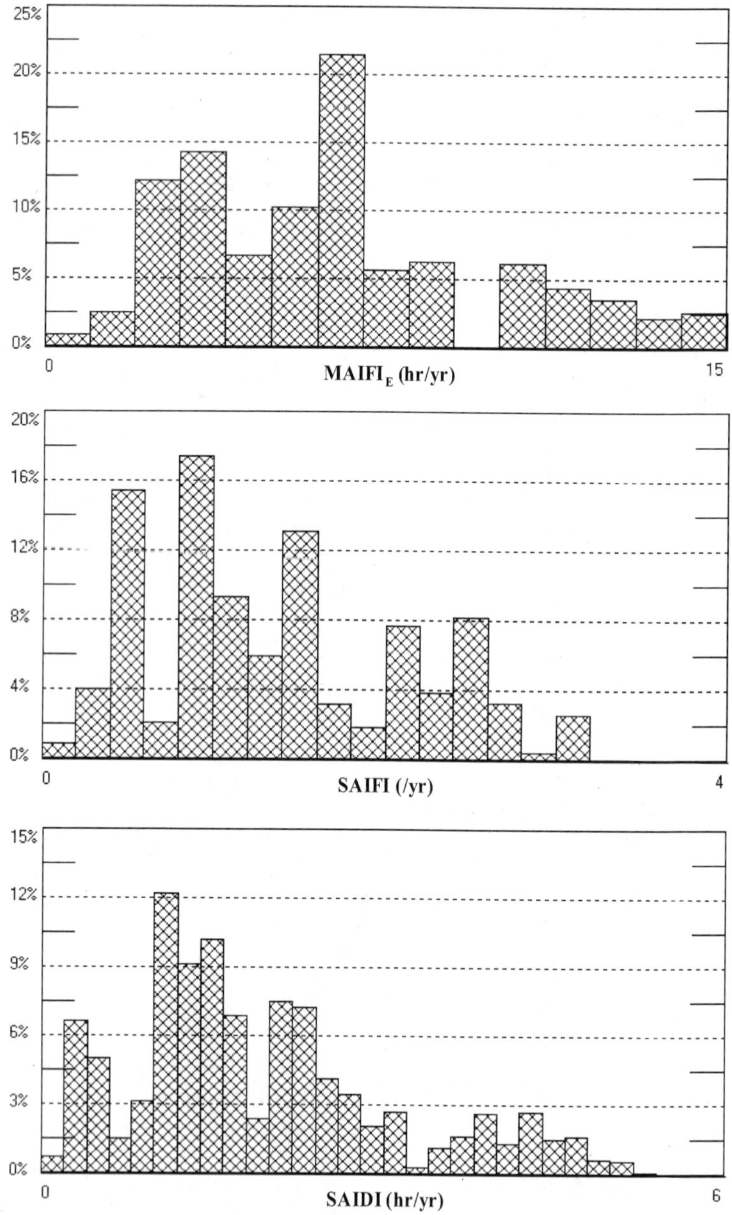

Figure 6.28. Histograms of customer reliability showing the distributions of customer MAIFI$_E$, SAIFI, and SAIDI. Histograms are helpful in identifying the number of customers that are experiencing substantially better than average reliability and substantially worse than average reliability. Oftentimes, reliability improvement projects should be focused on improving the distribution of reliability rather than improving average reliability.

At this point in the system analysis process, it is beneficial to compute the sensitivity of reliability indices to various component reliability parameters. Doing so can identify component parameters that should be carefully considered and others that do not substantially impact results, and can provide insight into the degree of improvement to expect from certain mitigation strategies.

A sensitivity analysis of $MAIFI_E$, SAIFI, and SAIDI to all major default component parameters has been performed. Sensitivity results for line and cable failures are shown in Table 6.17, sensitivity results for switching device failures are shown in Table 6.18, and sensitivity results for all component repair times are shown in Table 6.19.

$MAIFI_E$ values of several feeders are extremely sensitive to overhead line failure rates. On these feeders, temporary failure avoidance will be very effective at reducing $MAIFI_E$. Feeders with sensitivity values greater than 80% include A1, B1, C1, D2, F1, and F3.

Nearly all feeder SAIFI and SAIDI values are sensitive to overhead line failure rates. This reflects the primarily overhead nature of the system. Several feeders on substation E, however, have relatively low sensitivities to overhead line failure rates. For this reason, different reliability improvement strategies should be pursued on feeders E1, E4, E5, and E7.

Due to a large number of fused laterals, all feeders in this analysis area have reliability indices sensitive to fuse failures. Most, however, are not sensitive to switch failures. The exception is substation E. The feeders fed from this substation are relatively short, have a relatively large number of sectionalizing switches, and have reliability indices somewhat sensitive to switch failures.

Many feeder SAIDI values are very sensitive to overhead line repair time. For these feeders, reducing the time it takes to repair faults will be an effective way to improve reliability. Feeders that have high SAIDI sensitivities to line repair time include B1, B2, B3, B6, E1, F1, and F2.

Last, E1 and E7 have SAIDI values that are relatively sensitive to the repair time of switch failures. A summary of key sensitivity results is:

High sensitivity of $MAIFI_E$:	A1, B1, C1, D2, F1, F3
Low sensitivity to line failures:	E1, E4, E5, E7
High sensitivity to switch failures:	E1, E3, E4, E5, E7, E8
High sensitivity to line repair time:	B1, B2, B3, B6, E1, F1, F2
High sensitivity to switch repair time:	E1, E7

This sensitivity analysis gives good insight on how various parts of the system should be addressed differently in order to attain reliability improvements. It also identifies the components that are most likely to respond to improvement options that impact component reliability parameters such as failure rates and repair times. Further insight about specific components can be obtained through a root-cause analysis.

Table 6.17. Sensitivity of indices to failure rates of overhead lines and underground cables.

	Line Temporary Failures			Lines Sustained Failures			Cables Sustained Failures		
Feeder	dMAIFIe	dSAIFI	dSAIDI	dMAIFI	dSAIFI	dSAIDI	dMAIFIe	dSAIFI	dSAIDI
A1	81.83	3.52	3.96	16.14	58.79	65.98	0.69	0.35	0.48
E1	33.78	1.36	1.28	0.25	31.79	36.20	0.68	8.84	10.42
E2	38.93	0.61	0.73	37.61	46.26	47.60	9.27	2.97	4.03
E3	66.74	2.30	2.05	10.06	43.64	43.51	3.14	7.21	8.43
E4	54.47	1.68	1.85	7.21	31.48	38.21	4.85	15.89	12.59
E5	43.11	1.52	1.99	3.83	29.90	39.57	9.24	14.94	13.63
E6	51.65	0.98	1.22	29.63	46.52	58.37	6.58	5.23	3.49
E7	43.39	1.35	1.86	5.45	24.45	34.29	6.55	16.72	16.84
E8	63.44	1.01	1.22	26.36	42.59	49.28	3.15	4.74	6.02
F1	80.29	3.74	3.83	17.87	61.46	61.34	0.38	0.26	0.37
F2	72.53	3.69	3.81	14.96	61.01	61.78	3.36	17.76	16.49
F3	91.72	9.56	9.98	5.80	49.66	51.52	0.21	1.15	0.55
B1	83.14	3.30	3.30	15.47	55.08	55.08	0.15	0.85	0.96
B2	76.92	3.39	3.42	15.25	56.51	56.97	2.69	3.68	4.29
B3	72.79	3.12	3.46	17.01	52.05	57.79	2.06	4.41	3.27
B4	78.91	3.21	3.21	14.06	53.46	53.49	2.45	2.58	2.63
B5	74.26	2.86	3.21	13.98	47.67	53.51	1.73	6.74	6.24
B6	76.25	3.34	3.65	17.27	55.70	60.83	1.21	3.03	2.73
B7	53.70	2.71	2.71	10.70	45.16	45.11	8.21	6.70	5.48
B8	78.93	3.25	3.49	17.20	54.10	58.11	1.71	1.42	1.60
D1	72.28	2.79	2.84	15.01	46.49	47.40	2.90	4.05	4.49
D2	81.39	3.31	3.11	13.32	55.24	51.83	0.77	4.23	3.75
C1	82.26	3.22	3.10	11.58	58.93	56.60	1.02	1.07	1.60
C2	76.69	2.29	2.97	8.85	38.23	49.53	1.61	5.59	4.23
C3	65.51	2.76	2.91	11.00	39.37	41.08	3.80	6.19	5.60

Table 6.18. Sensitivity of indices to sustained failure rates of switchable components.

	BREAKERS			FUSES			SWITCHES		
Feeder	dMAIFIe	dSAIFI	dSAIDI	dMAIFIe	dSAIFI	dSAIDI	dMAIFIe	dSAIFI	dSAIDI
A1	0.00	0.87	0.96	0.57	24.78	21.82	0.00	10.70	5.80
E1	0.00	2.24	0.80	0.39	24.70	22.75	0.20	22.90	15.67
E2	0.00	0.19	0.16	6.31	36.76	33.70	1.39	12.51	11.14
E3	0.00	0.72	0.43	4.74	21.73	21.02	2.64	21.70	17.55
E4	0.00	0.58	0.38	0.91	20.29	21.81	1.42	29.31	20.47
E5	0.00	0.16	0.09	2.50	24.64	20.46	1.64	27.63	22.48
E6	0.00	0.26	0.08	3.48	31.19	26.73	1.02	15.49	9.61
E7	0.00	1.71	0.96	2.97	18.47	18.47	0.88	35.30	23.96
E8	0.00	0.24	0.17	1.90	24.66	25.08	0.82	25.95	15.44
F1	0.00	3.30	8.10	0.69	26.16	20.03	0.00	3.66	2.61
F2	0.00	3.25	6.95	2.03	25.18	20.08	0.54	3.81	2.64
F3	0.00	7.44	11.63	0.35	25.95	18.29	0.00	3.65	2.75
B1	0.00	9.37	11.56	0.75	24.78	19.96	0.00	3.97	3.10
B2	0.00	6.15	9.52	1.62	25.43	20.62	0.00	6.95	5.88
B3	0.00	3.85	5.15	2.01	31.76	25.06	0.57	6.17	5.12
B4	0.00	2.99	6.29	0.43	33.04	28.88	0.00	5.86	5.11
B5	0.00	3.21	4.02	1.64	26.93	22.32	0.40	12.92	10.41
B6	0.00	4.11	5.59	2.95	25.82	20.96	0.45	7.73	5.18
B7	0.00	8.65	14.96	2.44	29.21	22.74	0.79	9.45	6.40
B8	0.00	5.26	7.06	1.42	32.48	26.04	0.00	3.13	2.32
D1	0.00	1.34	3.05	2.68	29.80	26.86	0.46	15.00	12.07
D2	0.00	3.77	12.01	0.65	19.60	11.13	0.00	10.90	3.05
C1	0.00	0.33	0.64	0.56	26.75	25.50	0.00	9.00	9.28
C2	0.00	1.31	1.02	0.93	25.04	26.16	0.25	26.50	13.72
C3	0.00	1.72	1.88	2.13	35.15	31.94	0.49	11.09	6.94

Table 6.19. Sensitivity of SAIDI to component mean time to repair.

Feeder	LINES dSAIDI	CABLES dSAIDI	BREAKERS dSAIDI	FUSES dSAIDI	SWITCHES dSAIDI
A1	68.13	0.51	1.44	21.02	5.41
E1	37.49	10.98	1.18	22.74	15.66
E2	36.62	4.16	0.24	20.30	7.68
E3	33.30	6.27	0.64	16.65	10.95
E4	33.07	7.64	0.55	16.58	12.12
E5	32.03	8.61	0.13	10.77	11.31
E6	50.62	2.54	0.12	19.83	5.70
E7	35.52	14.23	0.42	16.24	20.33
E8	29.89	2.65	0.26	14.91	5.51
F1	55.00	0.40	9.45	14.35	1.42
F2	52.84	14.03	8.10	14.26	1.12
F3	53.37	0.41	13.57	14.07	2.23
B1	57.58	1.01	12.28	19.66	3.10
B2	54.31	4.55	10.11	17.60	4.53
B3	52.42	2.35	5.47	17.85	3.98
B4	32.78	2.40	6.68	12.55	2.19
B5	53.04	5.73	4.27	19.52	9.02
B6	55.33	1.50	5.94	15.74	3.03
B7	42.69	4.55	15.90	17.90	6.03
B8	49.68	1.42	7.50	18.17	1.48
D1	29.32	1.44	3.82	12.69	4.90
D2	35.78	3.74	15.01	7.34	2.57
C1	25.98	0.44	1.00	16.41	3.88
C2	43.17	2.18	0.81	19.58	7.68
C3	33.39	5.97	2.91	22.26	5.94

When attempting to improve reliability indices, it is helpful to know the greatest contributing factors to these indices, that is, the components with high root-cause scores. Data maps corresponding to the root-cause analyses for SAIFI and SAIDI for this analysis area are shown in Figure 6.29.

Sections with high root-cause scores are good candidates for increased inspection and maintenance. Feeders with sections of high SAIDI root cause scores include A1, B2, B3, B4, B5, B6, B8, F1, and F3.

Another beneficial way of examining root-cause scores is to graph values versus total exposure. Doing so reflects the percentage of exposure that is critical and the percentage of exposure that is less critical. A graph showing this relationship for SAIDI, broken down by substation, is shown in the left graph of Figure 6.30. The highest exposure values occur on 11 miles of substation A.

About 13 miles of substation F have very high root-cause scores. As mentioned earlier, this exposure occurs on feeders F1 and F3. Ensuring the health of these critical feeders is a good starting point for targeted reliability improvement efforts. An additional 30 miles of exposure on substation F has less, but significant contribution to high SAIDI values.

The other high contributors to SAIDI occur on about 45 miles of exposure on six feeders of substation B. Based on this observation, initiating a broad-based feeder inspection and vegetation management program on substation B should be investigated.

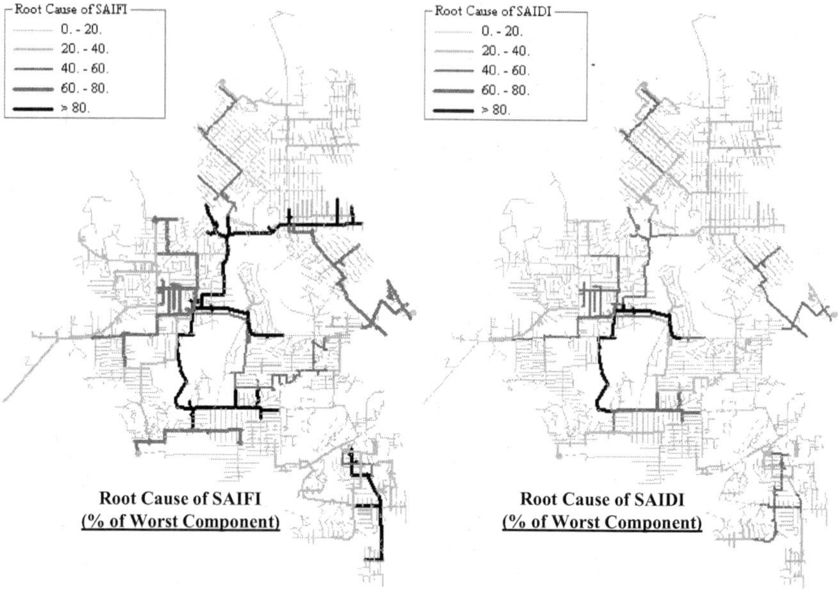

Figure 6.29. Root-causes of SAIFI and SAIDI. A root-cause score ranks each component according to its contribution to a predefined reliability measure. For this analysis, darkly shaded components contribute the most to poor reliability indices and can be investigated for potential reliability improvement projects.

The last aspect of root-cause analysis to be examined concerns switchable components. The right graph of Figure 6.30 represents the sorted SAIDI root-cause scores for breakers, fuses, and switches. In general, substation breaker failures have a high impact on reliability and should be maintained properly. Fuses tend to contribute more to SAIDI than switches and should be emphasized during feeder inspection activities.

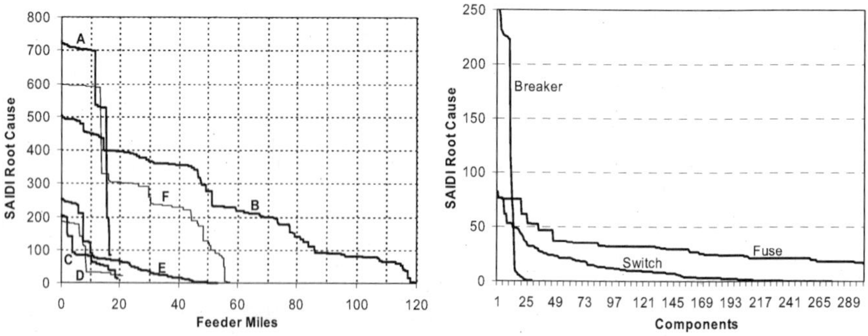

Figure 6.30. Sorted root causes of SAIDI for feeder sections (left) and switchable devices (right). The left graph shows the amount of exposure asociated with each substation that is contributing the most to high SAIDI values, allowing feeder inspection program to focus on critical areas. The right graph serves the same purpose for breakers, switches, and fuses.

6.9.4 Reliability Improvement Options

Now that a thorough analysis of the distribution system has been performed, it is appropriate to explore various reliability improvement options such as those described in Section 6.4. Normally, this is done by hand. A reliability engineer identifies possible reliability improvements based on experience and specific knowledge of the system, quantifies the impact of these improvements by incorporating them into the reliability model, and assigns them a cost according to the initial and future expenses that are associated with the project. Each project can then be ranked based on benefit-to-cost ratios, and selected in order of cost effectiveness until the reliability improvement goals are reached (in this case, a 20% reduction in $MAIFI_E$, SAIFI, and SAIDI).

In this case, a computer algorithm is used to automatically generate a large number of potential improvements.[48] This algorithm identifies potential high-value projects based on existing system results such as peak loading, reliability, root-cause scores, capacity constraints, quantified costs, and reliability benefits. It ranks them based on benefit-to-cost ratios. Several different classes of reliability improvement options are explored, allowing different approaches to reliability to be compared from a value perspective. Basic categories of the options explored include:

Transfer Path Upgrades – A transfer path is an alternate path to serve load after a fault occurs. If a transfer path is capacity constrained due to small conductor sizes, reconductoring may be a cost-effective way to improve reliability.

New Tie Points – A tie point is a normally open switch that allows a feeder to be connected to an adjacent feeder. Adding new tie points increases the number of possible transfer paths and may be a cost-effective way to improve reliability on feeders with low transfer capability.

Increased Line Sectionalizing – Increased line sectionalizing is accomplished by placing normally closed switching devices on a feeder. Adding fault interrupting devices (reclosers and fuses) improves reliability by reducing the number of customers interrupted by downstream faults. Adding switches without fault interrupting capability improves reliability by allowing more flexibility during post-fault system reconfiguration.

Feeder Automation – In this analysis, feeder automation refers to SCADA controlled switches on feeders. These automated switches allow post-fault system reconfiguration to occur much more quickly than with manual switches, allowing certain customers to experience a momentary interruption rather than a sustained interruption.

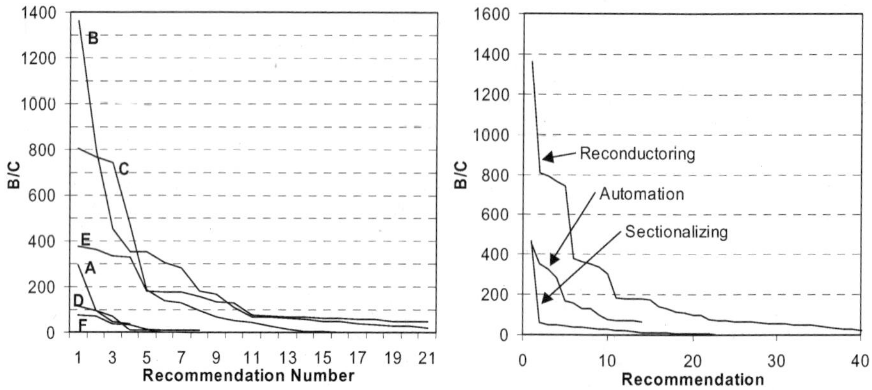

Figure 6.31. Cost effectiveness of computer-generated reliability improvement recommendations. Each of these graphs ranks recommendations based on a benefit-to-cost ratio, with the left graph organizing recommendation by substations and and right graph organizing recommendations by type. From these figures, it can be seen that the most cost-effective recommendations tend to be associated with substations B and C, and tend to involve re-conductoring projects.

The computer-generated reliability improvement projects are not a substitute for good human engineering. They are used, along with the system analysis results, as a starting point for manually generated reliability improvement recommendations. As a summary, the computer-generated reliability improvement options have been categorized by substation, sorted by their benefit-to-cost ratio, and plotted in the left graph of Figure 6.31. From this, it can be seen that substations B, C, and E have the most cost-effective possibilities for reliability improvement.

The right graph of Figure 6.31 shows a sorted list of of computer-generated recommendations broken down by recommendation type. Reconductoring (transfer path upgrades) seems to be the most cost-effective approach to reliability improvement, mostly due to overloaded feeder getaways. Feeder automation is the next most cost-effective way to improve reliability. Increasing line sectionalizing and adding new tie points offers a few attractive projects, but should not be used for broad-based reliability improvement strategies.

It is also of interest to compare the reliability characteristics associated with the use of fuse saving versus fuse clearing. The base system model, and all prior results have assumed the use of fuse saving, presumably resulting in a higher $MAIFI_E$ and a lower SAIFI and SAIDI when compared to fuse clearing strategies. To verify this prediction, the reliability of each feeder with a fuse clearing strategy is calculated by blocking the instantaneous relay on all feeder breakers. The results of this analysis is summarized in Table 6.20.

Table 6.20. Changes in reliability if fuse clearing is used instead of fuse saving.

Feeder	MAIFI	SAIFI	SAIDI	CAIDI	%Δ MAIFI	%Δ SAIFI	%Δ SAIDI	%Δ CAIDI
A1	1.60	1.54	6.10	3.97	-73.9	33.9	46.6	9.7
E1	0.06	0.22	0.34	1.58	-62.5	10.0	3.0	-4.8
E2	0.49	2.42	1.70	0.70	-86.9	5.7	6.2	0.0
E3	0.46	0.84	0.75	0.90	-71.4	37.7	23.0	-9.1
E4	0.70	0.92	0.78	0.85	-58.8	17.9	11.4	-5.6
E5	2.82	2.85	3.32	1.16	-52.4	5.2	10.3	4.5
E6	0.49	1.54	3.29	2.14	-82.3	10.8	23.7	11.5
E7	1.37	1.54	1.70	1.11	-56.6	10.0	12.6	2.8
E8	0.28	2.51	2.02	0.80	-89.1	34.2	34.7	0.0
F1	1.13	1.41	2.59	1.83	-82.3	54.9	75.0	12.3
F2	0.20	2.37	4.57	1.93	-96.8	157.6	164.2	3.2
F3	1.49	0.78	2.74	3.54	-74.0	95.0	166.0	38.3
B1	0.93	1.05	1.72	1.63	-69.7	23.5	24.6	0.6
B2	1.42	1.79	2.40	1.34	-77.3	37.7	42.9	3.9
B3	1.30	3.83	6.18	1.61	-90.8	84.1	98.7	8.1
B4	3.28	3.45	3.31	0.96	-70.4	28.7	30.3	1.1
B5	3.11	3.05	5.22	1.71	-69.9	22.5	30.8	6.9
B6	1.87	3.15	5.05	1.60	-85.1	61.5	76.6	8.8
B7	0.81	1.15	1.34	1.17	-82.0	23.7	25.2	0.9
B8	1.70	2.01	3.30	1.64	-79.3	32.2	45.4	10.1
D1	1.73	1.93	1.74	0.90	-77.5	29.5	32.8	2.3
D2	0.78	0.64	0.39	0.61	-61.8	20.8	18.2	-3.2
C1	3.40	2.71	1.61	0.59	-53.7	14.8	13.4	-1.7
C2	2.46	2.26	2.34	1.04	-46.4	7.6	12.0	4.0
C3	0.48	0.58	0.63	1.09	-77.4	26.1	28.6	2.8
TOTAL	**1.39**	**1.88**	**2.98**	**1.58**	**-78.0**	**37.2**	**53.6**	**12.1**

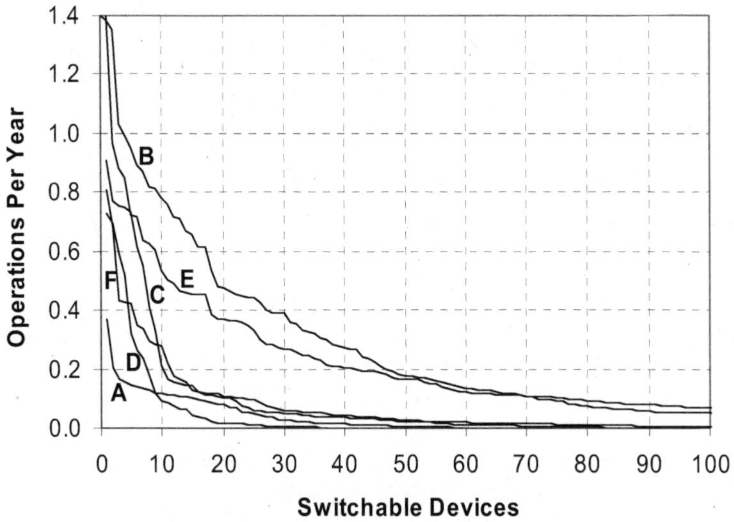

Figure 6.32. The expected operational frequency of switchable devices broken down by substation. In general, automating switches that are used frequently will be a more cost-effective way to improve reliability than automating switches that are used less frequently. In this case, there are several switches on substations B and E that operate substantially more than other switches on the system, and should be considered for possible automation.

When assessing the reliability of a system, many software programs keep track of the expected number of operations for switching devices including protection operation and sectionalizing operations. Gaining insight into the operational frequency of switching devices allows maintenance and other potential reliability improvement strategies to be examined. Figure 6.32 shows the switching frequency of devices for each substation. Switches on substations B and E experience a higher number of switching operations than other substations and are good candidates for distribution automation.

Tree trimming cycles have a large impact on cost and reliability. As such, it is important to select appropriate tree trimming cycles for each area of the system. Doing this effectively requires a model that links trimming cycles to overhead line failure rate. Using Figure 6.20 as a guide, the impact of increased vegetation management and reliablilty-centered maintenance on heavily treed areas can be explored. Other additional useful inputs are root-cause scores of overhead lines and sensitivity scores of overhead line failure rates. Feeder sections in heavily treed areas, with a high SAIDI and SAIFI root causes and high reliability index sensitivities to line failure rates, should be given tree trimming priority.

6.9.5 Recommendations

After generating a pool of reliability improvement options, it is desirable to select the combination that satisfies all objectives for the lowest possible cost. If projects are independent of each other, they can be approved according to their benefit-to-cost ratio. If they can be grouped into projects and project options, they can be approved based on a marginal benefit-to-cost analysis. Unfortunately, the benefit-to-cost ratios of projects often change depending upon which other projects are selected. Projects have the ability to both increase and decrease the effectiveness of other projects.

Consider a project that calls for adding a new tie switch between two feeders. Examined alone, this project is shown to reduce SAIDI by 10 min/yr. Now consider a second project that calls for reconductoring a section of small cable to reduce capacity constraints during load restoration. Examined alone, this project is shown to reduce SAIDI by 5 min/yr. Since reconductoring allows the tie switch to be more effective, executing both projects may result in a SAIDI reduction of 30 min/yr, doubling the cost effectiveness of each project.

Projects can also work against each other. Consider a project that calls for adding a midpoint recloser to a feeder. Examined alone, this project is shown to reduce $MAIFI_E$ by 2 per year. Now consider a second project that calls for replacing bare conductors with covered conductors in a heavily treed locations. Examined alone, this project is shown to reduce $MAIFI_E$ by 4 per year. Since using the covered conductor will reduce the number of temporary faults that the recloser sees, executing both projects may result in a $MAIFI_E$ reduction of 3 per year, reducing the cost effectiveness of each project by 50%.

Identifying the optimal set of interdependent projects can be achieved through computer optimization techniques (discussed in the next chapter), or can simply be addressed through good engineering judgement and some trial and error. Using the latter method, nine recommendations were found that cost-effectively reduce $MAIFI_E$, SAIFI, and SAIDI by 20%.

The first recommendation addresses feeder getaways since they constitute a large percentage of capacity constrained load transfers. To investigate this, all feeder getaway emergency ratings were set to infinity and a new reliability assessment was performed. Interestingly, the only indices that were substantially affected were associated with feeder B5. Other feeders have enough flexibility to switch around emergency rating constraints. To improve the reliability of B5, similar flexibility can be achieved by placing two new sectionalizing switches toward the end of the feeder.

Recommendation 1 – Add two new sectionalizing switches near the end of the main trunk on feeder B5.

The second recommendation addresses feeders with high SAIFI and SAIDI values. This includes three feeders on substation E, two feeders on substation C, and five feeders on substation B. For the most part, each of these three substations can be addressed separately.

The three feeders on substation E with high reliability indices are E2, E5, and E6, and are all located in an area of medium tree density south of a major highway. The first step in improving reliability to this area is to decrease the tree trimming cycle time by one year. For the purposes of this analysis, the resulting impact to overhead lines is accounted for by assuming a permanent failure rate reduction of 0.05/mi/yr and a temporary failure rate reduction of 0.15/mi/yr.

Recommendation 2 – Increase tree trimming cycle frequency by one year on substation E for areas south of the major highway.

Tree trimming can be supplemented by protection, sectionalizing, and automation. The south section of feeder E2 cannot easily be sectionalized for faults to the north. Adding a sectionalizing switch will increase switching flexibility. Upgrading this switch to a recloser improves reliability to the north by isolating it from downstream faults. The rest of substation E has sufficient sectionalizing, but effective SAIDI reductions can be realized through targeted automation.

Recommendation 3 – Add a line recloser near the end of feeder E2. This recloser should be able to be used as a SCADA controlled switch. In addition, implement targeted feeder automation on substation E. To do this, (1) automate the tie switches between E6 and E2, (2) automate the tie switch between E6 and E5, (3) automate two switches on the main trunk of E6, (4) automate one switch on the main trunk of E2, and (5) automate one switch on the main trunk of E5.

Feeder C2 does not have sufficient sectionalizing to ensure high levels of reliability. To increase SAIDI, adding two switches is recommended. To improve SAIFI, one of these switches should take the form of a recloser.

Recommendation 4 – Add a new sectionalizing switch and a new recloser to the main trunk of feeder C2.

The reliability indices of the substation B feeders are high when compared to surrounding substations. A recommendation to add two sectionalizing switches on feeder B5 has already been mentioned, but much more should be done. The first is to consolidate its service territory. The northern part of feeder B4 serves about 7 kVA of peak load that is more suitable to be served from substation A or substation F. It is recommended that these loads be sectionalized from feeder B4 by adding a new sectionalizing switch (normally open) and transferring them to a new feeder originating from substation A or substation F. This new feeder will be able to accept permanent load transfers as adjacent feeders become overloaded, and will increase reconfiguration flexibility.

Recommendation 5 – Connect the northern section of feeder B4 to a new feeder originating from substation A or substation F. This will require adding a new switch to the main trunk.

Most of the feeders on substation B are routed through moderate or heavily treed areas. Similar to substation E, tree trimming cycle frequency should be increased by one year to achieve assumed overhead line reductions of 0.05/mi/yr for permanent failure rates and 0.15/mi/yr for temporary failure rates.

Recommendation 6 – Reduce the tree trimming cycle frequency on all of the feeders on substation B by 1 year. This does not include the northern section of feeder B4 that will be transferred to a new feeder.

Substation B is also a good candidate for line reclosers and feeder automation. Presently, there is only one recloser—located to the northwest of feeder B1—and no feeder automation.

Recommendation 7 – Add reclosing and sectionalizing to feeder B8 including (1) two new line reclosers with SCADA control, (2) automate an existing normally closed switch, (3) automate the tie switch to feeder B1, and (4) automate the tie switch to feeder B4.

Recommendation 8 – Add reclosing and sectionalizing to feeder B3 including (1) two new line reclosers with SCADA control, (2) a new automated switch, (3) automate the tie switch to feeder B7, and (4) automate the tie switch to feeder B1.

Recommendation 9 – Add feeder automation to feeder B6 including (1) automate an existing normally closed switch, and (2) automate the tie switch to feeder C3.

After these nine reliability improvement recommendations are incorporated into the reliability model, $MAIFI_E$ improves by 22%, SAIFI improves by 21%, and SAIDI improves by 20%. Improvements seen by the worst performing individual feeders are more pronounced with several seeing reliability improvements greater than 40%.

The new reliability indices for each feeder, along with calculated improvements based on the original indices, are shown in Table 6.21. After modeling the reliability improvement, not only have aggregate reliability indices been reduced to target values, but the range of worst feeder indices to best feeder indices has been substantially reduced.

Table 6.21. Impact of the nine reliability improvement recommendations on reliability indices.

Feeder	MAIFIe	SAIFI	SAIDI	CAIDI	%Δ MAIFIe	%Δ SAIFI	%Δ SAIDI	%Δ CAIDI
A1	4.33	0.91	3.27	3.58	-29.5	-20.9	-21.4	-1.1
E1	0.13	0.18	0.31	1.67	-18.8	-10.0	-6.1	0.6
E2	2.1	1.92	1.31	0.68	-44.0	-16.2	-18.1	-2.9
E3	1.45	0.57	0.57	1	-9.9	-6.6	-6.6	1.0
E4	1.69	0.77	0.68	0.89	-0.6	-1.3	-2.9	-1.1
E5	6.49	1.94	2.55	1.32	9.4	-28.4	-15.3	18.9
E6	2.04	1.1	2.12	1.94	-26.4	-20.9	-20.3	1.0
E7	3.15	1.39	1.49	1.07	-0.3	-0.7	-1.3	-0.9
E8	1.99	1.77	1.4	0.79	-22.6	-5.3	-6.7	-1.3
F1	6.39	0.91	1.61	1.77	0.0	0.0	8.8	8.6
F2	6.17	0.92	1.73	1.87	0.0	0.0	0.0	0.0
F3	5.72	0.4	1.03	2.56	0.0	0.0	0.0	0.0
B1	1.4	0.58	0.97	1.66	-54.4	-31.8	-29.7	2.5
B2	5.28	1.13	1.46	1.3	-15.5	-13.1	-13.1	0.8
B3	7.89	1.52	2.49	1.64	-44.2	-26.9	-19.9	10.1
B4	6.93	1.5	1.34	0.89	-37.5	-44.0	-47.2	-6.3
B5	9.15	2.3	2.96	1.29	-11.5	-7.6	-25.8	-19.4
B6	11.9	1.35	2.24	1.67	-5.3	-30.8	-21.7	13.6
B7	3.32	0.73	0.86	1.18	-26.1	-21.5	-19.6	1.7
B8	4.36	0.8	1.39	1.73	-46.8	-47.4	-38.8	16.1
D1	7.7	1.49	1.31	0.88	0.0	0.0	0.0	0.0
D2	2.04	0.53	0.33	0.63	0.0	0.0	0.0	0.0
C1	3.87	1.24	0.89	0.72	-47.3	-47.5	-37.3	20.0
C2	3.59	1.56	1.73	1.11	-21.8	-25.7	-17.2	11.0
C3	2.12	0.41	0.39	0.96	0.0	-10.9	-20.4	-9.4
TOTAL	**4.92**	**1.08**	**1.55**	**1.44**	**-22.2**	**-21.2**	**-20.1**	**2.1**

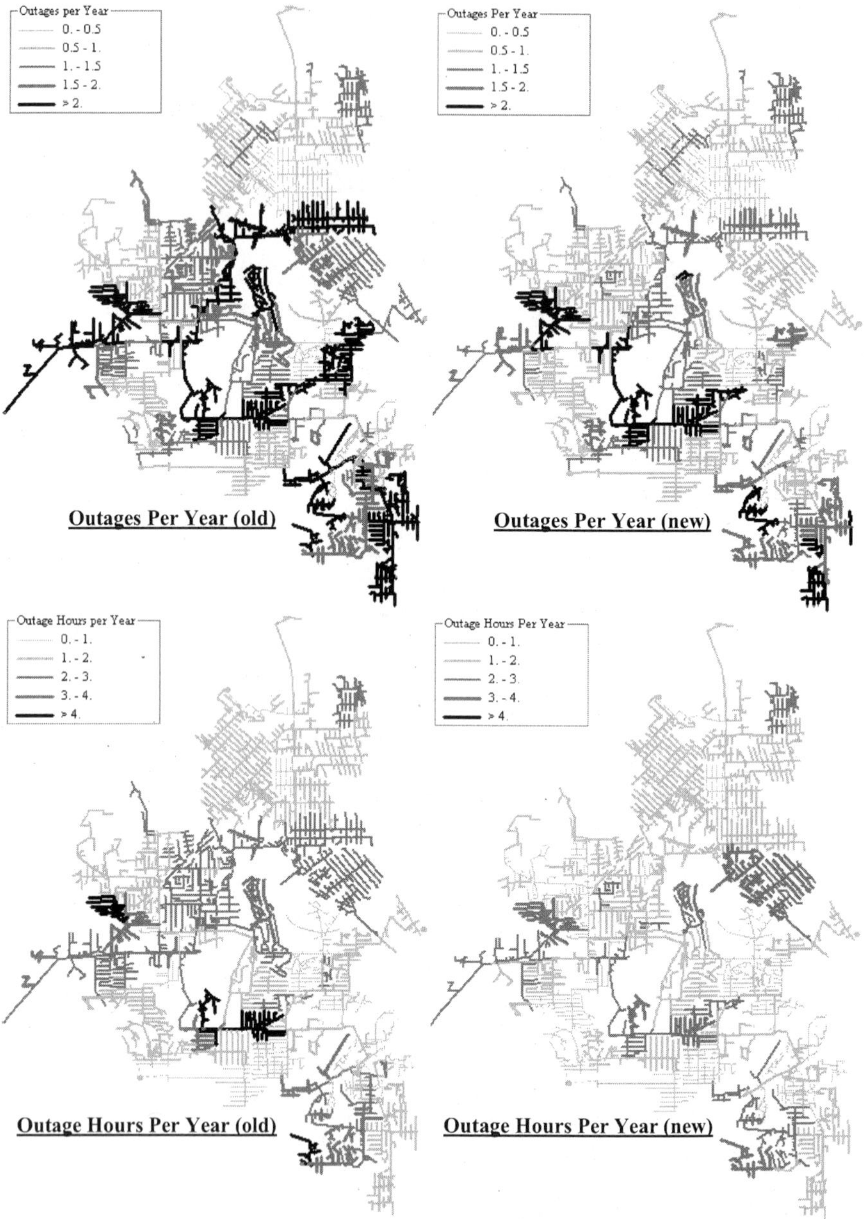

Figure 6.33. Visualization of reliability improvements after nine reliability improvement projects are modeled. The top two maps show the reduction in annual outage frequency, which corresponds to SAIFI. The bottom two maps show the reduction in annual outage hours, which corresponds to SAIDI. After improvements, SAIFI is reduced by 21% and SAIDI is reduced by 20%.

Geographic visualizations of SAIFI and SAIDI before and after the recommendations are implemented are shown in Figure 6.33. These data maps confirm that all reliability problems have been successfully mitigated. Further, they give geographic information corresponding to where the most reliability improvements have been achieved and where reliability has not substantially changed. An additional advantage to "before and after" data maps is that they can effectively present reliability improvement information to people unfamiliar with the associated technical jargon, perhaps increasing the possibility of the projects being approved for implementation.

Another key factor in securing permission for implementation of reliability improvement options is, of course, an economic analysis. This analysis should include the costs and benefits associated for each project, as well as a ranking based on benefit-to-cost ratios. For this analysis, the following two measures of benefit are used: (1) the number of avoided customer interruptions, and (2) the number of avoided customer interruption minutes. Using the ratio of dollars per customer interruption shows how cost effective a recommendation would be for improving frequency indices such as SAIFI. Using the ratio of dollars per customer interruption minute shows how cost effective a recommendation would be for improving duration indices such as SAIDI. A benefit-to-cost summary of recommendations is shown in Table 6.22.

The most cost-effective ways to improve SAIFI are recommendations 4, 7, and 9. Recommendation 4 adds a recloser to C2 and is easily implemented. Recommendations 7 and 9 add combinations of devices to feeders on substation B and utilize feeder automation. Feeder automation is effective for reducing sustained interruption frequency since large amounts of customers can be quickly restored after a fault occurs.

Table 6.22. Benefit and cost summary of reliability improvement recommendations.

Rec.	Description	Customers Affected	Cost ($)	Cust-Int Per $	Cust-Min Per $
1	Add 2 switches to B5	1235	10000	n/a	1.79
2	E tree trimming (27.1-mi)	3380	108400	0.003	0.56
3	E automation (7 switches)	3380	105000	0.006	0.39
4	Recloser on C2	628	20000	0.017	0.68
5	Split B4	1702	150000	0.013	0.81
6	B tree trimming (96.8-mi)	9273	387200	0.004	0.50
7	Add devices to B8	1898	75000	0.015	0.88
8	Add devices to B3	1369	80000	0.008	0.33
9	Add devices to B6	978	30000	0.016	0.63
All		**24839**	**965600**	**0.007**	**0.60**

The most cost-effective ways to improve SAIDI are recommendations 1, 5, and 7. Recommendation 1 adds two new sectionalizing switches near the end of the main trunk on feeder B5 and is easily implemented. Recommendation 5 requires splitting B4 into two feeders. This is more difficult to implement but is extremely cost-effective in reducing both SAIFI and SAIDI, and will provide additional benefits to surrounding feeders. Recommendation 7 adds new devices and applies automation to feeder B8. This recommendation is also cost-effective for reducing SAIFI and the same comments concerning automation apply.

For this analysis area, reconductoring is not a cost-effective option to improve reliability. Although there is a large amount of small conductor, most constrained transfer paths also include constrained large conductor, making reconductoring projects extremely expensive. Several reconductoring attempts were made on the most capacity constrained wires, but resulted in very small reliability gains that were not able to cost effectively compete with other reliability improvement alternatives. This result is interesting since reconductoring was identified by the computer as potentially the most cost-effective method of improving reliability (see Figure 6.31). This reinforces the message that at the present time, computer generated designs cannot fully replace good engineering judgment and experience and should be verified before being blindly implemented. As the next chapter will show, automated computer design and optimization algorithms can be a powerful way to explore millions of possibilities and achieve higher reliability for lower cost. They should, however, be used as a supplement and a complement to the system analysis techniques presented in this chapter, not as a replacement for them.

In summary, this exercise has applied rigorous reliability assessment techniques to a 25 feeder analysis area corresponding to an actual utility distribution system. The calibrated reliability model was used to identify reliability characteristics and concerns. Sensitivity and root-cause analyses were then performed to help identify potentially effective ways to address reliability concerns. An automated computer analysis automatically examined millions of reliability improvement projects and identified those with the best benefit-to-cost ratios. Finally, manual exploration of specific projects led to nine reliability improvement recommendations that successfully address specific reliability problems and improve overall reliability indices by more than 20%.

6.10 STUDY QUESTIONS

1. What are some of the advantages and drawbacks to predictive distribution reliability model reduction?
2. What is the goal of model calibration? Discuss the appropriateness of calibrating to (1) large system averages and to (2) individual customer reliability.

3. What are some of the techniques that can be used to explore the reliability characteristics of a system model?
4. What are some of the main methods to improve distribution reliability during nonstorm conditions? Which of these methods are typically the most cost-effective? Which of these methods are typically expensive compared to other methods?
5. How can system reconfiguration (changing the location of normally-open tie points) potentially result in improved reliability?
6. What are some of the main methods to improve distribution reliability during severe storms? Are these the same as for nonstorm conditions? How might a utility prioritize storm hardening activities?
7. Discuss some of the primary motivations for the conversion of overhead distribution facilities to underground. What are some of the potential disadvantages of undergrounding?
8. Is payback period a good measure to use for project approval? Explain.
9. Explain the potential impact of using net present value versus initial cost when computing the benefit-to-cost ratios of capital projects. How does this change when considering both capital projects and maintenance projects?
10. What is the largest potential problem associated with benefit-to-cost analysis? How can marginal benefit-to-cost analysis address this problem?

REFERENCES

1. R. Billinton and S. Jonnavitihula, "A Test System for Teaching Overall Power System Reliability Assessment," *IEEE Transactions on Power Systems* Vol. 11, No. 4, Nov. 1996, pp. 1670-1676.
2. R. E. Brown and T. M. Taylor, "Modeling the Impact of Substations on Distribution Reliability," *IEEE Transactions on Power Systems*, Vol. 14, No. 1, Feb. 1999, pp. 349-354.
3. J. J. Burke, "Using Outage Data to Improve Reliability," *IEEE Computer Applications in Power,* April 2000, pp. 57-60.
4. R. E. Brown and J. R. Ochoa, "Distribution System Reliability: Default Data and Model Validation," *IEEE Transactions on Power System,* Vol. 13, No. 2, May 1998.
5. E. R. Tufts, *The Visual Display of Information*, Edward R. Tuft (published by Graphics Press), 1983.
6. E. R. Tufts, *The Visual Display of Information*, Edward R. Tuft (published by Graphics Press), 1990.
7. R. E. Brown and J. J. Burke, "Managing the Risk of Performance Based Rates," *IEEE Transactions on Power Systems,* Vol. 15, No. 2, May 2000, pp. 893-898.
8. R. E. Brown and B. Howe, "Optimal Deployment of Reliability Investments," E-Source, Power Quality Series: PQ-6, March 2000, 37 pages.
9. R. E. Brown, A. P. Hanson, H. L. Willis, F. A. Luedtke, and M. F. Born, "Assessing the Reliability of Distribution Systems," *IEEE Computer Applications in Power*, Vol. 14, No. 1, Jan. 2001, pp. 44-49.
10. J. L. Blackburn, *Protective Relaying: Principles and Applications*, Marcel Dekker, 1987.
11. S. H. Horowitz and A. G. Phadke, *Power System Relaying*, Research Studies Press Ltd., 1992.
12. J. J. Burke, *Power Distribution Engineering: Fundamentals and Applications*, Marcel Dekker, Inc., 1994.

13. *Electrical Distribution System Protection (3rd Edition)*, Cooper Power Systems, 1990.
14. H. L. Willis, *Power Distribution Planning Reference Book*, Marcel Dekker, 1997.
15. Adly A. Girgis, David G. Hart, and William L. Peterson, "A New Fault Location Technique for Two- and Three-Terminal Lines," *IEEE Transactions on Power Delivery*, Vol. 7, No. 1, Jan. 1992, pp. 98-107.
16. A. A. Girgis, C. M. Fallon, and D. L. Lubkeman, "A Fault Location Technique for Rural Distribution Feeders," *IEEE Transactions on Industry Applications*, Vol. 29, No. 6, Nov. 1993, pp. 1170-1175.
17. Y-h Hsu, F. C. Lu, Y. Chien, J. P. Liu, J. T. Lin, H. S. Yu, and R. T. Kuo, "An Expert System for Locating Distribution System Faults," *IEEE Transactions on Power Delivery*, Vol. 6, No. 1, Jan. 1991, pp. 336-372.
18. P. Jarventausta, P. Verho, and J. Partanen, "Using Fuzzy Sets to Model the Uncertainty in the Fault Location Process of Distribution Networks," *IEEE Transactions on Power Delivery*, Vol. 9, No. 2, April 1994, pp. 954-960.
19. J. Zhu, D. L. Lubkeman, and A. A. Girgis, "Automated Fault Location and Diagnosis on Electric Power Distribution Feeders," *IEEE Transactions on Power Delivery*, Vol. 12, No. 2, April 1997, pp. 801-808.
20. P. A. Kuntz, *Optimal Reliability Centered Vegetation Maintenance Scheduling in Electric Power Distribution Systems*, Ph.D. Dissertation, University of Washington, 1999.
21. V. Borozan and N. Rajakovic, "Application Assessments of Distribution Network Minimum Loss Reconfiguration," *IEEE Transactions on Power Delivery,* Vol. 12, No. 4, Oct. 1997, pp. 1786-1792.
22. R. Takeski and D. Rajicic, "Distribution Network Reconfiguration for Energy Loss Reduction," *IEEE Transactions on Power Systems,* Vol. 12, No. 1, Feb. 1997, pp. 398-406.
23. Q. Zhou, D. Shirmohammadi, and W. H. E. Liu, "Distribution Reconfiguration for Operation Cost Reduction," *IEEE Transactions on Power Systems,* Vol. 12, No. 2, May. 1997, pp. 730-735.
24. B. W. Johnson, *Out of Sight, Out of Mind? A Study on the Costs and Benefits of Uundergrounding Overhead Power Lines*, Edison Electric Institute, 2004.
25. Navigant Consulting, *A Review of Electric Utility Undergrounding Policies and Practices*, Long Island Power Authority, 2005.
26. R. E. Brown, *Undergrounding Assessment Phase 1 Final Report: Literature Review and Analysis of Electric Distribution Overhead to Underground Conversion*, Florida Electric Utilities, 2007.
27. *Placement of Utility Distribution Lines Underground*, Commonwealth of Virginia State Corporation Commission, 2005.
28. North Carolina Natural Disaster Preparedness Task Force, *The Feasibility of Placing Electric Distribution Facilities Underground.* North Carolina Public Staff Utilities Commission, 2003.
29. Docket No. 98-026, *Inquiry into the Response by Public Utilities in Maine to the January 1998 Ice Storm*, State of Maine Public Utilities Commission, 1998.
30. Case No. 8977, *In the Matter of the Electric Service Interruptions due to Hurricane/Tropical Storm Isabel and the Thunderstorms of August 26-28, 2003*. Before the Public Service Commission of Maryland, 2003.
31. *Preliminary Analysis of Placing Investor-Owned Electric Utility Transmission and Distribution Facilities Underground in Florida.* Florida Public Service Commission, 2005.
32. Hi-Line Engineering, *Draft Report of Qualitative Advantages and Disadvantages of Converting Overhead Distribution Facilities to Underground Facilities within the Service Territory of the Fort Pierce Utilities Authority*. Fort Pierce Utilities Authority. 2005.
33. JLSD Consulting, *State Road Utility Study*, Town of Palm Beach, Florida, 2002.
34. T. G. Bradford, *Undergrounding Utilities Staff Report*, Town of Palm Beach, Florida, 2004.
35. R. W. Beck, *Conversion of Aerial to Underground Utilities Analysis*, Town of Palm Beach, Florida, 2006.
36. *Comparison of Impacts of Overhead versus Underground Transmission Line Construction*, Tallahassee, FL, 2003.

37. University of South Florida College of Engineering. *A Citizen's Initiative: Evaluating the Benefits of Underground Utility Distribution. Florida Department of Community Affairs*, 1999.
38. CVO Electrical Systems, *Undergrounding Feasibility Study*, Tahoe Donner Association, 2006.
39. P. Martin, *Undergrounding Public Utility Lines*, Legislative Reference Bureau, Honolulu, Hawaii, 1999.
40. James Lee Witt Associates, *Hurricane Isabel Response Assessment*, PEPCO Holdings, 2004.
41. T. W. Berrie, *Electricity Economics and Planning*, Peter Peregrinus Ltd., London, 1992.
42. D. G. Newnan, *Engineering Economic Analysis*, Engineering Press, Inc., 1983.
43. H. G. Stoll, *Least-Cost Electric Utility Planning*, John Wiley & Sons, 1989.
44. C. V. Jones, *Financial Risk Analysis of Infrastructure Debt: The Case of Water and Power Investments*, C. V. Jones (Quorum Books), 1991.
45. M. P. Reeser, *Introduction to Public Utility Accounting*, American Gas Association and the Edison Electric Institute, 1997.
46. R. E. Brown and M. M. Marshall, "Budget Constrained Planning to Optimize Power System Reliability," *IEEE Transactions on Power Systems*, Vol. 15, No. 2, May 2000, pp. 887-892.
47. A. J. Wood and B. F. Wollenberg, *Power Generation, Operation, and Control*, John Wiley & Sons, Inc., 1984.
48. R. E. Brown, "Distribution Reliability Modeling at Commonwealth Edison," *IEEE Transmission and Distribution Conference and Exposition*, Atlanta, GA, Oct. 2001.

7
System Optimization

Optimal, optimize, and optimization are terms prone to misuse. Though the concept of optimization and the techniques to achieve it are powerful, few people truly understand their implications and are able to apply them in a rigorous and systematic manner. System optimization has always been a favorite topic in academia, but until recently has had little applicability to distribution systems. With the exponential increase in computation power, applying optimization techniques to practical distribution reliability problems is becoming feasible, and a deeper understanding of optimization is warranted.

What does it imply when an engineer claims to have the optimal solution? Is it reasonable for top management to request a solution that optimizes both reliability and cost? Do the members of a distribution optimization team truly understand their mandate? The goal of this chapter is to allow the reader to understand and answer such questions. More importantly, it will provide a theoretical foundation so that optimization techniques can be properly applied to distribution systems and increase the probability of providing higher reliability for lower cost.

7.1 OVERVIEW OF OPTIMIZATION

The goal of system optimization is to minimize an objective function without violating any constraints. To do this, of course, an appropriate objective function must be defined and all applicable constraints must be identified. In addition, the objective function must be computable and all constraints must be testable for all possible solutions, which are characterized by a set of controllable variables. Mathematically, an optimization problem is expressed as follows:

Optimization Problem Formulation (7.1)

Minimize:

$$\Omega(x_1, x_2, \ldots, x_L) \qquad \text{Objective Function}$$

Subject to:

$$\begin{aligned}
C_1(x_1, x_2, \ldots, x_L) &= Y_1 && \text{Equality Constraints} \\
C_2(x_1, x_2, \ldots, x_L) &= Y_2 && \\
&\vdots && \\
C_M(x_1, x_2, \ldots, x_L) &= Y_M && \\
C_{M+1}(x_1, x_2, \ldots, x_L) &\le Y_{M+1} && \text{Inequality Constraints} \\
C_{M+2}(x_1, x_2, \ldots, x_L) &\le Y_{M+2} && \\
&\vdots && \\
C_N(x_1, x_2, \ldots, x_L) &\le Y_N &&
\end{aligned}$$

L = Number of Optimizable Variables
M = Number of Equality Constraints
N = Total Number of Constraints

From this formulation, it can be seen that the objective of an optimization problem is to identify a feasible set of variables that minimizes an objective function. The objective function can be any combination of formulae, algorithms, or processes as long as it returns a single value for each possible combination of parameters. Objective functions may also be formulated to be maximized, but distribution reliability optimization typically deals with cost and interruptions, which are generally more desirable as they decrease in magnitude. Feasible solutions are defined as solutions that satisfy all equality constraints and do not violate any inequality constraints.

For distribution system reliability, optimization problems generally take one of two forms. The first is to minimize cost while satisfying all reliability constraints. For example, minimize the net present value of reliability improvement projects while achieving a $MAIFI_E$ of 10 /yr, a SAIFI of 1 /yr, and a SAIDI of 60 min/yr. The second is to minimize customer interruptions subject to cost constraints. For example, minimize SAIDI without exceeding an annualized project cost of $5 million per year.

The same factors that appear in the objective function can also appear in constraints. For example, if the goal is to minimize the net present value of reliability improvement projects while achieving specific $MAIFI_E$, SAIFI, and SAIDI targets, there could also be a stipulation that net present value cannot exceed a certain budget limit. In this situation, it becomes much more likely that there will be no combination of optimization variables that will satisfy all constraints. If it is impossible to satisfy all constraints, the problem is not well-formulated and there is no feasible solution to the optimization problem.

Equality and inequality constraints, in general, make an optimization problem more difficult to solve. Even if they do not make all solutions infeasible, they may make the state space of feasible solutions difficult to search. For these reasons, constraints can be temporarily relaxed while performing an optimization, allowing the state space of infeasible solutions to be explored with the ultimate goal of finding a better solution. Alternatively, constraints can be incorporated into the objective criteria as penalty factors. If a constraint is violated, a penalty is added to the objective function based on the degree that the constraint is violated. In some cases, relaxed constraints in the final solution may be acceptable and small violations can translate into small penalties. In other cases, all constraints must be satisfied and care must be taken so that all solutions that violate constraints have a net objective function (base objective function plus all penalty factors) that is higher than all solutions that do not violate any constraints.

To illustrate the concept of penalty factors, consider an optimization problem that seeks to minimize cost while achieving a SAIDI of 4 hr/yr without overloading any equipment. In this case, the SAIDI target is set by state regulators and is considered a hard constraint—values greater than 4 hr/yr are not acceptable. Equipment overloads, however, are a soft constraint since small overloads are not strictly prohibited. Since the cost values for feasible solutions are expected to be in the $10 million range, all SAIDI violations are assigned a penalty factor of $1 billion plus an additional amount based on the degree of the violations. Doing this ensures that any solution that violates the SAIDI constraint will have an objective function higher than any solution that satisfies the SAIDI constraint. Loading violations, however, are assigned a penalty factor in strict proportion to the degree of each overload, allowing solutions with small loading violations to have a net objective function lower than other solutions without any loading violations.

The purpose of this chapter is to address system optimization in a rigorous manner. Doing so requires the consistent use of precisely defined terms. Definitions of terms corresponding to optimization problem formulation include:

Optimizable Variables – a set of variables that describe potential solutions to an optimization problem. One or more combinations of optimizable variables corresponds to each feasible solution, and all other combinations correspond to infeasible solutions.

Solution Space – all possible combinations of optimizable variables. A solution space can be divided into a feasible solution space and an infeasible solution space, each of which may be noncontiguous.

Objective Function – a mathematical expression, sometimes referred to as a criterion, that indicates the relative quality of a potential solution corresponding to a combination of optimizable variables. It is standard practice to define objective functions so that lower values are more desirable than higher values.

Constraint – a mathematical expression that corresponds to the feasibility of a potential solution in one particular aspect. Constraints are typically categorized into equality constraints and inequality constraints, with inequality constraints typically being formulated so that feasible values are less than or equal to a threshold value. Constraints can also be grouped into hard constraints, which cannot be violated, and soft constraints, which may be violated under certain conditions.

Relaxed Constraint – a soft constraint that is violated on purpose, typically to facilitate solution space investigation or to identify potential solutions that otherwise have desirable objective function values.

Binding Constraint – an inequality constraint that is at its threshold value for a particular combination of optimizable variables.

Penalty Factor – A number that is added to the objective function when a constraint is allowed to be violated. Penalty factors can be constant or can vary based on the degree of the violation. Large penalty factors are typically used for constraints that cannot be relaxed, and small penalty factors are typically used for constraints that can be relaxed.

As stated previously, the goal of system optimization is to minimize an objective function without violating any constraints. It would seem, then, that the optimal solution would correspond to the combination of optimizable parameters in the feasible solution space that results in the lowest objective function value. It may surprise the reader to learn that this is not the case, and that a feasible solution space may contain many optimal solutions depending upon which type of optimality is being considered.

7.1.1 Locally Versus Globally Optimal

Consider a reliability problem that seeks to find the optimal value of a single variable x based on an objective function f(x). To further simplify the problem, assume that x must have a value between x_{min} and x_{max}, and that f(x) is a continuous function for all values of x between x_{min} and x_{max}. A conceptual representation of this problem is shown in Figure 7.1.

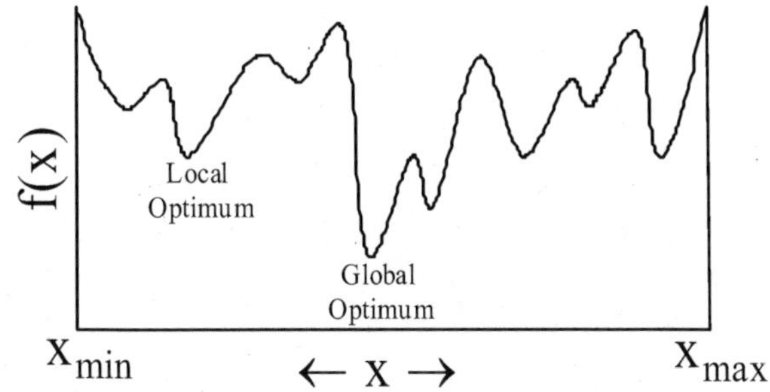

Figure 7.1. A conceptual representation of local and global optimality. Locally optimal solutions become worse for any small change in optimization variables. Globally optimal solutions have lower objective function values than all other feasible solutions. For large optimization problems such as distribution reliability, globally optimal solutions are not generally obtainable and locally optimal solutions must be pursued.

Now consider an optimization strategy that starts with $x = x_{min}$ and slowly increases x by infinitesimal values until it reaches x_{max}. After performing this exhaustive search of the solution space, the value of x that results in the lowest objective function is referred to as the *globally optimal solution.*

Global optimality is a fine concept, but is difficult to obtain for most reliability optimization problems since it is not typically feasible to test all possible solutions. Consider the impact of eliminating the x_{max} limit for the problem represented in Figure 7.1. As the value of x increases from x_{min}, it will periodically encounter troughs where increasing or decreasing x by a small amount will make the objective function worse. These points are referred to as local optima, and can be identified in twice differentiable functions by a zero first order derivative and a positive second order derivative. As the value of x increases towards infinity, the best locally optimal solution can be remembered, but there is no guarantee that higher values of x will not result in lower objective function values. Hence, most optimization algorithms can guarantee local optimality, but cannot guarantee global optimality. Precise definitions for these two types of optimality are:

Locally Optimum – a solution that will produce a worse objective function value if any optimization variable is perturbed.

Global Optimum – the solution that produces the best objective function value when compared to all other solutions in the feasible solution space.

The study of optimization problems is referred to as mathematical programming. There are two primary types of optimization problems that can be solved with guaranteed global optimality. The first are problems that are small enough for all possible solutions to be exhaustively tested. The second are problems that have a continuous and convex feasible solution space, such as linear programming problems and quadratic programming problems. A summary of the major classes of mathematical programming is:

Mathematical Programming – the study of problems that seek to optimize a function of variables subject to constraints. The term programming does not refer to computer programming, and was coined before the word became closely associated with computer software development.

Linear Programming – the study of optimization problems with a linear objective function and linear constraints. Linear programming problems have convex solution spaces, and well-known solution methods are generally able to guarantee global optimality.

Quadratic Programming – the study of optimization problems with a quadratic objective function and linear constraints. Quadratic programming problems have convex solution spaces, and well-known solution methods are generally able to guarantee global optimality, but are more computationally intensive that those used for linear programming.

Nonlinear Programming – the study of optimization problems where at least one of the functions (objective or constraints) does not vary linearly with all optimization variables. Unless convexity of the solution space can be demonstrated, solution methods cannot generally guarantee global optimality for nonlinear programming problems.

Up to this point, it has been implicitly assumed that optimization problems consist of a single objective function and utilize real numbers for variables. Though this is an appropriate way to introduce the concept of optimization, neither of these assumptions is necessary and, in fact, they are often not appropriate for distribution reliability optimization. For this reason, Sections 7.1.2 and 7.1.3 discuss the concepts of multi-objective optimization and discrete variable optimization, respectively.

7.1.2 Multi-Objective Optimization

The first step in approaching an optimization problem is to clearly define the metric that is being optimized. If this criterion can be represented as a single scalar value, the goal of the optimization process is clear—minimize this value. However, many utilities are not comfortable reducing their reliability problems to a single measure of goodness, and multiple objectives are required.

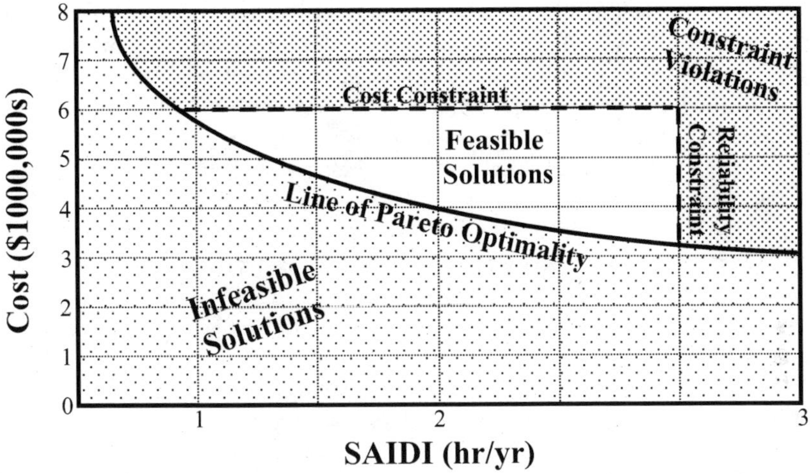

Figure 7.2. Classes of solutions for a multi-objective optimization problem considering reliability (SAIDI) and cost. For each value of cost, there exists a solution resulting in the lowest possible SAIDI. The set of these points is referred to as the line of Pareto optimality, and solutions on this line can only improve cost or reliability at the expense of the other.

To illustrate the concept of multi-objective optimization, consider a utility that is concerned about reliability and cost. Specifically, they would like to minimize net present value of reliability improvement projects while minimizing the expected value of SAIDI. Consider a specific cost, C. Theoretically, there is a best possible way to spend this money to result in the lowest possible SAIDI. Similarly, there is theoretically a lowest possible cost to achieve each specific SAIDI value. Each of these combinations of reliability and cost is said to be a Pareto optimum. This term is derived from the Italian economist and sociologist Vilfredo Pareto, who defined total societal welfare as an improvement in a person's condition that was not achieved at any other person's expense. The definition of Pareto optimality is:

> **Pareto Optimality** – a term used to describe a solution to a multi-objective optimization problem if any improvement to any objective function comes at the expense of one or more other objective functions. A solution with Pareto optimality is sometimes referred to as an *efficient solution*.

The concept of Pareto optimality is shown in Figure 7.2. The line of Pareto optimality represents the best possible combinations of reliability and cost. Any combination below this line cannot be achieved, and any combination above this line is suboptimal. In addition, this figure represents a maximum acceptable SAIDI value and a maximum acceptable cost value. These constraints result in infeasible combinations of cost and reliability on and above the line of Pareto optimality.

When first examining reliability and cost from a Pareto optimality perspective, most utilities will find that the cost and reliability of their system lies above the line of Pareto optimality. This means that it is possible to achieve the same level of reliability for lower cost, which is inarguably an improvement. Similarly, it is possible to achieve a higher level of reliability for the same cost, also an unambiguous improvement. There will also be designs that can improve reliability and reduce cost at the same time, which is often the desired outcome. These unambiguous improvements in reliability and cost can only continue until the line of Pareto optimality is reached. At this point, cost cannot be further reduced without increasing SAIDI, and SAIDI cannot be reduced without increasing cost. When Pareto optimality is reached, it becomes imperative for reliability engineers to be able to communicate the implications of budget reductions and reliability improvement goals to those that make these decisions. If the budget is cut by X%, a reliability engineer should be able to respond that this cut will result in a SAIDI increase of Y%. Similarly, if a corporate mandate states that SAIDI shall be reduced by Y%, a reliability engineer should respond that achieving this goal requires a budget increase of X%.

The concept of Pareto optimality is just as valid for optimization problems with more than two objective functions. An optimization problem with 1 objective function has a single (one-dimensional) optimal solution and an optimization problem with 2 objective functions has a two-dimensional curve of optimal solutions. Similarly, an optimization problem with 3 objective functions has a three-dimensional surface of optimal solutions and an optimization problem with N objective functions has an N-dimensional manifold of optimal solutions.

Consider a utility that wishes to optimize three values: cost, SAIDI, and SAIFI. For this scenario, there will be a lowest possible cost for any selected combination of SAIDI and SAIFI. Similarly, there will be a lowest possible SAIDI for and selected combination of cost and SAIFI and a lowest possible SAIFI for and selected combination of cost and SAIDI. The set of optimal combinations forms a three-dimensional surface of Pareto optimality (see Figure 7.3). This surface has all of the implications of the two-dimensional line of Pareto optimality discussed in the reliability versus cost scenario. Utilities will generally find themselves above this surface, and can improve any objective function without sacrificing others. After the surface of Pareto optimality is reached, the only way to improve one objective function is at the expense of one or both of the others.

Dealing with optimization problems with more than three objective functions becomes a bit more arcane since manifolds with more than three dimensions cannot generally be visualized. This is not a problem for computers, which can identify whether solutions are located on the manifold of Pareto optimality, what the trade-offs are between different objective functions if they are, and how far away they are if they are not. To present this information in a method comprehensible to humans, dimensionality must generally be reduced.

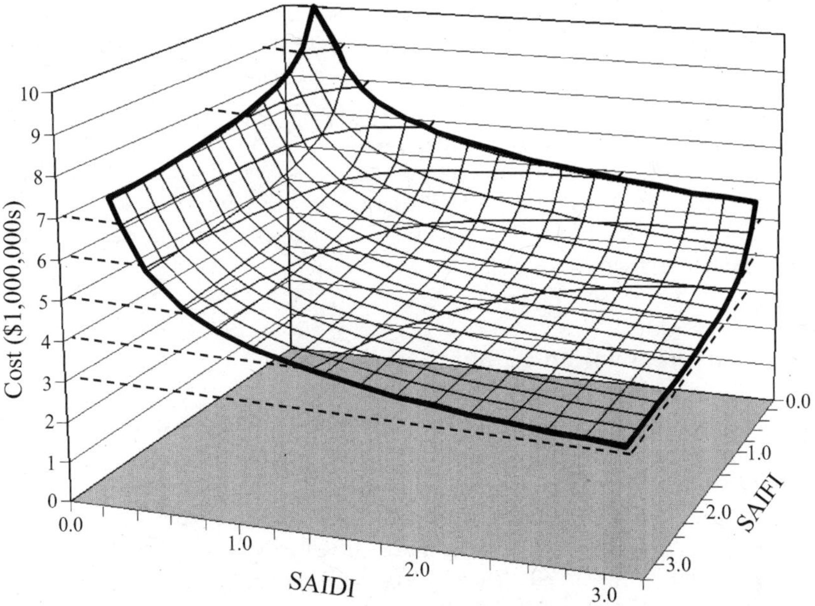

Figure 7.3. A surface of Pareto optimality for an optimization problem with three objective functions: cost, SAIDI, and SAIFI. If a solution is on the surface of Pareto optimality, no objective function can be improved without causing one or both of the others to become worse.

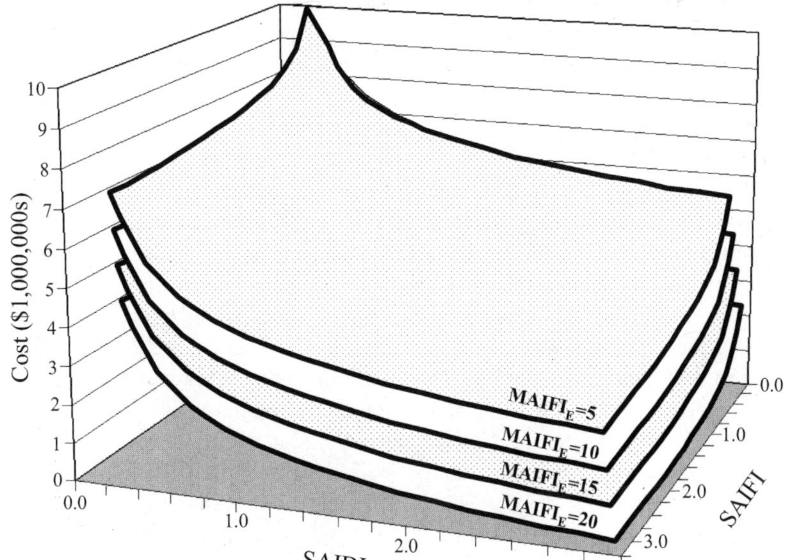

Figure 7.4. For optimization problems with more than three objective functions, Pareto optimality can be visualized by taking three-dimensional cross sections. In this case, optimizing $MAIFI_E$, SAIFI, SAIDI, and cost is shown by taking cross sections for different fixed values of $MAIFI_E$.

It is often beneficial to examine Pareto optimality of optimization problems with more than three objective functions by taking two-dimensional or three-dimensional cross sections. Consider a problem that seeks to optimize cost, $MAIFI_E$, SAIFI, and SAIDI. The surface of Pareto optimality will consist of a four-dimensional manifold that is impossible to visualize. A three-dimensional surface, however, can be generated by fixing one of the objective functions at a specific value and creating a manifold based on the remaining three objective functions. This can be done for several fixed values to create a set of three-dimensional cross sections similar to those shown in Figure 7.4.

7.1.3 Discrete Variable Optimization

Sections 7.1.1 and 7.1.2 have discussed optimization in terms of continuous optimization variables. Although beneficial for conceptual purposes, the vast majority of distribution reliability optimization problems consist of discrete choices rather than a continuum of choices. As an example, consider some of the choices that impact the reliability of a primary distribution feeder. There are an integer number of fuses and switches. These devices can only be placed at specific pole locations. These devices can only be purchased with specific ratings. Feeder sections can be either overhead or underground. Wire only comes in certain sizes. Switches are either automated or they are not. Optimization techniques that address discrete choices fall under the category of integer programming, and optimization techniques that address both continuous and discrete choices fall under the category of mixed integer programming:[1]

Integer Programming – the study of optimization problems with discrete optimization variables.

Mixed Integer Programming – the study of optimization problems with both continuous and discrete optimization variables.

In one sense, discrete choices are good for optimization problems. An optimal solution can be found by examining each combination of discrete possibilities and choosing the best combination. In reality, however, there are far too many possible solutions to examine each one separately. Discrete optimization suffers from a problem referred to as combinatorial explosion, which means that the number of possible solutions grows exponentially with problem size (i.e., the problem is NP hard). To illustrate, consider conductor sizing. If there are 5 possible conductor sizes and there is only one line section that is being built, the number of possible solutions is $5^1 = 5$. If there are two line sections, the number of possible solutions is $5^2 = 25$. If there are 20 line sections, the number of possible solutions is 5^{20} = 95 trillion. Clearly, even simple problems can quickly become intractable.

In the past, when computing power was much less than today, many discrete optimization problems were linearized so that computationally efficient linear programming techniques could be used. The final solution would then be selected by choosing discrete values nearest to the continuous values chosen by the continuous optimization process. For example, if the linear programming algorithm suggests placing a 567-kVAR capacitor bank 3.87 miles from the substation, the closest feasible solution may be placing a 600-kVAR capacitor bank (available for purchase) at 3.8 miles from the substation (the location of the nearest pole). Such techniques were an acceptable compromise in the past, but are generally not recommended since they can lead to suboptimal results when compared to solution methods specifically designed for discrete optimization problems.

Even with sophisticated algorithms and fast computers, discrete optimization problems can quickly become untenable if their solution space is allowed to become excessively large. For this reason, it is vital to restrict the domain space of discrete optimization problems wherever possible. This is a divide-and-conquer approach that breaks up a large problem into several subproblems that can be solved independent of one another. To illustrate the potential advantages, consider the conductor sizing problem with 20 conductors and 95 trillion possible solutions. If this problem can be broken down into 5 separate feeders with 4 sections each, the solution space is reduced to $5 \text{x} 5^4$ = 3125. This is a domain space reduction of 99.999999997%.

When dividing a reliability optimization problem into subproblems, it is important to insure that each subproblem is sufficiently independent of other subproblems. For example, when placing protection devices and normally closed switches on a distribution system for reliability improvement, each feeder can be treated separately since contingencies on one feeder will not affect the reliability of other feeders. Within a feeder, however, protection device and switch placement must be treated together since their placements are interdependent: protection device placement will impact optimal switch placement and switch placement will impact optimal protection device placement.

7.2 DISCRETE OPTIMIZATION METHODS

This section discusses the techniques necessary to implement a discrete optimization algorithm for distribution reliability problems. It starts by discussing problem formulation including objective functions, constraints, and penalty factors. It continues by discussing approaches to encoding potential solutions in a form that is usable by discrete optimization algorithms. It concludes by discussing the most common discrete optimization methods including local search, simulated annealing, and genetic algorithms.

7.2.1 Problem Formulation

Problem formulation, as previously discussed in the introduction of this section, consists of an objective function, a set of constraints and a set of optimizable variables.

For discrete optimization problems, defining the objective function is technically simple since there are no requirements for linearity, convexity, differentiability, or continuity. In fact, the objective function can usually be defined without knowing how it is computed. For example, SAIDI can be selected as the objective function without knowing precisely how SAIDI will be computed, a luxury not afforded by most continuous optimization algorithms.

In most situations, the distribution reliability optimization function and constraints will consist primarily of reliability and cost components. The three natural formulations are:

Common Distribution Reliability Problem Formulations

1. Optimize reliability subject to cost constraints
2. Optimize cost subject to reliability constraints
3. Optimize the total cost of reliability including the cost to provide reliability and the incurred costs associated with interruptions.

The first formulation is common for utilities with a fixed budget to spend on reliability improvement projects. Given a fixed amount of money, these utilities wish to achieve the best possible reliability, usually measured in terms of reliability indices. If multiple indices are of interest, it is generally appropriate to create an objective function equal to a weighted sum of multiple indices. For example, if a utility has present reliability indices of $MAIFI_{E0}$, $SAIFI_0$, and $SAIDI_0$, an appropriate objective function might be:

$$\text{Objective} = w_1\left(\frac{MAIFI_E}{MAIFI_{E0}}\right) + w_2\left(\frac{SAIFI}{SAIFI_0}\right) + w_3\left(\frac{SAIDI}{SAIDI_0}\right) \tag{7.2}$$

The weights of this objective function are selected according to the relative importance of reducing each index ratio. It is also convenient to make the sum of all weights equal to unity so that the base case objective function value is also equal to unity. For example, consider a system with reliability indices of $MAIFI_{E0}$ = 10 /yr, $SAIFI_0$ = 3 /yr, and $SAIDI_0$ = 240 min/yr. If an objective function is defined as simply the sum of these three indices, SAIDI improvements will tend to dominate results since these particular SAIDI values are much larger than the other indices. To solve this problem, each index can be normalized to initial index values so that a 10% increase in SAIDI is given the same value as a 10% increase in SAIFI or $MAIFI_E$. Last, weights are given to these normalized index values to reflect their relative importance. For example selecting the weights $w_1 = 0.25$, $w_2 = 0.25$, and $w_3 = 0.5$, will give SAIDI improvements twice as much importance as SAIFI or $MAIFI_E$ improvements. Further, the base case objective function value will be equal to one, allowing other scores to be viewed as per unit values.

The second formulation has a much more straightforward objective function since it is entirely based on cost. Once the cost basis is selected (e.g., net present value, annualized cost, initial cost), the objective function is simply equal to the sum of the cost of all reliability improvement projects. The third formulation is similar, but adds additional costs associated with the frequency and duration of interruptions. These incurred costs can include performance penalties, customer rebates, lawsuit awards, lost revenue, and a host of other possibilities.

Constraints in discrete optimization problems are generally handled in one of three ways: (1) do not allow search algorithms to consider solutions that violate constraints, (2) have a function that modifies all solutions until all constraints are satisfied, and (3) represent constraints as penalty factors. Of these, penalty factors are generally recommended as a first approach. Treating constrained solutions as impenetrable barriers in the solution space can greatly reduce the amount of solution space that is explored, potentially resulting in lower quality solutions. Dynamic solution modifications to satisfy constraints can be effective if implemented carefully, but can bias results and reduce the computational effectiveness of optimization algorithms.

Penalty factors can be implemented in a variety of ways, but generally consist of an initial penalty that occurs when the constraint is violated and a variable penalty that changes its value based upon the degree of the violation. A general form for a penalty factor for variable x with a constraint of x_{min} is:

$$\text{Penalty Factor} = C_1 + C_2\left(x - x_{min}\right)^{C_3} \tag{7.3}$$

Selection of C_1, C_2, and C_3 are important to both the type of solutions that will be identified as optimal and what their overall quality is likely to be. If constraints can be relaxed and slight violations are acceptable, C_1 should be small to nothing and the other parameters should be carefully chosen so that the tradeoffs being made between the objective function and the penalty factor are appropriate. If constraints cannot be relaxed, C_1 should be large when compared to the objective function of the worst feasible solution and the other parameters should be chosen so that solutions with a few small constraint violations are more desirable than solutions with many large constraint violations.

7.2.2 Solution Representation

An important step in discrete optimization is solution representation. This is simply a method of representing a solution in a manner that can be used by discrete optimization algorithms. The most popular way to represent a solution is to use an integer map, which is a string of bounded integer numbers that represent a solution. A common type of integer map restricts integer values to zero or one, and is called a *bitmap*. Consider a utility trying to decide where to place fault detectors on a system with 50 possible locations. Feasible solutions can be represented by a string of 50 binary numbers, with each number corresponding to a feasible location. A value of zero corresponds to not having a fault detector at the location and a value of one corresponds to having a fault detector at the location. For this scenario, the following bitmap represents a solution with fault detectors in locations 1, 5, 25 and 50.

	00000000011111111112222222222333333333344444444445
Location:	12345678901234567890123456789012345678901234567890
Value:	10001000000000000000000010000000000000000000000001

Sometimes discrete decisions are not binary in nature. An example is the selection of conductor size. Suppose there are five standard sizes to choose from when upgrading a conductor. For a bitmap to reflect this choice, each section has to be assigned a substring of three bits. The sequence 000 corresponds to no upgrade, the sequence 001 corresponds to the smallest upgrade, and so on until the sequence 101 corresponds to the largest possible upgrade. This type of representation is acceptable, but has several drawbacks. First, certain combinations may not correspond to meaningful solutions. In this case, the sequence 110 and 111 are not assigned to a conductor upgrade. Second, the changing of a single bit may result in a very different solution, making solution space exploration difficult. Consider the upgrades shown in Table 7.1, with solution 000 corresponding to no conductor upgrade. If the first bit changes, the resulting sequence of 100 corresponds to second largest possible upgrade, skipping all of the possibilities in between. Such problems can be reduced but not eliminated by using sequences of binary strings that only change one bit at a time, like the Gray code,[2] but it is often more convenient to use nonbinary maps.

Table 7.1. Representation of a conductor upgrade using a typical binary sequence and a Gray code sequence. Notice that the Gray code sequence only changes one bit at a time.

Typical Binary Representation		Gray Code Representation		
Decimal	Binary	Decimal	Binary	Upgrade
0	000	0	000	No Upgrade
1	001	1	001	Upgrade to 4/0
2	010	3	011	Upgrade to 250 kcmil
3	011	2	010	Upgrade to 500 kcmil
4	100	5	110	Upgrade to 750 kcmil
5	101	8	111	Upgrade to 1000 kcmil

A generalized integer map generally consists of a string of integers, with each integer representing each individual decision. For the conductor upgrade problem discussed earlier, solutions could be encoded by assigning to each line section an integer with a value from zero to five. The following integer map represents a solution with an upgrade of section 1 to 1000 kcmil (integer value of 5), an upgrade of sections 5 through 8 to 250 kcmil (integer value of 3) and an upgrade of sections 25 through 32 with an upgrade to 4/0 (integer value of 1).

	00000000011111111112222222222333333333344444444445
Location:	12345678901234567890123456789012345678901234567890
Value:	50003333000000000000000011111111000000000000000000

Integer maps can also be used to represent multiple decisions. Consider a solution that consists of both fault location devices and conductor upgrades. In this situation, two sequential integers can be used to represent each section: a binary integer to represent whether a fault location device is present (A) and a six-value integer to represent the conductor upgrade (B). A solution that places fault location devices on sections [1, 5, 10, 13], upgrades section 1 to 1000 kcmil, upgrades sections 5 through 8 to 250 kcmil and upgrades sections 16 through 17 with an upgrade to 4/0 is represented as:

	0	0	0	0	0	0	0	0	0	1	1	1	1	1	1	1	1
	1	2	3	4	5	6	7	8	9	0	1	2	3	4	5	6	7
Location:	AB	AB	AB	AB	AB	AB	AB	AB	AB	AB	AB	AB	AB	AB	AB	AB	AB
Value:	15	00	00	00	13	03	03	03	00	10	00	00	10	00	00	01	01

Solution representation is important since it is the interface between the optimization algorithms and the reliability assessment algorithms. As such, it is important to generate an integer map that is efficient for the optimization algorithms to manipulate and for the reliability assessment algorithms to translate into a model that can be used to compute objective functions and test for constraints. The solution representation can and will have an impact on the quality of solutions for complex problems, and enough time should be spent on developing an appropriate structure before selecting and implementing a solution methodology.

7.2.3 Local Search

A simple and robust approach to solving a discrete optimization problem is referred to as a *local search*. Essentially, a local search starts with a solution and makes small changes until stopping criteria are met. During the algorithm, the best solution is remembered and is used as the result when the algorithm terminates.

The most common form of local search is referred to as *hill climbing*. This method can either begin with the existing state of the system or with a random solution. After evaluating the objective function of this solution, a single integer is increased in value by one step and the new objective function is computed. If the new value is better than the initial value, the new solution is kept. If the new value is not better than the initial value, the original value is reduced in value by one step and kept if the objective function becomes better. This process is done for each variable sequentially, looping back to the first integer after the last integer has been checked, until the design cannot be further improved. The hill climbing algorithm can be summarized as follows:

Algorithm for a Hill Climbing Local Search

1. Randomly generate an initial integer string, $C_1, C_2, \ldots, C_n$.
2. Compute the initial objective function value, $\Omega(C_1, C_2, \ldots, C_n)$.
3. Set $x = 1$.
4. Increase the integer value of C_x by one increment (if possible) and compute the new objective function value, Ω_{test}.
5. Is Ω_{test} better than Ω? If yes, set $\Omega = \Omega_{test}$ and go to Step 10.
6. Decrease the integer value of C_x by one increment.
7. Decrease the integer value of C_x by one increment (if possible) and compute the new objective function value, Ω_{test}.
8. Is Ω_{test} better than Ω? If yes, set $\Omega = \Omega_{test}$ and go to Step 10.
9. Increase the integer value of C_x by one increment and set $x_{stop} = x$.
10. Set $x = x + 1$.
11. Is $x > n$? If yes, set $x = 1$.
12. Is $x = x_{stop}$. If yes, end.
13. Go to Step 4.

There are several variations to the hill climbing algorithm described above. The first variation is to alternate between tests for integer increments and integer decrements after each integer string pass rather than after each individual integer. This is a minor algorithmic change and will not tend to affect solution quality or algorithm performance by a substantial amount. The second variation is to examine all possible integer increments and integer decrements before selecting the single change that results in the best optimization function value. This type of discrete gradient search approach will tend to terminate in fewer accepted changes, but will also tend to be computationally much slower.

The hill climbing algorithm is important since it guarantees local optimality. After the algorithm terminates, a single increment or decrement to any integer is guaranteed not to produce a better solution. The hill climbing algorithm will also always result in the same solution if the same initial integer string is used. For this reason, it is often useful to run the hill climbing algorithm for many different randomly generated initial integer strings to see if solution quality or solution characteristics vary. If the solution space is convex, each trial, regardless of the initial solution, will result in the same answer. This, unfortunately, is rarely the case and many local minima are usually encountered. If most of the solutions are nearly equivalent, this is not a problem and hill climbing is a sufficient method. If solutions vary widely, more sophisticated methods such as simulated annealing or genetic algorithms may be required.

A local search technique designed to avoid being trapped in local minima is referred to as Tabu search.[3-4] This method essentially consists of a hill climbing algorithm in the gradient form. However, while making changes, the algorithm keeps a Tabu list of integers that have recently been changed. When a local minima is reached, the algorithm selects the movement that will minimize objective function degradation, but only considers movements that are not on the Tabu list. This technique is not only effective for escaping local minima, but for escaping discontinuous islands surrounded by infeasible solutions.

The size of the Tabu list is an important parameter since too short a list can result in search cycles and too long a list may restrict interesting moves that may lead to high-quality solutions.[5-6] Other features that can impact solution quality are aspiration, intensification, and diversification.[7] Aspiration allows the Tabu list to be ignored if an unusually good move is detected. Intensification uses a relatively short Tabu list to find the local optimum in the local neighborhood of solutions, and diversification uses a long Tabu list to escape the local neighborhood and examine unexplored areas of the solution space. The stopping criteria for a Tabu search is often set by the number of diversification phases performed.

Local search techniques are simple to implement, are robust, and do not have a large number of parameters to tune. As such, they are often sufficient for reliability optimization problems, especially when the problem is unique and the algorithm is not likely to be reused in the future. Perhaps the simplest method to implement is a hill climbing algorithm with a random initial solution. This algorithm can be repeated many times for many different initial solutions until a distribution of outcomes becomes apparent and there is confidence that the best encountered solution is near the global optimal. This method may not be as computationally efficient as some of the more sophisticated techniques that will be discussed in future sections, but is quick to implement and may produce answers that are nearly as good.

7.2.4 Simulated Annealing

Simulated annealing optimizes a function by performing a directed random search based on statistical mechanics. Consider the annealing process of metal. First, the metal is heated up so that the molecules have high energy and high mobility. Next, the metal is slowly cooled so that the atoms have a high probability of settling down in a low energy state. When the temperature becomes low enough, the metal molecules have very low mobility and the system is "frozen" in a low energy state. In a similar manner, a discrete optimization problem can be annealed to find high quality solutions. Initially, the solution has high mobility and is allowed to randomly move about the solution space regardless of whether solution quality improves or degrades. Gradually, the system is cooled, mobility is decreased, and the probability of moving to lower quality solutions is reduced. Eventually, the system is reduced to zero mobility, referred to as being frozen, and a high quality solution has hopefully been found. Simulated annealing is attractive since it can theoretically guarantee a globally optimal solution, but the annealing process is inherently slow and formulating an efficient cost function and annealing schedule has proven difficult.[8]

A suggested annealing schedule uses a single temperature parameter, T, for its annealing schedule. The schedule begins by creating a random solution and setting an initial value of T close to unity. A new solution is then generated by examining each integer in the integer map and randomly incrementing or decrementing its value with a probability of T (e.g., if T is 0.9, an integer will be reassigned 90% of the time). If this new solution is an improvement over the old solution, it is accepted. If this new solution is worse than the previous solution, it is accepted with a probability of T. The temperature is then reduced by multiplying it by an annealing rate, R, and the process is repeated. This procedure repeats until T has reduced in value a pre-specified number of times without any improvements in solution quality. At this point the solution is frozen and the annealing process is terminated. Since simulated annealing cannot guarantee that a local minimum has been reached, a hill climbing algorithm should be performed after a solution freezes. A summary of the algorithm is:

Algorithm for Simulated Annealing

1. Select a starting temperature, $T \in (0,1)$, and an annealing rate, $R \in (0,1)$.
2. Randomly generate an initial solution and compute the objective function, Ω.
3. Generate a new solution by changing each integer value with a probability of T. Increment or decrement the integer value with equal probability.
4. Compute the objective function of the new solution, Ω_{test}.
5. Is Ω_{test} better than Ω? If yes, keep the new solution and set $\Omega = \Omega_{test}$. If no, keep the new solution with a probability of T.
6. Has the solution changed since a preset number of temperature changes? If no, go to Step 8.
7. Multiply T by the annealing rate R and go to Step 3.
8. Perform a hill climbing local search on the frozen solution to guarantee local optimality.

The speed and solution quality of a simulated annealing algorithm is a strong function of both starting temperature and annealing rate, with higher quality solutions becoming much more likely with a slow annealing rate. In general, higher initial temperatures should be used with fast annealing rates, but lower starting temperatures are acceptable with slow annealing rates since the algorithm will still have sufficient time to explore the solution space. In the author's experience, a starting temperature of 0.3 with an annealing rate of 0.99 typically results in high quality solutions in a reasonable amount of time.

A common variation of simulated annealing determines the probability of accepting a worse solution based on both temperature and the difference between the objective function values of the candidate and present solution. This probability is typically formulated as follows:

$$\text{Probability of Acceptance} = \exp\left(\frac{\Omega - \Omega_{\text{test}}}{\Omega \cdot \text{T}}\right) \tag{7.4}$$

This probability of acceptance equation works well for certain objective functions, but may limit solution space exploration for problem formulations that utilize penalty factors that can be large when compared to the objective function. This is especially true for problems with hard constraints. Consider a problem where a hard constraint violation is reflected as a penalty factor approximately twice the value of the expected objective function. Even at a temperature of 0.9, the probability of accepting a solution that barely violates the constraint will be less than 10%.

Implementing a simulated annealing algorithm is somewhat more complicated than implementing a local search, and should actually include a local search so that solutions are guaranteed to be locally optimal. Therefore, simulated annealing is recommended for optimization problems that are not being handled well by a local search algorithm, or for optimization algorithms that will be required on a regular basis in the future.

7.2.5 Genetic Algorithms

In both simulated annealing algorithms and local search algorithms, each trial is independent. This means that each individual trial has the same probability of resulting in a good solution, and increasing the number of trials will not affect this probability. If a large number of trials are needed to ensure a high quality solution, it may be beneficial to use techniques in which trials are run in parallel, and each trial can share information with other trials in an attempt to improve overall solution quality. An example of such a technique is *genetic algorithms*.

A genetic algorithm is a form of directed random search based on natural selection and evolution. After an initial population of solutions is created, future generations are determined by probabilistically selecting high quality parents and combining them to create children. During this process, a small amount of random mutation is used to ensure genetic diversity. Genetic algorithms have proven to be robust and efficient when applied to difficult optimization problems and their application to distribution planning problems has increased exponentially in recent years.[9-12]

Genetic algorithms operate on a set of potential solutions. Using biological terminology, each individual member is referred to as a chromosome and the entire set is referred to as a population. A chromosome is usually encoded as an integer map, with each integer in the map referred to as a gene and each possible value of the genes being referred to as an *allele*. An example of two chromosomes, each with eight genes and binary alleles, is:

Chromosome A: `11010010`
Chromosome B: `10011011`

Each chromosome is characterized by a fitness function, which is typically equal to the objective function value plus all penalty factors. After all chromosomes are assigned a fitness value, the next generation is created through genetic operators such as crossover and mutation. Crossover is the fundamental genetic operator for genetic algorithms and consists of probabilistically selecting two fit parents and combining their chromosomal information to produce offspring. Mutation consists of randomly altering a small percentage of genes to ensure genetic diversity.

There are two common methods of parental selection: roulette wheel and tournament selection. The roulette wheel method allocates wheel bins based on parental fitness, with each parent receiving at least one bin and fitter parents receiving more bins. Parents are chosen by randomly selecting a bin, with each bin having an equal chance of being selected.

A simple way of assigning roulette wheel bins to parents is to allocate a small number of bins to the worst parent and a large number of bins to the best parents. The number of bins allocated to other parents can be assigned by linear interpolation. For example, consider a population with a best fitness score of Ω_{best} and a worst fitness score of Ω_{worst}. If the best member is allocated B_{max} bins and the worst member is allocated B_{min} bins, members with a fitness of Ω are allocated the following number of bins:

$$\text{Bins} = B_{min} + \left(B_{max} - B_{min}\right)\frac{\left(\Omega - \Omega_{best}\right)}{\left(\Omega_{worst} - \Omega_{best}\right)} \tag{7.5}$$

An example of bin allocation for a population size of eight, a minimum of one bin per member and a maximum of ten bins per member is shown in Figure 7.5.

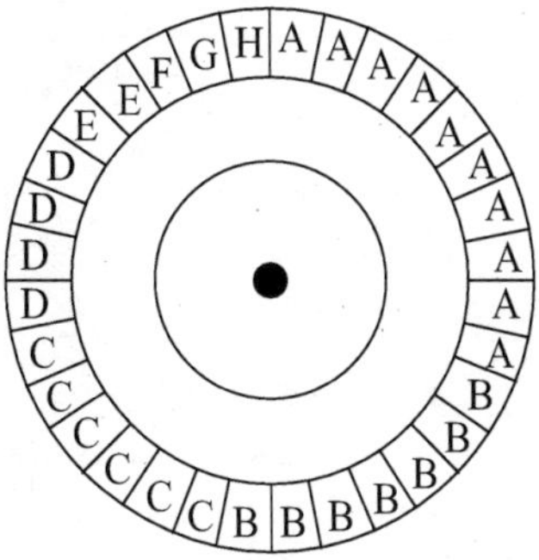

	Fitness	# Bins
A	22,142	10
B	25,389	7
C	26,203	6
D	28,489	4
E	30,532	2
F	31,581	1
G	31,908	1
H	32,189	1

Figure 7.5. Example of roulette wheel parental selection. In this case, the roulette wheel is formed from a population of eight members with the best member (A) receiving ten bins and the worst member (H) receiving a single bin. The number of bins allocated to other members is determined by linear interpolation.

A second popular method of parental selection is referred to as *tournament selection*. In tournament selection, a parent is selected by winning a tournament consisting of two or more randomly selected members. The winner is chosen in a similar manner to roulette wheel selection, with fitter members having a higher probability of winning the tournament than weaker members do. If a tournament consists of just two members, the stronger member can simply be given a fixed probability of being selected, such as 70% or 80%.

Once two parents are selected, the genetic operator crossover is used to create two new children. The simplest form of crossover, referred to as single point crossover, is performed by selecting a random crossover point and creating two new children by swapping genes on either side of the crossover point. An example of single point crossover is shown below (the selected crossover point is indicated by an "x"):

Parent A: **1101** x **0010**
Parent B: 1001 x 1011

Child 1: **1101** 1011
Child 2: 1001 **0010**

The theory addressing the computational efficiency of genetic algorithms is based on schema, which is defined as a set of alleles in a chromosome. The most important aspect of a schema is its length, which is equal to the separation of the first encountered allele and the last encountered allele. An example of several different alleles with various lengths is shown in Table 7.2.

Table 7.2. Examples of schema and their corresponding length.

Schema	Alleles	Length
A	-123----	3
B	-321----	3
C	-1-2-3--	5
D	3--3--33	8

There are several important points regarding schema as illustrated in Table 7.2. First, different schema can consist of the exact same genes with different allele values as shown by Schema A and Schema B. Second, the alleles associated with a schema do not have to be sequential as shown by Schema C and Schema D (a dash indicates that the gene is not associated with the schema).

The efficiency of genetic algorithms using roulette wheel parental selection and single point crossover has been mathematically proven in what is referred to as the schema theorem,[13] which states that schema with short lengths that are correlated with high fitness functions are exponentially reinforced in subsequent generations. It has been further shown that genetic algorithms are good at avoiding local minima and have inherent parallelism in computation.

The schema theorem proves the inherent computational efficiency of genetic algorithms, but has an important implication that is often overlooked. The probability of good allele combinations being reinforced is directly related to how closely these gene combinations are grouped together in the integer map. Consider a highly desirable two-gene combination. If these two genes are far apart, they have a good chance of being separated during the crossover process. Therefore, it is vitally important to group highly correlated genes as close as possible when creating a chromosome to represent a solution.

A common variation of single point crossover is referred to as *two-point crossover*. In two-point crossover, two crossover points are randomly chosen and the bits between these crossover points are exchanged to create children. An example of two-point crossover is:

Parent A:	**0010101101** x **001000** x **000010101101**
Parent B:	1100010100 x 101101 x 001010001011
Child 1:	**0010101101** x 101101 x **000010101101**
Child 2:	1100010100 x **001000** x 001010001011

In single or two-point crossover, crossover points can be chosen randomly or be restricted to a subset of gene locations. The restriction of crossover points is used to keep closely associated genes from being split during the crossover process. Sets of integers located between allowed crossover points are referred to as *co-adapted alleles*.

The second commonly used genetic operator is *mutation*. After a child is created through crossover, a small chance is given (typically less than 1%) for any particular integer to change its value. Mutation serves as a constant supply of genetic diversity and prevents the permanent loss of any particular allele. High mutation values may be necessary for small populations without a large amount of genetic information, but this reduces the computational effectiveness associated with the schema theorem.[14] Essentially, high mutation rates result in undirected random searches that will hopefully stumble upon a good solution. If high mutation rates seem to increase solution quality, it is probable that the population is too small for genetic algorithms to be effective.

The third feature commonly used in genetic algorithms is an *elitist class*. When a new generation is being created, a certain percentage of the future population is reserved for the fittest members of the present population. Instead of crossover, these elite members are copied directly into the new population and are shielded from mutation. At least one elite member should always be used since this ensures that the best encountered solution will always be remembered. Larger elite populations will tend to reinforce good schema before they are corrupted from crossover, but will tend to reduce genetic diversity over time and may lead to premature algorithm convergence.

The last feature of a genetic algorithm is the *stopping criterion*, which terminates the algorithm after certain conditions are met. The simplest method is to stop after a fixed number of generations, but this may cause the algorithm to stop while solutions are continuing to improve or stop well after solutions have stopped improving. Other possible methods are to stop after a preset number of generations have elapsed without the best member improving, the average of the elite population improving, the average of the entire population improving, or the convergence of average solution quality to best solution quality.

It is difficult to know *a priori* how to best apply genetic algorithms to a specific problem—genetic operators must be selected, parameter values must be chosen, stopping criteria must be defined, and so on. All choices are ultimately interrelated and, therefore, finding the best combination can be an arduous and elusive task. Although optimal choices will vary based on the specific problem formulation, some general recommendations are as follows:

- Use a population size equal to 25% of the number of genes in the solution (e.g., a population size of 150 for a solution is represented by a chromosome with 600 genes).
- Use an elitist class equal to 2% of the total population.
- Use a two-member tournament to select parents with the more fit parent being chosen 70% of the time and the less fit parent being chosen 30% of the time.
- Use two-point crossover instead of single-point crossover.
- Avoid the use of co-adapted alleles.

- Do not perform mutation on elite members and use a mutation rate of 0.05% per gene per generation for all other members.
- Terminate the genetic algorithm when the best elitist member has not improved for 10 generations.
- Perform a hill climbing algorithm after the genetic algorithm is complete to ensure local optimality.

Genetic algorithms can come in many forms depending upon various choices that are made. The following algorithm is provided as an example for optimization problems related to distribution reliability:

<u>Genetic Algorithm with 2-Point Crossover</u>

1. Design solution chromosome with highly related genes being located in close proximity to each other.
2. Randomly generate an initial population of solutions referred to as generation zero.
3. Compute the fitness of each member in the population.
4. From Generation G, create Generation G + 1.
   - 4.1. Copy the elite members of G directly into Generation G + 1.
   - 4.2. Create the remaining members for Generation G + 1.
     - 4.2.1. Select 2 parents using a parental selection method.
     - 4.2.2. Randomly select 2 crossover points.
     - 4.2.3. Create 2 children by swapping the parental genes located between the crossover points.
     - 4.2.4. Perform mutation on these two children.
     - 4.2.5. Place the 2 children in Generation G + 1.
5. Compute the fitness of each member in Generation G + 1.
6. Has the stopping criterion been satisfied? If no, go to Step 4.
7. Perform hill climbing on all members to ensure local optimality.
8. Select the member with the best objective function.

For complicated distribution reliability optimization problems of moderate size and larger, genetic algorithms will almost always outperform local search and simulated annealing, with performance gains being reflected in both solution quality and in computation time. These gains, however, come at a price. Implementing genetic algorithms is substantially more difficult, and tuning the genetic algorithm parameters for optimal performance can be difficult and time consuming. Regardless, a good genetic algorithm remains one of the most effective methods of solving discrete optimization problems and the added performance will often be worth the added effort.

7.3 KNOWLEDGE-BASED SYSTEMS

The discrete optimization methods described in the previous section are powerful methods of exploring a complex solution space, but do not take advantage of human expertise that may be available from distribution planners and engineers. This section describes several techniques that are able to make reliability improvement decisions based on expert human knowledge represented as rules. The first method, referred to as an expert system, represents rules as a set of conditional statements in the form of if-then rules. The second method, referred to as a fuzzy expert system, extends this concept by accommodating vague human semantics such as small, medium and large.

7.3.1 Expert Systems

Expert systems are algorithms that perform tasks historically reserved for human experts. This includes the representation of expert domain knowledge, the ability to infer an answer based on this knowledge, and the ability to explain why a particular inference has been made.[15] Expert systems can be quite valuable since they allow expert-like decisions to be made by nonexperts. Further, those using the expert system are taught to think and reason like experts since all decisions made by the expert system are explained in detail.

Expert system domain knowledge is typically captured by a set of expert rules collectively referred to as a knowledge base. These rules are generally represented as if-then conditional statements with the "if clause" referred to as the antecedent and the "then clause" referred to as the consequence. The consequence can also contain a measure of confidence reflecting the percentage of time that the satisfaction of the antecedent ultimately corresponds to a recommendation of the consequence. An example of an expert rules is:

Rule	[R1]
If	[A1] the lateral tap is overhead, and
	[A2] the lateral tap is single phase, and
	[A3] the lateral tap is longer than 500 feet
Then	[C1] the lateral tap should be fused (0.95 confidence)

This rule (R1) consists of three antecedent clauses (A1-A3) and a single consequence (C1). If all of the antecedent clauses are true, then the expert that created the rule has a 95% confidence that the consequence should be recommended.

A typical expert system can consist of thousands of rules with many rules associated with the same consequences. This creates the possibility of conflicting rules. Consider the recommendation associated with rule [R1] in the above example. There may be many other rules that recommend that the lateral tap be fused, leading to a question of what the confidence of the recommendation should be. There may also be rules that recommend that the lateral tap not be fused, leading to a question of which recommendation is preferred.

To perform rule resolution, each expert system is equipped with an inference engine that identifies active rules (rules with true antecedents) and resolves rules with conflicting consequences. A simple inference approach is to group together all active rules that have the same or opposite consequence (e.g., "the lateral tap should be fused" and "the lateral tap should not be fused"). The net consequence is taken to be the sum of all active rules with the same consequence minus the sum of all active rules with the opposite consequence divided by the total number of active rules. This is essentially an average value, with a negative result indicating that the opposite consequence should be recommended. A sample rule inference is:

Rule	Consequence	Confidence
[R1]	the lateral tap should be fused	0.95
[R5]	the lateral tap should be fused	0.70
[R12]	the lateral tap should be fused	0.55
[R47]	the lateral tap should not be fused	-0.30
[R81]	the lateral tap should not be fused	-0.10
[Final]	the lateral tap should be fused	0.36

For the above example, the final recommendation is that the lateral tap should be fused. This recommendation has a 36% confidence and is based on five active rules—three in concurrence with the final recommendation and two in contradiction with the final recommendation.

Inferring final recommendations by taking the average of active rules assumes that all rules are of equal importance. This may not be the case. Consider the following two rules: (1) if a wasp is flying towards me from the right, then run to the left with 95% confidence, and (2) if a crocodile is running towards me from the left, then run to the right with 95% confidence. If both rules are active, a simple inference engine will recommend standing still—not good expert advice. This problem can be overcome by assigning a weight to each rule based in its relative importance. The inference engine can then base the final confidence of the consequence based on the weighted average of the active rule confidence values rather than the unweighted average.

Expert systems are well suited for analyzing very large systems[16-17] and producing high level analyses that can be used as a starting point for more in-depth analyses. To illustrate, consider the expert system analysis of Commonwealth Edison's 15-kV distribution system, which consists of 3,300 feeders. This assessment was a result of the major reliability problems experienced in the summer of 1999,[18] and had a goal of identifying the quickest and most cost-effective ways of improving customer reliability while avoiding similar problems in the future.[19]

The expert system rules for this project were based on the anticipated benefit-to-cost ratio of potential reliability improvement projects, with benefit being defined as reduction in customer kVA-minutes per year and cost being defined as annualized project cost. These rules were then applied to automatically generated reliability improvement projects, with an average of 1000 projects being considered per feeder (over 3 million projects in all). Done by hand, expert analysis of each potential project is obviously not feasible. Using an expert system, recommendations for each potential project (after the system was modeled and calibrated) was accomplished in less than three weeks.

The expert system makes extensive use of system topology for its recommendations. In addition, rules make use of reliability analysis results such as peak loading, reliability, root cause scores, and capacity constraints. Using this wide array of information allows complicated rules to be generated such as:

Rule	[R1]
If	[A1] the lateral tap is 3ϕ overhead, and
	[A2] the number of lateral tap customers is more than 500, and
	[A3] the lateral tap is longer than 100 feet, and
	[A4] the lateral tap SAIDI is greater than 120 min/yr, then
Then	[C1] a recloser should be placed on the lateral tap (0.9 confidence)

Rule	[R2]
If	[A1] the switch operates more than once per year, and
	[A2] the number of upstream customers is more than 1000, and
	[A3] there is a downstream tie switch, and
	[A4] the transfer path has no capacity constraints, then
Then	[C1] the switch should be automated (0.8 confidence)

Several different categories of reliability improvement projects were explored, corresponding to consequences of the expert system rules. This allowed different reliability improvement approaches to be compared on the same basis. Basic categories of reliability improvement options examined include:

- **Transfer Path Upgrades** – A transfer path is an alternate path to serve load after a fault occurs. If a transfer path is capacity constrained due to small conductor sizes, reconductoring may be a cost-effective way to improve reliability.

- **New Tie Points** – Adding new tie points increases the number of possible transfer paths and may be a cost-effective way to improve reliability on feeders with low transfer capability.

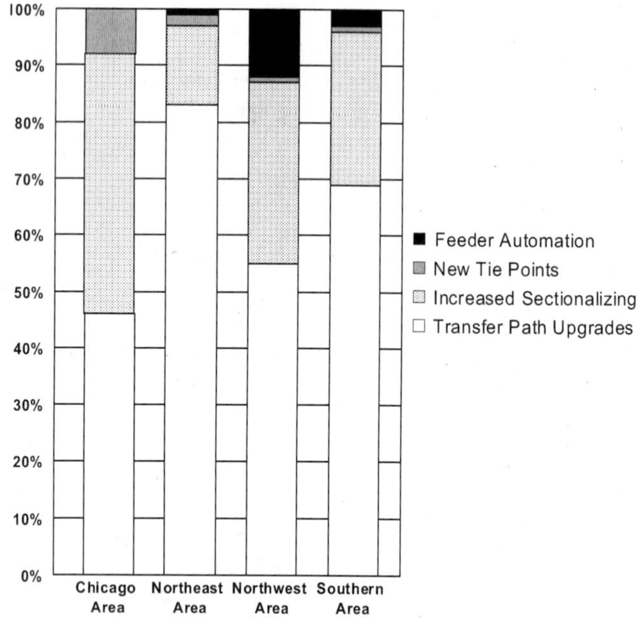

Figure 7.6. A breakdown of reliability improvement recommendations made by an expert system for the four Commonwealth Edison operating areas. This figure clearly shows that reliability improvement strategies will vary for different areas.

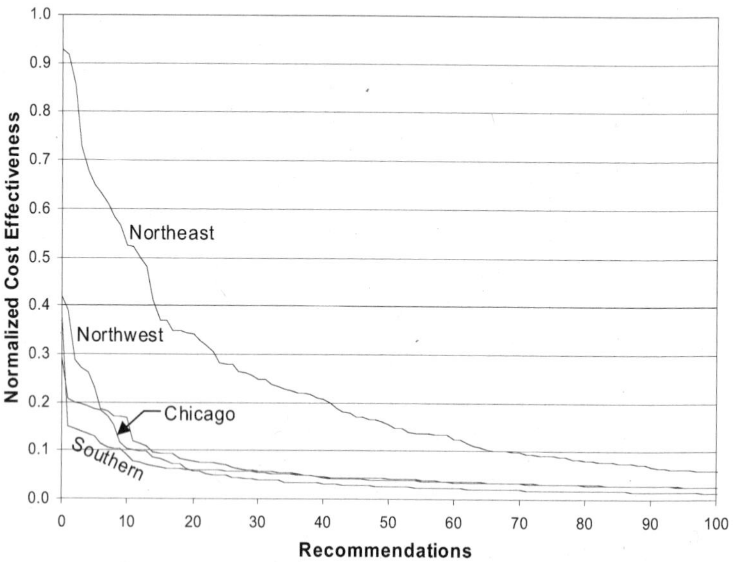

Figure 7.7. The range of project cost-effectiveness broken down by operating region. The expert system results indicate that the most cost-effective reliability gains can be made in the Northeast region and the least in the Southern region.

- **Increased Line Sectionalizing** – Increased line sectionalizing is accomplished by placing normally closed switching devices on a feeder. Adding fault interrupting devices (reclosers and fuses) improves reliability by reducing the number of customers interrupted by downstream faults. Adding switches without fault interrupting capability improves reliability by allowing more flexibility during post-fault system reconfiguration.

- **Feeder Automation** – In this study, feeder automation refers to SCADA-controlled switches on feeders. These automated switches allow post-fault system reconfiguration to occur much more quickly than with manual switches, allowing certain customers to experience a momentary interruption rather than a sustained interruption.

Recommendations were generated for each feeder and grouped into the four utility operating areas: Chicago, Northeast, Northwest, and Southern. A breakdown of the type of recommendations made for each area is shown in Figure 7.6, and the ranking of projects by cost effectiveness is shown in Figure 7.7. These results show that reliability improvement recommendations vary widely for each region, and that the expected cost-effectiveness of reliability improvement projects also varies widely from project to project and from region to region. For example, the highest ranked project for the Northeast region is more than three times as cost-effective as the highest ranked project for the Southern region. Within the Northeast region, the highest ranked project is more than 5 times as cost-effective as the fiftieth ranked project.

Expert systems are an excellent tool for capturing expert knowledge and using this knowledge to obtain tentative recommendations for large systems. In a sense, they can fulfill a triage function to quickly identify problem areas and quickly identify potential cost-effective solutions. Final recommendations, however, should still be based on rigorous cost and reliability model analysis so that details not considered in the expert rules, such as project interaction, can be properly considered.[20]

7.3.2 Fuzzy Expert Systems

Traditional expert systems can sometimes have difficulty capturing expert knowledge since expert knowledge is often vague whereas expert rules must be precise. Consider an expert who states that, "long feeders in areas of high tree density should have frequent tree trimming." In this statement, the imprecise modifiers "long," "high" and "frequent" are difficult to translate into an expert rule. Many expert systems handle this difficulty by assigning numerical values that approximate the intent of the rule such as, "feeders longer than 10 miles in areas of more than 100 trees per mile should perform tree trimming at least twice per year." Although this statement may accurately reflect expert opinion for a specific case, it may not reflect expert opinion for other cases such as feeders slightly less than 10 miles long or feeders in areas of slightly less than 100 trees per mile.

Fuzzy set theory is a framework capable of representing vague human semantics and incorporating them into expert rules.[21-22] In classical set theory, membership is crisp; an item is either a full member or it is a nonmember. This is practical in many scenarios. For example, the set of "protection devices" is crisp—any fuse is a full member while any capacitor is a nonmember. In fuzzy sets, each item in the domain has a grade of membership for each set—a number between zero and one that represents how much the item belongs to a set (one indicates absolute membership and zero indicates nonmembership). For example, a feeder 8 miles in length might have a membership of 0.5 in the set of "long feeders." Possible fuzzy sets for short, medium, long and very long are shown in Figure 7.8.

The four sets of Figure 7.8 have a domain (every possible member) of all feeders, and each feeder belongs to each fuzzy set to a certain degree. For example, the statement "a feeder 4 miles long is short" has a truth of 0.5, and the statement "a feeder 20 miles long is very long" has a truth of 1.0.

Fuzzy set theory, like crisp set theory, becomes more useful if the union, intersection, and implication of sets are defined. These operators allow the truth of more complex statements using "and," "or" and "if" to be handled. Typical definitions for these three operators are (A∧B refers to the minimum value of A and B and A∨B refers to the maximum value of A and B):

Intersection (A and B) $\quad A \cap B = A \wedge B \quad$ (7.6)

Union (A or B) $\quad A \cup B = A \vee B \quad$ (7.7)

Implication (B if A) $\quad A \rightarrow B = 1 \wedge (1 - A + B) \quad$ (7.8)

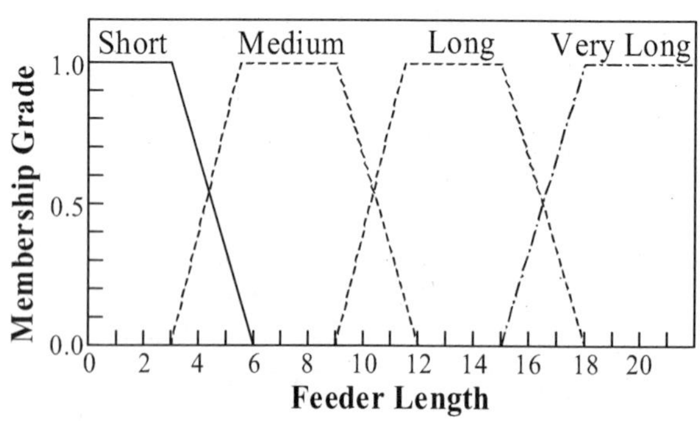

Figure 7.8. Examples of fuzzy sets related to feeder length. Each feeder, depending upon its length, belongs to each fuzzy set to a varying degree, called its grade of membership. The above example shows fuzzy sets as trapezoids, but other shapes, such as triangular and Gaussian, are also common.

Fuzzy sets can be used to create expert rules similar to those used in basic expert systems. The difference is that rules can contain modifiers corresponding to predefined fuzzy sets. An example fuzzy rule with three inputs and one output is:

Rule	[R1]
If	[A1] the lateral tap is long, and
	[A2] the tree density is high, and
	[A3] the lateral tap has many customers
Then	[C1] the lateral tap should be fused (high confidence)

In addition to fuzzy sets in the antecedent clauses (i.e., long, high, and many), a fuzzy set is used for the consequence. This consequence, unlike those in traditional expert system rules, is not restricted to being either active or inactive. Since the antecedent clause of a fuzzy rule can be partially true, the consequences of fuzzy rules are allowed to be partially active. The degree of activeness of a fuzzy rule is determined by fuzzy inference, which returns the mean value of all consequences weighted by their respective antecedent truths. To illustrate, consider the following two fuzzy rules, each with two fuzzy inputs and one fuzzy output:

$$\text{[R1]:} \quad \text{IF } x_1 \text{ is } A_{11} \text{ and } x_2 \text{ is } A_{12} \text{ THEN } y_1 \text{ is } B_1 \tag{7.9}$$

$$\text{[R2]:} \quad \text{IF } x_1 \text{ is } A_{21} \text{ and } x_2 \text{ is } A_{22} \text{ THEN } y_1 \text{ is } B_2 \tag{7.10}$$

The first step of fuzzy inference is to compute the compatibility, C, of each antecedent, which is equivalent to the truth of the antecedent:

$$C_i = A_{i1}(x_1) \wedge A_{i2}(x_2) \tag{7.11}$$

In this equation, $A_{ij}(x_k)$ refers to the truth of "x_k is A_{ij}". The consequence of rule i, referred to as Y_i, is now computed by multiplying the set B_i by the truth of its respective antecedent. This is written as follows:

$$Y_i = C_i \cdot B_i \tag{7.12}$$

The final solution is found by combining the results of all fuzzy rules. This is done by dividing the sum of all Y_i sets by the sum of all antecedent truths. This resulting fuzzy output set, Y_{out}, is mathematically expressed as:

$$Y_{out} = \sum_{i=1}^{N} Y_i \Big/ \sum_{i=1}^{N} C_i \tag{7.13}$$

The result of a set of fuzzy rules is a fuzzy output set, Y_{out}. This fuzzy set result can be translated into a single numerical value by computing its mean value, which is equal to the integral of the moments of the function divided by the integral of the function. This process, sometimes referred to as *defuzzification*, is mathematically expressed as follows:

$$\bar{y} = \frac{\int_{-\infty}^{\infty} x \cdot Y_{out}(x)dx}{\int_{-\infty}^{\infty} Y_{out}(x)dx} \tag{7.14}$$

Fuzzy expert systems that utilize defuzzified outputs can be applied in exactly the same manner as traditional expert systems. This includes adding weights to rules to reflect their relative importance. The drawback to fuzzy expert systems is that rules can sound semantically correct, but will not actually represent expert knowledge unless the fuzzy sets are chosen properly. Regardless, fuzzy set theory is a powerful concept that, when applied properly, can aid in capturing expert knowledge, help to identify the reliability characteristics of large systems, and generate reliability improvement options that have a high likelihood of cost-effectively improving distribution reliability.

7.4 OPTIMIZATION APPLICATIONS

The optimization techniques described previously in this chapter are not just abstract academic concepts. They are powerful computational frameworks that can be applied to practical distribution system reliability optimization problems with computing power available on a typical personal computer.[23-29] To reinforce this point and to give the reader a better feel for practical reliability optimization, this section presents several actual optimization applications and corresponding results. It concludes with a comparison of optimization methods so that their relative effectiveness can be examined.

7.4.1 Feeder Automation

A simple application of reliability optimization is the selection of existing manual feeder switches to be automated. The objective function and constraints are easily defined in terms of cost and reliability, potential solutions are easily represented as an integer map, the reliability improvements associated with potential solutions are easily modeled, and locally optimal solutions are easily generated by using hill climbing techniques. In addition, this is a practical problem of interest to many utilities considering the use of automated feeder switches to improve the reliability of their distribution system.

In this application, the goal of feeder automation is to achieve a SAIDI target using the fewest possible number of automated switches. As previously mentioned, this will be done by adding automation equipment to existing manual switches (both normally closed and normally open). The area under consideration is the service territory of an eight feeder substation in the midwestern US. This system serves approximately 12,000 customers with a peak demand of 50 MVA. There are a total of 40 normally closed switches and 36 normally open switches (all of the switches are manual), and the expected SAIDI value is 120 minutes per year. The geographic layout of this system (including the location of switches) is shown in Figure 7.9.

Each feasible solution for this optimization project is represented by a bitmap, with each bit corresponding to an existing switch on the system. A bit value of zero indicates that the corresponding switch is not automated, and a bit value of one indicates that the corresponding switch is automated.

The objective function is equal to the arithmetic sum of all bits in the bitmap, which is equal to the number of automated switches in the solution. The only constraint considered is a SAIDI target, which is represented as a penalty factor that linearly increases as the degree of the constraint violation increases. For a solution with N_a automated switches, a reliability target $SAIDI_T$ and a reliability value of SAIDI, the objective is to minimize the following:

$$\text{Minimize:}\ N_a + \begin{cases} 0 & ;\text{SAIDI} <= \text{SAIDI}_T \\ \dfrac{10^9\left(\text{SAIDI} - \text{SAIDI}_T\right)}{\text{SAIDI}_T} & ;\text{SAIDI} > \text{SAIDI}_T \end{cases} \tag{7.15}$$

The solution methodology is a simple hill climbing algorithm starting with a random bitmap and using analytical simulation to compute system reliability. This problem was solved for a wide range of SAIDI targets to determine what happens to automation requirements as reliability targets become more aggressive. The required number of automated switches to achieve SAIDI targets from 60 minutes to 120 minutes is shown in Figure 7.10. Solutions for each SAIDI target were computed multiple times to improve the probability of finding high quality solutions.

Multiple trials for the same target SAIDI sometimes result in substantial solution variation. To demonstrate, 20 solutions for a SAIDI target of 90 min/yr were generated using random initial solutions. A histogram of outcomes is shown in Figure 7.10, with the worst solution requiring 23 switches and the best solution requiring 13 switches (the switches corresponding to the best solution are shown Figure 7.9). Since each of these solutions is quite different in terms switch number and location, multiple trials are needed to help ensure that a near globally optimal solution is ultimately selected for implementation.

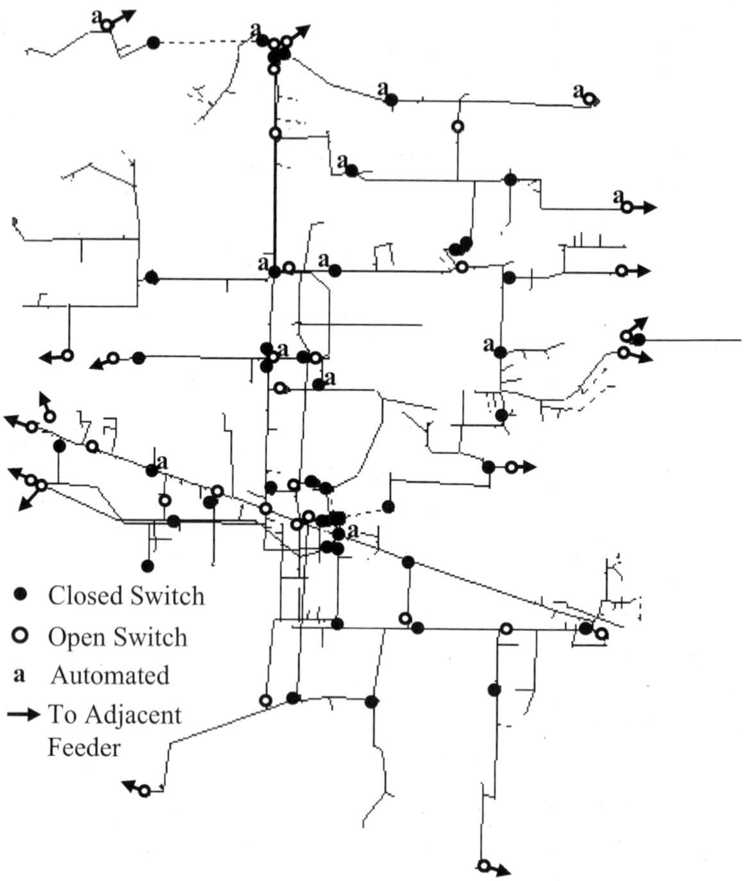

Figure 7.9 Eight-feeder system to demonstrate feeder automation optimization. Using a hill climbing algorithm, it is shown that the original SAIDI of 120 min/yr can be reduced to 90 min/yr by automating 3 of the 36 normally open switches and 10 of the 40 normally closed switches.

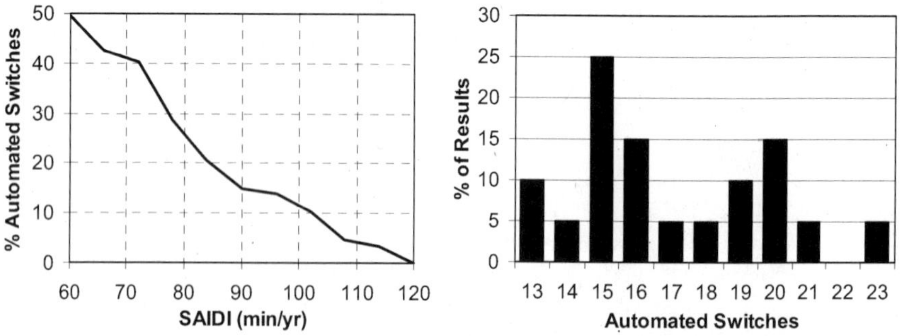

Figure 7.10. The left graph shows the number of automated switches required to improve SAIDI to various levels for the system shown in Figure 7.9. The right graph demonstrates the natural variation of multiple trials by showing a histogram of 20 solutions found for a target SAIDI of 90 min/yr.

This application of optimization techniques to feeder automation demonstrates several points. First, it is not necessarily difficult to formulate an optimization problem. In this case, the objective was simply to meet a SAIDI target using the least possible number of automated switches. Second, it is not necessarily difficult to represent potential solutions as an integer map. In this case, a simple string of bits corresponds to switches being manual or automated. Third, it is not necessarily difficult to implement an effective optimization algorithm. In this case, a simple hill climbing approach achieves good results, especially if multiple trials are performed with different random starting points.

Optimization can be an effective way to approach practical distribution reliability problems. This section has applied optimization to the problem of feeder automation, but many other problems are just as appropriate for optimization and are limited only by the creativity of the distribution engineer. With practice, more sophisticated algorithms can be implemented and more complicated problems (like switching and protection optimization, discussed in the next section) can be addressed.

7.4.2 Reclosing and Switching

Utilities usually have reliability targets for several reliability indices such as $MAIFI_E$, SAIFI, and SAIDI. One way to achieve these targets is to place reclosers and switches on the primary distribution system. Reliability can be further improved by automating existing devices and automating new devices. Identifying the optimal combinations and locations of switches, reclosers, and automation to achieve multiple reliability targets is a complicated optimization problem that is difficult to solve using simple optimization techniques.

This section applies a genetic algorithm to improve the reliability of an actual utility feeder located in the southern US. Reliability improvement projects can include the installation of new reclosers and the installation of new sectionalizing switches. For an additional cost, these devices can be fitted with automation equipment so that they can be remotely switched by dispatchers. Last, reliability can be improved by retrofitting existing reclosers and sectionalizing switches with automation so that they can be remotely switched by dispatchers. The existing reliability indices of this feeder are: $MAIFI_E$ = 8.2 /yr, SAIFI = 1.2 /yr, and SAIDI = 133 min/yr. The objective of the optimization algorithm is to achieve the following reliability index targets: $MAIFI_E$ = 4 /yr, SAIFI = 0.5 /yr, and SAIDI = 45 min/yr. The algorithm attempts to do this for the least possible cost with switches costing \$1000, reclosers costing \$15,000, and installing automation on any new or existing switch or recloser costing \$5000. The formal problem formulation can be written as follows:

Reclosing and Switching Problem Formulation

$$
\begin{aligned}
&\text{Minimize:} \\
&\quad 1000 \cdot N_s + 15000 \cdot N_r + 5000 \cdot N_a \\
&\text{Subject to:} \\
&\quad \text{MAIFI}_E < 4.0 \text{ /yr} \\
&\quad \text{SAIFI} \quad < 0.5 \text{ /yr} \\
&\quad \text{SAIDI} \quad < 45 \text{ min/yr}
\end{aligned}
\tag{7.16}
$$

N_s = Number of new switches
N_r = Number of new reclosers
N_a = Number of devices equipped with new automation equipment

The feeder used in this example, shown in Figure 7.11, serves 1900 customers and has a peak load of nearly 10 MW. It has no line reclosers, no automation equipment, one normally closed switch, and six ties to other feeders. It is divided into 300 line segments and 15 cable segments, with each segment being a candidate for a new device. Solutions are represented by using three bits for each existing switch and three bits for each line or cable section as follows:

Bit 1: Is a switch or protection device present? A 0 corresponds to "yes" and a 1 corresponds to "no."
Bit 2: If a device is present, a 0 corresponds to a switch and a 1 corresponds to a recloser.
Bit 3: If a device is present, a 0 corresponds to no automation and a 1 corresponds to automation.

For existing switches, Bit 1 is always equal to 1, a Bit 2 value of 1 represents an upgrade to a recloser, and a Bit 3 value of 1 represents the installation of automation equipment. For existing reclosers, Bit 1 and Bit 2 are always equal to 1, and a Bit 3 value of 1 represents the installation of automation equipment. For line and cable sections, Bit 2 and Bit 3 are only relevant if Bit 1 has a value of 1, which indicates that a new device has been installed. Since there are a total of 322 possible device locations (7 existing switches + 300 line segments + 15 cable segments), the total solution representation requires 322 x 3 = 966 bits.

This optimization problem has been solved by using analytical simulation (as described in Chapter 5) to compute system reliability and genetic algorithms (as described in Section 7.2.5) to identify the optimal solution. This algorithm uses random initial population generation, a total population of 500, an elite population of 10, a mutation rate of 0.5% per bit per generation, and a stopping criterion of 10 generations without improvement of the best solution. The best solution has a cost of $66,000 and results in the indices shown in Table 7.3.

Table 7.3. Reliability indices associated with the reclosing and switching optimization problem.

Index	Present	Target	Optimized
$MAIFI_E$	8.2	4.0	3.8
SAIFI	1.2	0.5	0.4
SAIDI	133	45	45

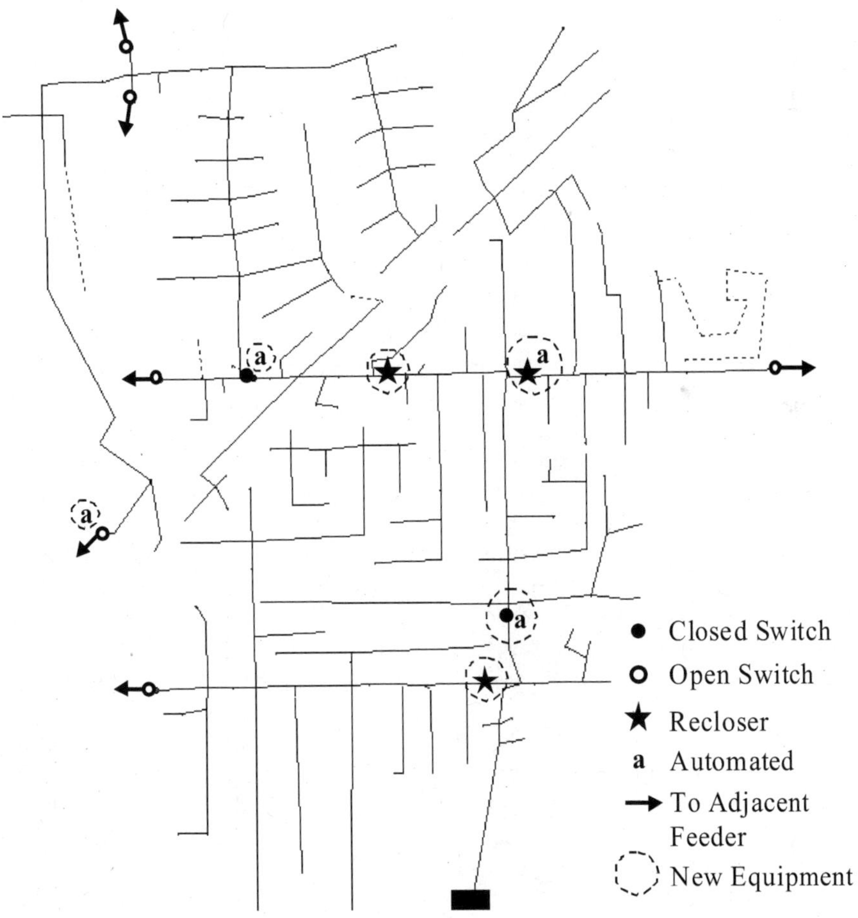

Figure 7.11. Feeder used for the reclosing and switching optimization problem. The present system has no line reclosers, one normally closed switch, and six ties to adjacent feeders. The least cost solution that meets all reliability index targets, identified by a genetic algorithm, automates the existing normally closed switch, automates an existing tie switch, adds two new reclosers, adds one new automated recloser, and adds one new automated switch.

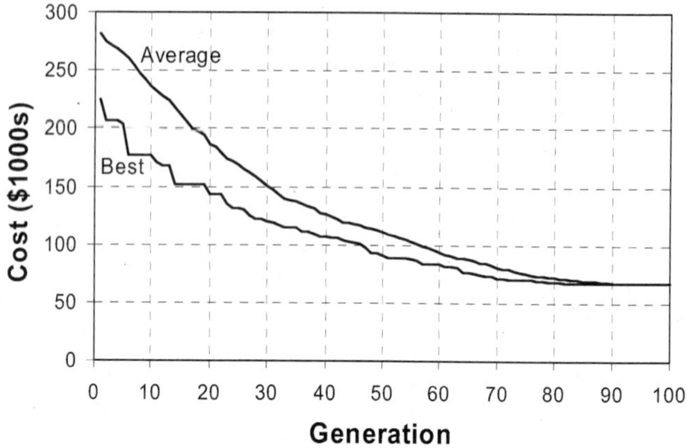

Figure 7.12. Convergence of the genetic algorithm for reclosing and switching optimization. As the average cost of all solutions approaches the cost of the best solution, genetic diversity is lost and further improvements can only be achieved through random mutations.

Convergence of a genetic algorithm can be examined by tracking the objective function score of both the best solution and the average of all solutions for each generation. The genetic algorithm is considered to have converged when average solution quality approaches the quality of the best solution. At this point, it is likely that most genetic diversity has been lost and additional improvements will only occur due to random mutations. The curves for this example, shown in Figure 7.12, indicate that convergence occurs after approximately 90 generations.

The least cost solution identified by the genetic algorithm is shown in Figure 7.11. It consists of automating the existing normally closed switch, automating the tie switch furthest away from the substation, adding two new reclosers, adding one new automated recloser, and adding one new automated switch. These actions just meet the SAIDI target and slightly exceed the SAIFI and $MAIFI_E$ targets.

The solution identified by the genetic algorithm is nonobvious and would be nearly impossible to identify through manual methods. In fact, the solution identified by the genetic algorithm significantly outperformed simple hill climbing. While the genetic algorithm identified a solution with a cost of $66K, ten hill climbing runs with random initial bitmaps resulted in costs of $117K, $123K, $139K, $167K, $182K, $192K, $267K, $270K, $272K, and $360K. Each hill climbing run resulted in a different locally optimal solution, with the best solution nearly twice as expensive as the genetic algorithm solution. For difficult optimization problems with complex solution spaces, sophisticated techniques such as genetic algorithms clearly outperform more simple methods such as hill climbing.

7.4.3 System Reconfiguration

A distribution system can be reconfigured by changing the location of normally open switches. This changes the allocation of customers and the flow of power for the affected feeders. Feeder reconfiguration methods have been developed to minimize losses[30-31] and operational costs.[32] This section demonstrates how to use the feeder reconfiguration concept to improve system reliability by using an annealed local search.

The algorithm described in this section is designed to optimize radially operated distribution systems. A radially operated system may be highly interconnected, but at any particular time, a component will have a unique electrical path back to a distribution substation. In a radial system, the direction of power flow is well defined. Figure 7.13 shows an example of a radial tree structure consisting of a component (A) with a parent (B) and two children (C, D). The parent/child structure is important since its modification is the goal of the feeder reconfiguration process.

A radial system is modeled by representing components as objects and parent/child relationships as pointers between objects. Each component has a pointer to its parent, and a list of pointers to its children. This data structure allows for the efficient implementation of power flow and reliability algorithms using tree navigation techniques. Upstream components such as protection devices and sources can be found by following parent pointers. Downstream components such as interrupted customers can be found using depth-first and breadth-first search techniques.

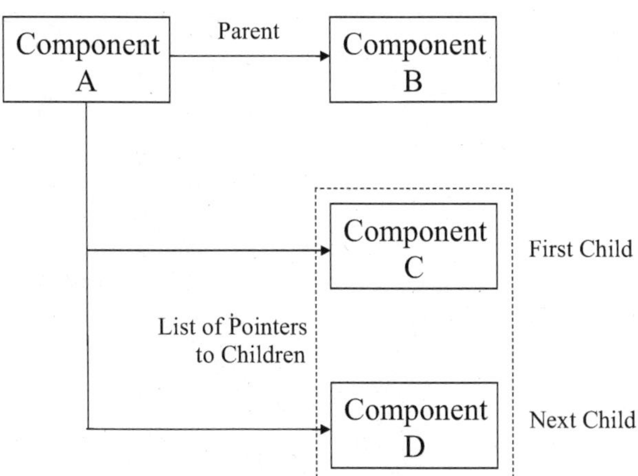

Figure 7.13. In a radial tree structure, each component is described by a parent and a set of children. The parent is the component next closest to the source of power and the children are the next furthest components from the source of power. System reconfiguration is accomplished by modifying parent/child relationships of components.

The optimization function for feeder reconfiguration is defined as a weighted sum of reliability indices. The goal of the feeder reconfiguration is to minimize this function subject to equipment overloads and voltage violations:

System Reconfiguration Problem Formulation

Minimize:

$$w_1\text{SAIDI} + w_2\text{SAIFI} + w_3\text{MAIFI} \tag{7.17}$$

w_x: Reliability Index Weight

Subject to:

No equipment overloads
No voltage drop violations

The reliability index weights are chosen to reflect utility targets, performance based rate structures or relative cost to customers. Constraints are implemented as penalty factors added to the optimization function. The penalty factor formulation for loading violations and voltage violations is:

$$pf_{lv} = \sum_i \max\left[\frac{kVA_i - Rated\_kVA_i}{Rated\_kVA_i}, 0\right] \tag{7.18}$$

$$pf_{vv} = \sum_i \max\left[\frac{Min\_Voltage_i - Voltage_i}{Min\_Voltage_i}, 0\right] \tag{7.19}$$

pf_{lv} = Penalty factor due to loading violations
pf_{vv} = Penalty factor due to voltage violations

Feeder reconfiguration using existing switch locations is a discrete optimization problem. Simulated annealing and genetic algorithms, however, are problematic to apply to radial structure reconfiguration problems for two reasons. First, most of the generated switch position combinations will not represent feasible solutions. Second, generating a new radial tree structure for each combination of switch positions is computationally intensive. A local search avoids the problems of simulated annealing and genetic algorithms by making small changes in the radial structure by closing a normally open switch and opening a nearby upstream switch. This process, referred to hereafter as a tie switch shift, allows incremental changes to the tree structure as opposed to complete reformulation of the structure for each switch position change.

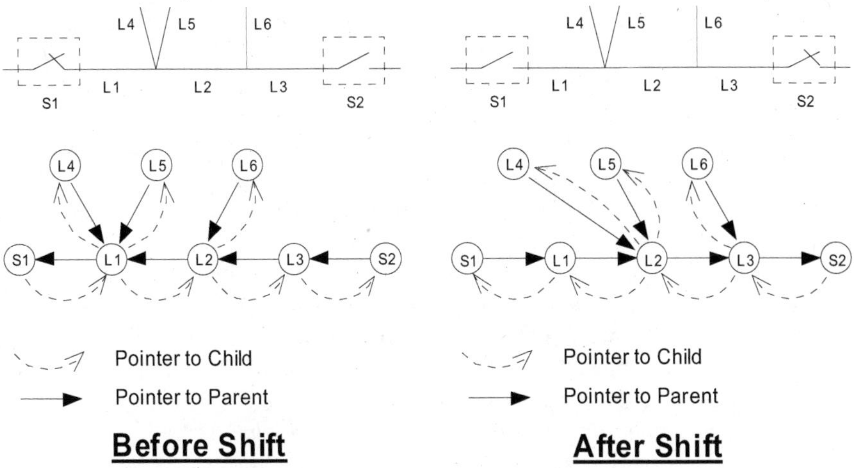

Figure 7.14. Radial tree structure before and after a tie switch shift. Before the shift, power flows from switch S1 to switch S2. After the open point is shifted from switch S2 to switch S1, the parent and child relationships associated with lines L1-L3 are reversed to reflect the shift in power flow. Similarly, the locations at which lateral branches L4-L6 are connected to the main trunk are shifted in what was a downstream direction before the shift.

To illustrate a tie switch shift, consider the simple system in Figure 7.14. It consists of a normally closed switch (S1), a normally open switch (S2), and six line segments (L1-L6). The radial tree structure is shown by solid arrows (pointers to parents) and dotted arrows (pointers to children). **P** is defined as the set of components on the path from S1 to S2 and **D** is defined as the set of components directly connected to **P**. In Figure 2, **P** $\in$ (L1,L2,L3) and **D** $\in$ (L4,L5,L6).

Consider a tie switch shift from S2 to S1. Prior to the shift, power flows from S1 to S2. After the shift, a reversal occurs and power flows from S2 to S1. This requires the following changes to the radial tree structure:

Radial Tree Structure Changes After a Tie Switch Shift

1. For P_1 and P_2 in **P** and a pointer $P_1 \rightarrow P_2$, reverse the connection to become $P_2 \rightarrow P_1$.
2. For P_1 in **P**, D_1 in **D** and a child pointer $P_1 \rightarrow D_1$, alter the pointer to become $P_2 \rightarrow D_1$ where P_2 is the new parent of P_1.
3. For P_1 in **P**, D_1 in **D** and a parent pointer connecting $D_1 \rightarrow P_1$, change the pointer to become $D_1 \rightarrow P_2$ where P_2 is the new parent of P_1.

After performing a tie switch shift, the system shown on the left side of Figure 7.14 is reconfigured to become the system shown on the right side of Figure 7.14. Power now flows in the reverse direction and terminates at the new normally open point.

The tie switch shift is well suited for hill climbing, which could simply test tie switch shifts and keep changes if the solution improves; continuing until a local optimum is reached. Unfortunately, hill climbing has difficulty identifying solutions that differ substantially from the initial condition since there is a high probability of becoming trapped in a local optimum. The concept of a local annealed search overcomes the limitations of hill climbing, allowing a greater search area to be explored by probabilistically accepting certain solutions with worse objective functions.

The primary variable that drives a local annealed search is temperature, T: $T \in [0,1]$. The temperature of a system represents the probability that a worse solution will be accepted. A temperature of 1 generalizes to an undirected random search and a temperature of zero reduces to simple hill climbing.

A local annealed search is characterized by two parameters: initial temperature T_0 and annealing rate R. The initial temperature determines how freely the algorithm will initially traverse the solution space, and the annealing rate determines the rate at which the temperature is reduced. When the temperature keeps dropping without the solution changing, the process is frozen and the algorithm terminates. The annealed local search algorithm is outlined below.

Annealed Local Search

1. Set $T = T_0$: $T_0 \in [0,1]$.
2. Compute the objective function Ω_{best}.
3. Select a tie switch, shift the tie switch in Direction 1, and then compute the objective function Ω_{test}.
4. Let R be a uniform random number: $R \in [0,1]$.
5. If $\Omega_{test} < \Omega_{best}$ or $T > r$, set $\Omega_{best} = \Omega_{test}$ and go to step 8.
6. Shift the tie switch back to its previous position.
7. Repeat steps 3-7 for Direction 2.
8. Repeat steps 3-8 for all untested tie switches.
9. Set T = R x T.
10. If Ω_{best} has changed since the last change in T, go to step 3.
11. End.

The annealed local search probabilistically allows system reconfigurations that result in worse reliability and/or violate constraints. Occasionally accepting worse solutions allows a greater area of the search space to be explored and increases the likelihood of a near-optimal solution being discovered. Local optimality is assured since the annealed local search reduces to hill climbing at a zero temperature.

The characteristics of an annealed local search have been examined on an eight-feeder test system. Each feeder is a multibranch layout[33] and the total system consists of 576 load points, 1865 components, 208 fused laterals, and 200 switches. The test system is shown in Figure 7.15.

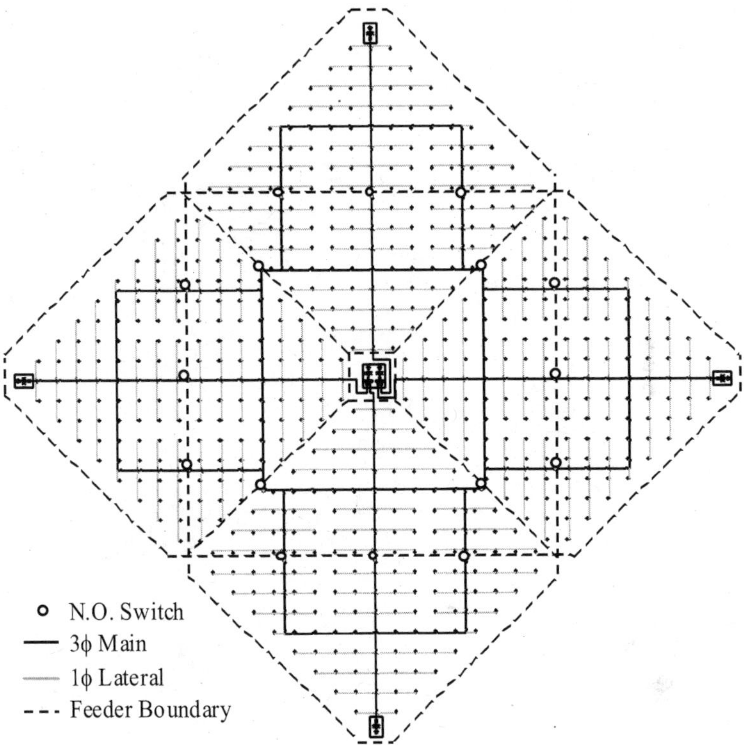

Figure 7.15. Eight-feeder test system used to demonstrate an annealed local search to optimize the location of normally open switches. This system consists of 16 open points and more than 200 switch locations, resulting in more than 10^{36} possible system configurations.

The feeder boundaries shown in Figure 7.15 represents the most reliable configuration for this system if all feeders are topologically identical and all loads are equal. The optimal system configuration changes if the loading patterns are altered sufficiently. To create a system for optimization, each feeder was assigned a random average loading level and each load point assigned a random demand ranging from zero to twice the average demand. The randomization resulted in two of the feeders being overloaded and provides an opportunity for optimization.

The impact of annealing rate on solution quality was assessed by performing ten trials for annealing rates between 0 and 0.99. The initial temperature was set equal to the annealing rate so that fast annealing rates tended to explore solutions similar to the initial conditions, and slow annealing rates tended to explore a wide range of solutions. Results for optimizing SAIDI (w_1 = 1, w_2 = 0, and w_3 = 0 in Eq. 7.17) are shown in Figure 7.16.

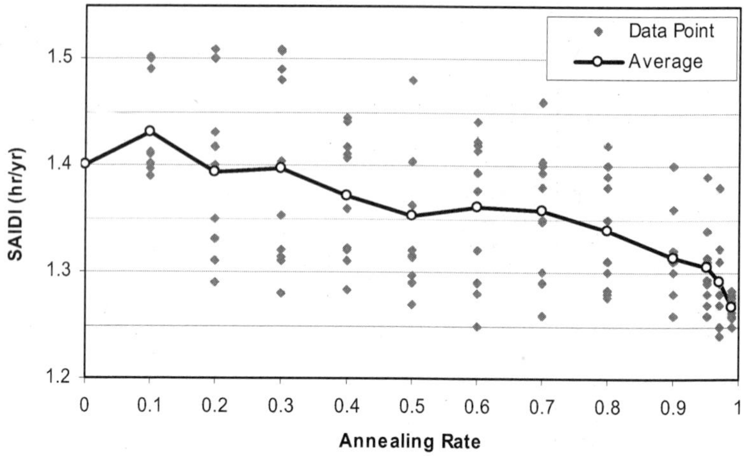

Figure 7.16. Solution quality versus annealing rate for the 8-feeder test system. Both solution quality and solution variation become better as the annealing rate becomes slower. Based on this graph, an annealing rate of 0.99 can be used to ensure a high quality solution with a single trial.

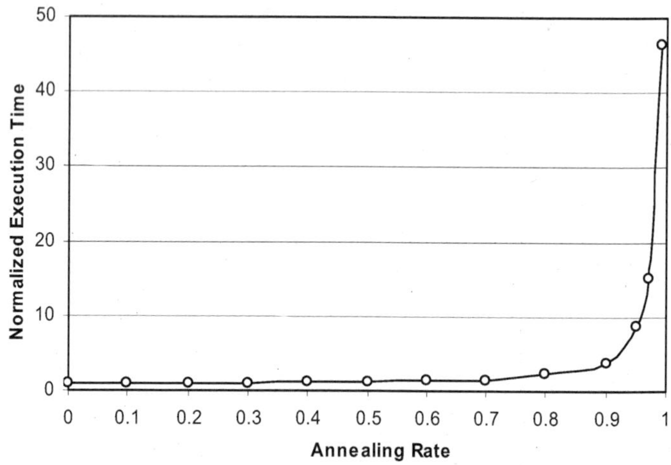

Figure 7.17. A graph of execution time versus annealing rate for the 8-feeder test system shows that the time required to produce a solution rises hyperbolically as the annealing rate approaches unity.

Figure 7.16 shows that average solution quality improves as the annealing rate increases and shows that the variance of solution quality remains high until very slow annealing rates are used. The disadvantage of high annealing rates is increasing execution time. A normalized plot of the average execution time versus annealing rate is shown in Figure 7.17. The plot is normalized such that the execution time for an annealing rate of zero is equal to unity. The plot shows a hyperbolic relationship as the annealing rate approaches unity.

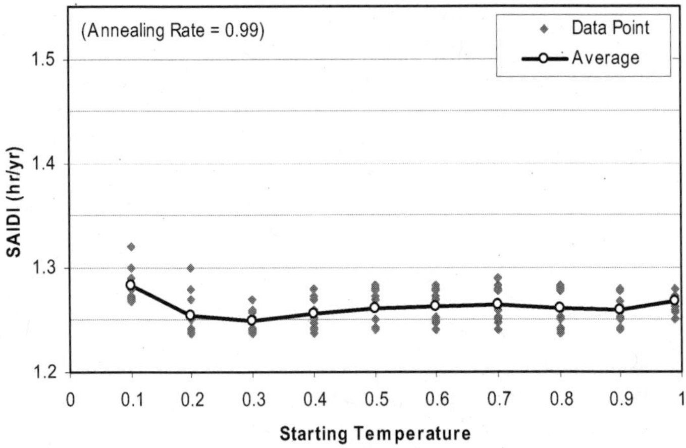

Figure 7.18. A graph of solution quality versus starting temperature for the 8-feeder test system shows that a starting temperature of 0.3 is sufficient to produce high quality results when an annealing rate of 0.99 is used.

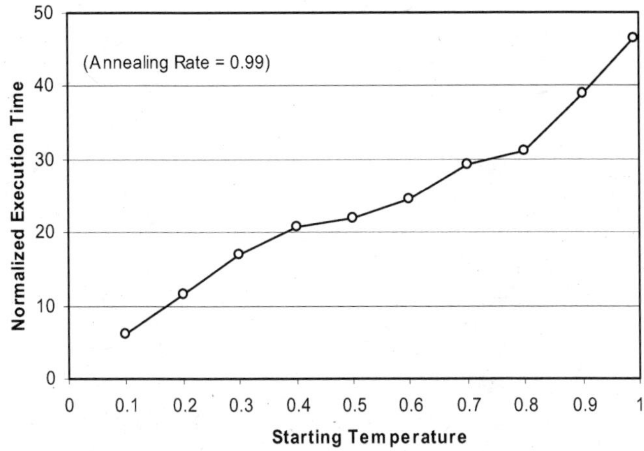

Figure 7.19. A graph of execution time versus starting temperature shows that the time required to produce a solution rises linearly with starting temperature. By choosing a starting temperature of 0.3 rather than 0.9, execution time can be reduced by approximately 67%.

Based on Figure 7.16, an annealing rate of 0.99 was selected to provide reasonable assurance of a high quality solution with a single trial. Similar trials were performed to determine the minimum required initial temperature, T_0, which consistently produces a quality solution (a low initial temperature being desirable to reduce execution time). Ten trials were performed for initial temperatures between 0.1 and 0.99. Solution quality versus T_0 is shown in Figure 7.18 and execution time versus T_0 is shown in Figure 7.19. Solution quality and solution variance do not significantly improve for initial temperatures greater than 0.3. Execution time increases linearly for increasing starting temperatures.

The annealed local search methodology has been applied to seven actual distribution systems in order to test the ability of the algorithm to optimize reliability through feeder reconfiguration. When applying the annealed local search to actual systems, optimization parameters must be selected *a priori*. Based on the test system results, an annealing rate of 0.99 and a starting temperature of 0.3 were selected for the analyses on actual systems. These values are shown to result in high quality solutions for a single trial while avoiding unnecessary computation time. Brief descriptions of these topologically diverse systems are now provided.

Urban 1 – This system serves a commercial and hotel district in a large southeastern coastal city. It is served by an aging underground system and is moderately loaded.

Urban 2 – Serves an industrial and commercial district in a large midwestern city. It is served by an aging underground system and is heavily loaded.

Suburban 1 – This system serves a newly developing suburban area in the Southeast. There are several moderately sized industrial customers served by this system.

Suburban 2 – This system serves a quickly growing suburban area in the Southeast that is primarily residential.

Suburban 3 – This system serves a suburban area in the Southeast. The infrastructure is lightly loaded in most areas, but is heavily loaded in an area that serves several industrial customers.

Rural – This system serves a rural area in the Southeast. Load density is low and most of the circuit exposure is single phase overhead construction.

Institutional – This system serves a large institutional campus in a coastal area of the Southeast. All buildings are served from a single substation and the site is served through underground feeders and pad-mounted equipment.

These seven test systems are diverse in terms of topology, load density, construction type, and customer type. The number of normally open and normally closed switches also varies considerably, resulting in varying solution space sizes (solution space grows as the number of switches grows). Larger solution spaces require more time to explore, but increase the likelihood of finding higher quality solutions. Key characteristics of the test systems are summarized in Table 7.4.

Table 7.4. Test system characteristics for optimizing system configuration.

System	Circuits	Load (MVA)	Circuit Miles	Switches
Urban 1	29	223	20.4	295
Urban 2	31	193	86.3	242
Suburban 1	15	212	545.7	102
Suburban 2	21	267	545.6	151
Suburban 3	13	85	204.9	61
Rural	8	144	496.0	46
Institutional	7	108	52.1	345

Table 7.5. Test system results for optimizing system configuration.

System	Original SAIDI (hr/yr)	Optimized SAIDI (hr/yr)	Reduction
Urban 1	1.81	1.55	14.4%
Urban 2	0.86	0.74	13.8%
Suburban 1	4.31	4.08	5.3%
Suburban 2	4.24	4.05	4.5%
Suburban 2	10.84	10.31	4.9%
Rural	5.72	5.68	0.7%
Institutional	1.23	1.20	2.4%

For each test system, the algorithm is instructed to optimize SAIDI (corresponding to weights of $w_1 = 1$, $w_2 = 0$, and $w_{3=} = 0$). The results of the reliability optimization for the seven test systems are summarized in Table 7.5.

The annealed local search is able to find configurations with higher reliability for each of the seven test systems without violating voltage and loading constraints. This improvement ranges from less than 1% to nearly 15%. In general, higher density systems result in greater reliability improvement. This is true for load density (MVA per circuit mile), switch density (switches per circuit mile), and switches per MVA of load. With the exception of the institutional system, the relationship appears to be nonlinear for the first two measures and linear with respect to switches per MVA of load. The sample set is small and further investigation is required to determine whether the relationship is evidenced in a broader array of systems.

Feeder reconfiguration can improve reliability for little or no capital expenditure. As such, feeder reconfiguration for reliability optimization presents electric utilities with an opportunity to become more competitive by providing higher levels of reliability at a lower cost. The results seen on these seven test systems are encouraging. In an era of deregulation and increasing customer reliability needs, utilities are under pressure to improve reliability without increasing cost. For certain types of systems, the annealed local search is able to improve reliability by nearly 15% without any capital spending. Assuming a conservative capital reliability improvement cost of $10/kVA-hr, reconfiguration optimization results in a reliability improvement that could cost nearly $580,000 for the “Urban 1” system.

7.4.4 Comparison of Methods

When selecting an optimization algorithm for distribution reliability problems, it is best to select the simplest method that will produce reasonably good answers in an acceptable amount of time.[26] To do this, a basic feel for the performance differences between various optimization methods is required. This section applies various algorithms to the same optimization problem so that these types of comparisons can be made. It begins by defining an optimization problem and a test system, continues by comparing hill climbing, simulated annealing, and genetic algorithms, and concludes by examining the impact of probabilistic initialization using fuzzy rules.

The problem that will be used to benchmark algorithms is to optimize the detailed design of a pre-routed primary distribution system.[34] The solution must identify protection device locations, switch locations, tie point locations, automation strategies, and underground versus overhead strategies so that overall cost is minimized. This overall cost, referred to as the total cost of reliability (TCR), is equal to the sum of the utility cost of reliability (UCR) and the customer cost of reliability (CCR):

$$\text{TCR} = \text{UCR} + \text{CCR} \tag{7.20}$$

The utility cost of reliability consists of annual operation and maintenance costs and annualized capital costs. The customer cost of reliability is equal to the sum of all customer incurred costs ($C_{incurred}$) due to momentary and sustained interruptions. This is based on both the customer's cost of interrupted power (C_{kW}, in \$/kW) and the customer's cost of interrupted energy (C_{kWh}, in \$/kWh). For a customer with an average load of P (in kW), CCR can be modeled as:

$$\text{CCR} = \sum_{i=1}^{N} P\left(C_{kW} + t_i \cdot C_{kWh}\right) \tag{7.21}$$

N: the number of interruptions
t_i : the duration of interruption i

The total cost of reliability is used as the objective function to be minimized subject to design constraints. The constraint considered in this test-case problem is the coordination of protection devices—only three protection devices can be placed in series while maintaining coordination. If a design violates a coordination constraint, a penalty factor is added to the TCR to make it more expensive than the TCR of any valid design.

Since distribution system design is a discrete optimization problem, it is convenient to represent potential solutions as bitmaps. The bitmap for this problem is created by dividing the distribution system into line sections and assigning 7 bits to represent information concerning each section. Each of these bits is characterized by a question. If the bit value is zero, the answer to the question is "no." If the bit value is one, the answer to the question is "yes." The following bits and their corresponding questions are used:

Bit 1: Is the section underground?
Bit 2: Is a normally closed switch present?
Bit 3: Is the normally closed switch automated?
Bit 4: Is a protection device present?
Bit 5: Is the protection device a recloser?
Bit 6: Is a normally open switch present?
Bit 7: Is the normally open switch automated?

It is important to note that there is a hierarchy in the above bits. For example, Bit 7 will have no impact on a design unless Bit 6 has a value of one. Similarly, Bit 5 will have no impact on a design unless both Bit 2 and Bit 4 have a value on one.

This benchmark optimization problem requires feeder routing information. In addition, feasible tie points must be established. These are points on the feeder where it is feasible to build tie lines and install a normally open switch so that loads can be back fed during contingencies. To ensure that the benchmarking results are based on a realistic system, the routing and feasible tie points of the test system are based on an actual distribution system in the Pacific Northwest, shown in Figure 7.20.

In addition to routing information, both utility cost information and customer cost information are needed. The utility cost data used is summarized in Table 7.6 and the customer cost data used is summarized in Table 7.7. Reliability data used are taken from the recommended values presented in Chapter 4.

This optimization problem is now solved using hill climbing, simulated annealing, and genetic algorithms. A description of the details of each methodology is now summarized:

Hill Climbing – The hill climbing algorithm uses random initial bit values, tests each bit sequentially and immediately accepts changes that improve the objective function score. It is executed 1000 times to obtain a distribution of results.

Simulated Annealing – The simulated annealing algorithm uses random initial bit values, an initial temperature of 0.9, an annealing rate of 0.99 and a probability of accepting detrimental changes equal to the current temperature. Hill climbing is performed on all frozen solutions to ensure local optimality. It is executed 1000 times to obtain a distribution of results.

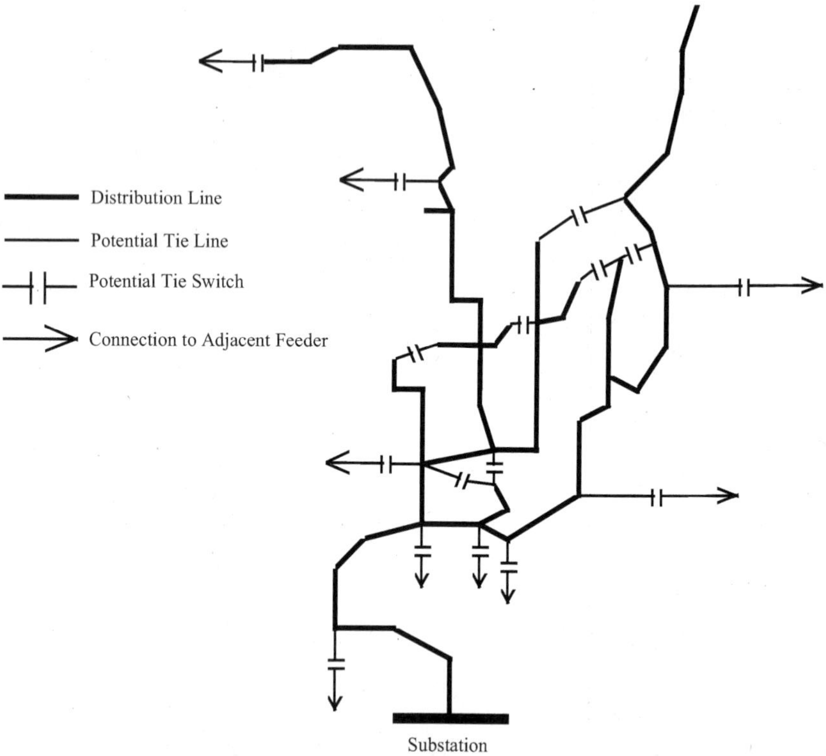

Figure 7.20. Test system primary feeder routing and feasible tie points. This test system is based on an actual utility feeder in the Pacific Northwest. The goal of the optimization problem is to identify protection device locations, switch locations, tie point locations, automation strategies, and underground versus overhead strategies so that the sum of utility cost and incurred customer cost is minimized.

Table 7.6. Annualized equipment cost data.

Device	Base Cost ($/year)	Automation Cost ($/year)
Sectionalizing Switch	1,358	4,939
Tie Switch	1,358	2,488
Recloser	3,771	3,071
Fuse	1,553	4,939
OH Line (per mile)	33,937	--
UG Line (per mile)	238,918	--

Table 7.7. Incurred customer cost data.

Customer Type	% of Load	C_{kW} ($/kW)	C_{kWh} ($/kWh)
Residential	50	0.01	2.68
Commercial	20	1.19	25.61
Office Buildings	20	6.44	20.09
Small Users	10	1.00	5.08
Total System	**100**	**1.63**	**10.99**

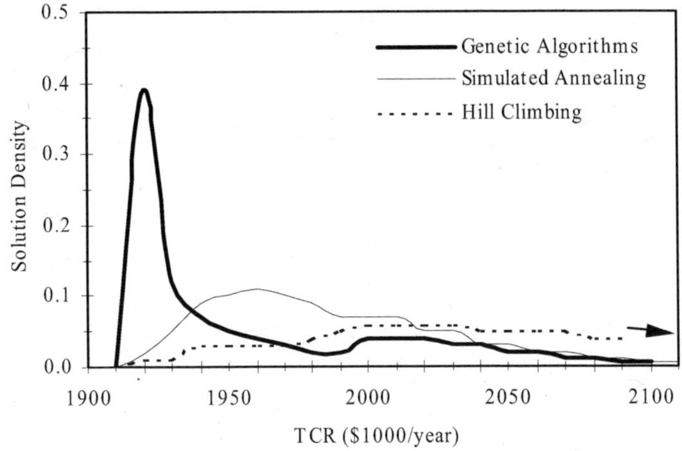

Figure 7.21. Comparison of results for three optimization methodologies. Hill climbing results in a broad array of locally optimal solutions that vary widely in total cost. Simulated annealing is less sensitive to initial conditions than integer programming, but still results in a wide range of costs. Genetic algorithms take advantage of multiple test cases by sharing information and results in better overall solution quality when compared to the other two techniques.

Genetic Algorithms – The genetic algorithm uses random initial bit values, a total population of 1000, an elite population of 50, a mutation rate of 1% per bit per generation and stops when the best member does not change for 50 generations. Hill climbing is performed on all members of the final generation to ensure local optimality. Each of the members in the final population is kept to show a distribution of results.

A summary of the solutions resulting from hill climbing, simulated annealing, and genetic algorithms are shown in Figure 7.21. Integer Programming, when presented with 1000 different initial conditions, resulted in 1000 different locally optimal solutions. This illustrates the complexity of the search space, and shows that integer programming is highly sensitive to initial conditions. Simulated annealing is less sensitive to initial conditions than integer programming, but still results in a wide range of costs, requiring many trials to be performed before a distribution engineer would be confident that a near-optimal solution has been found. Genetic algorithms take advantage of multiple test cases by simultaneously running trials and constantly sharing information among trials through genetic crossover. This improves computational efficiency and results in better overall solution quality when compared to hill climbing and simulated annealing.

Up to this point, all optimization techniques have used random initialization, which essentially gives an algorithm a random starting solution from which to begin its search. This begs the question, "Can solution quality be improved by providing starting locations that are more likely to contain elements of an optimal design when compared to random starting locations? If so, how can these new starting locations be generated?"

To date, distribution system design is performed almost exclusively by distribution planners and engineers using a few analytical tools and a large number of heuristic rules. This method cannot optimize a design, but the heuristic rules used certainly contain information about good designs. In this situation, where optimal designs are desired and large amounts of expert knowledge exist, it is often beneficial to combine optimization techniques and expert rules into a hybrid optimization method.[35-36] Since the knowledge relating to distribution system reliability optimization tends to be vague in nature, fuzzy rules are selected and used to probabilistically create distribution system designs that correspond to heuristic knowledge.[37-38] These designs are then used as initial solutions in an attempt to improve the performance of hill climbing, simulated annealing, and genetic algorithms.

Before fuzzy rules can be generated, it is necessary to define the fuzzy sets that will be used by the fuzzy rules. For this particular problem, five triangular fuzzy sets will be used: very small, small, medium, large, and very large. These sets are shown in Figure 7.22. Fuzzy rules will then be used to create a design on a bit by bit basis. Since each bit corresponds to a particular section, information about this section is needed to accomplish this task.

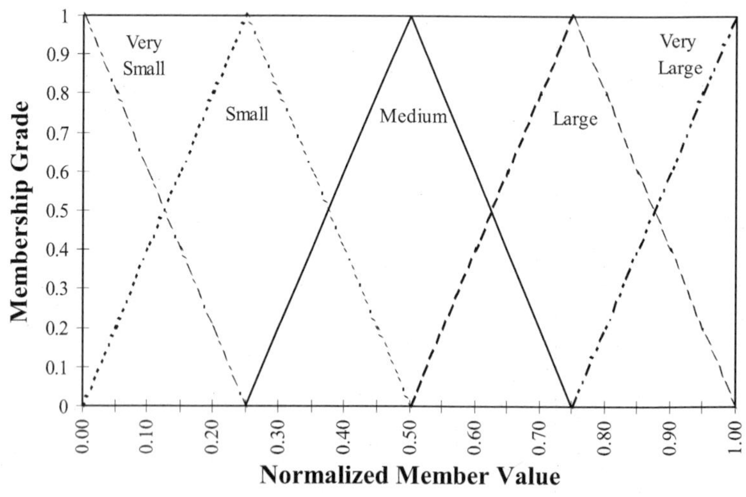

Figure 7.22. Fuzzy sets used for probabilistic initialization of optimization problems. In this case, five triangular functions are used to represent normalized values between 0 and 1. The membership grade of a number x in fuzzy set **S** corresponds to the truth of the statement "x is **S**." For example, the statement "0.6 is Small" has a truth of 0.0, the statement "0.6 is Medium" has a truth of 0.6, and the statement "0.6 is Large" has a truth of 0.4.

After examining many low-cost designs for various system and reliability data, the following features have been chosen as inputs for fuzzy rules (all features are represented by values between zero and one and can be described using the fuzzy sets shown in Figure 7.22):

Downstream Load (DSL) – This value is a measure of how much connected load is downstream of the section being considered. A per unit value is obtained by dividing the total load downstream of the section (including the load on the section) by the total feeder load.

Number of Siblings (SIB) – This is a measure of the branching that is occurring at the beginning of the section being considered. Since the maximum number of branches stemming from a node on the system being considered is 3, a per-unit value is obtained by dividing the number of other sections stemming from the upstream node of the section by 2.

Overhead Line Failure Rate (FR) – This is a measure of the per-mile failure rate of a section if it were to be of overhead construction. A normalized measure is obtained by dividing the overhead line failure rate of the section (in failures/mile/year) by 0.3.

Load Density (LD) – This is a measure of the overall loading of the system. A normalized measure is obtained by dividing the load density of the system (in kW/mile) by 2000.

Tie Point Quality (TPQ) – This is a measure of how often a tie point will be able to connect an interrupted section to a source via an alternate path. A per-unit measure is obtained by dividing the length of the overlapping part of the normal path and backfeedable path by the normal path length.

Not all of the above characteristics are needed when examining each bit. For example, deciding whether or not to place a tie switch on a section requires different information than deciding whether to make the section overhead or underground. In addition, the number of fuzzy inputs should be kept to a minimum since the number of fuzzy rules required increases exponentially with the number of inputs (if there are n fuzzy input sets and x fuzzy inputs, the number of rules required is n^x). The following is a description of each bit and the corresponding fuzzy inputs that have been chosen:

Bit 1 – Is the section underground? Low cost designs tend use underground cable if overhead line failure rate is high and if the system load density is high. Because of this reason, FR and LD are the fuzzy inputs used to determine Bit 1.

Bit 2 – Is a normally closed switch present? The number of normally closed switches on low cost designs increases as load density increases. In addition, there is a tendency to place normally closed switches on branching sections. Because of these observations, SIB and LD are the fuzzy inputs used to determine Bit 2.

Bit 3 – Is the normally closed switch automated? Low cost designs tend to automate normally closed switches that are close to the substation and to not automate normally closed switches that are far away from the substation. The number of automated switches also tends to increase as load density increases. Because of these observations, DSL and LD are the fuzzy inputs used to determine Bit 3.

Bit 4 – Is a protection device present? Low cost designs tend to place protection devices at points of branching. In addition, the number of protection devices increases with load density. Because of these observations, SIB and LD are the fuzzy inputs used to determine Bit 4.

Bit 5 – Is the protection device a recloser? Low cost designs tend to place reclosers near the substation. In addition, the number of reclosers increases with load density. Because of these observations, DSL and LD are the fuzzy inputs used to determine Bit 5.

Bit 6 – Is a normally open switch present? Low cost designs tend to place normally open switches towards the end of the feeder. It is also important for this location to have a high probability of connecting the section to a source via an alternate path during a fault. Because of these observations, DSL and TPQ are the fuzzy inputs used to determine Bit 6.

Bit 7 – Is the normally open switch automated? The most important determiner of whether a tie switch should be automated or not is tie point quality. In addition, more tie switches are automated as load density increases. For these reasons, TPQ and LD are the fuzzy inputs used to determine Bit 7.

Since there are 5 fuzzy sets and two fuzzy inputs corresponding to each bit, there are 25 rules associated with each bit. This rule will have an antecedent (the "if" portion of the rule) and a consequence (the "then" portion of the rule). An example of a fuzzy rule corresponding to a section being underground is:

Example fuzzy rule – If the failure rate of the section is medium and the load density of the system is very large, then the probability that the section should be underground is medium.

To determine the value assigned to a particular bit, the truth of each of its associated rule antecedents is computed. The fuzzy set corresponding to each rule consequence is then weighted by its antecedent truth. All 25 weighted consequences are then added together and the center of mass is computed. This number, a value between 0 and 1, can then be used to probabilistically determine the state of a bit. For example, if the fuzzy output number of a bit is 0.8, then that bit will have a probability of 0.8 of having a value of 1 and a probability of 0.2 of having a value of 0. Using this process, many design solutions can be quickly generated and used for initial solutions in hybrid optimization methods. The fuzzy rules corresponding to Bit 1 of this problem are shown in Table 7.8. A complete list of rules can be found in Reference 39.

A comparison of hill climbing solutions for random versus fuzzy initialization is shown in the top graph of Figure 7.23. Solution quality does improve for fuzzy initialization, but only slightly. The solution variation is still large and many trials will be needed to insure that a near-optimal solution has been found.

A comparison of simulated annealing solutions for random versus fuzzy initialization is shown in the middle graph of Figure 7.23. As with integer programming, solution quality improves, but only marginally. This result is somewhat expected since simulated annealing is relatively insensitive to initial conditions.

A comparison of genetic algorithm solutions for random versus fuzzy initialization is shown in the bottom graph of Figure 7.23. Unlike integer programming and simulated annealing, fuzzy initialization has a profound impact on genetic algorithms. By providing the genetic algorithm good initial genetic information, lower cost solutions are more likely to be found.

Assessment method performances are summarized in Table 7.9. It is evident that genetic algorithms result in lower cost solutions when compared to integer programming and simulated annealing. It is further shown that using fuzzy initialization in genetic algorithms improves solution quality, reduces the standard deviation of solutions, and even reduces computation time. (The average CPU time taken to perform integer programming with random initialization is given a relative CPU time of 1. All other methods are compared to this norm.)

In summary, this section has compared the effectiveness of optimization techniques for a complex distribution system reliability design problem. It initially compared the performance of hill climbing, simulated annealing, and genetic algorithms and showed that genetic algorithms result in lower cost solutions and lower variability in results. It then examined the impact on solution quality of replacing random initialization with probabilistic initialization based on heuristic knowledge encoded as fuzzy rules.

Table 7.8. Rules for Bit 1, which determines the probability of a section initially being specified as underground. In this case, rules are based on the failure rate of the section (FR) and the load density of the overall system (LD). These features are described by the following five fuzzy sets: very small (VS), Small (S) Medium (M), Large (L), and Very Large (VL). Since there are two features and five fuzzy sets, there are twenty-five total rules corresponding to this bit (5^2). The probability of Bit 1 being equal to one, P(1), is equal to center of mass of the 25 fuzzy consequences weighted by their respective antecedent truths.

Rule 1-1:	If	FR	is	VS	and	LD	is	VS	then	P(1)	is	VS
Rule 1-2:	If	FR	is	VS	and	LD	is	S	then	P(1)	is	VS
Rule 1-3:	If	FR	is	VS	and	LD	is	M	then	P(1)	is	VS
Rule 1-4:	If	FR	is	VS	and	LD	is	L	then	P(1)	is	VS
Rule 1-5:	If	FR	is	VS	and	LD	is	VL	then	P(1)	is	S
Rule 1-6:	If	FR	is	S	and	LD	is	VS	then	P(1)	is	VS
Rule 1-7:	If	FR	is	S	and	LD	is	S	then	P(1)	is	VS
Rule 1-8:	If	FR	is	S	and	LD	is	M	then	P(1)	is	VS
Rule 1-9:	If	FR	is	S	and	LD	is	L	then	P(1)	is	S
Rule 1-10:	If	FR	is	S	and	LD	is	VL	then	P(1)	is	S
Rule 1-11:	If	FR	is	M	and	LD	is	VS	then	P(1)	is	VS
Rule 1-12:	If	FR	is	M	and	LD	is	S	then	P(1)	is	VS
Rule 1-13:	If	FR	is	M	and	LD	is	M	then	P(1)	is	VS
Rule 1-14:	If	FR	is	M	and	LD	is	L	then	P(1)	is	S
Rule 1-15:	If	FR	is	M	and	LD	is	VL	then	P(1)	is	M
Rule 1-16:	If	FR	is	L	and	LD	is	VS	then	P(1)	is	VS
Rule 1-17:	If	FR	is	L	and	LD	is	S	then	P(1)	is	VS
Rule 1-18:	If	FR	is	L	and	LD	is	M	then	P(1)	is	S
Rule 1-19:	If	FR	is	L	and	LD	is	L	then	P(1)	is	M
Rule 1-20:	If	FR	is	L	and	LD	is	VL	then	P(1)	is	L
Rule 1-21:	If	FR	is	VL	and	LD	is	VS	then	P(1)	is	VS
Rule 1-22:	If	FR	is	VL	and	LD	is	S	then	P(1)	is	S
Rule 1-23:	If	FR	is	VL	and	LD	is	M	then	P(1)	is	M
Rule 1-24:	If	FR	is	VL	and	LD	is	L	then	P(1)	is	L
Rule 1-25:	If	FR	is	VL	and	LD	is	VL	then	P(1)	is	VL

Table 7.9. Comparison of optimization algorithm performance. Genetic algorithms result in lower cost solutions when compared to integer programming and simulated annealing. In addition, fuzzy initialization in genetic algorithms improves solution quality, reduces the standard deviation of solutions, and slightly reduces computation time. Relative CPU time is the ratio of CPU time to the average CPU time taken to perform integer programming with random initialization.

Algorithm	Mean TCR ($1000/yr)	Standard Deviation (% of mean)	Relative CPU Time
Hill climbing			
Random Initialization	2,061	4.57	1.00
Fuzzy Initialization	2,039	3.50	1.09
Simulated Annealing			
Random Initialization	1,970	2.09	1.62
Fuzzy Initialization	1,968	2.11	1.60
Genetic Algorithms			
Random Initialization	1,944	2.50	1.51
Fuzzy Initialization	1,915	1.37	1.43

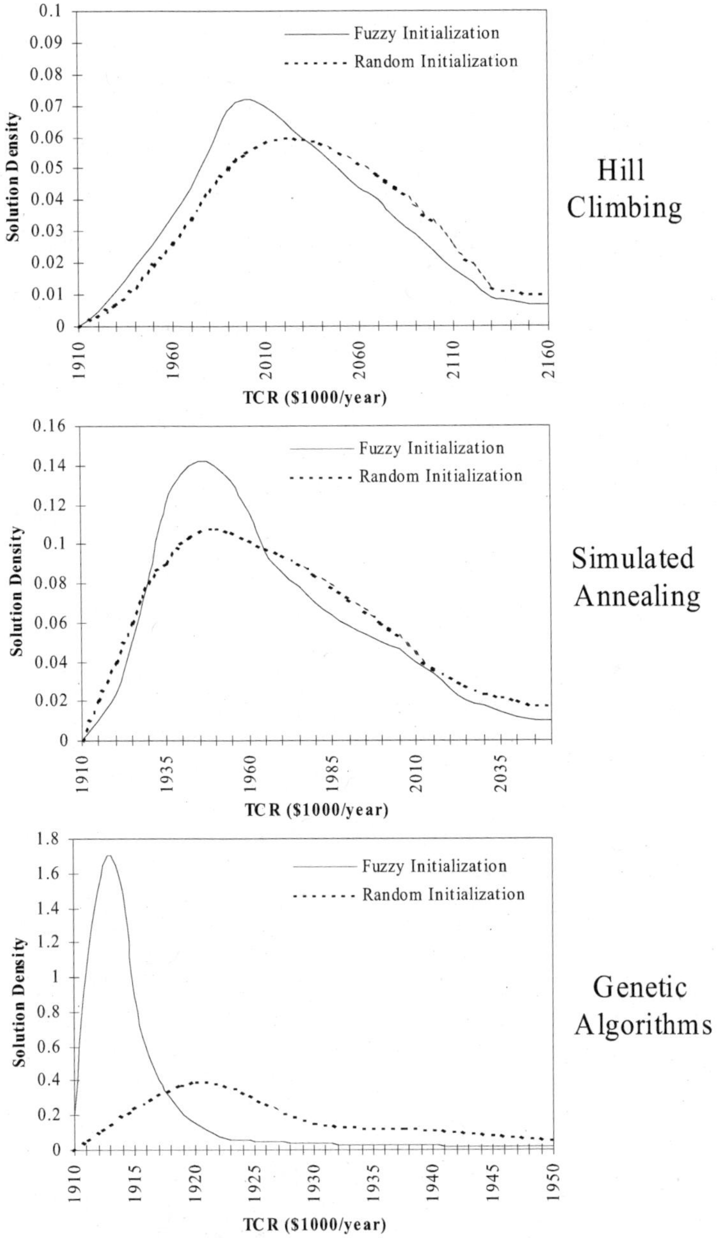

Figure 7.23. Impact of fuzzy initialization on hill climbing, simulated annealing, and genetic algorithms. Fuzzy initialization has modest impact on hill climbing and simulated annealing, but substantially improves the performance of genetic algorithms. By providing the genetic algorithm good initial genetic information, lower cost solutions are more likely to be found.

The performance of all three optimization methods improved when random initialization was replaced with fuzzy initialization. This improvement was relatively small for integer programming and simulated annealing, but was substantial for genetic algorithms. Providing the genetic algorithm with good initial genetic material resulted in superior average solutions, reduced solution variance, and improved computation time.

7.5 FINAL THOUGHTS ON OPTIMIZATION

Armed with a basic theoretical knowledge of distribution reliability assessment and optimization, the reader should begin to consider the practical issues associated with applying these concepts on a broad basis. Small applications such as identifying optimal recloser locations on a few feeders may be straightforward. Large applications such as total capital and operational expenditure allocation, however, may prove difficult due to organizational barriers and conflicting agendas that may not be directly related to cost or reliability—at least in a technical sense.

Perhaps the most fundamental barrier to optimization is the unwillingness of distribution companies to define broad-based objectives and problem scopes that span many divisions, departments, budgets, and functions. Doing so creates inherent conflict between the affected groups. Each will tend to view reliability in a particular way and will be extremely reluctant to reduce its budget so that others can be increased. Reliability optimization can be addressed separately in each department, but doing so leads to many suboptimal solutions and is not in the long-term best interest of the distribution company.

Achieving a broad problem definition that encompasses most of the large reliability and cost factors facing a distribution company requires a proponent at the executive level who has formal authority or influence over all affected departments. For investor owned utilities, people at this level in the organizational hierarchy will be interested in objectives that seek to maximize market capitalization, which is ultimately based on the free cash flow associated with reliability decisions. For publicly owned utilities, executives will be interested in objectives that seek to maximize societal welfare, which is generally approached by meeting all customer expectations for the lowest possible rates.

A distribution company maximizes its chances of optimizing reliability when it can continuously reallocate all capital and operational budgets based on well-defined reliability objectives. For example, it should be possible to compare the cost effectiveness of looping an underground distribution lateral through a new subdivision to increasing tree trimming on the main overhead trunk. If tree trimming results in higher reliability for lower cost, the loop should not be built and a part of the construction budget should be transferred to the vegetation management budget. Similarly, the value of substation equipment maintenance should be able to be directly compared to feeder automation. If feeder automation results in higher marginal value, the substation maintenance budget should be reduced and the feeder automation budget should be increased.

A mental barrier that commonly manifests itself when applying optimization to distribution reliability is a fixation on risk minimization. When a distribution company has a risk-averse culture, the results of an optimization process are often altered to minimize the possibility of reliability ending up worse than promised. For example, an optimization process might conclude that the best size for a transformer is 15 MVA, but a size of 20 MVA is selected "just to make sure." If this increase in capacity is genuinely intended to reduce the risk of an unacceptable reliability outcome, this risk should be explicitly included in the objective function.[40-41] Otherwise, the additional spending undermines the optimization process by increasing cost to satisfy an emotional need.

Other emotional issues that can undermine distribution reliability optimization are personal attachments to the present way of doing business. If an optimization process recommends changes, people responsible for the status quo will tend to become defensive and attack the credibility of the optimization results. This problem can occur at all levels of the organization, from an engineer responsible for equipment ratings to a top executive responsible for setting reliability targets. An engineer who has been recommending loading levels lower than those recommended by the optimization process may warn of impending doom if ratings are adjusted upwards. An executive who has been chanting a mantra of "reliability must improve" for the last three years may entrench and become an opponent of the optimization process when results show that reliability should, in fact, not improve.

As this book is technical in nature, it is not appropriate to address issues concerning leadership, team building, negotiation, and influence, which are precisely the skills required to address ego barriers to reliability optimization. For information on these topics, the reader is referred elsewhere.[42-44] It must suffice to say that before implementing reliability optimization, a thorough stakeholder analysis should be performed. Once all people capable of impeding a successful implementation are identified, they can be included in the reliability optimization process from its inception.

Another common argument used against reliability optimization is "These results cannot be trusted due to a lack of good data." It is generally easy to find shortcomings in assumptions relating to cost, failure rates, the impact on failure rates due to reliability improvement options, and so forth. If successful, the insufficient data argument can lead to a "chicken and egg syndrome" where reliability data collection is not justified because there is no use for it and reliability optimization is not justified because there is a lack of required data.

The best way to defend against insufficient data arguments is to carefully select default data, to meticulously calibrate system models based on conservative and justifiable assumptions, and to only use results within the context of these assumptions. Once a reliability optimization process of limited scope is in place, there will be an incentive to increase reliability data collection efforts to both improve the confidence in results and to gradually increase the optimization scope in both breadth and depth.

Most US electricity customers are served from distribution systems that are operated as regulated monopolies under the governing authority of state regulatory commissions. Regulators approve rate cases and therefore determine the revenue stream of distribution companies. Regulators set performance targets and performance-based rates. Their job is to be an advocate for the people and to insure their best interest, which is to have high levels of reliability for the lowest possible cost without jeopardizing the financial health of the regulated companies. As such, the last reliability optimization difficulties that will be discussed are related to regulation.

The first potential issue with regulators is their tendency to transfer cost savings to rate payers. Consider a reliability improvement optimization effort that results in a 5% reduction in operational expenditures without impacting reliability. In a fully regulated situation, regulators would consider this a reduction in recoverable costs and would require the distribution company to adjust rates so that these savings are completely passed on to customers. If the optimization effort results in a 5% reduction in the rate base, regulators would force a 5% reduction in earnings so that the ratio of profits to rate base remains the same. Such anticipated regulatory responses pose the question, "If regulators take reliability optimization benefits away as soon as they are realized, why should distribution companies bother to make the effort?" The answer to this question is not simple, but involves a regulatory shift away from guaranteed return-on-assets and towards rate freezes, performance-based rates, and reliability guarantees.

There are other potential difficulties with regulators. Many of them view reliability as something that must always improve, arguing that technology and process efficiencies should allow for higher reliability at a lower cost. Others view electricity as an entitlement and embed cross-subsidies in rate structures. In each case, the regulatory view of optimization may be at odds with the utility view of optimization, potentially resulting in conflict and wasted effort.

Rather than performing a broad-based reliability optimization initiative and having it undermined by regulatory action, distribution companies are encouraged to include regulators in the optimization process. If distribution companies and regulatory commissions work together, the probability of implementing a fair combination of rate structure, reliability incentives, reliability targets, and customer choice increases greatly. Once the rules of the game are set, reliability can be optimized knowing it is in the best interest of all parties including distribution regulators, distribution customers, and equity investors.

Reliability optimization is imperative to the survival of distribution companies. Although distribution systems are a natural monopoly, there are emerging nonregulated technologies that are able to bypass the distribution system function at increasingly lower costs. If distribution companies are not able to provide higher reliability for a lower cost than these technologies, they will vanish along with their infrastructure and market valuation. Up to this point, this book has presented all of the classical issues related to distribution reliability including system design, system operation, reliability metrics, reliability modeling, reliability analysis, and reliability optimization. Using these concepts and techniques, distribution companies can remain competitive by providing increasingly higher value to customers. At the same time, customer satisfaction will increase through rate choices and reliability guarantees, regulatory volatility will stabilize as commissioners see that they are acting in the best public interest, and investor expectations will be exceeded by providing high levels of reliability while improving profitability and minimizing risk. Today, these issues are somewhat complicated by aging infrastructure and a business shift towards asset management, which are discussed in the following two chapters.

7.6 STUDY QUESTIONS

1. Describe what is meant by the term "optimal." What are the main components of an optimization problem?
2. What is the difference between locally-optimal solution and a globally-optimal solution? In what situations is it possible to guarantee global optimality?
3. What is the difference between continuous optimization problems and discrete optimization problems? List some distribution decision variables that fall into these two categories.
4. What is the role of solution representation in the overall optimization process? Why is solution representation of particular importance for genetic algorithms?
5. What are some typical methods to solve a discrete optimization problem? Describe some of the advantages and disadvantages associated with each.
6. Name the basic components of an expert system rule.
7. How can expert systems handle conflicting rules and rules of differing importance?
8. How does a fuzzy expert system differ from a traditional expert system? What are some of the potential advantages and disadvantages of a fuzzy expert system?
9. What are some potential practical applications for computer optimization?
10. Can computer optimization replace human engineering with regards to distribution system reliability improvements? Explain.

REFERENCES

1. G. L. Nemhauser and L. A. Wolser, *Integer Programming and Combinatorial Optimization*, Wiley, 1988.
2. F. J. Hill and G. R. Peterson, *Introduction to Switching Theory and Logical Design*, John Wiley and Sons, Inc., 1968.
3. F. Glover, "Tabu Search: Part I," *ORSA Journal on Computing*, Vol. 1, 1990.
4. F. Glover, "Tabu Search: Part II," *ORSA Journal on Computing*, Vol. 1, 1990.
5. E. L. da Silva, J. M. A. Ortiz, G. C. de Oliveira, and S. Binato, "Transmission Network Expansion Planning Under a Tabu Search Approach," *IEEE Transactions on Power Systems,* Vol. 16, No. 1, Feb. 2001, pp. 62-68.
6. R. A. Gallego, R. Romero, and A. J. Monticelli, "Tabu Search Algorithm for Network Synthesis," *IEEE Transactions on Power Systems,* Vol. 15, No. 2, May 2000, pp. 490-495.
7. K. Nara, Y. Hayashi, Y. Yamafuji, H. Tanaka, J. Hagihara, S. Muto, S. Takaoka, and M. Sakuraoka, "A Tabu Search Algorithm for Determining Distribution Tie Lines," *Proceedings of the International Conference on Intelligent Applications to Power Systems,* IEEE Catalog 96TH8152, Jan. 1996.
8. L. Davis, *Genetic Algorithms and Simulated Annealing*, Morgan Kauffmann Publishers, Inc., 1987.
9. V. Miranda, J. V. Ranito, and L. M. Proenca, "Genetic Algorithms in Optimal Multistage Distribution Network Planning," *IEEE Transactions on Power Systems,* Vol. 9, No. 4, Nov. 1994, pp. 1927-1933.
10. S. Sundhararajan and A. Pahwa, "Optimal Design of Secondary Distribution Systems Using a Genetic Algorithm," *Proceedings of the 27th Annual North American Power Symposium,* IEEE, October 1995, pp. 681-684.
11. E. C. Yeh, S. S. Venkata, and Z. Sumic, "Improved Distribution System Planning Using Computational Evolution," *Proceedings of 1995 IEEE Power Industry Computer Applications Conference,* IEEE, May, 1995.
12. D. Srinivasan, F. Wen, C. S. Chang, and A. C. Liew, "A Survey of Applications of Evolutionary Computing to Power Systems," *Proceedings of the International Conference on Intelligent Systems Applications to Power Systems,* IEEE Catalog 96TH8152, Jan. 1996.
13. D. E. Goldberg, *Genetic Algorithms in Search, Optimization, and Machine Learning*, Addison-Wesley Publishing Company, 1989.
14. W. M. Spears, "Crossover or Mutation?" *Foundations of Genetic Algorithms 2*, Morgan Kauffmann Publishers, Inc., 1993.
15. E. Rich and K. Knight, *Artificial Intelligence*, 2nd Edition, McGraw-Hill Int., 1991.
16. G. Brauner and M. Zobel, "Knowledge Based Planning of Distribution Networks," *IEEE Transactions on Power Systems,* Vol. 9, No. 2, May 1994, pp. 942-948.
17. J. Shao, N. D. Rao, and Y. Zhang, "An Expert System for Secondary Distribution System Design," *IEEE Transactions on Power Delivery,* Vol. 6, No. 4, Oct. 1991, pp. 1607-1615.
18. *Report of the U.S. Department of Energy's Power Outage Study Team: Findings and Recommendations to Enhance Reliability from the Summer of 1999,* U.S. Department of Energy, March 2000.
19. C. Segneri, "Reversal of Fortunes: ComEd Rebuilds Its System and Its Reputation," *Transmission and Distribution World,* Vol. 53, No. 6, May 2001, pp. 6-13.
20. R. E. Brown, A. P. Hanson, H. L. Willis, F. A. Luedtke, and M. F. Born, "Assessing the Reliability of Distribution Systems," *IEEE Computer Applications in Power*, Vol. 14, No. 1, Jan. 2001, pp. 44-49.
21. T. Terano, K. Asai, and M. Sugeno, *Fuzzy Systems Theory and Fuzzy Systems Theory and Its Applications,* Academic Press, Inc., 1987.
22. N. Kagan and R. N. Adams, "Electrical Power Systems Planning Using Fuzzy Mathematical Programming," *International Journal of Electrical Power and Energy Systems,* Vol. 16, No. 3, 1994, pp. 191-196.

23. R. E. Brown, "Distribution Reliability Assessment and Reconfiguration Optimization," *2001 IEEE/PES Transmission and Distribution Conference*, Atlanta, GA, Oct. 2001.
24. R. E. Brown, S. S. Gupta, R. D. Christie, and S. S. Venkata, "A Genetic Algorithm for Reliable Distribution System Design," *International Conference on Intelligent Systems Applications to Power Systems*, Orlando, FL, January 1996, pp. 29-33.
25. R. E. Brown and B. Howe, "Optimal Deployment of Reliability Investments," E-Source, Power Quality Series: PQ-6, March 2000, 37 pages.
26. H. L. Willis, H. Tram, M. V. Engel, and L. Finley, "Selecting and Applying Distribution Optimization Methods," *IEEE Computer Applications in Power,* Vol. 9, No. 1, IEEE, January 1996, pp. 12-17.
27. W. F. Horton, S. Goldberg, and R. A. Hartwell, "A Cost/Benefit Analysis in Feeder Reliability Studies," *IEEE Transactions on Power Delivery,* Vol. 4, No. 1, Jan. 1989, pp. 446-452.
28. Y. Tang, "Power Distribution System Planning with Reliability Modeling and Optimization," *IEEE Transactions on Power Systems,* Vol. 11, No. 1, Feb. 1996, pp. 181-189.
29. I. J. Ramírez-Rosado and J. L. Bernal-Agustín, "Reliability and Costs Optimization for Distribution Networks Expansion Using an Evolutionary Algorithm," *IEEE Transactions on Power Systems,* Vol. 16, No. 1, Feb. 2001, pp. 111-118.
30. V. Borozan and N. Rajakovic, "Application Assessments of Distribution Network Minimum Loss Reconfiguration," *IEEE Transactions on Power Delivery,* Vol. 12, No. 4, Oct. 1997, pp. 1786-1792.
31. R. Takeski and D. Rajicic, "Distribution Network Reconfiguration for Energy Loss Reduction," *IEEE Transactions on Power Systems,* Vol. 12, No. 1, Feb. 1997, pp. 398-406.
32. Q. Zhou, D. Shirmohammadi, and W. H. E. Liu, "Distribution Reconfiguration for Operation Cost Reduction," *IEEE Transactions on Power Systems,* Vol. 12, No. 2, May. 1997, pp. 730-735.
33. H.L. Willis, *Power Distribution Reference Book*, Marcel Dekker, 1997.
34. R. E. Brown, S. Gupta, R. D. Christie, S. S. Venkata, and R. D. Fletcher, "Automated Primary Distribution System Design: Reliability and Cost Optimization," *IEEE Transactions on Power Delivery*, Vol. 12, No. 2, April 1997, pp. 1017-1022.
35. Y. Fukuyama, H. Endo, and Y. Nakanishi, "A Hybrid System for Service Restoration Using Expert System and Genetic Algorithm," *Proceedings of the International Conference on Intelligent Systems Applications to Power Systems,* Orlando, FL, IEEE, January 1996.
36. A. H. Mantawy, Y. L. Abdel-Magid, and S. Z. Selim, "Integrating Genetic Algorithms, Tabu Search and Simulated Annealing for the Unit Commitment Problem," *IEEE Transactions on Power Systems,* Vol. 14, No. 3, Aug. 1999, pp. 829-836.
37. R. E. Brown, "The Impact of Heuristic Initialization on Distribution System Reliability Optimization," *International Journal of Engineering Intelligent Systems for Electrical Engineering and Communications*, Vol. 8, No. 1, March 2000, pp. 45-52.
38. R. E. Brown, S. S. Venkata, and R. D. Christie, "Hybrid Reliability Optimization Methods for Electric Power Distribution Systems," *International Conference on Intelligent Systems Applications to Power Systems*, Seoul, Korea, IEEE, July 1997.
39. R. E. Brown, *Reliability Assessment and Design Optimization for Electric Power Distribution Systems,* Ph.D. Dissertation, University of Washington, Seattle, WA, June, 1996.
40. V. Miranda and L. M. Proenca, "Probabilistic Choice vs. Risk Analysis–Conflicts and Synthesis in Power System Planning," *IEEE Transactions on Power Systems,* Vol. 13, No. 3, Aug. 1998, pp. 1038-1043.
41. V. Miranda and L. M. Proenca, "Why Risk Analysis Outperforms Probabilistic Choice as the Effective Decision Support Paradigm for Power System Planning," *IEEE Transactions on Power Systems,* Vol. 13, No. 2, May 1998, pp. 643-648.
42. J. P. Kotter, "Leading Change: Why Transformation Efforts Fail," *Harvard Business Review,* March-April 1995.
43. D. Carnegie, *How to Win Friends and Influence People,* Mass Market Paperback, 1936.
44. L. A. Wortman, *Effective Management for Engineers and Scientists,* John Wiley & Sons, 1981.

8
Aging Infrastructure

Many utilities in the United States and around the world have distribution systems with a large percentage of very old equipment. The amount of very old equipment is increasing, and age-related deterioration is, in many cases, beginning to have a detrimental impact on distribution system reliability. In the future, issues surrounding aging infrastructure will increasingly become more critical for distribution systems in terms of cost and reliability. Addressing aging infrastructure issues requires a large corporate and regulatory commitment, and will often require rate increases. As such, it is important to understand aging infrastructure not only in terms of reliability impact, but in terms of its broader context.

8.1 EQUIPMENT AGING

At face value, equipment aging seems to be a simple concept. A piece of equipment can be brand new, five years old, ten years old, fifty years old, and so forth. Unfortunately it is not as simple as this. It is possible for a piece of equipment to be purchased fifty years ago, installed forty five years ago, taken out of service thirty years ago, put back into service twenty five years ago, and had most of its critical parts replaced ten years ago. What age should be assigned to this piece of equipment?

Before answering this question, it is beneficial to consider another question, "Why do we care about the age of equipment?" There are many answers, but the following are most important:

Key Considerations of Equipment Aging

- The likelihood of failure tends to increase
- Maintenance costs tend to increase
- Replacement parts can become difficult to obtain
- Old equipment may become technologically obsolete

Although all of the above considerations are important, the first bullet point is the most difficult to address. This is because issues such as maintenance cost, spare parts availability, and technological obsolescence can be addressed in a business case. If it is cheaper to buy new equipment than maintain old equipment, the new equipment should be purchased. If the value proposition of new technology is compelling, old technology should be replaced with new technology.

But what about equipment that is neither expensive to maintain nor obsolete? It is simply old and getting older. It should be emphasized that old equipment is not undesirable simply because it is old. It is undesirable because it may have a high likelihood of failure.[1] In fact, any piece of equipment that has a high likelihood of failure is also undesirable. Utilities are not so much interested in equipment age. Rather, utilities are interested in the condition of the equipment, and often use age as an indicator of equipment condition.[2] Typically the best measure of age when used for this purpose is based on purchase date or installation date.

How Should Equipment Aging be Viewed?

- Age can often be used as an indication of equipment condition
- Age should generally be based on purchase date or installation date

Since the age of each piece of equipment is important, it is also important to understand the age characteristics of a population of equipment. This is typically done through equipment age profiles.

8.2 EQUIPMENT AGE PROFILES

Consider a utility that has installed one transformer each year for the last fifty years. This utility will have one transformer that is one year in age, one that is two years in age, one that is three years in age, and so forth up until fifty years in age. If you are interested in the age characteristics of these transformers, you could look at a list of all the transformers and their corresponding ages. In this case, you would see the numbers one through fifty. This type of list is difficult to interpret, especially for large populations. You could also use statistical measures like average age and standard deviation. In this case, the average age is about twenty five years and the standard deviation is about fifteen years. These measures are also difficult to interpret with regards to aging infrastructure.

A better way to look at the data is to use equipment age profiles and survivor curves. An equipment age profile is simply a histogram showing the percentage of equipment within various age categories. For large populations, it is useful to have a separate age category for each year. For smaller populations, it may make sense to group equipment based on five-year age categories (1-5, 6-10, 11-15, ...) or ten-year age categories (1-10, 11-20, 21-30, ...).

A sample age profile for underground cable is shown in Figure 8.1. The squares represent the percentage of cable that is each year in age, and correspond to the left vertical axis. For instance, about 2.7% of all cable is one year in age, as shown by the left-most square. Similarly, about 3.5% of all cable is twenty years in age, as shown by the square directly above the twenty on the horizontal axis.

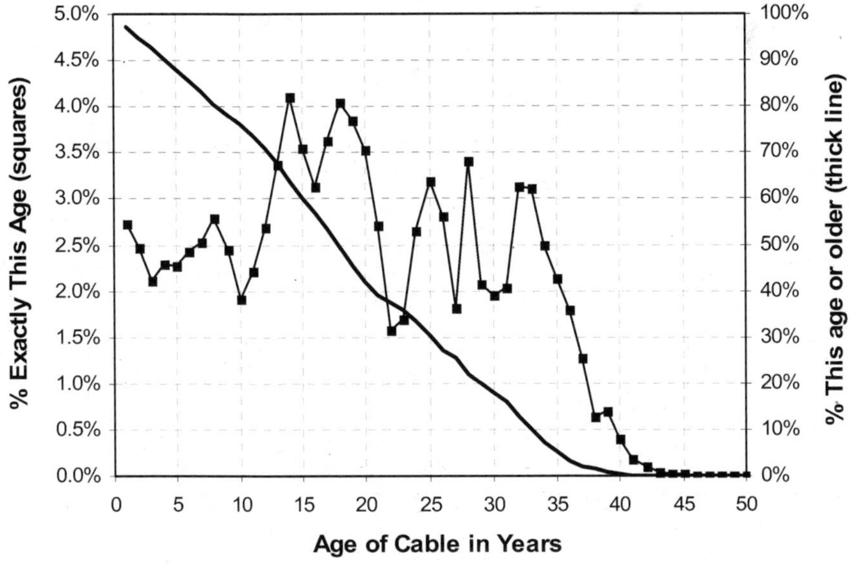

Figure 8.1. The squares connected by the thin line represent the percentage of underground cable at each specific age. For example, the left-most square shows that 2.7% of all underground cable is between zero and one year old. The thick line is a survivor curve and shows the percentage of cable that is older than each age level. For example, the thick line crosses the point "Age 25" at a y-axis value of 30%. This means that 30% of all cable is older than 25 years in age.

Equipment age profiles contain a lot of data, but can sometimes be difficult to interpret since equipment purchases and installations can vary widely from year to year, resulting in an erratic histogram. For this reason, it is often times beneficial to look at a "survivor curve," which represents the percentage of equipment that has survived up to any given age. In the case that 5% of all cable is between zero and one year in age, it follows that 95% of all cable has survived passed the age of one. Similarly, if another 5% of cable is between one and two years in age, it follows that 90% of all cable has survived past the age of two. The calculation of the amount of cable that has survived past all ages can be calculated in a similar manner, resulting in a survivor curve.

The survivor curve corresponding to the histogram in Figure 8.1 is shown as a thick line in the same figure, which corresponds to the right vertical axis. This curve shows that about 50% of all cable is older than 18 years in age. This is determined by following the 50% line on the right vertical axis over to where it crosses the thick line at a value of about 18 years. In a similar way, it can be seen that about 20% of all cable is older than 29 years and about 10% of cable is older than 33 years. Very little cable is older than 40 years in this case.

The following sections of this chapter use equipment age histograms and survivor curves similar those of Figure 8.1 to illustrate a range of aging infrastructure issues.

8.3 POPULATION AGING BEHAVIOR

Each class of equipment is constantly aging and being rejuvenated. Each year, the entire population of existing equipment becomes one year older. Each year, new equipment is installed that makes the average equipment age for the entire population lower. Each year, some equipment will typically fail and be replaced. This will also make the average equipment age lower. These three dominant population factors are summarized as follows:

Dominant Factors Affecting Average Population Age

- Chronological aging of existing equipment
- Installation of new equipment
- Replacement of old equipment with new equipment (typically after a failure or technological obsolescence)

Consider a population of 1000 new transformers. Initially, the average age of this population is 1 year. As long as there are no failures, the average age will increase each year. The second year will have an average age of 2 years, the third year will have an average age of 3 years, and so forth. Eventually, older equipment will start to fail and be replaced by new equipment. This will slow the rate of increase for the average population age. Eventually, the rate of rejuvenation due to replacement will fully offset the rate of chronological aging and the average equipment age will stabilize. However, this process may take decades and involve oscillations where average population age rises and falls dramatically as large percentages of old equipment fail together and are replaced within a short period of time.

Consider again the example that starts with a population of 1000 new transformers. Also assume that the failure rate of transformers is 0.03 per year. After the first year, there will be 30 transformers that fail and are replaced with new units. This is determined by multiplying the failure rate of the transformers by the population size. In year two, there will be 30 transformers 1-year in age and 970 transformers that are 2-years in age. The resulting average age in the second year is 1.97 years, which is slightly less than what it would be with no failures. Assuming a constant failure rate, this process can be repeated to compute average population over time. This curve is shown in Figure 8.2.

Figure 8.2 shows that the population of transformers is initially characterized by an average age that is quickly increasing. Over time, the rate-of-change in average age slows and gradually converges on an asymptotic value. In this case, this asymptotic value for average population age can easily be computed by taking the reciprocal of the failure rate, or 33.33 years.

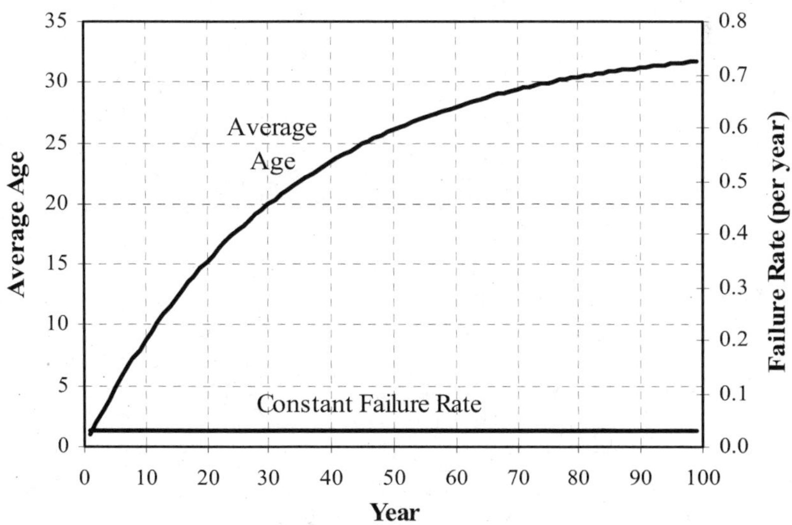

Figure 8.2. This figure shows the average age of a population of transformers with a constant failure rate of 0.03 per year. Initially the average population age is 1 year. As years pass, chronological aging is increasingly offset by failures (and replacement with new equipment) until the average population reaches an asymptotic value. In this case the steady-state population average age is 33 years.

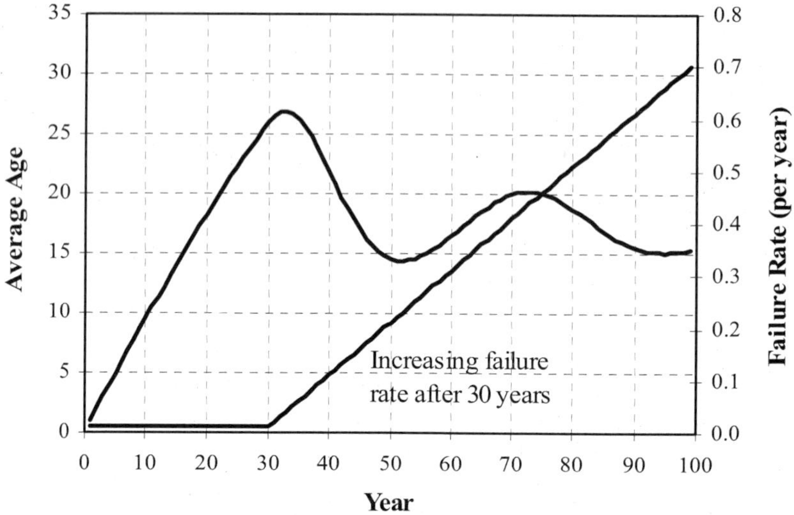

Figure 8.3. This figure shows the average age of a population of transformers with a failure rate of 0.01 per year for the first 30 years and a linearly increasing failure rate afterwards. As a large percentage of transformers fail and are replaced in years 30 through 50, the average population age reduces significantly. The average age then gradually rises until the first wave of replacement transformers reaches 30 years in age, resulting in a "failure echo" starting just past year 50.

Population aging dynamics become more complicated if failure rates are assumed to increase with equipment age. If failure rates tend to increase rapidly after an expected useful life, large amounts of equipment can fail in a relatively short period of time. Widespread replacement will then have a rejuvenating effect on the system, but will eventually result in a "failure echo" when the replacement equipment reaches the end of its useful life.

An example of population aging with nonconstant failure rates is shown in Figure 8.3. This case also starts with an initial population of 1000 new distribution transformers. These transformers are assumed to have a failure rate of 0.01 per year during the first 30 years. After this period of useful life, failure rates are assumed to grow linearly by 0.01 per year. In this scenario, relatively few failures occur during the first thirty years due to the low failure rate. From years thirty to fifty, most of the original transformers fail and are replaced. This results in a naturally rejuvenated system by year fifty. By year seventy, the second generation of equipment is beginning to near the end of its useful life, resulting in the previously mentioned "failure echo."

An important point that is demonstrated in these examples is that population dynamics occur over many decades, but dramatic increases in equipment failures can occur over a short period of time depending upon the amount of equipment nearing the end of the useful life period, and the magnitude of failure rate increase that occurs.

These examples are provided to give the reader a feel for basic population aging dynamics, but hold the population size constant. In reality, most classes of distribution equipment within electric utilities have growing populations. Growth tends to mask the signs of population aging. Consider a population of 1000 transformers with an average age of 30 years. In ten years, if no transformers are replaced, the average age will be 40 years. In the same ten years, if growth is about 7% per year, the number of transformers will typically double. Therefore, the final population will consist of the initial 1,000 transformers plus 1,000 new transformers. The initial transformers have an average age of 40 years, and the new transformers have an average age of 5 years. Therefore, the final population will have an average age of 22.5 years, which is lower than the average age of the original 1,000 transformers.

In the situation described in the previous paragraph, it may not seem like there is an aging infrastructure problem since average equipment age is going down. However, there is a large population of equipment that is all getting old at the same time, which is a major concern. Therefore, it is typically better to view aging infrastructure by looking at the amount of very old equipment on the system rather than looking at average equipment age.

Important Points Regarding Population Age

- Average age is not as important as the amount of very old equipment.
- Low average age may be a result of high equipment population growth. This does not eliminate the obligation of a utility to address aging equipment issues in older parts of the system.

Since average age is not as important as the amount of very old equipment, it is often beneficial to use metrics focused on the amount of old equipment. One can always look at histograms and survivor curves, but this becomes unwieldy when examining trends over many years. A better approach is to select a threshold age and measure the percentage of equipment that exceeds the threshold age. To illustrate, consider again Figure 8.3. This example shows the average age of a population of equipment that begins to deteriorate quickly after thirty years. As such, it is of interest to know the percentage of equipment that exceeds thirty years of age. This information is shown in Figure 8.4. For the first thirty years, none of the equipment meets this criterion. At year 31, about 73% of the transformers have survived and now exceed the thirty-year threshold. As these begin to rapidly fail and are replaced, the amount of equipment older than thirty years drops to less than 10% at year sixty, and then begins to rise again.

In the same way that high load growth can mask the effects of aging infrastructure, low load growth can have the opposite effect. In addition to making

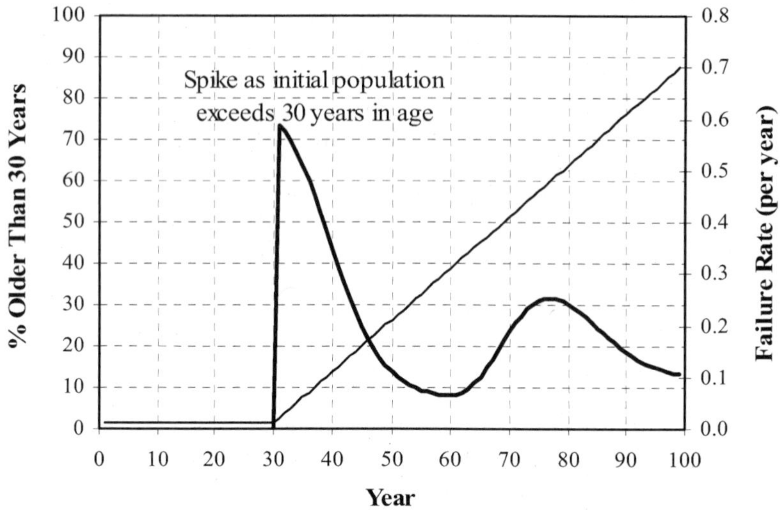

Figure 8.4. This figure shows the percentage of transformers older than thirty years in age. Like the previous figure, failure rates are 0.01 per year for the first 30 years and a linearly increasing afterwards. At year 31, 73% of the initial population has survived. A large percentage of transformers then fail and are replaced in years 30 through 50, reducing the percentage of old transformers significantly.

aging statistics more pronounced, a higher percentage of utility resources must focus on deteriorating equipment as opposed to new construction. This type of shift can be disruptive because spending related to new construction typically corresponds to increases in revenue. In contrast, efforts related to aging infrastructure typically increase spending without corresponding increases in revenue, at least without any associated rate adjustment.[3]

8.4 AGE AND INCREASING FAILURE RATES

As the useful life of a piece of equipment comes to an end, the previously constant failure rate will start to increase as the components start to wear out. This time is referred to as the wearout period of the equipment, as previously discussed in Section 4.2. During the wearout period, the failure rates of many categories of distribution equipment tend to increase, typically at an increasing rate.

For electrical equipment with insulation subject to thermal aging (such as transformers and cables), failure rates tend to increase exponentially with age. This means that after an initial infant mortality period, failure rates tend to be low during the design life of the equipment. After a certain point, failure rates will tend to dramatically rise as the electrical and mechanical strength of the insulation becomes insufficient to withstand physical or electrical disruptions.

Two primary methods have been used for empirically determining age-versus-failure models for equipment with organic insulation. One is based on stress tests and the other is based on historical outage data. Examples of stress tests include condition assessments and accelerated aging experiments to correlate the voltage stress, aging temperature and ac breakdown strength.[4-7] Although potentially useful in a laboratory setting, these types of methods are not practical for use on installed distribution equipment.

Examples of failure modeling based on historical data include fitting models[8-9] and applying the nonhomogeneous Poisson process to establish the age-related failure functions.[10] Weibull modeling is a common approach to capture equipment failure patterns,[11] but is often selected due to its tradition in reliability engineering rather than its ability to best describe actual age-versus-failure data.

In order to model failure rates, both population data and outage data must be obtained. Although equipment installation records may be incomplete, purchasing records in accounting systems can always be used as a proxy for population data. The extent of the details in outage records also affects the quality of failure rate models. When detail data is not available, assumptions can often be used as a substitute.

The remainder of Section 8.4 presents a method for age-based failure modeling that identifies the important equipment features based on significance tests. These features are then used to categorize equipment into several major types, whose age-related failure rate trends are given respectively. This method is described through the use of cable population and outage data from a large US utility.[12]

8.4.1 Historical Data

Although many power delivery companies are presently keeping complete and accurate cable installation data, such records are often missing, especially for cable that has been in the ground for several decades or more. In these cases, purchase records in utility accounting system can often be used to approximate the missing installation data. An example of purchasing data is shown in Figure 8.5, which gives the approximate age distribution of cross-linked polyethylene (XLP) nonconcentric neutral (NCONC) 25-kV cables for the distribution system considered in this section.

Ideally, both population data and outage data will include information about equipment characteristics that might impact age-versus-failure characteristics. For outage data, at a minimum this includes the age of each piece of equipment that fails. The availability of features in the data sets allows for the development of different models for different cable types. It is desirable to have as much outage data as possible so that more feature combinations can be examined without losing statistical significance.

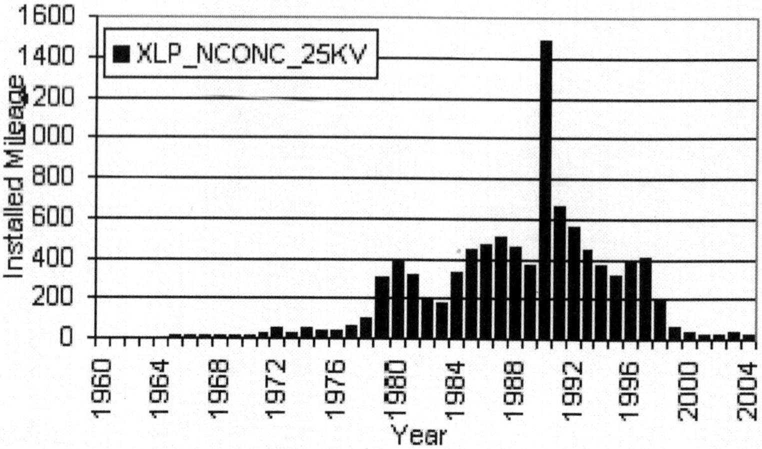

Figure 8.5. An approximate age distribution of XLP nonconcentric neutral 25-kV cable for a large US utility. This data came from purchasing records.

8.4.2 Feature Identification

Equipment features may significantly affect age-versus-failure rate characteristics. As such, it is important to identify important features to include. It is also important to identify unimportant features so that they can be excluded and the number of failure rate models kept to a minimum. For the cable example of this section, some of the potential important features include insulation type, rated voltage, and the presence of a waterproof jacket.

For the cable population, insulation types include Paper Insulated Lead Covered (PILC), High Molecular Weighted Polyethylene (HMWPE), Polyethylene (PE), Crosslink Polyethylene (XLP), and Ethylene Propylene Rubber (EPR). For this utility PILC cable has shown a very long expected life, but is becoming very old. HMWPE cable is problematic and has being gradually phased since 1990. EPR cables seem to have good performance, but have only been installed in recent years.

Rated voltages for this cable population include 12 kV and 25 kV. Rated voltage is an important factor since voltage stress is known to be strongly correlated to expected cable life. In this case, 25-kV cables are often operated at 15 kV with the hope of increasing cable life.

Concentric neutrals often corrode over time. Therefore, cables with concentric neutrals may show higher failure rates than those without concentric neutrals. Jackets can protect cables from moisture and corrosion, and cables without jackets may less reliable than those with jackets.

Table 8.1. Significance tests on cable features.

Feature	# of Failures	Cable Miles	Failure Rate (/yr/mile)	Difference (%)
Jacket	148	5838	0.025	3.1
Non-jacket	104	3979	0.026	
12 kV	34	812	0.042	98
25 kV	299	14127	0.021	
Concentric	148	5838	0.025	261
Non-concentric	62	677	0.092	

Some other features may also be considered such as metallic shields, construction types, environmental moisture levels, etc. However, the final failure rate models will not necessarily address all of these features. First, these features may not be recorded in outage data and/or population data. Second, an exhaustive inclusion of features may lead to data insufficiency in model developments. Third, many features may not have a significant influence on failure rates.

In order to reduce the feature dimensions and retain enough data for the failure rate modeling, significance tests can be used to filter out the relatively unimportant features. This compares the failure rate difference between all components sharing a single feature. For example, to examine the significance of jackets, the average failure rate of all cable with a jacket is compared to the average failure rate of all cable without a jacket. Results for each feature significance test are shown in Table 8.1.

For this cable population, the difference in average failure rates between jacketed and non-jacketed cables are small, only about 3.10%. Thus, it is reasonable to not consider this feature in developing the cable failure rate models. However, voltage levels and concentric neutral are features corresponding to large differences in average failure rates. Cables with these features should therefore be addressed separately in different failure rate models.

8.4.3 Failure Rate Models

After identifying important features, both population data and outage data can be assigned to categories of equipment sharing common important features. Each category will ultimately be characterized by an age-versus-failure rate model.

Ideally, the failure rates of each category at each age are obtained from statistical analysis. For example, the failure rate of 27-year-old 25-kV XLPE cable would ideally be determined by examining the number of actual failures for cable having this description. Unfortunately, outage data is often limited and this approach results in many years having a zero corresponding outage records. The large number of zero-failure rates makes it problematic to fit a regression model directly. To circumvent this problem, age bins can used and regression models cat be fitted to the average failure rates within age bins.

After population and failure statistics are obtained, it is desirable to identify the model that best describes age-versus failure characteristics. Three forms of models are commonly tested for goodness-of-fit: exponential, power, and linear (t corresponds to age in years).

$$\lambda = ae^{bt} \quad \text{; exponential model} \tag{8.1}$$

$$\lambda = at^{b} \quad \text{; power model} \tag{8.2}$$

$$\lambda = a + bt \quad \text{; linear model} \tag{8.3}$$

A good model selection criterion is based on the minimum of the root mean squared error between the historical failure rates and the predicted failure rates at the median points within age bins. This criterion is used both to find the best coefficients in each model and to ultimately select the best model.

Figure 8.6 shows the fitted exponential model, power model and linear model for historical failure rates of 12-kV XLP cables with concentric neutrals. In this case, the exponential model has the best fit.

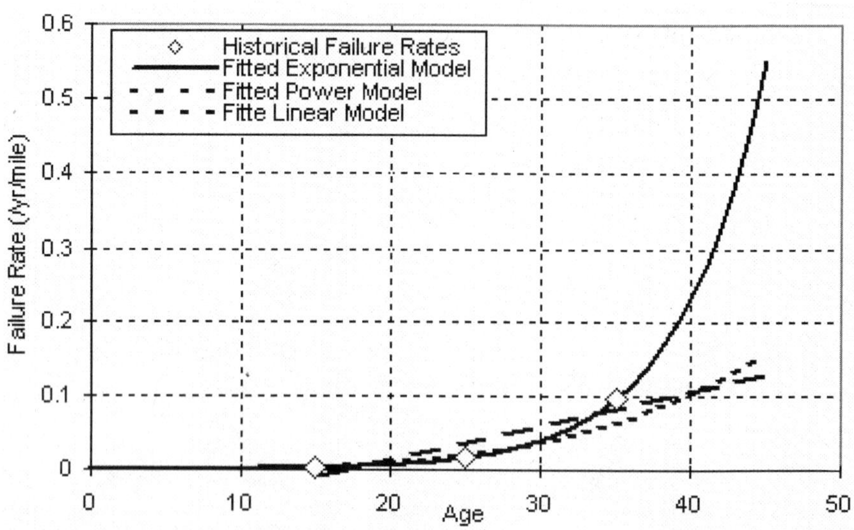

Figure 8.6. Fitted age-versus-failure models for historical failure rates of XLP concentric neutral 12-kV cables. The models correspond to an exponential, power, and linear model. In this case, the exponential model is best able to fit the historical data.

8.4.4 System Reliability Models

After age-versus-failure rate models have been developed for various classes of equipment, it is often desirable to examine the impact of aging equipment on system reliability.[13] This can be done for a base case and for various proactive replacement strategies. Typical metrics of interest are the total number of cable failures, SAIFI, and SAIDI. To model SAIFI and SAIDI, it is necessary to determine the average number of customer interruptions and customer interruption minutes associated with various categories of equipment.

<u>System Modeling Process</u>

1. Determine the current amount of equipment in each category at each age (bin). Each bin has a corresponding failure rate, λ.
2. Determine the expected number of failures in each bin, which is equal to the bin λ multiplied by the amount of equipment in the bin.
3. Determine the amount of equipment that is replaced in each bin due to failures. This is equal to the expected number of failures times the average amount of replaced equipment per failure. Reduce the amount of equipment in the bin accordingly and add the same amount in a zero-age bin of the category of replacement equipment.
4. Determine the amount of SAIFI and SAIDI contribution due to the equipment failures.
5. Perform proactive replacement (if any). To do this, identify the equipment to be replaced, remove the desired amount from the corresponding bin, and add the same amount in a zero-age bin of the category of replacement equipment.
6. Age each bin by one year and increase the bin λ accordingly.
7. Go to step 2 and repeat until the desired number of years has been simulated.

To illustrate the system modeling process, consider again the example of 1000 aging transformers. Assume that today all transformers have a failure rate of 0.01 per year, but that failure rates increase exponentially as the new transformers age. Also assume that today, system SAIDI is 100 min/yr and that 10% of this (10 min/yr) is due to transformer failures. Figure 8.7 shows how the SAIDI of this system changes over time, assuming no proactive replacement of aging transformers. Over a period of fifty years, the system with all-new transformer gradually deteriorates from a SAIDI of 100 min/yr to nearly 120 min/yr. Other analyses can then be performed to illustrate the impact of proactive inspection, maintenance, and replacement on SAIDI.

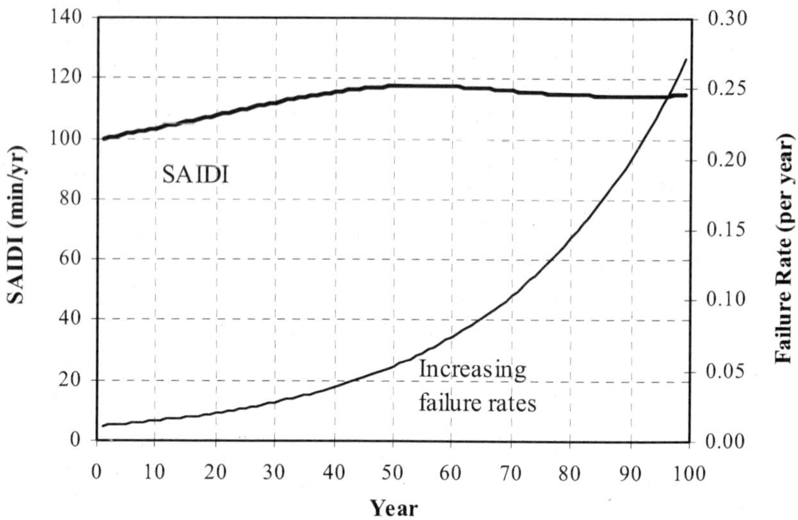

Figure 8.7. This graph shows how SAIDI deteriorates over time as an initially new population of transformers fail. Initially, transformer failures contribute to 10% of SAIDI. Over a period of fifty years, the system with all new transformers gradually deteriorates from a SAIDI of 100 min/yr to nearly 120 min/yr.

The example shown in Figure 8.7 is simplified in the sense that all transformers are initially the same age and all transformers have identical age-versus-failure rate models. For actual systems, this is not the case and the initial system model will have a mix of equipment ages with a mix of failure rate models. For example, an initial system may have different failure rate models for transformers of different voltages, different sizes, different manufacturers, and so forth. Multiple types of equipment can also be considered such as transformers, cables, and poles. System models should not be more complicated than necessary, but should have sufficient detail so that the benefits viable aging infrastructure mitigation scenarios can be modeled and quantified.

8.5 INSPECTION, REPAIR, AND REPLACEMENT

Many types of distribution equipment require inspection, testing and/or maintenance to ensure proper operation and minimize the probability of failures. Maintenance strategies can be broadly categorized as run-to-failure, periodic, condition based, and reliability centered.

Run-to-failure is the simplest maintenance strategy. After installation, equipment is not inspected or maintained until a failure occurs. This is cost effective for noncritical components with minimal maintenance requirements. It is also commonly used for equipment that is difficult or expensive to obtain meaningful condition data.

Periodic maintenance is the next simplest maintenance strategy. At specific time intervals, certain maintenance procedures are performed on equipment regardless of equipment condition or equipment criticality. Intervals are usually time-based, but other measures such as number of operations are also used. This has historically been the most common form of maintenance for major pieces of equipment, with periodic inspection and maintenance activities often being performed according to manufacturer recommendations. However, many types of equipment should not be maintained until a critical level of deterioration has occurred. This is because the very act of maintenance can cause damage to otherwise healthy equipment (i.e., don't fix it if it isn't broken). Many utilities recognize this and are moving away from periodic maintenance and towards condition-based maintenance.

Condition-based maintenance (CBM) monitors equipment and only requires maintenance when signs of deterioration become evident. Condition assessment techniques are too numerous to list exhaustively, but include techniques such as visual inspection, acoustical inspection, infrared thermography, voltage withstand testing, partial discharge testing, laboratory testing, and many other techniques. Pure CBM will perform a basic inspection on a piece of equipment and perform additional testing and maintenance as a result of this inspection. In addition, the future date and extent of the next inspection is based on the present condition.

There is always a danger that a piece of equipment will fail between inspections, especially if the inspections occur far apart in time. For this reason, there is an increasing trend towards the use of real-time condition monitoring for critical pieces of equipment such as large substation transformers. These devices will monitor key indicators of condition and report and signs of incipient failure to dispatchers over the SCADA system (a real-time communications system). Although far from perfect, such systems allow utilities to move closer to "just-in-time" maintenance and replacement.

Reliability-centered maintenance (RCM) is a process used to determine maintenance requirements based on equipment condition, equipment criticality, and cost. Effective applications of RCM result in maintenance schedules that maximize system reliability by maintaining components that are most likely to fail, have a high impact to customers when they fail and can be maintained for a reasonable cost. This holistic approach makes intuitive sense, but is difficult to implement. RCM has been successfully applied to power plants, is beginning to be applied to substations, and is in its infancy for feeder maintenance and vegetation management.

When determining the optimal maintenance strategy for an aging piece of major equipment, it is often possible to justify replacement based on escalating failure costs. For example, consider an old substation transformer with an expected remaining life of five years. Due to its deteriorated condition, this transformer has to be frequently inspected and maintained. The NPV of these inspection and maintenance costs can be computed and compared to the cost of replacement. In this case, the appropriate comparison is to five years worth of interest on the replacement cost of the transformer. The transformer will have to be replaced in any case, but the money spent on replacement today could be put in the bank earning interest until replacement in the future. If the NPV of five years worth of inspection and maintenance costs is more than five years worth of interest on the cost of a new transformer, the transformer should be replaced today.

Of course, it is impossible to know with certainty when a piece of equipment will fail in the future. Many times good condition data is simply not available. Even if it is, it is difficult to convert this condition data into failure probabilities, which are required for a NPV analysis. For many engineers within utilities, there is an aversion to replacing equipment when it may have remaining life, partly due to this uncertainty. However, this aversion is in large part unwarranted.

Equipment inspection, maintenance, and replacement decisions are not made in isolation. Many thousands of decisions are typically made each month. These decisions represent a portfolio that has an inherent performance in terms of cost, reliability, and risk. If professional investors purchase an individual stock, they are not surprised or disappointed if the value of the stock declines. Instead, these investors purchase a portfolio of stocks, and are only interested in the performance of the portfolio. In a similar manner, many inspection, maintenance, and replacement decisions will end up being suboptimal due to a lack of perfect knowledge. This is perfectly natural as long as the overall portfolio of inspection, maintenance, and replacement decisions has good performance.

It is always justifiable to spend money now to save money later, as is sometimes the case when replacing deteriorated but functioning equipment. However, there are also performance and risk issues associated with equipment failures. Even if it is the cheapest option to never replace old equipment, doing so may result in unacceptable reliability to customers. Regulations require that utilities provide adequate levels of service, and utilities must therefore spend money to achieve these levels of service even if life cycle cost is not minimized. Stated differently, utilities must strive to provide adequate levels of service for the minimum life cycle cost, and increasingly this will involve the proactive replacement of old equipment.

The last difficulty with equipment inspection, maintenance, and replacement decisions relates to risk. In addition to technical risk, this might include other aspects of risk including social, regulatory, and political. Many old systems serve central business districts. Failures occurring in these systems can often lead to "headline events" such as interruptions to major public facilities that may not contribute substantially to overall customer interruptions. However, this is a high profile situation that may result in an adverse response from city officials and, ultimately, strained regulatory relations. The costs associated with this outcome are real and substantial, but difficult to quantify. Therefore, it is difficult to decide how much to spend in consideration of these types of risk. In many cases, qualitative risk factors play an important role in spending decisions. This situation is not necessarily inappropriate, but makes it difficult to prioritize spending and to know how much risk-based spending is justifiable.

Key Points Regarding Inspection, Maintenance, and Replacement

- The basic ways to inspect and maintain equipment are run-to-failure, periodic maintenance, condition-based maintenance (CBM), and reliability-centered maintenance (RCM).
- Sometimes it is less expensive to replace deteriorated equipment today rather than incur high maintenance costs.
- Sometimes it is justified to spend more money in inspection and maintenance in order to ensure adequate service to customers.
- Sometimes it is justified to spend more money in inspection and maintenance in order to mitigate unacceptable risks.

8.6 STATE OF THE INDUSTRY

This section provides a general overview of the state of the electric utility industry with an emphasis on infrastructure in the United States. The approach is to discuss the history of the industry and the impact of this history on equipment age, equipment condition, and equipment investment trends. Also included are discussions on several classes of assets that are of particular interest to distribution utilities with regards to aging infrastructure.

8.6.1 Historical Overview

The electric power industry began in the late 1800s as a component of the electric lighting industry. At this time, lighting was the only application for electricity, and homes had other methods of illumination if the electricity supply was interrupted. Electricity was essentially a luxury item and reliability was not an issue.

As electricity became more common, new applications began to appear. Examples include electric motors, electric heating, irons, and phonographs. People began to grow accustomed to these new electric appliances, and their need for reliable electricity increased. This trend culminated with the invention of the radio. No nonelectrical appliance could perform the same function as a radio. If a person wanted to listen to the airwaves, electricity was required. As radio sales exploded in the 1920s, people found that reliable electricity was a necessity. By the late 1930s, electricity was regarded as a basic utility.

From roughly 1935 to the early 1970s, distribution investment was massive, generally between 7% and 15% per year as a percentage of electricity revenues. This has several implications. First, overall utility growth was consistent and predictable. This led to an attractive combination of increasing revenue, increasing efficiencies of scale, and generally decreasing rates. In this situation everybody is generally happy, and the primary focus of utilities is on system capacity expansion.

In addition, the massive distribution investment from 1935 to 1975 resulted in a doubling of installed distribution equipment every five to ten years. Essentially the distribution systems were self-rejuvenating, with the amount of old equipment always being small as a percentage of total equipment population.

The industry reached an inflection point with the oil crisis of the early 1970s. Load growth declined from over 7% to less than 2%. In some areas, energy conservation efforts actually resulted in a decline in electricity consumption. For the first time in more than forty years, utilities were faced with poor growth prospects, both in terms of revenue and infrastructure expansion. The effect of these pressures was a gradual decrease in distribution capital spending as a percentage of utility revenue.

In the middle of the 1980s, many utilities began to complete major capital projects that were initiated in the 1970s in the anticipation of robust load growth (primarily power plants). When this growth did not materialize, many available resources could not be put to work on new construction. At the same time, many utilities had high allowed rates-of-return due to the high interest rates of the 1970s. This combination led many investor-owned utilities to self-imposed rate freezes and to aggressively pursue cost reduction programs.

In many ways, the electric utility trends starting in the 1980s can be summarized as a shift from an engineering-focused industry towards a business-focused industry. Prior to the 1980s, utilities executives commonly had engineering backgrounds. Since the 1980s, utility executives are more likely to have business or legal backgrounds. The end results tended towards employee downsizing, reductions in spending, operational efficiency initiatives.

Industry Trends in the 1980s

- A massive reduction in personnel, especially with regards to higher-paid technical employees who best understand the technical issues related to aging equipment and system issues.
- Efforts to reduce capital spending by loading existing equipment to higher levels than historically were allowed. This higher loading accelerates condition deterioration, and may limit the ability of dispatchers to restore customers after a fault occurs through system reconfiguration.

- Efforts to reduce maintenance spending by extending maintenance cycles and adopting condition-based maintenance for certain classes of equipment.
- Efforts to improve operational efficiency through the use of sophisticated software tools (financial, project management, maintenance management, etc.), performance management, outsourcing, benchmarking, and so forth.

The cost-cutting trends of the 1980s were reinforced by the passage of the National Electric Policy Act in 1992 (NEPA). NEPA indicated that the industry was headed towards "de-regulation" and competition. During this time, each state also took its own approaches with regards to electric distribution regulation. Results varied by state, but the general effect was a re-doubled emphasis on cost cutting and efficiency improvement initiatives in order to maximize the chances of survival in this new environment.

Nearly every aspect of utility efficiency improvement efforts since the 1980s has had a deleterious effect on aging infrastructure and the ability of utilities to address aging infrastructure. Much of the distribution equipment installed in the 1960s and early 1970s is reaching the end of its expected life. Due to NEPA, this equipment has been pushed harder and has been maintained less in the last decade. Many utilities have achieved substantial improvements in business performance, but are now seeing signs of increasing equipment failure rates. Many of these same utilities are finding it difficult to address this difficult issue due to a lack of senior technical staff and a heavily loaded workforce.

8.6.2 Present Situation in the United States

Today, most large investor-owned utilities have either begun to increase spending on aging infrastructure, have announced intentions to increase spending on aging infrastructure, or are beginning to experience signs of increased failure and are deciding how to address the problem. Since this is a direct result of electricity growth patterns since the 1960s, the situation is similar for many utilities, and is most problematic for utilities serving older cities.

Perhaps the first signs of systematic aging infrastructure problems were witnessed in the summer of 1999, when the country experienced major electricity outages in New York City, Long Island, New Jersey, the Delmarva Peninsula, and Chicago. These events were investigated by the U.S. Department of Energy through a Power Outage Study Team (POST). The resulting POST report concludes the following:

> *The POST investigations found that the aging infrastructure and increased demand for power have strained many transmission and distribution systems to the point of interrupting service.*

Table 8.2. US rate case filings and aggregate amount requested from 1995 through 2004.

Year (As of Dec. 1)	Number of Rate Case Filings		Aggregate Amount Requested (MUSD)	
	Electric	Gas	Electric	Gas
2004	21	21	2,171.8	661.8
2003	20	23	2,400.3	970.5
2002	16	20	1,692.7	351.4
2001	21	12	1,816.8	560.4
2000	12	8	1,003.1	161.7
1999	10	13	759.0	562.9
1998	14	8	894.8	546.1
1997	13	11	624.9	161.9
1996	16	18	429.8	266.7

In the case of Commonwealth Edison in Chicago (now Exelon Corporation), these outages resulted in immediate and large increases in capital spending–more than one billion dollars from 1999 to 2004.

When aging infrastructure began to manifest itself in 1999, many utilities were under rate freezes, and had a limited ability to fund large increases in capital spending in the short term. However, the pressure to proactively address the aging infrastructure problem resulted in a substantial increase in rate case activity. From 1996 to 2000, the average amount requested from US utilities was $742 million. From 2001 through 2004, the average amount requested per year rose to more than $2 billion (see Table 8.2).

In many situations, utilities have been subject to a rate freeze or have otherwise not filed a rate case in more than a decade. Many utilities are now beginning to file new rate cases, in large part to address aging infrastructure issues, after long periods without base rate adjustments.

8.6.3 General Industry Trends

Although the specific approaches to aging infrastructure are different for each utility, there are some general trends in the industry. These trends generally relate to increased condition monitoring, increased efforts at life extension, more rigorous approaches to replacement decisions, and efforts to optimize spending in a rigorous manner. Not all utilities are making progress in these areas, but best-in-class performers are effectively utilizing these approaches to help address aging infrastructure. A brief summary of industry trends in each of these areas is now provided.

Condition monitoring. Most medium and large utilities are placing more emphasis on condition monitoring, and are increasingly using condition monitoring data to help make operational and maintenance decisions. Often these efforts focus on more aggressive inspection and testing. Techniques that are becoming more popular include infrared inspections, dissolved gas-in-oil analysis, frequency response signatures, and many others. Online condition monitoring is also becoming more popular, especially for substation equipment that can be cost-effectively monitored through the SCADA systems. Online techniques range from simple alarms (e.g., temperature, pressure) to continuous monitoring of dissolved gases in oil. Some of the newest equipment can be equipped by the vendor with self-diagnostic capability that will interpret monitoring data and automatically notify operators if significant equipment deterioration or an incipient failure has been detected.

Equipment Life Extension. Equipment replacement is often expensive, resource intensive, and operationally disruptive. For this reason, utilities are increasingly looking at life extension strategies as a way to defer replacement. Often times an equipment life extension program is coupled with a condition monitoring program. When the condition of a piece of equipment reaches a certain degree of deterioration, life extension options are examined for economic attractiveness. An example is a wooden pole program that requires periodic inspections. If a pole is likely to fail before the next inspection, the pole is treated, reinforced, or replaced depending upon the nature of deterioration. Similar strategies can be applied to transformers, circuit breakers, underground cables, and many other types of equipment.

Repair versus Replacement Decisions. Utility practices vary widely in this area, with some utilities making explicit repair-versus-replace decisions, while others will only replace equipment if it fails or becomes overloaded. In general, only the largest utilities in the US are moving towards specific replacement criteria for major equipment. Most utilities are basing replacement decision on a combination of condition data and qualitative judgment.

Economic Optimization Strategies. In terms of aging infrastructure, few companies are making rigorous optimization decisions, fewer are making purely economic decisions, and none that the author is aware of are employing strict optimization approaches. The rigor of optimization is simply not appropriate given the lack of data and uncertainty in assumptions. Pragmatic utilities are moving down the learning curve by gradually incorporating economic analysis into aging infrastructure assessment, and moving toward a more rigorous process that incorporates factors such as cost-to-benefit ratios and risk assessments.

8.6.4 Substations

From an aging infrastructure perspective, the most critical pieces of substation equipment are substation power transformers and distribution circuit breakers, with most aging infrastructure problems occurring in old substations located in older cities.

Several survivor curves for substation power transformers are shown in Figure 8.8, and similar curves for distribution circuit breakers are shown in Figure 8.9. These curves were assembled through a combination of public data and author surveys. Figure 8.8 shows the survivor curves of substation transformers for eight large US utilities. All but one utility have at least 50% of their transformers older than 30 years in age. Three out of the eight utilities have more than 30% of their transformers older than 50 years in age.

There are a lot of old transformers, but lightly loaded transformers can last many years. However, utilities have generally been loading transformers to higher levels in the past 15 years, and the combination of old chronological age and increased thermal aging has created significant deterioration in many transformers. Mainly because of this, substations were the first concern with regards to aging infrastructure, with emphasis starting after the national reliability problems in the summer of 1999. Many utilities have done a good job of assessing the condition of their older transformers and increasing condition monitoring for those at risk, but proactive replacement efforts have been generally modest and the overall population of transformers continues to get older.

At this point, it is important to remember that having an old population of equipment is not necessarily good or bad. In fact, good life-extension practices and condition-based maintenance intentionally strive to allow equipment to become older. However, older equipment in general will be more costly to maintain and more likely to fail than younger equipment.

Figure 8.9 shows the survivor curves of substation circuit breakers (distribution class) for five large US utilities. There is wide variation in this graph, primarily because some utilities have already replaced many of their old breakers. Oftentimes this is due to a lack of spare part availability. If an old circuit breaker fails, many utilities will replace the unit rather than custom-machine replacement parts.

From a criticality perspective, distribution circuit breakers are of special concern. This is because they are often the beginning of the radial distribution system. If a substation transformer fails, other transformers in the station can generally serve the load entire load with only a short interruption in service (often times with no interruptions at all). If a feeder circuit breaker experiences an internal fault, the entire feeder is interrupted. Worse, if a feeder circuit breaker is supposed to interrupt a downstream fault and fails to do so due to its deteriorated condition, the bus associated with the feeder will have to be de-energized. When this happens, all feeders served by this bus will experience a complete outage.

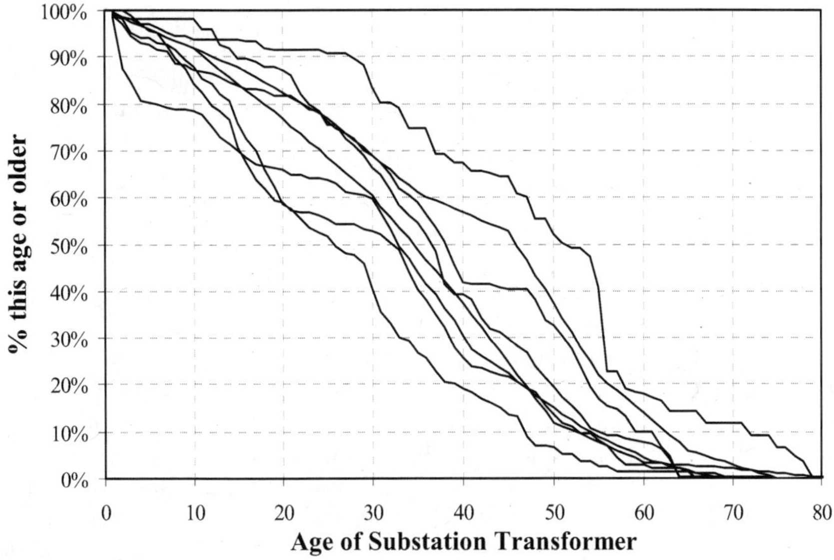

Figure 8.8. This figure shows the survivor curves of substation transformers for eight large US utilities. All but one utility have at least 50% of their transformers older than 30 years in age.

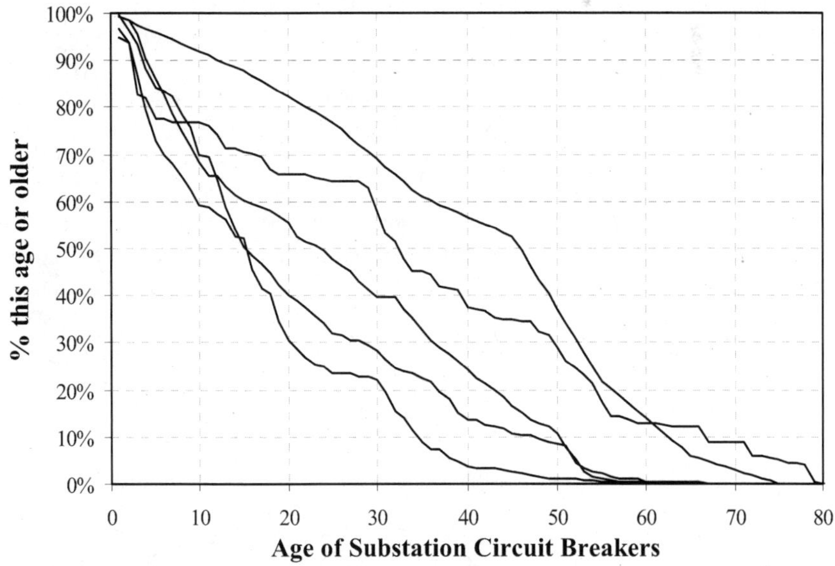

Figure 8.9. This figure shows the survivor curves of substation circuit breakers (distribution class) for five large US utilities. There is wide variation in this graph, indicating that some utilities have already replaced many of their older circuit breakers. For example, one utility only has 2% of circuit breakers older than 50 years, while others have corresponding values of 8%, 10%, 30%, and 37%.

8.6.5 Underground Cables

Aging underground cable and the resulting increase in cable failures is one of the most important issues for all utilities in the US that have significant underground distribution systems. However, the situation is not easy to characterize since underground cables can have many different characteristics that impact aging and failures.

Figure 8.10 shows the cable age distributions for five utilities with extensive underground distribution systems. Since urban areas are typically served by a large percentage of underground distribution systems, it is common for utilities serving large cities to have the oldest populations of cable.

The author has performed a cable survey with responses from twenty large US utilities. Of these, nearly 70% have formal cable replacement programs that proactively replace old cable that is expected to be prone to failure in the future. The most common criterion for replacement is simply insulation type. For example, if a utility is experiencing a large number of failures on XLPE cable installed in the early 1980s, all cable of this type may be targeted for replacement. The second most common method is for utilities to track specific failures on each cable section. If multiple failures occur on a section, the section is replaced.

Cable rejuvenation programs are also becoming more common. About 25% of surveyed utilities have cable rejuvenation programs that inject a gel into cables to restore deteriorated insulation properties. This technology allows the condition of direct buried cable to be improved without any excavation activity.

Aging cables are a significant problem for many utilities since they can adversely affect customer reliability and they are costly to replace. A few utilities have aggressively addressed this problem and, as a result, have already succeeded in replacing most of their first-generation extruded cables. In fact, about 60% of surveyed utilities feel that their aging cable problems will improve in the near future rather than get worse. The remainder feels that their present rate of cable replacement is not keeping up with the rate of cable deterioration due to aging.

8.6.6 Wooden Poles

Poles support overhead distribution equipment and are an important part of all overhead systems. Most poles are made of treated wood, but concrete, steel, and composite materials are also used. Wood poles represent some of the oldest assets for many utilities. It is common for large utilities to have more than one million distribution poles, and deteriorated poles can be a safety concern as well as a reliability concern. As such, wood poles are becoming a significant aging infrastructure concern at many utilities.

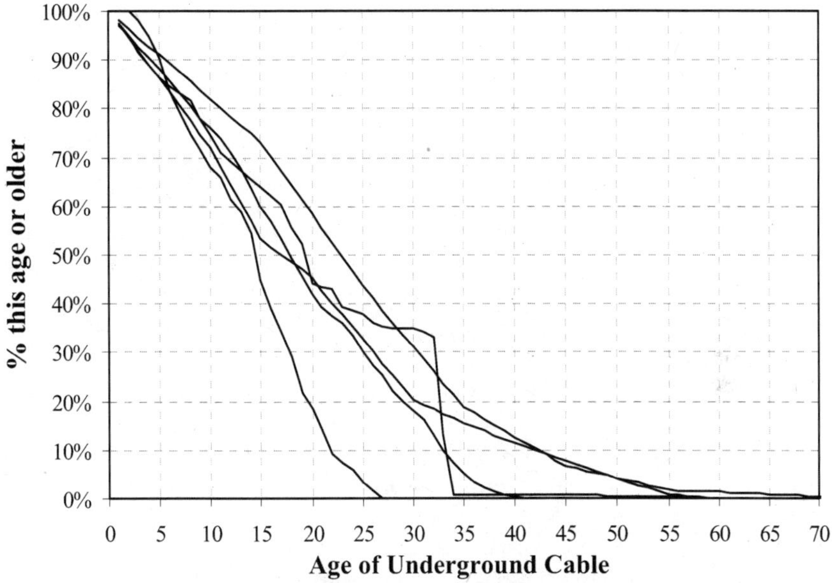

Figure 8.10. This figure shows the survivor curves of underground cable (distribution class) for five large US utilities. Often times the oldest cable in a system is PILC (paper insulation), which can often have better reliability characteristics at a very old age than certain types of cable that are much younger (e.g., first-generation polyethylene insulation).

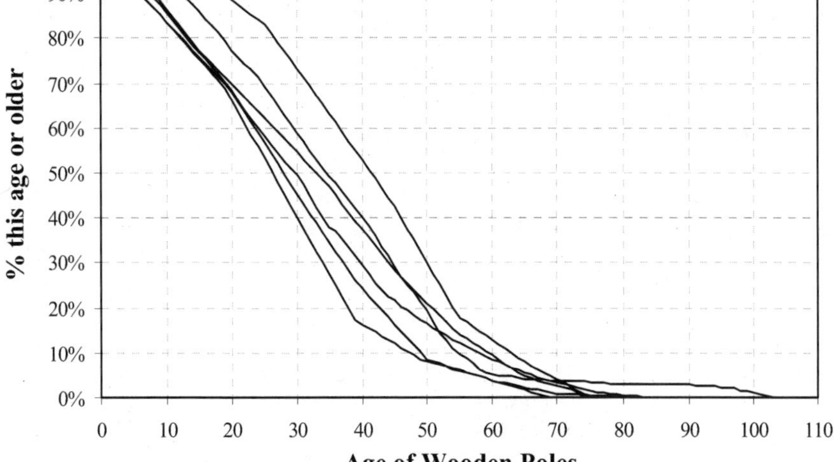

Figure 8.11. This figure shows the survivor curves of wooden poles (distribution class) for six large US utilities. For poles older than fifty years, results generally range from 9% to 20%, with only one utility scoring higher than 30%.

Older poles are susceptible to rot and decay, but periodic pole inspections typically can identify such problems and determine whether problem poles should be treated, reinforced, or replaced. A typical target inspection cycle is ten years, with ten percent of poles being inspected every year. Many utilities have found it difficult to maintain their target cycle times, and are more commonly resorting to outsourcing this function.

Figure 8.11 shows the age distribution of wooden poles for six large US utilities. All of these utilities have between 5% and 15% of their poles greater than sixty years in age. The utility with the highest percentage of extremely old poles has about 3% of its wood poles being older than 80 years in age. It is interesting to note that this system is located in a mild climate that is not susceptible to extreme weather events that tend to damage poles. Compare this to a typical utility in Florida, where a major hurricane is likely to occur every ten years or so. Since wooden poles are not designed to withstand hurricane-force winds, many wooden poles will be damaged during hurricanes and will require replacement. It is unlikely that a large percentage of wooden poles will survive 80 years in this situation.

For most utilities, aging wood poles have not yet resulted in significant decreases in reliability performance. In part this is because of aggressive inspection and treatment programs. Also, wooden poles under the right circumstances can last a long time. Last, deteriorated wood poles will often be supported by adjacent poles until the deterioration can be identified and addressed. Regardless, many utilities with an old population of poles are only replacing 0.5% of poles or less each year. For this amount to be sustainable, average pole life must be at least 200 years or more. The author does not believe this to be the case, and utilities will eventually have to increase the rate of pole replacement to sustainable levels.

8.7 FINAL THOUGHTS

For most US utilities, distribution equipment is getting older both in terms of average age and the percentage of very old equipment. This situation is a direct result of the business environment. The oil crisis of the early 1970s resulted in low load growth, drastically reducing the amount of new equipment being installed. The National Electric Policy Act of 1992 created the threat of deregulation and competition, resulting in efficiency efforts leading to increased equipment loading and less frequent maintenance activities. Details vary for each utility, but the effects of aging on certain classes of distribution equipment have begun to manifest themselves in recent years. Mitigating the adverse reliability impact of aging infrastructure and equipment deterioration will require increased levels of investment.

The problems associated with aging infrastructure are compounded by increased equipment loading. Not only does increased loading accelerate equipment deterioration. Increased equipment loading also limits the ability of a utility to restore service to customers after a fault occurs. Typically, many interrupted customers can be transferred to another feeder through normally open tie points. Presently, many utilities are not able to operate in this manner since performing these transfers, as a result of higher loading, will now result in equipment overloading. Fixing this problem will generally require an increase in the number of feeders and/or feeder tie locations, with both options requiring increased levels of investment.

Utilities are recognizing that aging distribution infrastructure will require increased levels of investment. This is one reason that the number of US rate case filings, and the aggregate amounts requested, have significantly increased in recent years. Ratepayer advocates can and should insist that spending related to aging infrastructure be reasonably justified, but limiting a utility's ability to manage its aging equipment will eventually result in lower levels of reliability and higher cost.

In a sense, aging infrastructure is changing the entire business paradigm for many distribution utilities. In the past, the focus was on building new stuff. In the future, the focus must shift to managing old stuff. In the past, the core competence was efficient execution of construction projects. In the future, the core competence must shift to rigorous reliability and cost management based asset-level data and detailed reliability models. In the past, capacity plans implicitly built reliability into the system; reliability problems were addressed reactively. In the future, reliability planning and capacity planning must be considered simultaneously; reliability must be addressed proactively.

Aging infrastructure has also resulted in increased regulatory and media attention on reliability performance. Customers are expecting higher levels of performance, but aging infrastructure is making reliability worse every year. To address this paradox, successful distribution utilities must address reliability in a rigorous and comprehensive manner so that all stakeholders can be assured that reliability targets are being achieved for the lowest possible cost. The contents of this book will hopefully help to guide and instruct those attempting to meet this significant challenge.

8.8 STUDY QUESTIONS

1. Why should a utility care if a piece of equipment is getting old?
2. Is average population age a good way to track a group of equipment as it ages over time? Explain.
3. What is the impact of system growth on aging infrastructure statistics?

4. If the expected life of a distribution pole is fifty years, what percentage of poles, in the long run, must be replaced annually so that replacements offset failures?
5. Besides outage data, what other data is required to develop age-versus-failure rate models?
6. Why is it important to consider features when developing age-versus-failure rate models? Can there be too many features?
7. Most utilities would like to improve reliability, or at least keep reliability from getting worse. Why might this be difficult?
8. Explain why spending on new construction has a lower approval standard when compared with spending on aging infrastructure
9. Describe how aging infrastructure is fundamentally changing the role of a distribution utility. How does reliability fit into this new role?
10. How must a distribution utility change so that it can better address challenged related to aging infrastructure and reliability?

REFERENCES

1. R. E. Brown, G. Frimpong and H. L. Willis, "Failure Rate Modeling Using Equipment Inspection Data," IEEE Transactions on Power Systems, Vol. 19, No. 2, May 2004, pp. 782-787.
2. H. L. Willis, G. Welch, and R. R. Schrieber, *Aging Power Delivery Infrastructures,* Marcel-Dekker, 2001.
3. R. E. Brown and H. L. Willis, "The Economics of Aging Infrastructure," *IEEE Power and Energy Magazine*, Vol. 4, No. 3, May/June 2006.
4. K. Abdolall, G. L. Halldorson, and D. Green, "Condition Assessment and Failure Modes of Solid Dielectric Cables in Perspective," IEEE Transactions on Power Delivery, Vol. 17, No. 1, Jan 2002, pp. 18-24.
5. M. D. Walton, B. S. Bernstein, W. A. Thue and J. T. Smith III, "Dielectric Breakdown Strength Correlation with Time-To-Failure During Wet Aging of XLP-Insulated Cables," IEEE Transactions on Power Delivery, Vol. 14, No. 3, Jul 1999, pp. 750-755.
6. M.D. Walton, B.S. Bernstein, W.A. Thue, and J.T. Smith, JIII, "Dielectric breakdown strength correlation with time-to-failure during wet aging of XLPE-insulated cables," *IEEE Transactions on Power Delivery*, Vol. 14, No. 3, July 1999, pp. 750 – 755.
7. L.E. Lundgaard, W. Hansen, D. Linhjell, and T.J. Painter, "Aging of oil-impregnated paper in power transformers," *IEEE Transactions on Power Delivery*, Vol. 19, No. 1, Jan. 2004, pp. 230 – 239.
8. R. M. Bucci, R. V. Rebbapragada, A. J. McElroy, E. A. Chebli and S. Driller, "Failure Prediction of Underground Distribution Feeder Cables," IEEE Transaction on Power Delivery, Vol. 9, No. 4, Oct 1994, pp. 1943-1955.
9. L. Bertling, R. Allan, and R. Eriksson, "A Reliability-Centered Asset Management Method for Assessing the Impact of Maintenance in Power Distribution Systems," *IEEE Transactions on Power Systems*, Vol. 20, No. 1, Feb 2005, pp. 75-82.
10. V. I. Kogan and T. L. Jones, "An Explanation for the Decline in URD Cable Failures and Associated Nonhomogeneous Poisson Process," IEEE Transactions on Power Delivery, Vol. 9, No. 1, Jan 1994, pp. 534-541.
11. D. Chinh, J.L. Parpal, and J.P. Crine, "Electrical aging of extruded dielectric cables: review of existing theories and data," *IEEE Transactions on Dielectrics and Electrical Insulation*, Vol. 3, No. 2, April 1996, pp. 237 – 247.

12. Y. Zhou and R. E. Brown, "A Practical Method for Cable Failure Rate Modeling," *IEEE/PES Transmission and Distribution Conference and Exposition*, New Orleans, LA, Sept. 2005.
13. W. Li, "Incorporating aging failures in power system reliability evaluation," *IEEE Transactions on Power Systems*, Vol. 17, No. 3, Aug. 2002, pp. 918 – 923.

Index

A

B

D

F

G

H

I

N

O

P

Q

R

S

T

U

X

Y

Z